Artificial Intelligence
and Soft Computing

Behavioral and Cognitive Modeling
of the Human Brain

Artificial Intelligence
and Soft Computing
Behavioral and Cognitive Modeling
of the Human Brain

Amit Konar
Department of Electronics and Tele-communication Engineering
Jadavpur University, Calcutta, India

CRC Press
Boca Raton   London   New York   Washington, D.C.

Library of Congress Cataloging-in-Publication Data

Konar, Amit.
    Artificial intelligence and soft computing : behavioral and cognitive modeling of the human brain / Amit Konar.
        p.   cm.
    Includes bibliographical references and index.
    ISBN 0-8493-1385-6 (alk. paper)
    1. Soft computing. 2. Artificial intelligence. 3. Brain—Computer simulation. I. Title.
    QA76.9.S63 K59 1999
    006.3--dc21   99-048018
                                                                                                    CIP

© 2000 by CRC Press

No claim to original U.S. Government works
International Standard Book Number 0-8493-1385-6
Library of Congress Card Number 99-048018
Printed in the United States of America   3 4 5 6 7 8 9 0
Printed on acid-free paper

PREFACE

The book, to the best of the author's knowledge, is the first text of its kind that presents both the traditional and the modern aspects of 'AI and Soft Computing' in a clear, insightful and highly comprehensive writing style. It provides an in-depth analysis of the mathematical models and algorithms, and demonstrates their applications in real world problems of significant complexity.

1. About the book

The book covers 24 chapters altogether. It starts with the behavioral perspective of the 'human cognition' and covers in detail the tools and techniques required for its intelligent realization on machines. The classical chapters on search, symbolic logic, planning and machine learning have been covered in sufficient details, including the latest research in the subject. The modern aspects of soft computing have been introduced from the first principles and discussed in a semi-informal manner, so that a beginner of the subject is able to grasp it with minimal effort. Besides soft computing, the other leading aspects of current AI research covered in the book include non-monotonic and spatio-temporal reasoning, knowledge acquisition, verification, validation and maintenance issues, realization of cognition on machines and the architecture of AI machines. The book ends with two case studies: one on 'criminal investigation' and the other on 'navigational planning of robots,' where the main emphasis is given on the realization of intelligent systems using the methodologies covered in the book.

The book is unique for its diversity in contents, clarity and precision of presentation and the overall completeness of its chapters. It requires no mathematical prerequisites beyond the high school algebra and elementary differential calculus; however, a mathematical maturity is required to follow the logical concepts presented therein. An elementary background of data structure and a high level programming language like Pascal or C is helpful to understand the book. The book, thus, though meant for two semester courses of computer science, will be equally useful to readers of other engineering disciplines and psychology as well as for its diverse contents, clear presentation and minimum prerequisite requirements.

In order to make the students aware of the applied side of the subject, the book includes a few homework problems, selected from a wide range of topics. The problems supplied, in general, are of three types: i) numerical, ii) reflexive and iii) provocative. The numerical problems test the students' understanding of the subject. The reflexive type requires a formulation of the problem from its statement before finding its solution. The provocative type includes the well-known problems of modern AI research, the solution to some of which are known, and some are open ended. With adequate hints supplied with the problems, the students will be able to solve most of the numerical and reflexive type problems themselves. The provocative type, however, requires some guidance from the teacher in charge. The last type of problems is included in the text to give the research-oriented readers an idea of the current trend in AI research. Graduate students of AI will also find these problems useful for their dissertation work.

The book includes a large number of computer simulations to illustrate the concepts presented in logic programming, fuzzy Petri nets, imaging and robotics. Most of the simulation programs are coded in C and Pascal, so that students without any background of PROLOG and LISP may understand them easily. These programs will enhance the students' confidence in the subject and enable them to design the simulation programs, assigned in the exercise as homework problems. The professionals will find these simulations interesting as it requires understanding of the end results only, rather than the formal proofs of the theorems presented in the text.

2. Special features

The book includes the following special features.

i) Unified theme of presentation: Most of the existing texts on AI cover a set of chapters of diverse thoughts, without demonstrating their inter-relationship. The readers, therefore, are misled with the belief that AI is merely a collection of intelligent algorithms, which precisely is not correct. The proposed book is developed from the perspective of cognitive science, which provides the readers with the view that the psychological model of cognition can be visualized as a cycle of 5 mental states: sensing, acquisition, perception, planning and action, and there exists a strong interdependence between each two sequential states. The significance of search in the state of perception, reasoning in the state of planning, and learning as an intermediate process between sensing and action thus makes sense. The unified theme of the book, therefore, is to realize the behavioral perspective of cognition on an intelligent machine, so as to enable it act and think like a human being. Readers will enjoy the book especially for its totality with an ultimate aim to build intelligent machines.

ii) Comprehensive coverage of the mathematical models: This probably is the first book that provides a comprehensive coverage of the mathematical models on AI and Soft Computing. The existing texts on "mathematical modeling in AI" are beyond the scope of undergraduate students. Consequently, while taking courses at graduate level, the students face much difficulty in studying from monographs and journals. The book, however, bridges the potential gap between the textbooks and advanced monographs in the subject by presenting the mathematical models from a layman's understanding of the problems.

iii) Case studies: This is the only book that demonstrates the realization of the proposed tools and techniques of AI and Soft Computing through case studies. The readers, through these case studies, will understand the significance of the joint usage of the AI and Soft Computing tools and techniques in interesting problems of the real world. Case studies for two distinct problems with special emphasis to their realization have been covered in the book in two separate chapters. The case study I is concerned with a problem of criminal investigation, where the readers will learn to use the soft computing tools in facial image matching, fingerprint classification, speaker identification and incidental description based reasoning. The readers can build up their own systems by adding new fuzzy production rules and facts and deleting the unwanted rules and facts from the system. The book thus will serve the readership from both the academic and the professional world. Electronic and computer hobbyists will find the case study II on mobile robots very exciting. The algorithms of navigational planning (in case study II), though tested with reference to "Nomad Super Scout II robot," have been presented in generic form, so that the interested readers can code them for other wheel-based mobile robots.

iv) Line Diagrams: The book includes around 190 line diagrams to give the readers a better insight to the subject. Readers will enjoy the book for they directly get a deeper view of the subject through diagrams with a minimal reading of the text.

3. Origin of the book

The book is an outgrowth of the lecture materials prepared by the author for a one semester course on "Artificial Intelligence," offered to the graduate students in the department of Electronics and Telecommunication Engineering, Jadavpur University, Calcutta. An early version of the text was also used in a summer-school on "AI and Neural Nets," offered to the faculty members of various engineering colleges for their academic development and training. The training program included theories followed by a laboratory course, where the attendees developed programs in PROLOG, Pascal and C with the help of sample programs/toolkit. The toolkit is included in the book on a CD and the procedure to use it is presented in Appendix A.

4. Structural organization of the book

The structural organization of the book is presented below with a dependency graph of chapters, where Ch. 9 → Ch. 10 means that chapter 10 should be read following chapter 9, for example.

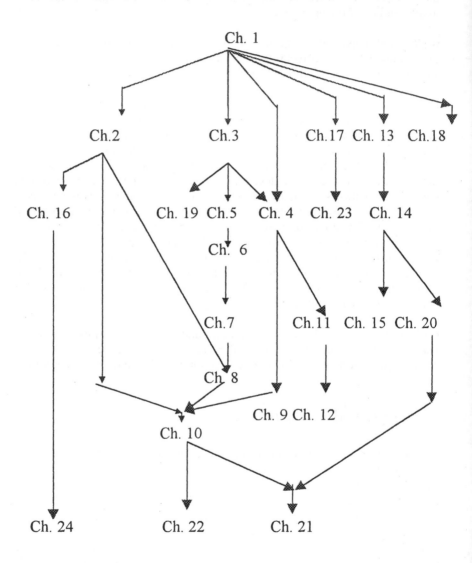

July 12, 1999
Jadavpur University   Amit Konar

ABOUT THE AUTHOR

Amit Konar is a Reader in the Department of Electronics and Telecommunication Engineering, Jadavpur University, Calcutta. He received a Ph.D. (Engineering) degree in Artificial Intelligence from the same university in 1994 and has been teaching the subject of Artificial Intelligence to the graduate students of his department for the last 10 years. Dr. Konar has supervised a number of Ph.D. and M.E. theses on different aspects of machine intelligence, including logic programming, neural networks, cognitive systems, stochastic and fuzzy models of uncertainty, fuzzy algebra, image understanding, architecture of intelligent machines and navigational planning of mobile robots. He has published more than 60 papers in international journals and conferences. He is an invited contributor of a book chapter in an edited book published by Academic Press. Dr. Konar is a recipient of the 1997 Young Scientist Award, offered by the All India Council for Technical Education (AICTE) for his significant contributions in Artificial Intelligence and Soft Computing.

ACKNOWLEDGMENT

The author gratefully acknowledges the contributions of many people, who helped him in different ways to complete the book. First and foremost, he wishes to thank his graduate students attending the course entitled "AI and Pattern Recognition" in ETCE department, Jadavpur University during the 1993-1999 sessions. Next, he would like to thank the scholars working for their Ph.D. degree under his supervision. In this regard, the author acknowledges the contribution of Ms. Jaya Sil, a recipient of the Ph.D. degree in 1996, for spending many of her valuable hours on discussion of the Bayesian and Markov models of knowledge representation. The other scholars, to whom the author is greatly indebted for sharing their knowledge in different areas of AI, are Mr. Srikant Patnaik, Mr. Biswajit Paul, Mrs. Bijita Biswas, Ms. Sanjukta Pal, Ms. Alakananda Bhattacharya and Ms. Parbati Saha. The contributions of Mr. Patnaik in chapter 24, Mr. Paul in chapter 14, Ms. Biswas in chapter 23, Ms. Pal in chapter 16, Ms. Bhattacharya in chapter 22 and Ms. Saha in chapter 10 need special mention. Among his scholars, the author wants to convey his special thanks to Mr. Patnaik, who helped him in many ways, which simply cannot be expressed in a few sentences.

The author acknowledges the contribution of his friend Mr. Dipak Laha, a faculty member of the Mechanical Engineering department, Jadavpur University, who helped him in understanding the many difficult problems of scheduling. He also would like to thank his friend Dr. Uday Kumar Chakraborty, a faculty member of the Computer Science department, Jadavpur University, for teaching him the fundamentals in Genetic Algorithms. The author gives a special thanks to Ms. Sheli Murmu, his student and now a colleague, who helped him in correcting many syntactical errors in the draft book. He also wants to thank his graduate students including Mr. Diptendu Bhattacharya, Ms. Bandana Barmn, and Mr. Srikrishna Bhattacharya for their help in drawing many figures and in the technical editing of this book. The author also wishes to thank his ex-student Ms. Sragdhara Dutta Choudhury, who helped him draw a very simple but beautiful sketch of the 'classroom' figure in chapter 6.

The architectural issues of knowledge based systems, which is the main theme of chapter 22, is the summary of the M.E. thesis (1991-1992) of Mr. Shirshendu Halder, who critically reviewed a large number of research papers and interestingly presented the pros and cons of these works in his thesis.

The author owes a deep gratitude to Prof. A. K. Mandal of the department of Electronics and Telecommunication Engineering, Jadavpur University, for teaching him the subject of AI and providing him both technical and moral support as a teacher, Ph.D. thesis adviser and colleague.

He is also indebted to Prof. A.K. Nath of the same department for encouraging him to write a book and spending long hours in valuable discussion. The author would like to thank his teacher Prof. A. B. Roy of the department of Mathematics, Jadavpur University, who inspired his writing skill, which later enabled him to write this book. He remembers his one-time project supervisor Prof. T. K. Ghosal of the Department of Electrical Engineering, Jadavpur University, for his constructive criticism, which helped him develop a habit of checking a thought twice before deliberating. The author also gratefully acknowledges his unaccountable debt to his teacher Mr. Basudeb Dey, who taught him the basis to uncover the mysteries from the statement of arithmetic problems, without which the author could never have been able to reach his present level of maturity in mathematics.

The author wants to convey a special vote of thanks to his colleagues Prof. S. K. Choudhury and Dr. B. Gupta for their kind gesture of attending his classes on AI for a complete semester, which helped him to make necessary corrections in the book.

Among his friends and well-wishers, the author would like to mention Mr. Gourishankar Chattopadhyay, Mr. Bisweswar Jana, Mrs. Dipa Gupta, Mr. P. K. Gupta and Prof. P.K. Sinha Roy, without whose encouragement and inspiration the book could not have taken its present shape. His ex-students Ms. Sanghamitra Sinha of Sun Microsystems, USA, Ms. Indrani Chakraborty of MIE University, Japan, Mr. Ashim Biswas of HCL Technologies, NOIDA, India and Dr. Madhumita Dasgupta of Jadavpur University, India helped him in many ways improve the book.

The author would like to thank Ms. Nora Konopka, Acquisition Editor, and staff members of CRC Press LLC for their kind cooperation in connection with writing this book. He would also like to thank Prof. L. C. Jain of the University of South Australia, Adelaide, for active cooperation and editorial guidance on this book.

Lastly, the author wishes to express his deep gratitude to his parents, who always stood by him throughout his life and guided him in his time of crisis. He also wishes to thank his wife Srilekha for her tolerance of his indifference to the family life and her assistance in many ways for the successful completion of the book. The author is equally grateful to his in-laws and especially his brother-in-law, Mr. Subrata Samanta, for their inspiration and encouragement in writing this book.

September 17, 1999

Jadavpur University   Amit Konar

To my parents, Mr. Sailen Konar and Mrs. Minati Konar, who brought me up despite the stress and complexities of their lives and devoted themselves to my education;

To my brother Sanjoy, who since his childhood shouldered the responsibility of running our family smoothly;

To my wife Srilekha, who helped me survive and inspired me in many ways to write and complete this book in the present form;

To my students in various parts of the world, who through their forbearance allowed me to improve my teaching skills;

To my teachers, who taught me the art of reacting to a changing environment; and

To millions of the poor and down-trodden people of my country and the world, whose sacrifice and tolerance paved the royal road of my education, and whose love and emotion, smile and tears inspired me to speak their thoughts in my words.

Amit Konar

Contents

Chapter 1: Introduction to Artificial Intelligence and Soft Computing

Chapter 2: The Psychological Perspective of Cognition

Chapter 3: Production Systems

Chapter 4: Problem Solving by Intelligent Search

Chapter 5: The Logic of Propositions and Predicates

Chapter 8: Structured Approach to Knowledge Representation

Chapter 9: Dealing with Imprecision and Uncertainty

Chapter 10: Structured Approach to Fuzzy Reasoning

Chapter 11: Reasoning with Space and Time

Chapter 12: Intelligent Planning

Chapter 13: Machine Learning Techniques

Chapter 14: Machine Learning Using Neural Nets

Chapter 15: Genetic Algorithms

Chapter 16: Realizing Cognition Using Fuzzy Neural Nets

Chapter 17: Visual Perception

Chapter 18: Linguistic Perception

Chapter 19: Problem Solving by Constraint Satisfaction

Chapter 20: Acquisition of Knowledge

Chapter 21: Validation, Verification and Maintenance Issues

Chapter 22: Parallel and Distributed Architecture for Intelligent Systems

Chapter 23: Case Study I: Building a System for Criminal Investigation

Chapter 24: Case Study II: Realization of Cognition for Mobile Robots

Chapter 24[+]: The Expectations from the Readers,

Appendix A: How to Run the Sample Programs?

Appendix B: Derivation of the Back-propagation Algorithm,

Appendix C: Proof of the Theorems of Chapter 10,

Index,

1

Introduction to Artificial Intelligence and Soft Computing

This chapter provides a brief overview of the disciplines of Artificial Intelligence (AI) and Soft Computing. It introduces the topics covered under the heads of intelligent systems and demonstrates the scope of their applications in real world problems of significant complexity. It also highlights the direction of research in this broad discipline of knowledge. The historical development in AI and the means by which the subject was gradually popularized are briefly outlined here. The chapter addresses many new tools and techniques, commonly used to represent and solve complex problems. The organization of the book in light of these tools and techniques is also presented briefly in this chapter.

1.1   Evolution of Computing

At the beginning of the Stone Age, when people started taking shelters in caves, they made attempts to immortalize themselves by painting their images on rocks. With the gradual progress in civilization, they felt

1

interested to see themselves in different forms. So, they started constructing models of human being with sand, clay and stones. The size, shape, constituents and style of the model humans continued evolving but the man was not happy with the models that only looked like him. He had a strong desire to make the model 'intelligent', so that it could act and think as he did. This, however, was a much more complex task than what he had done before. So, he took millions of years to construct an 'analytical engine' that could perform a little arithmetic mechanically. Babbage's analytical engine was the first significant success in the modern era of computing. Computers of the first generation, which were realized following this revolutionary success, were made of thermo-ionic valves. They could perform the so-called 'number crunching' operations. The second-generation computers came up shortly after the invention of transistors and were more miniaturized in size. They were mainly used for commercial data processing and payroll creation. After more than a decade or so, when the semiconductor industries started producing integrated circuits (IC) in bulk, the third generation computers were launched in business houses. These machines had an immense capability to perform massive computations in real time. Many electromechanical robots were also designed with these computers. Then after another decade, the fourth generation computers came up with the high-speed VLSI engines. Many electronic robots that can see through cameras to locate objects for placement at the desired locations were realized during this period. During the period of 1981-1990 the Japanese Government started to produce the fifth generation computing machines that, besides having all the capabilities of the fourth generation machines, could also be able to process *intelligence*. The computers of the current (fifth) generation can process natural languages, play games, recognize images of objects and prove mathematical theorems, all of which lie in the domain of Artificial Intelligence (AI). But what exactly is AI? The following sections will provide a qualitative answer to this question.

1.2 Defining AI

The phrase AI, which was coined by John McCarthy [1] three decades ago, evades a concise and formal definition to date. One representative definition is pivoted around the comparison of intelligence of computing machines with human beings [11]. Another definition is concerned with the performance of machines which "historically have been judged to lie within the domain of intelligence" [17], [35]. None of these definitions or the like have been universally accepted, perhaps because of their references to the word "intelligence", which at present is an abstract and immeasurable quantity. A better definition of AI, therefore, calls for formalization of the term "intelligence". Psychologist and Cognitive theorists are of the opinion that intelligence helps in identifying the right piece of knowledge at the appropriate instances of decision making [27], [14].The phrase "AI" thus can be defined as the *simulation of human intelligence on a machine, so as to make the machine efficient to identify and use the right piece of "Knowledge" at a given step of solving a problem.* A system capable of planning and executing the right task at the right time is generally called **rational** [36]. Thus, AI alternatively may be stated as *a subject dealing with computational models that can think and act rationally* [18][1], [47][2], [37][3], [6][4]. A common question then naturally arises: Does rational thinking and acting include all possible characteristics of an intelligent system? If so, how does it represent behavioral intelligence such as machine learning, perception and planning? A little thinking, however, reveals that a system that can reason well must be a successful planner, as planning in many circumstances is part of a reasoning process. Further, a system can act rationally only after acquiring adequate knowledge from the real world. So, perception that stands for building up of knowledge from real world information is a prerequisite feature for rational actions.   One step further thinking envisages that a machine without learning capability cannot possess perception. The rational action of an agent (actor), thus, calls for possession of all the elementary characteristics of intelligence. Relating AI with the computational models capable of thinking and acting rationally, therefore, has a pragmatic significance.

1.3 General Problem Solving Approaches in AI

To understand what exactly AI is, we illustrate some common problems. Problems dealt with in AI generally use a common term called 'state'. A state represents a status of the solution at a given step of the problem solving procedure. The solution of a problem, thus, is a collection of the problem states. The problem solving procedure applies an operator to a state to get the next state. Then it applies another operator to the resulting state to derive a new state. The process of applying an operator to a state and its subsequent

---

1.The branch of computer science that is concerned with the automation of intelligent behavior.
2. The study of computations that make it possible to perceive, reason and act.
3. A field of study that seeks to explain and emulate intelligent behavior in terms of computational processes.
4. The study of mental faculties through the use of computational models.

transition to the next state, thus, is continued until the goal (desired) state is derived. Such a method of solving a problem is generally referred to as **state-space approach**. We will first discuss the state-space approach for problem solving by a well-known problem, which most of us perhaps have solved in our childhood.

Example 1.1:   Consider a 4-puzzle problem, where in a 4-cell board there are 3 cells filled with digits and 1 blank cell. The initial state of the game represents a particular orientation of the digits in the cells and the final state to be achieved is another orientation supplied to the game player. The problem of the game is to reach from the given initial state to the goal (final) state, if possible, with a minimum of moves. Let the initial and the final state be as shown in figures 1(a) and (b) respectively.

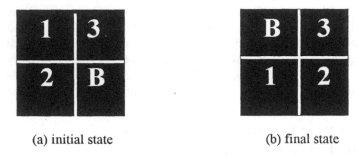

(a) initial state   (b) final state

Fig. 1.1: The initial and the final states of the Number Puzzle game, where B denotes the blank space.

We now define two operations, blank-up (BU) / blank-down (BD) and blank-left (BL) / blank-right (BR) [9], and the state-space (tree) for the problem is presented below (vide figure 1. 2) using these operators.

The algorithm for the above kind of problems is straightforward. It consists of three steps, described by steps 1, 2(a) and 2(b) below.

Algorithm for solving state-space problems
Begin
      1. state: = initial-state; existing-state:=state;
      2. **While** state ≠ final state **do**
      Begin

          a.   Apply operations from the set {BL, BR, BU, BD} to each state so as to generate new-states;
          b.   **If** new-states ∩ the existing-states ≠ φ
              Then do

Begin state := new-states – existing-states;
Existing-states := existing-states ∪ {states}
End;
End while;

End.

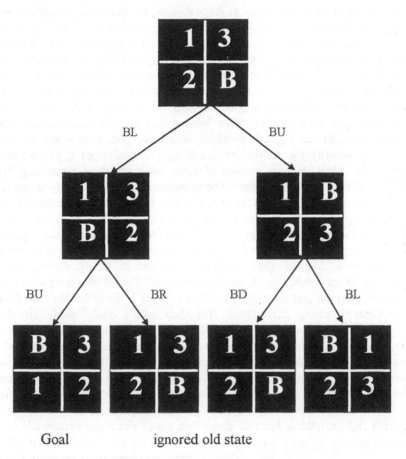

Goal   ignored old state

Fig.1. 2: The state-space for the Four-Puzzle problem.

It is thus clear that the main trick in solving problems by the state-space approach is to determine the set of operators and to use it at appropriate states of the problem.

Researchers in AI have segregated the AI problems from the non-AI problems. Generally, problems, for which straightforward mathematical / logical algorithms are not readily available and which can be solved by intuitive approach only, are called AI problems. The 4-puzzle problem, for instance, is an ideal AI Problem. There is no formal algorithm for its realization, i.e., given a starting and a goal state, one cannot say prior to execution of the tasks the sequence of steps required to get the goal from the starting state. Such problems are called the ideal *AI problems*. The well-known water-jug problem [35], the Travelling Salesperson Problem (TSP) [35], and the n-Queen problem [36] are typical examples of the classical AI problems. Among the non-classical AI problems, the diagnosis problems and the pattern classification problem need special mention. For solving an AI problem, one may employ both AI and non-AI algorithms. An obvious question is: what is an *AI algorithm*? Formally speaking, an AI algorithm generally means a non-conventional intuitive approach for problem solving. The key to AI approach is intelligent search and matching. In an intelligent search problem / sub-problem, given a goal (or starting) state, one has to reach that state from one or more known starting (or goal) states. For example, consider the 4-puzzle problem, where the goal state is known and one has to identify the moves for reaching the goal from a pre-defined starting state. Now, the less number of states one generates for reaching the goal, the better is the AI algorithm. The question that then naturally arises is: how to control the generation of states. This, in fact, can be achieved by suitably designing some control strategies, which would filter a few states only from a large number of legal states that could be generated from a given starting / intermediate state. As an example, consider the problem of proving a trigonometric identity that children are used to doing during their schooldays. What would they do at the beginning? They would start with one side of the identity, and attempt to apply a number of formulae there to find the possible resulting derivations. But they won't really apply all the formulae there. Rather, they identify the right candidate formula that fits there best, such that the other side of the identity seems to be closer in some sense (outlook). Ultimately, when the decision regarding the selection of the formula is over, they apply it to one side (say the L.H.S) of the identity and derive the new state. Thus they continue the process and go on generating new intermediate states until the R.H.S (goal) is reached. But do they always select the right candidate formula at a given state? From our experience, we know the answer is "not always". But what would we do if we find that after generation of a few states, the resulting expression seems to be far away from the R.H.S of the identity. Perhaps we would prefer to move to some old state, which is more promising, i.e., closer to the R.H.S of the identity. The above line of thinking has been realized in many intelligent search problems of AI. Some of these well-known search algorithms are:

    a)   Generate and Test
    b)   Hill Climbing
    c)   Heuristic Search
    d)   Means and Ends analysis

(a) Generate and Test Approach: This approach concerns the generation of the state-space from a known starting state (root) of the problem and continues expanding the reasoning space until the goal node or the terminal state is reached. In fact after generation of each and every state, the generated node is compared with the known goal state. When the goal is found, the algorithm terminates. In case there exist multiple paths leading to the goal, then the path having the smallest distance from the root is preferred. The basic strategy used in this search is only generation of states and their testing for goals but it does not allow filtering of states.

(b) Hill Climbing Approach: Under this approach, one has to first generate a starting state and measure the total cost for reaching the goal from the given starting state. Let this cost be f. While f ≤ a predefined utility value and the goal is not reached, new nodes are generated as children of the current node. However, in case all the neighborhood nodes (states) yield an identical value of f and the goal is not included in the set of these nodes, the search algorithm is trapped at a hillock or local extrema. One way to overcome this problem is to select randomly a new starting state and then continue the above search process. While proving trigonometric identities, we often use Hill Climbing, perhaps unknowingly.

(c) Heuristic Search: Classically *heuristics* means rule of thumb. In heuristic search, we generally use one or more *heuristic functions* to determine the better candidate states among a set of legal states that could be generated from a known state. The heuristic function, in other words, measures the fitness of the candidate states. The better the selection of the states, the fewer will be the number of intermediate states for reaching the goal. However, the most difficult task in heuristic search problems is the selection of the heuristic functions. One has to select them intuitively, so that in most cases hopefully it would be able to prune the search space correctly. We will discuss many of these issues in a separate chapter on Intelligent Search.

(d) Means and Ends Analysis: This method of search attempts to reduce the gap between the current state and the goal state. One simple way to explore this method is to measure the distance between the current state and the goal, and then apply an operator to the current state, so that the distance between the resulting state and the goal is reduced. In many mathematical theorem- proving processes, we use Means and Ends Analysis.

Besides the above methods of intelligent search, there exist a good number of general problem solving techniques in AI. Among these, the most common are: Problem Decomposition and Constraint Satisfaction.

Problem Decomposition: Decomposition of a problem means breaking a problem into independent (de-coupled) sub-problems and subsequently sub-problems into smaller sub-problems and so on until a set of decomposed sub-problems with known solutions is available. For example, consider the following problem of integration.

$$I = \int (x^2 + 9x + 2) \, dx,$$

which may be decomposed to

$$\int (x^2 \, dx) + \int (9x \, dx) + \int (2 \, dx) ,$$

where fortunately all the 3 resulting sub-problems need not be decomposed further, as their integrations are known.

Constraint Satisfaction: This method is concerned with finding the solution of a problem by satisfying a set of constraints. A number of constraint satisfaction techniques are prevalent in AI. In this section, we illustrate the concept by one typical method, called hierarchical approach for constraint satisfaction (HACS) [47]. Given the problem and a set of constraints, the HACS decomposes the problem into sub-problems; and the constraints that are applicable to each decomposed problem are identified and propagated down through the decomposed problem. The process of re-decomposing the sub-problem into smaller problems and propagation of the constraints through the descendants of the reasoning space are continued until all the constraints are satisfied. The following example illustrates the principle of HACS with respect to a problem of extracting roots from a set of inequality constraints.

Example 1.2: The problem is to evaluate the variables X_1, X_2 and X_3 from the following set of constraints:

$$\{ X_1 \geq 2; \quad X_2 \geq 3 ; \quad X_1 + X_2 \leq 6; \quad X_1 , X_2 , X_3 \in I \}.$$

For solving this problem, we break the ' $\geq$ ' into ' $>$ ' and ' $=$ ' and propagate the sub-constraints through the arcs of the tree. On reaching the end of the arcs, we attempt to satisfy the propagated constraints in the parent constraint and reduce the constraint set. The process is continued until the set of constraints is minimal, i.e., they cannot be broken into smaller sets (fig. 1.3).

There exists quite a large number of AI problems, which can be solved by *non-AI approach*. For example, consider the Travelling Salesperson Problem. It is an optimization problem, which can be solved by many non-AI algorithms. However, the *Neighborhood search AI method* [35] adopted for this problem is useful for the following reason. The design of the AI algorithm should be such that the time required for solving the problem is a *polynomial (and not an exponential) function* of the size (dimension) of the problem. When the computational time is an exponential function of the dimension of the problem, we call it a **combinatorial exploration** problem. Further, the number of variables to be used for solving an AI problem should also be minimum, and should not increase with the dimension of the problem. A non-AI algorithm for an AI problem can hardly satisfy the above two requirements and that is why an AI problem should be solved by an AI approach.

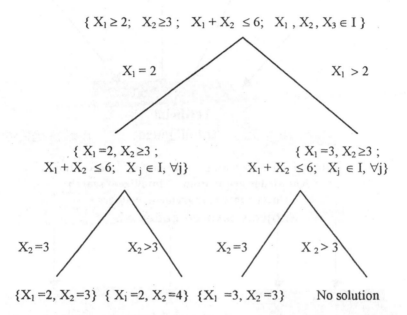

Fig. 1.3: The constraint tree, where the arcs propagate the constraints, and the nodes down the tree hold the reduced set of constraints.

1.4 The Disciplines of AI

The subject of AI spans a wide horizon. It deals with the various kinds of knowledge representation schemes, different techniques of intelligent search, various methods for resolving uncertainty of data and knowledge, different schemes for automated machine learning and many others. Among the application areas of AI, we have Expert systems, Game-playing, and Theorem-proving, Natural language processing, Image recognition, Robotics and many others. The subject of AI has been enriched with a wide discipline of knowledge from Philosophy, Psychology, Cognitive Science, Computer

Science, Mathematics and Engineering. Thus in fig. 1.4, they have been referred to as the parent disciplines of AI. An at-a-glance look at fig. 1.4 also reveals the subject area of AI and its application areas.

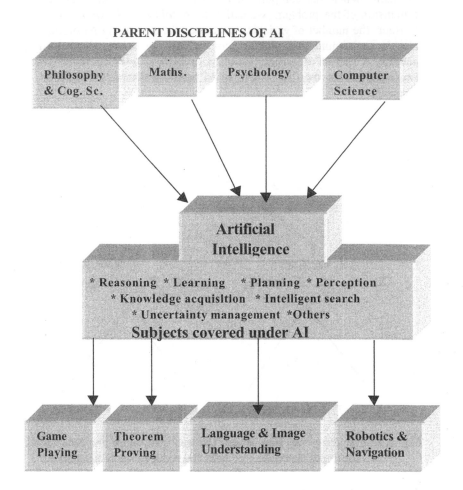

Fig. 1.4: AI, its parent disciplines and application areas.

1.4.1 The Subject of AI

The subject of AI was originated with game-playing and theorem-proving programs and was gradually enriched with theories from a number of parent disciplines. As a young discipline of science, the significance of the topics covered under the subject changes considerably with time. At present, the topics which we find significant and worthwhile to understand the subject are outlined below:

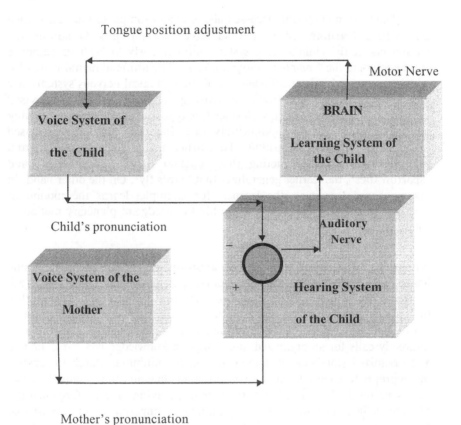

Fig. 1. 5: Pronunciation learning of a child from his mother.

Learning Systems: Among the subject areas covered under AI, learning systems needs special mention. The concept of learning is illustrated here with reference to a natural problem of learning of pronunciation by a child from his mother (vide fig. 1.5). The hearing system of the child receives the pronunciation of the character "A" and the voice system attempts to imitate it. The difference of the mother's and the child's pronunciation, hereafter called the error signal, is received by the child's learning system through the auditory nerve, and an actuation signal is generated by the learning system through a motor nerve for adjustment of the pronunciation of the child. The adaptation of the child's voice system is continued until the amplitude of the error signal is insignificantly low. Each time the voice system passes through an adaptation cycle, the resulting tongue position of the child for speaking "A" is saved by the learning process.

The learning problem discussed above is an example of the well-known **parametric learning**, where the adaptive learning process adjusts the parameters of the child's voice system autonomously to keep its response close enough to the "*sample training pattern*". The artificial neural networks, which represent the electrical analogue of the biological nervous systems, are gaining importance for their increasing applications in supervised (parametric) learning problems. Besides this type, the other common learning methods, which we do unknowingly, are inductive and analogy-based learning. In inductive learning, the learner makes generalizations from examples. For instance, noting that "cuckoo flies", "parrot flies" and "sparrow flies", the learner generalizes that "birds fly". On the other hand, in analogy-based learning, the learner, for example, learns the motion of electrons in an atom analogously from his knowledge of planetary motion in solar systems.

Knowledge Representation and Reasoning: In a reasoning problem, one has to reach a pre-defined goal state from one or more given initial states. So, the lesser the number of transitions for reaching the goal state, the higher the efficiency of the reasoning system. Increasing the efficiency of a reasoning system thus requires minimization of intermediate states, which indirectly calls for an organized and complete knowledge base. A complete and organized storehouse of knowledge needs minimum search to identify the appropriate knowledge at a given problem state and thus yields the right next state on the leading edge of the problem-solving process. Organization of knowledge, therefore, is of paramount importance in knowledge engineering. A variety of knowledge representation techniques are in use in Artificial Intelligence. Production rules, semantic nets, frames, filler and slots, and predicate logic are only a few to mention. The selection of a particular type of representational scheme of knowledge depends both on the nature of applications and the choice of users.

Example 1. 3: A semantic net represents knowledge by a structured approach. For instance, consider the following knowledge base:

Knowledge Base: A bird can fly with wings. A bird has wings. A bird has legs. A bird can walk with legs.

The bird and its attributes here have been represented in figure 1.6 using a graph, where the nodes denote the events and the arcs denote the relationship between the nodes.

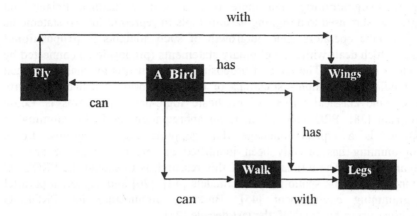

Fig. 1.6: A semantic net representation of " birds".

Planning: Another significant area of AI is planning. The problems of reasoning and planning share many common issues, but have a basic difference that originates from their definitions. The reasoning problem is mainly concerned with the testing of the satisfiability of a goal from a given set of data and knowledge. The planning problem, on the other hand, deals with the determination of the methodology by which a successful goal can be achieved from the known initial states [1]. Automated planning finds extensive applications in robotics and navigational problems, some of which will be discussed shortly.

Knowledge Acquisition: Acquisition (Elicitation) of knowledge is equally hard for machines as it is for human beings. It includes generation of new pieces of knowledge from given knowledge base, setting dynamic data structures for existing knowledge, learning knowledge from the environment and refinement of knowledge. Automated acquisition of knowledge by machine learning approach is an active area of current research in Artificial Intelligence [5], [20].

Intelligent Search: Search problems, which we generally encounter in Computer Science, are of a deterministic nature, i.e., the order of visiting the elements of the search space is known. For example, in depth first and breadth first search algorithms, one knows the sequence of visiting the nodes in a tree. However, search problems, which we will come across in AI, are

non-deterministic and the order of visiting the elements in the search space is completely dependent on data sets. The diversity of the intelligent search algorithms will be discussed in detail later.

Logic Programming: For more than a century, mathematicians and logicians were used to designing various tools to represent logical statements by symbolic operators. One outgrowth of such attempts is **propositional logic,** which deals with a set of binary statements (propositions) connected by Boolean operators. The logic of propositions, which was gradually enriched to handle more complex situations of the real world, is called **predicate logic.** One classical variety of predicate logic-based programs is **Logic Program** [38]. PROLOG, which is an abbreviation for **PRO**gramming in **LOG**ic, is a typical language that supports logic programs. Logic Programming has recently been identified as one of the prime area of research in AI. The ultimate aim of this research is to extend the PROLOG compiler to handle spatio-temporal models [42], [20] and support a parallel programming environment [45]. Building architecture for PROLOG machines was a hot topic of the last decade [24].

Soft Computing: Soft computing, according to Prof. Zadeh, is "an emerging approach to computing, which parallels the remarkable ability of the human mind to reason and learn in an environment of uncertainty and imprecision" [13]. It, in general, is a collection of computing tools and techniques, shared by closely related disciplines that include fuzzy logic, artificial neural nets, genetic algorithms, belief calculus, and some aspects of machine learning like inductive logic programming. These tools are used independently as well as jointly depending on the type of the domain of applications. The scope of the first three tools in the broad spectrum of AI is outlined below.

♦   **Fuzzy Logic:** Fuzzy logic deals with fuzzy sets and logical connectives for modeling the human-like reasoning problems of the real world. A fuzzy set, unlike conventional sets, includes all elements of the universal set of the domain but with varying membership values in the interval [0,1]. It may be noted that a conventional set contains its members with a value of membership equal to one and disregards other elements of the universal set, for they have zero membership. The most common operators applied to fuzzy sets are AND (minimum), OR (maximum) and negation (complementation), where AND and OR have binary arguments, while negation has unary argument. The logic of fuzzy sets was proposed by Zadeh, who introduced the concept in systems theory, and later extended it for approximate reasoning in expert systems [45]. Among the pioneering contributors on fuzzy logic, the work of Tanaka in stability analysis of control systems [44], Mamdani in cement kiln control

[19] , Kosko [15] and Pedrycz [30] in fuzzy neural nets, Bezdek in pattern classification [3], and Zimmerman [50] and Yager [48] in fuzzy tools and techniques needs special mention.

♦ **Artificial Neural Nets:** Artificial neural nets (ANN) are electrical analogues of the biological neural nets. Biological nerve cells, called neurons, receive signals from neighboring neurons or receptors through dendrites, process the received electrical pulses at the cell body and transmit signals through a large and thick nerve fiber, called an axon. The electrical model of a typical biological neuron consists of a linear activator, followed by a non-linear inhibiting function. The linear activation function yields the sum of the weighted input excitation, while the non-linear inhibiting function attempts to arrest the signal levels of the sum. The resulting signal, produced by an electrical neuron, is thus bounded (amplitude limited). An artificial neural net is a collection of such electrical neurons connected in different topology. The most common application of an artificial neural net is in machine learning. In a learning problem, the weights and / or non-linearities in an artificial neural net undergo an adaptation cycle. The adaptation cycle is required for updating these parameters of the network, until a state of equilibrium is reached, following which the parameters no longer change further. The ANN support both supervised and unsupervised types of machine learning. The supervised learning algorithms realized with ANN have been successfully applied in control [25], automation [31], robotics [32] and computer vision [31]. The unsupervised learning algorithms built with ANN, on the other hand, have been applied in scheduling [31], knowledge acquisition [5], planning [22] and analog to digital conversion of data [41].

♦ **Genetic Algorithms:** A genetic algorithm (GA) is a stochastic algorithm that mimics the natural process of biological evolution [35]. It follows the principle of *Darwinism*, which rests on the fundamental belief of the *"survival of the fittest"* in the process of natural selection of species. GAs find extensive applications in intelligent search, machine learning and optimization problems. The problem states in a GA are denoted by chromosomes, which are usually represented by binary strings. The most common operators used in GA are crossover and mutation. The processes of execution of *crossover* and *mutation* are illustrated in fig.1.7 and 1.8 respectively. The evolutionary cycle in a GA consists of the following three sequential steps [23].

    a) Generation of population (problem states represented by chromosomes).

    b) Genetic evolution through crossover followed by mutation.

      c) Selection of better candidate states from the generated
         population.

In step (a) of the above cycle, a few initial problem states are first
identified. The step (b) evolves new chromosomes through the process of
crossover and mutation. In step (c) a fixed number of better candidate states
are selected from the generated population. The above steps are repeated a
finite number of times for obtaining a solution for the given problem.

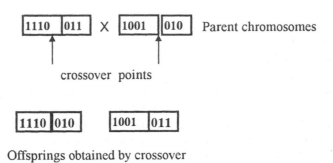

Fig.1.7: Exchange of genetic information by crossover operation.

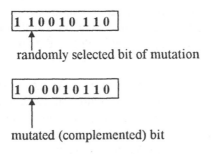

Fig. 1. 8: The mutation operation: randomly selected
           bits are complemented.

Management of Imprecision and Uncertainty: Data and knowledge-
bases in many typical AI problems, such as reasoning and planning, are often
contaminated with various forms of incompleteness. The incompleteness of
data, hereafter called **imprecision**, generally appears in the database for i)
lack of appropriate data, and ii) poor authenticity level of the sources. The
incompleteness of knowledge, often referred to as **uncertainty**, originates in
the knowledge base due to lack of certainty of the pieces of knowledge.

Reasoning in the presence of imprecision of data and uncertainty of knowledge is a complex problem. Various tools and techniques have been devised for reasoning under incomplete data and knowledge. Some of these techniques employ i) **stochastic** ii) **fuzzy** and iii) **belief network** models [16]. In a stochastic reasoning model, the system can have transition from one given state to a number of states, such that the sum of the probability of transition to the next states from the given state is strictly unity. In a fuzzy reasoning system, on the other hand, the sum of the membership value of transition from the given state to the next state may be greater than or equal to one. The belief network model updates the stochastic / fuzzy belief assigned to the facts embedded in the network until a condition of equilibrium is reached, following which there would be no more change in beliefs. Recently, fuzzy tools and techniques have been applied in a specialized belief network, called a **fuzzy Petri net,** for handling both imprecision of data and uncertainty of knowledge by a unified approach [14].

1.4.2 Applications of AI Techniques

Almost every branch of science and engineering currently shares the tools and techniques available in the domain of AI. However, for the sake of the convenience of the readers, we mention here a few typical applications, where AI plays a significant and decisive role in engineering automation.

Expert Systems: In this example, we illustrate the reasoning process involved in an expert system for a weather forecasting problem with special emphasis to its architecture. An expert system consists of a **knowledge base**, **database** and an **inference engine** for interpreting the database using the knowledge supplied in the knowledge base. The reasoning process of a typical illustrative expert system is described in Fig. 1.9. PR 1 in Fig. 1.9 represents i-th production rule.

The inference engine attempts to match the antecedent clauses (IF parts) of the rules with the data stored in the database. When all the antecedent clauses of a rule are available in the database, the rule is fired, resulting in new inferences. The resulting inferences are added to the database for activating subsequent firing of other rules. In order to keep limited data in the database, a few rules that contain an explicit consequent (THEN) clause to delete specific data from the databases are employed in the knowledge base. On firing of such rules, the unwanted data clauses as suggested by the rule are deleted from the database.

Here PR1 fires as both of its antecedent clauses are present in the database. On firing of PR1, the consequent clause "it-will-rain" will be added to the database for subsequent firing of PR2.

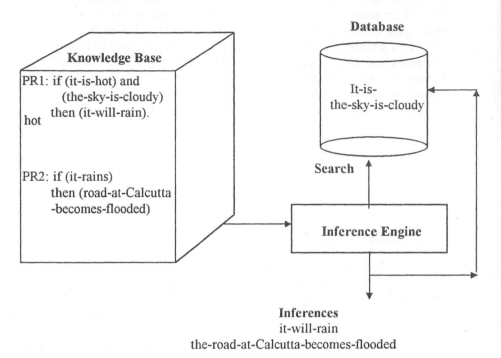

Fig. 1. 9: Illustrative architecture of an expert system.

Image Understanding and Computer Vision: A **digital image** can be regarded as a two-dimensional array of pixels containing gray levels corresponding to the intensity of the reflected illumination received by a video camera [6]. For interpretation of a scene, its image should be passed through three basic processes: low, medium and high level vision (fig.1.10). The importance of **low level vision** is to pre-process the image by filtering from noise. The **medium level vision** system deals with enhancement of details and segmentation (i.e., partitioning the image into objects of interest). The **high level vision** system includes three steps: recognition of the objects from the segmented image, labeling of the image and interpretation of the scene. Most of the AI tools and techniques are required in high level vision systems. Recognition of objects from its image can be carried out through a process of pattern classification, which at present is realized by supervised learning algorithms. The interpretation process, on the other hand, requires knowledge-based computation.

Navigational Planning for Mobile Robots: Mobile robots, sometimes called automated guided vehicles (AGV), are a challenging area of research, where AI finds extensive applications. A mobile robot generally has one or more camera or ultrasonic sensors, which help in identifying the obstacles on its trajectory. The navigational planning problem persists in both static and dynamic environments. In a static environment, the position of obstacles is fixed, while in a dynamic environment the obstacles may move at arbitrary directions with varying speeds, lower than the maximum speed of the robot. Many researchers using spatio-temporal logic [7-8] have attempted the navigational planning problems for mobile robots in a static environment. On the other hand, for path planning in a dynamic environment, the genetic algorithm [23], [26] and the neural network-based approach [41], [47] have had some success. In the near future, mobile robots will find extensive applications in fire-fighting, mine clearing and factory automation. In accident prone industrial environment, mobile robots may be exploited for automatic diagnosis and replacement of defective parts of instruments.

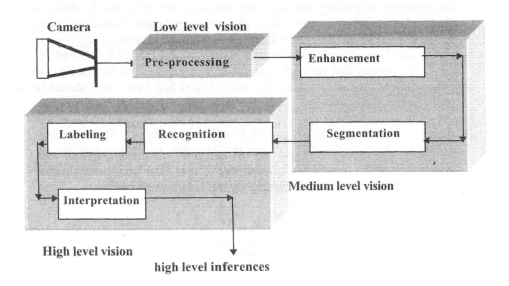

Fig. 1.10: Basic steps in scene interpretation.

Speech and Natural Language Understanding: Understanding of speech and natural languages is basically two classical problems. In speech analysis, the main problem is to separate the syllables of a spoken word and determine features like amplitude, and fundamental and harmonic frequencies of each syllable. The words then could be identified from the extracted features by pattern classification techniques. Recently, artificial neural networks have been employed [41] to classify words from their features. The problem of understanding natural languages like English, on the other hand, includes syntactic and semantic interpretation of the words in a sentence, and sentences in a paragraph. The syntactic steps are required to analyze the sentences by its grammar and are similar with the steps of compilation. The semantic analysis, which is performed following the syntactic analysis, determines the meaning of the sentences from the association of the words and that of a paragraph from the closeness of the sentences. A robot capable of understanding speech in a natural language will be of immense importance, for it could execute any task verbally communicated to it. The phonetic typewriter, which prints the words pronounced by a person, is another recent invention where speech understanding is employed in a commercial application.

Scheduling: In a scheduling problem, one has to plan the time schedule of a set of events to improve the time efficiency of the solution. For instance in a class-routine scheduling problem, the teachers are allocated to different classrooms at different time slots, and we want most classrooms to be occupied most of the time. In a flowshop scheduling problem [42], a set of jobs J_1 and J_2 (say) are to be allocated to a set of machines M_1, M_2 and M_3 (say). We assume that each job requires some operations to be done on all these machines in a fixed order say, M_1, M_2 and M_3. Now, what should be the schedule of the jobs (J_1-J_2) or (J_2 –J_1), so that the completion time of both the jobs, called the make-span, is minimized? Let the processing time of jobs J_1 and J_2 on machines M_1, M_2 and M_3 be (5, 8, 7) and (8, 2, 3) respectively. The gantt charts in fig. 1.11 (a) and (b) describe the make-spans for the schedule of jobs J_1 - J_2 and J_2 - J_1 respectively. It is clear from these figures that J_1-J_2 schedule requires less make-span and is thus preferred.

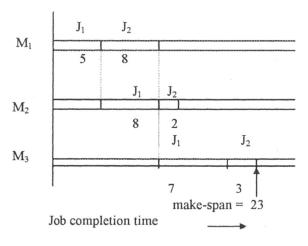

(a) The J_1 - J_2 schedule.

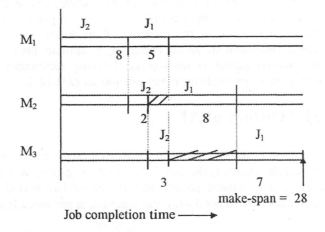

Job completion time ⟶

(b): The J_2 - J_1 schedule where the hatched lines indicate waiting
time of the machines.

Fig. 1.11: The Gantt charts for the flowshop scheduling problem
with 2 jobs and 3 machines.

Flowshop scheduling problems are a **NP complete problem** [1] and determination of optimal scheduling (for minimizing the make-span) thus requires an exponential order of time with respect to both machine-size and job-size. Finding a sub-optimal solution is thus preferred for such scheduling problems. Recently, artificial neural nets and genetic algorithms have been employed to solve this problem. The heuristic search, to be discussed shortly, has also been used for handling this problem [34].

Intelligent Control: In process control, the controller is designed from the known models of the process and the required control objective. When the dynamics of the plant is not completely known, the existing techniques for controller design no longer remain valid. Rule-based control is appropriate in such situations. In a rule-based control system, the controller is realized by a set of production rules intuitively set by an expert control engineer. The antecedent (premise) part of the rules in a rule-based system is searched against the dynamic response of the plant parameters. The rule whose antecedent part matches with the plant response is selected and fired. When more than one rule is firable, the controller resolves the conflict by a set of strategies. On the other hand, there exist situations when the antecedent part of no rules exactly matches with the plant responses. Such situations are handled with fuzzy logic, which is capable of matching the antecedent parts of rules partially/ approximately with the dynamic plant responses. Fuzzy control has been successfully used in many industrial plants. One typical application is the power control in a nuclear reactor. Besides design of the controller, the other issue in process control is to design a plant (process) estimator, which attempts to follow the response of the actual plant, when both the plant and the estimator are jointly excited by a common input signal. The fuzzy and artificial neural network-based learning techniques have recently been identified as new tools for **plant estimation** [25], [43].

1.5   A Brief History of AI

Professor Peter Jackson of the University of Edinburgh classified the history of AI into three periods namely i) the classical period (of game playing and theorem proving), ii) the romantic period, and iii) the modern period [12]; the major research work carried out during these periods is presented below.

1.5.1 The Classical Period

This period dates back to 1950. The main research works carried out during this period include game playing and theorem proving. The concept of state-space approach for solving a problem, which is a useful tool for intelligent problem-solving even now, was originated during this period [27].

The period of classical AI research began with the publication of Shannon's paper on chess (1950) [35] and ended with the publication by Feigenbaum and Feldman [10]. The major area of research covered under this period is intelligent search problems involved in game-playing and theorem- proving. Turing's "test", which is a useful tool to test machine intelligence, originated during this period.

1.5.2 The Romantic Period

The romantic period started from the mid 1960s and continued until the mid 1970s. During this period, people were interested in making machines "understand", by which they usually mean the understanding of natural languages. Winograd's (1972) SHRDLU system [46], a program capable of understanding a non-trivial subset of English by representing and reasoning about a restricted domain (a world consisting of toy blocks), in this regard needs special mention. The knowledge representation scheme using special structures like "semantic nets" was originated by Quillian [33] during this period. Minisky (1968) also made a great contribution from the point of view of information processing using semantic nets. Further, knowledge representation formalisms using frames, which was another contribution of Minisky during this period, also need special mention [28].

1.5.3 The Modern Period

The modern period starts form the latter half of the 1970s to the present day. This period is devoted to solving more complex problems of practical interest. The MYCIN experiments of Stanford University [4], [39] resulted in an expert system that could diagnose and prescribe medicines for infectious bacteriological diseases. The MECHO system for solving problems of Newtonian machines is another expert system that deals with real life problems. It should be added that besides solving real world problems, researchers are also engaged in theoretical research on AI including heuristic search, uncertainty modeling and non-monotonic and spatio-temporal reasoning. To summarize, this period includes research on both theories and practical aspects of AI.

1.6 Characteristic Requirements for the Realization of the Intelligent Systems

The AI problems, irrespective of their type, possess a few common characteristics. Identification of these characteristics is required for designing a common framework for handling AI problems. Some of the well-known characteristic requirements for the realization of the intelligent systems are listed below.

1.6.1 Symbolic and Numeric Computation on Common Platform

It is clear from the previous sections that a general purpose intelligent machine should be able to perform both symbolic and numeric computations on a common platform. Symbolic computing is required in automated reasoning, recognition, matching and inductive as well as analogy-based learning. The need for symbolic computing was felt since the birth of AI in the early fifties. Recently, the connectionist approach for building intelligent machines with structured models like artificial neural nets is receiving more attention. The ANN based models have successfully been applied in learning, recognition, optimization and also in reasoning problems [29] involved in expert systems. The ANNs have outperformed the classical approach in many applications, including optimization and pattern classification problems. Many AI researchers, thus, are of the opinion that in the long run the connectionist approach will replace the classical approach in all respects. This, however, is a too optimistic proposition, as the current ANNs require significant evolution to cope with the problems involved in logic programming and non-monotonic reasoning. The symbolic and connectionist approach, therefore, will continue co-existing in intelligent machines until the latter, if ever, could replace the former in the coming years.

1.6.2   Non-Deterministic Computation

The AI problems are usually solved by state-space approach, introduced in section 1.3. This approach calls for designing algorithms for reaching one or more goal states from the selected initial state(s). The transition from one state to the next state is carried out by applying appropriate rules, selected from the given knowledge base. In many circumstances, more than one rule is applicable to a given state for yielding different next states. This informally is referred to as non-determinism. Contrary to the case, when only one rule is applicable to a given state, this system is called **deterministic**. Generally AI problems are **non-deterministic**. The issues of determinism and non-determinism are explained here with respect to an illustrative knowledge-based system. For instance, consider a knowledge base consisting of the following production rules and database.

Production Rules

    PR1: IF (A) AND (B) THEN (C).
    PR2: IF (C) THEN (D).
    PR3: IF (C) AND (E) THEN (Y).
    PR4: IF (Y) THEN (Z).

Database: A, B, E.

    The graph representing the transition of states for the above reasoning problem is presented in fig.1. 12. Let A and B be starting states and Z be the goal state. It may be noted that both PR2 and PR3 are applicable at state (C) yielding new states. However, the application of PR3 at state (C) can subsequently lead to the goal state Z, which unfortunately remains unknown until PR4 is applied at state Y. This system is a typical example of non-determinism. The dropping of PR2 from the knowledge base, however, makes the system deterministic. One formal approach for testing determinism / non-determinism of a reasoning system can be carried out by the following principle:

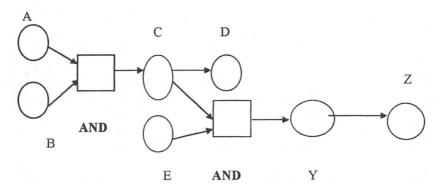

Fig. 1.12: A Petri-like net representing non-determinism in a reasoning system
with initial states A and B and goal state Z.

Principle for testing determinism: After deriving the goal state from the
initial states, continue marking (backtracking) the parents of each node
starting from the goal node, until the initial states are reached. If unmarked
nodes are detected, then the system is non-deterministic; otherwise it is
deterministic. It may be noted that testing of determinism in a knowledge-
based system, for any set of starting and goal states, is a distinct problem and
no conclusion about determinism can be drawn for modified initial or goal
states.

The principle for testing determinism in the proposed knowledge-based
system is illustrated here with reference to the dependence graph (Petri-like
net) of fig.1. 12. It may be noted that while backtracking on the graph, node
D is not marked and thus the system is non-deterministic.

Besides reasoning, non-determinism plays a significant role in many
classical AI problems. The scope of non-determinism in heuristic search has
already been mentioned. In this section, we demonstrate its scope in
recognition problems through the following example.

Example 1.3: This example illustrates the differences of deterministic and
non-deterministic transition graphs [9], called automata. Let us first consider
the problem of recognition of a word, say, "robot". The transition graph (fig.
1.13(a)) for the current problem is deterministic, since the arcs emerging out
from a given state are always distinct. However, there exist problems, where
the arcs coming out from a state are not always distinct. For instance,
consider the problem of recognizing the words "robot" and "root". Here,
since more than one outgoing arc from state B (fig. 1.13(b)) contains the
same label (o), they are not distinct and the transition graph is non-
deterministic.

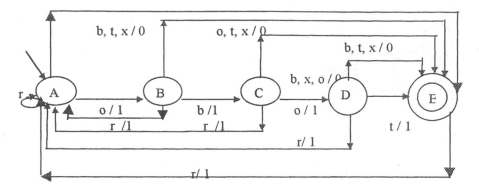

System states: A,B,C,D,E;   Received symbols : r, o, b, o, t, x where x ∉ {r, o, b, t}.
x / 1: transition on x to next state with output =1 (success) ; x/ 0: transition on x to
next state with output 0 (failure).

Fig.1. 13 (a) : A deterministic automata used for the recognition of the word
"robot".

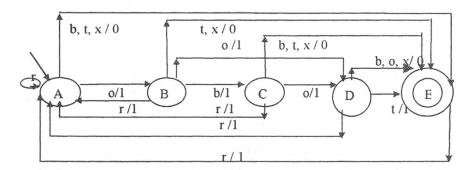

Fig.1. 13 (b): A non-deterministic automata that recognizes words "robot"
and "root".

1.6.3 Distributed Computing

Because of too much non-determinism in the AI problems, distributed
computing with shared resources is given much importance in the current
computational models of intelligence. Data structures like Petri nets [14],
and computational models like "AND-parallelism", "OR-parallelism",
"Stream-parallelism" [45] etc., have been recently emerging to realize
distributed computing on intelligent machines. Distributed computing has
twofold advantages in AI: i) to support massive parallelism and ii) to

improve reliability by realizing fragments of computational models onto a number of software or hardware modules.

1.6.4 Open System

The design of an intelligent system should be made, so that it can be readily extended for spatio-temporal changes in the environment. For example, one should use a dynamic knowledge base, which can automatically acquire knowledge from its environment. Further, the learning systems should adapt their parameters to take into account the new problem instances. An open system also allows hardware/software interfacing with the existing system.

1.7 Programming Languages for AI

Generally relational languages like PROLOG [38] or functional languages like LISP are preferred for symbolic computation in AI. However, if the program requires much arithmetic computation (say, for the purpose of uncertainty management) then procedural language could be used. There is a dilemma between the choice of programming languages for solving AI problems to date. A procedural language that offers a call for a relational function or a relational language that allows interface with a procedural one is probably the best choice. Currently, a number of shells (for ES) are available, where the user needs to submit knowledge only and the shell offers the implementation of both numeric as well as symbolic processing simultaneously.

1.8 Architecture for AI Machines

During the developmental phase of AI, machines used for conventional programming were also used for AI programming. However, since AI programs deal more with relational operators than number crunching, the need for special architectures for the execution of AI programs was felt. Gradually, it was discovered that due to non-determinism in the AI problems, it supports a high degree of concurrent computing. The architecture of an AI machine thus should allow symbolic computation in a concurrent environment. Further, for minimizing possible corruption of program resources (say variables or procedures), concurrent computation may be realized in a fine grain distributed environment. Currently PROLOG and LISP machines are active areas of AI research, where the emphasis is to incorporate the above issues at the hardware and software levels. Most of these architectures are designed for research laboratories and are not available in the open commercial market to date. We hope for a better future for AI, when these special architectures will find extensive commercial exploitation.

1.9 Objective and Scope of the Book

The objective of the book is to bridge the potential gap between the existing textbooks of AI and research papers/monographs. The available texts on AI usually do not cover mathematical issues. Further, soft computing, which plays a significant role in modeling intelligent problems, has rarely been included in a textbook on AI. The book, thus, to the best of the author's knowledge, is the first text of its kind that covers all modern aspects of AI and soft computing and their applications in a single easy-to-read volume.

The book has been organized in the following manner. Chapter 2 introduces the behavioral aspects of cognition [21]. Problem solving by production system is presented in chapter 3. A detailed review on intelligent search is presented in chapter 4. Various aspects of knowledge representation and reasoning are covered in chapters 5-11. The chapters included are predicate logic (chapter 5), logic programming (chapter 6), default and non-monotonic logic (chapter 7), structured approach to knowledge representation (chapter 8), and dealing with imprecision and uncertainty (chapter 9). A separate chapter on structured approach to fuzzy reasoning systems is included in chapter 10 of the book for both its increasing demand and scarcity of the literature on the topic. The spatio-temporal models for reasoning are covered in chapter 11. Some aspects of intelligent planning are covered in chapter 12. The principles of machine learning are presented in chapter 13 and some of their realizations with ANN are covered in chapter 14. One complete chapter (chapter 15) is devoted to genetic algorithm. These are the major areas covered in the book. The other issues outlined in the book, for the sake of completeness are: realizing cognition with fuzzy neural nets (chapter 16), visual perception (chapter 17), linguistic perception (chapter 18), constraint satisfaction (chapter 19), knowledge acquisition (chapter 20), verification and validation models (chapter 21) and architecture of AI machines (chapter 22). These are relatively growing topics and thus will help the readers learn the frontier areas of the subject. Chapter 23 and 24 cover two case studies: one on criminal investigation and the other on navigational planning of robots.

1.10 Summary

The subject of AI deals more with symbolic processing than numeric computation. Knowledge representation, reasoning, planning, learning, intelligent search and uncertainty management of data and knowledge are the common areas covered under AI. Some of the applications areas of AI are speech and image understanding, expert systems, pattern classification and navigational planning of mobile robots. LISP and PROLOG are the usual languages for programming AI problems. Because of severe non-determinism in AI problems, it supports massive parallelism. Specialized parallel processing architecture for PROLOG and LISP machines is in development and may be available shortly in the open market.

Exercises

1. Determine the starting state, goal state, and legal moves and draw the state-space for the well-known missionaries-cannibals problem, listed below:

   There are three missionaries, three cannibals and a boat on the left bank of a river. All of the six persons need to be transported to the right bank using a boat. The boat carries only two persons and at least one person should bring the boat back. If the cannibals outnumber the missionaries on either bank, they will eat the missionaries.

2. Design a semantic net, describing relationships among different modules of an aircraft. Write a program in Pascal/C to realize the net for answering the following questions:

   Queries: a) Does the aircraft have seats?
                 b) If answer to (a) is yes, then how many seats does it have?

3. Consider the following knowledge base and database.

   Knowledge base:

   PR1: IF ((X is a man and Y is a woman) AND
            (X and Y are lovers))
       THEN (X is a friend of Y).

   PR2: IF ((X is a man and Y is a woman) AND
            (X married Y))
       THEN (X loves Y).

   PR3: IF ((X is a man and Y is a woman) AND
            (Y married X))
       THEN (Y loves X).

   PR4: IF ((X loves Y) AND
            (Y loves X))
       THEN (X and Y are lovers).

Database:

   1. Ram is a man.
   2. Sita is a woman.
   3. Ram married Sita.
   4. Sita married Ram.

a)   Show the sequence of selection of PRs to prove that "Ram is a friend of Sita". What other (intermediate) inferences does the inference engine derive to prove the above conclusion?

b)   What additional piece of knowledge should one add to the system to eliminate the fourth data element, which is redundant in presence of the third data element?

4.   Design a deterministic automata for the recognition of the word "apple". Extend your design to recognize the words "apple" and "ape" by a non-deterministic automata.

References

[1] Bender, Edward A., *Mathematical Methods in Artificial Intelligence*, IEEE Computer Society Press, Los Alamitos, CA, chapter 1, pp. 26, 1996.

[2] Besnard, P., *An Introduction to Default Logic*, Springer-Verlag, Berlin, pp.27-35, 1989.

[3] Bezdek. J. C., Ed., *Pattern Recognition with Fuzzy Objective Function Algorithms*, Kluwer Academic Press, 1991.

[4] Buchanan, B. G. and Shortliffe, E. H., Eds., *Rule-Based Expert Systems*, Addison-Wesley, Reading, MA, 1984.

[5] Buchanan, B.G. and Wilkins, D.C., Eds., Readings in Knowledge Acquisition and Learning: Automating the Construction and Improvement of Expert Systems, Morgan Kaufmann, San Mateo, CA, 1993.

[6] Charniak, E. and McDermott, D., *Introduction to Artificial Intelligence*, Addison-Wesley, Reading, MA, 1985.

[7] Dean, T., Allen, J. and Aloimonds, Y., *Artificial Intelligence: Theory and Practice*, Addison-Wesley, Reading, MA, 1995.

[8] Dickmans, E.D., Mysliwetz, B. and Christians, T., "An integrated spatio-temporal approach to vehicles," *IEEE Trans. on Systems, Man and Cybernetics*, vol. 20, no. 6, Dec. 1990.

[9] Dougherty, E. R. and Giardina, C.R., *Mathematical Methods for Artificial Intelligence and Autonomous Systems*, Prentice-Hall, Englewood Cliffs, NJ, 1988.

[10] Feigenbaum, E. A. and Feldman, J., Eds., *Computers and Thought*, McGraw-Hill, New York, 1963.

[11] Haugeland, J., Ed., *Artificial Intelligence: The Very Idea*, MIT Press, Cambridge, 1985.

[12] Jackson, P., *Introduction to Expert Systems*, Addison-Wesley, Reading, MA, 1986.

[13] Jang, J. R., Sun, C. and Mizutani, E., *Neuro-Fuzzy and Soft Computing: A Computational Approach to Learning and Machine Intelligence*, Prentice-Hall, Englewood Cliffs, NJ, pp. 7-9, 1997.

[14] Konar A., *Uncertainty Management in Expert Systems using Fuzzy Petri Nets*, Ph.D. thesis, Jadavpur University, Calcutta, 1994.

[15] Kosko, B., *Neural Networks and Fuzzy Systems*, Prentice-Hall, Englewood Cliffs, NJ, 1994.

[16] Kruse, R., Schwecke, E. and Heinsohn, J., *Uncertainty and Vagueness in Knowledge Based Systems*, Springer-Verlag, Berlin, 1991.

[17] Kurzweil, R., *The Age of Intelligent Machines*, MIT Press, Cambridge, 1990.

[18] Luger, G. F. and Stubblefield, W. A., *Artificial Intelligence: Structures and Strategies for Complex Problem Solving*, Benjamin/Cummings, Menlo Park, CA, 1993.

[19] Mamdani, E. H., Application of fuzzy set theory to control systems, in *Fuzzy Automata and Decision Processes*, Gupta, M. M., Saridies, G. N. and Gaines, B. R., Eds., Oxford University Press, Amsterdam, New York, pp. 77-88, 1977.

[20] Mark, S. *Introduction to Knowledge Systems*, Morgan Kaufmann, San Mateo, CA, Chapter 5, pp. 433-458, 1995.

[21] Matlin, M. W., *Cognition*, Original U.S. edition by Harcourt Brace Publishers, Indian reprint by Prism Books Pvt. Ltd., Bangalore, 1994.

[22] McDermott, D. and Davis, E., "Planning routes through uncertain territory," *Artificial Intelligence*, vol. 22, 1984.

[23] Michalewicz, Z., *Genetic algorithms + Data Structures = Evolution Programs*, Springer-Verlag, Berlin, 1992.

[24] Naganuuma, J., Ogura, T., Yamada, S. I. and Kimura, T., "High speed CAM-based architecture for a PROLOG machine," *IEEE Trans. on Computers*, vol. 37, no. 11, November 1997.

[25] Narendra, K. S. and Parthasarathi, K., "Identification and control of dynamical system using neural networks," *IEEE Trans. on Neural Networks*, vol. 1. , pp. 4-27, 1990.

[26] Xiao, J., Michalewicz, Z., Zhang, L. and Trojanowski, K., "Adaptive Evolutionary Planner/Navigator for Mobile Robots," *IEEE Trans. on Evolutionary Computation*, vol. 1, no.1, April 1997.

[27] Newell, A .and Simon, H.A., *Human Problem Solving*, Prentice-Hall, Englewood Cliffs, NJ, 1972.

[28] Nilson, N. J., *Principles of Artificial Intelligence*, Morgan Kaufmann, San Mateo, CA, pp. 6-7, 1980.

[29] Patterson, D. W., *Introduction to Artificial Intelligence and Expert Systems*, Prentice-Hall, Englewood Cliffs, NJ, pp. 345-347, 1990.

[30] Pedrycz, W., *Fuzzy Sets Engineering*, CRC Press, Boca Raton, FL, 1995.

[31] Proceedings of First Int. Conf. on Automation, Robotics and Computer Vision, Singapore, pp. 1-133, 1990.

[32] Proceedings of Fourth Int. Conf. on Control, Automation, Robotics and Computer Vision, Singapore, 1-376, 1996.

[33] Quillian, M. R., Semantic Memory, In *Semantic Information Processing*, Minisky, M., Ed., MIT Press, Cambridge, MA, pp. 227-270, 1968.

[34] Rajendra, C. and Chaudhury, D., "An efficient heuristic approach to the scheduling of jobs in a flowshop," *European Journal of Operational Research*, vol. 61, pp. 318-325, 1991.

[35] Rich, E. and Knight, K., *Artificial Intelligence*, McGraw-Hill, New York, 1996.

[36] Russel, S. and Norvig, P., *Artificial Intelligence: A Modern Approach*, Prentice-Hall, Englewood Cliffs, NJ, 1995.

[37] Schalkoff, J., Culberson, J., Treloar, N. and Knight, B., "A world championship caliber checkers program," *Artificial Intelligence*, vol. 53, no. 2-3, pp. 273-289, 1992.

[38] Shobam, Y., *Artificial Intelligence Techniques in PROLOG*, Morgan Kaufmann, San Mateo, CA, 1994.

[39] Shortliffe, E. H., *Computer-based Medical Consultations*: MYCIN, Elsevier, New York, 1976.

[40] Spivey, M., *An Introduction to Logic Programming through PROLOG*, Prentice-Hall, Englewood Cliffs, NJ, 1996.

[41] Sympson, P., *Artificial Neural Nets*: *Concepts, Paradigms and Applications*, Pergamon Press, Oxford, 1988.

[42] Szats, A., Temporal logic of programs: standard approach, In *Time & Logic*, Leonard, B. and Andrzej, S., Eds., UCL Press, London, pp. 1-50, 1995.

[43] Takagi, T.and Sugeno, M., "Fuzzy identification of systems and its application to modeling and control," *IEEE Trans. on Systems, Man and Cybernetics*, pp. 116-132, 1985.

[44] Tanaka, K., "Stability and Stabilizability of Fuzzy-Neural-Linear Control Systems," *IEEE Trans. on Fuzzy Systems*, vol. 3, no. 4, 1995.

[45] Wah, B.W. and Li, G.J., "A survey on the design of multiprocessing systems for artificial intelligence," *IEEE Trans. On Systems, Man and Cybernetics*, vol. 19, no.4, July/Aug. 1989.

[46] Winograd, T., "Understanding natural language," *Cognitive Psychology*, vol. 1, 1972.

[47] Winston, P. H., *Artificial Intelligence*, Addison-Wesley, 2nd ed., Reading, MA, 1994.

[48] Yager, R.R., "Some relationships between possibility, truth and certainty," Fuzzy Sets and Systems, Elsevier, North Holland, vol. 11, pp. 151-156, 1983.

[49] Zadeh. L. A., " The role of fuzzy logic in the management of uncertainty in Expert Systems," *Fuzzy Sets and Systems*, Elsevier, North Holland, vol. 11, pp. 199-227, 1983.

[50] Zimmermann, H. J., *Fuzzy Set Theory and its Applications*, Kluwer Academic, Dordrecht, The Netherlands, 1991.

2

The Psychological Perspective of Cognition

The chapter provides an overview of the last thirty years' research in cognitive psychology with special reference to the representation of sensory information on the human mind. The interaction of visual, auditory and linguistic information in memory for understanding instances of the real world scenario has been elucidated in detail. The construction of mental imagery from the visual scenes and interpretation of the unknown scenes with such imagery have also been presented in this chapter. Neuro-physiological evidences to support behavioral models of cognition have been provided throughout the chapter. The chapter ends with a discussion on a new 'model of cognition' that mimics the different mental states and their inter-relationship through reasoning and automated machine learning. The scope of AI in building the proposed model of cognition has also been briefly outlined.

2.1 Introduction

The word 'cognition' generally refers to a faculty of mental activities dealing with abstraction of information from a real world scenario, their representation and storage in memory and automatic recall [27]. It also includes construction of higher level percepts from primitive/low level information/knowledge, hereafter referred to as perception. The chapter is an outgrowth of the last thirty years' research in neurocomputing and cognitive science. It elucidates various models of cognition to represent different types of sensory information and their integration on memory for understanding and reasoning with the real world context. It also outlines the process of construction of mental imagery from the real instances for subsequent usage in understanding complex three-dimensional objects.

The chapter starts with the cognitive perspective of pattern recognition. It covers elementary matching of 'sensory instances' with identical 'templates' saved in memory. This is referred to as the 'template matching theory'. The weakness of the template matching theory is outlined and the principle to overcome it through matching problem instances with stored minimal representational models (prototypes) is presented. An alternative feature- based approach for pattern recognition is also covered in this chapter. A more recent approach to 3-dimensional object recognition based on Marr's theory is also presented here.

The next topic, covered in the chapter, is concerned with cognitive models of memory. It includes the Atkinson-Shiffring's model, the Tulving's model and the outcome of the Parallel Distributed Processing (PDP) research by Rumelhart and McClelland [30].

The next section in the chapter deals with mental imagery and the relationship among its components. It includes a discussion on the relationship between object shape versus imagery and ambiguous figures versus their imagery. Neuro-physiological support to building perception from imagery is also outlined in this section. Representation of spatial and temporal information and the concept of relative scaling is also presented here with a specialized structure, called cognitive maps.

Understanding a problem and its representation in symbolic form is considered next in the chapter. Examples have been cited to demonstrate how a good representation serves an efficient solution to the problem.

The next section proposes a new model of cognition based on its behavioral properties. The model consists of a set of mental states and their possible inter-dependence. The state transition graph representing the model of cognition includes 3 closed cycles, namely, i) the sensing-action cycle, ii) the acquisition-perception cycle, and iii) the sensing-acquisition-perception-planning-action cycle. Details of the functionaries of the different states of cognition will be undertaken in this section.

The scope of realization of the proposed model of cognition will be briefly outlined in the next section by employing AI tools and techniques. A concluding section following this section includes a discussion on the possible direction of research in cognitive science.

2.2 The Cognitive Perspective of Pattern Recognition

The process of recognizing a pattern involves 'identifying a complex arrangement of sensory stimuli' [20], such as a character, a facial image or a signature. Four distinct techniques of pattern recognition with reference to both contexts and experience will be examined in this section.

2.2.1 Template-Matching Theory

A 'template' is part or whole of a pattern that one saves in his memory for subsequent matching. For instance, in template matching of images, one may search the template in the image. If the template is part of an image, then matching requires identifying the portion (block) of the image that closely resembles the template. If the template is a whole image, such as the facial image of one's friend, then matching requires identifying the template among a set of images [4]. Template matching is useful in contexts, where pattern shape does not change with time. Signature matching or printed character matching may be categorized under this head, where the size of the template is equal to the font size of the patterns.

Example 2.1: This example illustrates the principle of the template-matching theory. Fig. 2.1 (a) is the template, searched in the image of a boy in fig. 2.1(b). Here, the image is partitioned into blocks [5] equal to the size of the template and the objective is to identify the block in the image (fig. 2.1(b)) that best matches with the template (fig. 2.1 (a)).

(a)   (b)

Fig. 2.1: Matching of the template (a) with the blocks in (b).

The template-matching theory suffers from the following counts.

i)   **Restriction in font size and type:** Template-matching theory is not applicable to cases when the search domain does not include the template of the same size and font type. For instance, if someone wants to match a large-sized character, say Z, with an image containing a different font or size of letter Z, the template-matching theory fails to serve the purpose.

ii) **Restriction due to rotational variance:** In case the search space of the template contains a slightly rotated version of the template, the theory is unable to detect it in the space. Thus, the template-matching theory is sensitive to rotational variance of images.

It may be added here that the template-matching theory was framed for exact matching and the theory as such, therefore, should not be blamed for the reason for which it was meant. However, in case one wants to overcome the above limitations, he/she may be advised to use the Prototype-matching theory outlined below.

2.2.2 Prototype-Matching Theory

'Prototypes are idealized / abstract patterns' [20] in memory, which is compared with the stimulus that people receive through their sensory organs. For instance, the prototype of stars could be an asterisk (*). A prototype of a letter 'A' could be a symbol that one can store in his memory for matching with any of the patterns in fig. 2.2 (a) or the like.

Fig. 2.2 (a): Various fonts and size of 'A'.

Prototype-matching theory works well for images also. For example, if one has to identify his friend among many people, he should match the prototype of the facial image and his structure, stored in memory, with the visual images of individuals. The prototype (mental) image, in the present context, could include an approximate impression of the face under consideration. How exactly the prototype is kept in memory is unknown to the researchers still today.

An alternative approach to pattern recognition, to be discussed shortly, is the well-known feature-based matching.

2.2.3 Feature-based Approach for Pattern Recognition

The main consideration of this approach is to extract a set of primitive features, describing the object and to compare it with similar features in the sensory patterns to be identified. For example, suppose we are interested to identify whether character 'H' is present in the following list of characters (fig. 2.2 (b)).

$$A \quad H \quad F \quad K \quad L$$

Fig. 2.2 (b): A list of characters including H.

Now, first the elementary features of 'H' such as two parallel lines and one line intersecting the parallel lines roughly at half of their lengths are detected. These features together are searched in each of the characters in fig. 2.2 (b). Fortunately, the second character in the figure approximately contains similar features and consequently it is the matched pattern.

For matching facial images by the feature-based approach, the features like the shape of eyes, the distance from the nose tip to the center of each eye, etc. are first identified from the reference image. These features are then matched with the corresponding features of the unknown set of images. The image with the best matched features is then identified. The detailed scheme for image matching by specialized feature descriptors such as **fuzzy moments** [5] will be presented in chapter 23.

2.2.4 The Computational Approach

Though there exist quite a large number of literature on the computational approach for pattern recognition, the main credit in this field goes to David Marr. Marr [19] pioneered a new concept of recognizing 3-dimensional objects. He stressed the need for determining the edges of an object and constructed a 2 ½-D model that carries more information than a 2-D but less than a 3-D image. An approximate guess about the 3-D object, thus, can be framed from its 2 ½-D images.

Currently, computer scientists are in favor of a neural model of perception. According to them, an electrical analogue of the biological neural net can be trained to recognize 3-D objects from their feature space. A number of training algorithms have been devised during the last two decades to study the behavioral properties of perception. The most popular among them is the well-known back-propagation algorithm, designed after Rumelhart in connection with their research on Parallel Distributed Processing (PDP) [30]. The details of the neural algorithms for pattern recognition will be covered in chapter 14.

2.3 Cognitive Models of Memory

Sensory information is stored in the human brain at closely linked neuronal cells. Information in some cells can be preserved only for a short duration. Such memory is referred to as **Short Term Memory** (STM). Further, there are cells in the human brain that can hold information for a quite long time, of the order of years. Such memory is called **Long Term Memory** (LTM). STMs and LTMs can also be of two basic varieties, namely iconic memory and echoic memory. The **iconic memories** can store visual information, while the **echoic memories** participate in storing audio information. These two types of memories together are generally called sensory memory. Tulving alternatively classified human memory into three classes, namely episodic, semantic and procedural memory. Episodic memory saves facts on their happening, the semantic memory constructs knowledge in structural form, while the procedural ones help in taking decisions for actions. In this section, a brief overview of memory systems will be undertaken, irrespective of the type/ variety of memory; these memory systems together are referred to as cognitive memory. Three distinct classes of cognitive memory models such as Atkinson-Shiffrin's model, Tulving's model and the PDP model will be outlined in this section.

2.3.1 The Atkinson-Shiffrin's Model

The Atkinson-Shifrin's model consists of a three layered structure of memory (fig. 2.3). Sensory information (signals) such as scene, sound, touch, smell, etc. is received by receptors for temporary storage in sensory registers (memory). The **sensory registers** (Reg.) are large capacity storage that can save information with high accuracy. Each type of sensory information is stored in separate (sensory) registers. For example, visual information is saved in iconic registers, while audio information is recorded in echoic registers. The sensory registers decay at a fast rate to keep provisions for entry of new information. Information from the sensory registers is copied into short term memory (STM). STMs are fragile but less volatile than sensory registers. Typically STMs can hold information with significant strength for around 30 seconds, while sensory registers can hold it for just a fraction of a second. Part of the information stored in STM is copied into long term memory (LTM). LTMs have large capacity and can hold information for several years.

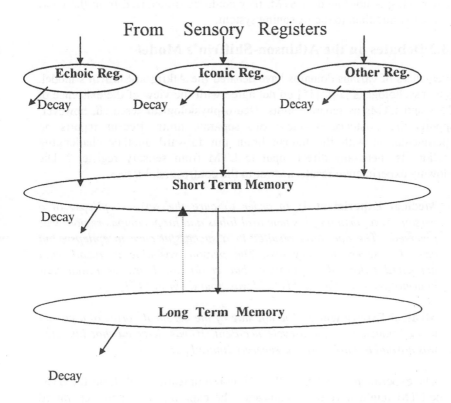

Fig. 2.3: Three-level hierarchical model of cognitive memory.

STMs have faster access time [1] than LTMs. Therefore, for the purpose of generating inferences, useful information from LTMs is copied into STMs. This has been shown in fig. 2.3 by a feedback path from LTM to STM. Because of its active participation in reasoning, STMs are sometimes called **active memory**.

The hierarchical structure of Atkinson-Shiffrin's cognitive memory model can be compared with the memory systems in computers. The STM is similar with cache, while the LTM could be compared with main (RAM) memory. The reasoning system in the human brain is analogous with the central processing unit (CPU) of the computer. The reasoning system fetches information from the STM, as CPU receives information from cache, when the addressed words are available in the cache. In case the addressed words are not available in cache, they are fetched from the main and transferred to the cache and the CPU as well. Analogously, when the information required for reasoning is not found in STM, they could be transferred from the LTM to the STM and then to the reasoning system.

2.3.2 Debates on the Atkinson-Shiffrin's Model

Researchers of various domains have debated the Atkinson-Shiffrin's model. Many psychologists [14], [15] do not agree with the view of the existence of STMs and LTMs as separate units. Neuro-physiological research, however, supports the existence of these two separate units. Recent reports on experimentation with the human brain put forward another challenging question: Is there any direct input to LTM from sensory registers? The following experimental results answer to the controversial issues.

> **Medical Experiment 1:** *In order to cure the serious epilepsy of a person X, a portion of his temporal lobes and hippocampus region was removed. The operation resulted in a successful cure in epilepsy, but caused a severe memory loss. The person was able to recall what happened before the operation, but could not learn or retain new information, even though his STM was found normal [21].*

> **Medical Experiment 2:** *The left side of the cerebral cortex of a person was damaged by a motorcycle accident. It was detected that his LTM was normal but his STM was severely limited [2].*

The experiment 1 indicates that the communication link from the STM to the LTM might have been damaged. The experiment 2, however, raised the question: how does the person input information to his LTM when his STM is damaged? The answer to this question follows from Tveter's model, outlined below.

Without referring to the Atkinson-Shiffrin's model, Tveter in his recent book [34] considered an alternative form of memory hierarchy (fig. 2.4). The sensory information, here, directly enters into the LTM and can be passed on to the STM from the LTM. The STM has two outputs leading to the LTM. One output of the STM helps in making decisions, while the other is for permanent storage in the LTM.

2.3.3 Tulving's Model

The Atkinson-Shiffrin's model discussed a flow of control among the various units of memory system. Tulving's model, on the other hand, stresses the significance of abstracting meaningful information from the environment by cognitive memory and its utilization in problem solving. The model comprises of three distinct units namely **episodic, semantic and procedural memory.**

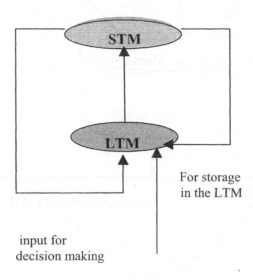

Fig. 2.4: Tveter's Model showing direct entry to the LTM.

Episodic memory stores information about happened events and their relationship. The semantic memory, on the other hand, represents knowledge and does not change frequently. The procedural memory saves procedures for execution of a task [33]. A schematic architecture of Tulving's model is presented in fig. 2.5.

The episodic memory in fig. 2.5 receives an information "the sky is cloudy" and saves it for providing the necessary information to the semantic memory. The semantic memory stores knowledge in an antecedent-consequent form. The nodes in the graph denote information and the arcs denote the causal relationship. Thus the graph represents two rules: Rule 1 and Rule 2, given below.

Rule 1: *If the sky is cloudy*
          Then it will rain.

Rule 2: *If it rains*
          Then the roads will be flooded.

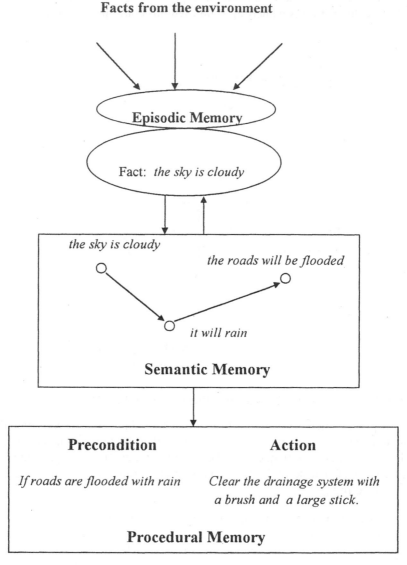

Fig. 2.5: Schematic architecture of Tulving's cognitive memory.

After execution of these two sequential rules, the semantic memory derives that 'the roads will be flooded'. The procedural memory first checks the pre-condition: "the road will be flooded" and then implements the action of cleaning the drainage system.

Tulving's model bridges a potential gap between the Atkinson-Shiffrin's model and the modern aspects of knowledge representation. For instance, the episodic memory contains only facts like data clauses in a PROLOG program (vide chapter 6). The semantic memory is similar with 'semantic nets' used for structured knowledge representation and reasoning (vide chapter 8). The procedural memory may be compared with a frame (vide chapter 8) that provides methods to solve a problem.

2.3.4 The Parallel Distributed Processing Approach

The outcome of the research on Parallel Distributed Processing (PDP) by Rumelhart, McClelland and their associates opened up a new frontier in machine learning. Unlike the other models of cognitive memory, the PDP approach rests on the behavioral characteristics of single cellular neurons. The PDP team considered the cognitive system as an organized structure of neurons, which together forms a neural net. Such a network has an immense capability to learn and save the learned information / knowledge for subsequent usage. The PDP approach thus supports the behavioral features of cognitive memory but cannot explain the true psychological perspectives of cognition. For instance, it cannot differentiate the STM with the LTM, but experimental evidences support their co-existence. However, irrespective of any such issues, the PDP approach undoubtedly has significance from the point of view of realization of cognition on machines. The fundamental characteristics of this approach, which gave it a unique status in cognitive science, are presented below.

♦ It is one of the pioneering works on cognition that resembled the biological memory as a distributed collection of single cellular neurons that could be trained in a time parallel manner.

♦ It demonstrated a possible realization of cognition on machines.

♦ For similar input patterns, the neural net should respond similarly, while for distinct input patterns with sufficient difference, the resulting responses of the neural net will also be different sufficiently.

In fact, this is a significant observation that led to the foundation of a completely new class of pattern recognition by **supervised learning**. In a supervised learning scheme there is a trainer that can provide the desired output for the given input patterns. The details of a supervised learning paradigm will be covered in a separate chapter on neural nets [35].

♦ The PDP approach supports the content addressable features of memory, rather than addressable memory.

In conventional computers we use random access memory, where we find the contents, once the address is supplied. But in our biological brain, we may sometimes remember part of an information and retrieve the whole after a while. Retrieving the whole information from its part is usually done by **content addressable memory (CAM)**.

2.4 Mental Imagery

How do people remember scenes? Perhaps they represent scenes in some form of image in their brain. The mental representation of scenes is informally called mental imagery [17] in cognitive science. This section explores two important areas in cognitive science. First, it covers mental imagery, its rotation and part-whole relationship. Next it presents **cognitive maps**[1] that denote the spatial relationship among objects. People form an idea of distances between any two places by their cognitive map.

2.4.1 Mental Representation of Imagery

Psychologists have a decade-long controversy on the mental representation of images of physical objects. One school of psychologists [17] believes that the images are stored in human memory in analog patterns, i.e., a similar prototype image of the object is recorded in memory. The other school [29] argues that we store images by symbolic logic-based codes. Symbolic logics are currently used in Artificial Intelligence to represent spatial, temporal and relational data and knowledge but are inadequate to describe complex shape of imagery. We are, however, trained to remember the shape of objects / animals with high accuracy, which probably could not be reproduced (decoded) from symbolic codes. Therefore, without going into the controversy, we may favor the opinion of the first school of psychologists.

2.4.2 Rotation of Mental Imagery

Psychologists are of the view that people can rotate their mental imagery, as we physically rotate the objects around a point or an axis. As experimental evidence, let us consider the character images in fig. 2.6.

---

[1] Cognitive maps in Artificial Intelligence, however, have a more general meaning. It stands for networks capable of acquiring, learning, encoding and decoding information / knowledge.

(a)   (b)

Fig. 2.6: The character 'A' and its 180 degree rotated
        view around its top vertex point.

It is clear from fig. 2.6 that (b) is the inverted view of that in (a). Based on the experimental evidences on rotation of mental imagery, the following points can be envisaged.

- More complex is the shape of the original image, more **(reaction) time** [20] is required to identify its rotated view [31].

- More is the angle of rotation, more is the reaction time to identify the uniqueness of the mental imagery [31].

- Non-identical but more close images and their rotated view require large reaction time to identify their non-uniqueness [32].

- Familiar figures can be rotated more quickly than the unfamiliar ones [8].

- With practice one can improve his / her reaction time to rotate a mental image [13].

2.4.3 Imagery and Size

In this section, the different views on imagery and size will be outlined briefly.

Kosslyn's view: Stephen Kosslyn was the pioneering researcher to study whether people make faster judgements with large images than smaller images. Kosslyn has shown that a mental image of an elephant beside that of a rabbit would force people to think of a relatively small rabbit. Again, the same image of the rabbit seems to be larger than that of a fly. Thus, people have their own relative judgement about mental imageries. Another significant contribution of Kosslyn is the experimental observation that people require more time to create larger mental imageries than smaller ones [18]. The results, though argued by the contemporary psychologists, however, follow directly from intuition.

Moyer's view: Robert Moyer provided additional information on the correspondence between the relative size of the objects and the relative size of their mental imagery. Moyer's results were based on psychophysics, the branch of psychology engaged in measuring peoples' reactions to perceptual stimuli [24]. In psychophysics, people take a longer time to determine which of the two almost equal straight lines is larger. Moyer thus stated that the reaction time to identify a larger mental image between two closely equal images is quite large.

Peterson's view: Unlike visual imagery, Intons-Peterson [28] experimented with auditory signals (also called images). She asked her students to first create a mental imagery of a cat's purring and then the ringing tone of a telephone set. She then advised her students to move the pitch of the first mental imagery up and compare it with the second imagery. After many hours of experimentation, she arrived at the conclusion that people require quite a large time to compare two mental imageries, when they are significantly different. But they require less time to traverse the mental imagery when they are very close. For instance, the imagery of purring, being close enough to the ticking of a clock, requires less time to compare them.

2.4.4 Imagery and Their Shape

How can people compare two similar shaped imageries? Obviously the reasoning process looks at the boundaries and compares the closeness of the two imageries. It is evident from commonsense reasoning that two imageries of an almost similar boundary require a longer time to determine whether they are identical. Two dissimilar shaped imageries, however, require a little reaction time to arrive at a decision about their non-uniqueness.

Paivio [26] made a pioneering contribution in this regard. He established the principle, stated above, by experiments with mental clock imagery. When the angle between the two arm positions in a clock is comparable with the same in another imagery, obviously the reaction time becomes large to determine which angle is larger. The credit to Paivio lies in extending the principle in a generic sense.

2.4.5 Part-whole Relationship in Mental Imagery

Reed was interested in studying whether people could determine a part-whole relationship of their mental image [20]. For instance, suppose one has saved his friend's facial image in memory. If he is now shown a portion of his friend's face, would he be able to identify him? The answer in many cases was in the affirmative.

Reed experimented with geometric figures. For instance, he first showed a Star of David (vide fig. 2.7) and then a parallelogram to a group of people and asked them to save the figures in their memory. Consequently, he asked them whether there exists a part-whole relationship in the two mental imageries. Only 14% of the people could answer it correctly. Thus determining part-whole relationship in mental imagery is difficult. But we do it easily through practicing.

Fig. 2.7: The Star of David.

2.4.6 Ambiguity in Mental Imagery

Most psychologists are of the opinion that people can hardly identify ambiguity in mental imagery, though they can do it easily with paper and pencil [20]. For instance consider the following imagery (vide fig. 2.8).

Fig. 2.8: The letter X topped by the letter H is difficult to extract from the mental imagery but not impossible by paper and pencil.

Peterson and her colleagues, however, pointed out that after some help, people can identify ambiguity also in mental imagery [28], [20].

2.4.7 Neuro Physiological Similarity between Imagery and Perception

The word 'perception' refers to the construction of knowledge from the sensory data for subsequent usage in reasoning. Animals including lower class mammals generally form perception from visual data. Psychologists, therefore, have long wondered: does mental imagery and perception have any neuro-physiological similarity? An answer to this was given by Farah [12] in 1988, which earned her the Troland award in experimental psychology. Goldenbarg and his colleagues [9] noted through a series of experiments that there exists a correlation between accessing of mental imagery and increased blood flow in the visual cortex. For example, when people make judgements with visual information, the blood flow in the visual cortex increases.

2.4.8 Cognitive Maps of Mental Imagery

Cognitive maps are the internal representation of real world spatial information. Their exact form of representation is not clearly known to date. However, most psychologists believe that such maps include both propositional codes as well as imagery for internal representation. For example, to encode the structural map of a city, one stores the important places by their imagery and the relationship among these by some logical codes. The relationship in the present context refers to the distance between two places or their directional relevance, such as place A is north to place B and at a distance of ½ Km.

How exactly people represent distance in their cognitive map is yet a mystery. McNamara and his colleagues [22] made several experiments to understand the process of encoding distance in the cognitive maps. They observed that after the process of encoding the road maps of cities is over, people can quickly remember the cities closely connected by roads to a city under consideration. But the cities far away by mileage from a given city do not appear quickly in our brain. This implicates that there must be some mechanisms to store the relative distances between elements in a cognitive map.

Besides representing distance and geographical relationship among objects, the cognitive maps also encode shapes of the pathways connecting the objects. For example, if the road includes curvilinear paths with large straight line segments, vide fig. 2.9, the same could be stored in our cognitive map easily [7]. However, experimental evidences show that people cannot easily remember complex curvilinear road trajectories (fig. 2.10). Recently, Moar and Bower [23] studied the encoding of angle formation by two non-collinear road trajectories. They observed experimentally that people have a general tendency to approximate near right angles as right angles in their cognitive map. For example, three streets that form a triangle in reality may not appear so in the cognitive map. This is due to the fact that the sum of the internal angles of a triangle in any physical system is 180 degrees; however, with the angles close to 90 degrees being set exactly to 90 degrees

in the cognitive map, the sum need not be 180 degrees. Thus a triangular
path appears distorted in the cognitive map.

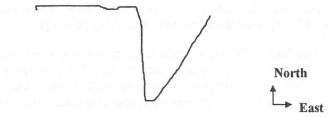

North

East

Fig. 2.9: A path with large straight-line segments.

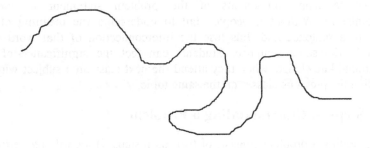

Fig. 2.10: A complex curvilinear path, difficult for encoding in
             a cognitive map.

2.5 Understanding a Problem

According to Greeno [10], understanding a problem involves constructing an
internal representation of the problem statement. Thus to understand a
sentence we must have some representation of the words and phrases and
some semantic links between the words, so that the construct resembles the
original sentence in meaning. The understanding of a problem, therefore,
calls for understanding the meaning of the words in the problem in more
elementary forms. Greeno stressed the need for three issues: *coherence,
correspondence and relationship to background knowledge* in connection
with understanding a problem.

Coherence: A coherent representation is a pattern, so that all its
components (fragments) make sense. Readers with a background of wave
propagation theory, of course, will wonder: why the term 'coherence'!
Coherence in wave propagation theory corresponds to wavelets (small
waves) with the same phase. In the present context, coherence stands for
equal emphasis on each component of the pattern, so that it is not biased to

one or more of its components. For example, to create a mental representation of the sentence "Tree trunks are straws for thirsty leaves and branches", one should not pay more emphasis on straws than trunks (stems of the trees). Formally, coherence calls for a mental representation of the sentence with equal emphasis on each word / fact / concept.

Correspondence: Correspondence refers to one-to-one mapping from the problem statement to the mental representation. If the mapping is incorrect or incomplete, then a proper understanding is not feasible. Correspondence, in most cases, however, is determined by the third issue, presented below.

Relationship to background knowledge: Background knowledge is essential to map components of the problem statement to mental representation. Without it people fail to understand the meaning of the words in a sentence and thus lose the interconnection of that word with others in the same sentence. Students can feel the significance of the background knowledge, when they attend the next class on a subject without attending the previous classes on the same topic.

2.5.1 Steps in Understanding a Problem

Understanding a problem consists of two main steps: i) identifying pertinent information from the problem description, deleting many unnecessary ones and ii) a well-organized scheme to represent the problem. Both the steps are crucial in understanding a problem. The first step reduces the scope of generality and thus pinpoints the important features of the problem. It should, of course, be added here that too much specialization of the problem features may sometimes yield a subset of the original problem and thus the original problem is lost. Determining the problem features, thus, is a main task in understanding the problem. Once the first step is over, the next step is to represent the problem features by an internal representation that truly describes the problem. The significance of the second step lies in the exact encoding of the features into the mental representation, so that the semantics of the problem and the representation have no apparent difference. The second step depends solely on the type of the problem itself and thus differs for each problem. In most cases, the problem is represented by a specialized data structure such as matrices, trees, graphs, etc. The choice of the structure and organization of the data / information by that structure, therefore, should be given priority for understanding a problem. It should be mentioned here that the time-efficiency in understanding a problem depends largely on its representation and consequently on the selection of the appropriate data structures. A few examples are presented below to give the readers some idea about understanding and solving a problem.

Example 2.2: This example demonstrates how a graphic representation can help in solving a complex problem. The problem is with a monk. He started climbing up a tall mountain on one sunny morning and reached the top on the same evening. Then he started meditating for several days in a temple at the hilltop. After several days, on another sunny morning, he left the temple and started climbing down the hill through the same road surrounding the mountain. The road is too narrow and can accommodate only one passenger at one time.

The problem is to prove that *there must be a point on the hill that the monk will visit at the same time of the day both in his upward and downward journey*, irrespective of his speed. This problem can be best solved by assuming that there are two monks, one moving up, while the other is climbing down the hill. They started moving at the same time of the day. Since the road is narrow, they must meet at some spot on the road (vide fig. 2.11).

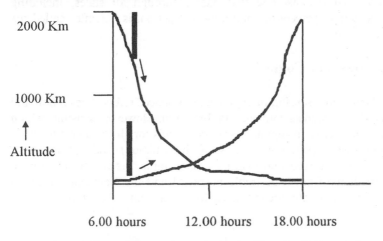

The vertical bars denote the monks. The upward (downward) arrow indicates the upward (downward) journey of the monks. Note that the two monks must meet at some point on the road.

Fig. 2.11: Representation of the monk problem.

It may be noted that the main stress of the problem should be given to the meeting of the monks only and should not be confused with their meeting time. The solution of the given problem is a simple extension of the modified problem with two monks, which the reader can guess easily.

There exist quite a large number of interesting problems (see exercises) that can be efficiently represented by specialized data structures. For instance, the 4-puzzle problem can be described by a matrix; the water-jug problem by a graph and the missionaries-cannibals problem by a tree. There exist also a different variety of problems that could be formulated by propositional codes or other knowledge representation techniques. Identifying the best representation of a problem is an art and one can learn it through intuition only. No hard and fast rules can be framed for identifying the appropriate representation of the problem space.

2.6 A Cybernetic View to Cognition

An elementary model of cognition (vide fig. 2.12) is proposed in this section based on its foundation in cognitive psychology [7], [11]. The model consists of a set of 5 mental states, denoted by ellipses, and their activation through various physiological and psychological processes. The model includes feedback of states and is thus cyclic. The model [16] in fig. 2.12, for instance, contains three cycles, namely *the perception-acquisition cycle, the sensing-action cycle,* and the last one that passes through all states, including sensing, acquisition, perception, planning and action, is hereafter called *the cognition cycle* [16].

2.6.1 The States of Cognition

Sensing: Apparently, sensing in engineering sciences refers to reception and transformation of signals into measurable form. However, sensing, which has a wider perspective in cognitive science, stands for all the above together with pre-processing (filtering from stray information) and extraction of features from the received information. For example, visual information on reception is filtered from undesirable noise [5] and the elementary features like size, shape, color, etc. are extracted for storing into short term memory (STM).

Acquisition: The acquisition state compares the response of the STM with already acquired and permanently stored information of the LTM. The content of LTM, however, changes occasionally, through feedback from the perception state. This process, often called refinement of knowledge, is generally carried out by a process of unsupervised learning. The learning is unsupervised since such refinement of knowledge is an autonomous process and requires no trainer for its adaptation.

Perception: This state constructs high level knowledge from acquired information of relatively lower level and organizes it, generally, in a structural form for the efficient access of knowledge in subsequent phases.

The construction of knowledge and its organization is carried out through a process of automated reasoning that analyzes the semantic (meaningful) behavior of the low-level knowledge and their association. The state of perception itself is autonomous, as the adaptation of its internal parameters continues for years long until death. It can be best modeled by a semantic net [3], [6] .

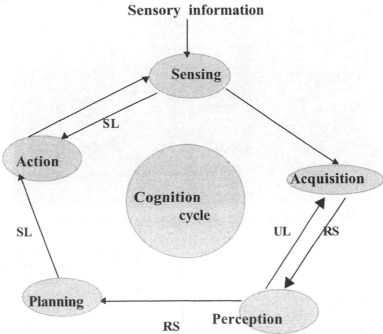

Sensory information

SL= Supervised learning, UL= Unsupervised learning, RS= Reasoning
Fig. 2.12: The different mental states of cognition and their relationship.

Planning: The state of planning engages itself to determine the steps of action involved in deriving the required goal state from known initial states of the problem. The main task of this state is to identify the appropriate piece of knowledge for application at a given instance of solving a problem. It executes the above task through matching the problem states with its perceptual model, saved in the semantic memory.

It may be added here that planning and reasoning, although sharing much common formalism, have a fundamental difference that originates from their nomenclature. The reasoning may be continued with the concurrent execution of the actions, while in planning, the schedule of actions are derived and executed in a later phase. In our model of cognition, we, thus, separated the action state from the planning state.

Action: This state determines the control commands for actuation of the motor limbs in order to execute the schedule of the action-plan for a given problem. It is generally carried out through a process of supervised learning, with the required action as input stimulus and the strength of the control signals as the response.

Example 2.3: This example demonstrates the various states of cognition with reference to a visual image of a sleeping cat in a corridor (Fig. 2.13).

Fig. 2.13: Digital image of a sleeping cat in a corridor.

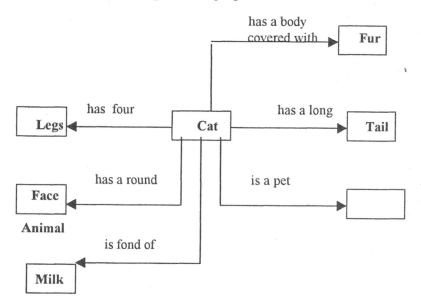

Fig. 2.14: The state of perception about a cat in the semantic
memory.

Here, the sensing unit is a video camera, which received the digital image of the cat in the corridor. The image was then pre-processed and its bit-map was saved in a magnetic media, which here acts as an acquisition unit. The Acquisition State of the model of cognition, thus, contains only a pixel-wise intensity map of the scene. Human beings, however, never store the bit-map of a scene; rather, they extract some elementary features, for instance, the shape of the face (round / oval-shaped), length of the tail (long / too long / short), texture of the fur and the posture of the creature. The extracted features of a scene may vary depending on age and experience of the person. For example, a baby of 10 months only extracts the (partial) boundary edges of the image, while a child of 3 years old can extract that "the face of the creature is round and has a long tail". An adult, on the other hand, gives priority to postures, saves it in STM and uses it as the key information for subsequent search in the LTM model of perception. The LTM in the present context is a semantic net, which keeps record of different creatures with their attributes. A typical organization of a semantic net, representing a cat (fig.2.14) and a corridor (fig. 2.15) is presented below.

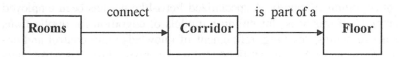

Fig. 2.15 : The state of perception about a corridor in the semantic memory.

Now, for illustrating the utility of the perception, planning and action states, let us consider the semantic net for the following sentence in fig. 2.16.

"The milkman lays packets full of milk in the corridor."

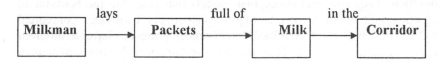

Fig. 2.16: Semantic net of a typical fact in the state of perception.

Combining all the above semantic nets together, we form a composite model of the entire scene, which together with the additional piece of knowledge: *"If a living creature is fond of something and it is kept away from him, then he cannot access it,"* helps someone to generate the following schedule of plans through a process of backward reasoning [16].

Plan: *Keep the packets full of milk away from the cat.*

Further, for execution of the above plan, one has to prepare the following schedule of actions:

    1.   Move to the corridor.
    2.   Pick up the packets of milk.
    3.   Keep them in a safe place beyond the reach of the cat.

It may be noted that the generation of the above schedule of actions for a given plan by human beings requires almost no time. This, perhaps, is due to the supervised learning scheme that helps speeding up the generation of such a schedule.

The semantic net that serves as a significant tool for knowledge representation and reasoning requires further extension for handling various states of cognition efficiently. A specialized Petri-like net has been employed in chapter 16 of this book, for building models of cognition for applications in inexact reasoning, learning, refinement of knowledge and control and co-ordination of tasks by an artificial cognitive map. The Petri-like nets mentioned here, however, have structural resemblance only with ideal Petri nets [25] but are distinct with respect to their properties.

2.7 Scope of Realization of Cognition in Artificial Intelligence

'Cognition' being an interdisciplinary area has drawn the attention of peoples of diverse interest. The psychologists study the behavioral aspects of cognition. They construct conceptual models that resemble the behavior of cognition and interpret the biological phenomena with their conceptual models. Researchers of Artificial Intelligence, however, have a different attitude towards cognition. They observe the biological behavior of human beings and attempt to realize such behavior on an **intelligent agent** by employing intelligent tools and techniques. A robot, for example, could be such an agent, which receives sensory signals from its environment and acts on it by its actuators and motor assemblies to execute a physical task. A question then naturally arises: should we call all the sensing-action cycle executing agents artificially intelligent? If so, where is their intelligence?

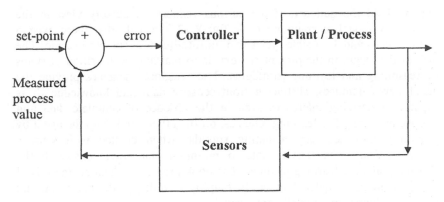

Fig. 2.17: A process control loop that executes the sensing-action cycle.

A little thinking, however, reveals that most of the closed loop control systems sense signals from their environment and act on it to satisfy a desired goal (criterion). There is a controller in the control loop that receives the deviation (error) of the measured signal from the desired signal (set-point), and generates a control command for the plant. The plant in turn generates an output signal, which is fed back to the controller through the error detector module (vide fig. 2.17). The controller could be analog or digital. An analog controller is a lag/ lead network, realized with R-C circuits. A digital controller, on the other hand, could be realized with a microcomputer that recursively executes a difference equation. Such controllers were called intelligent two decades back. But, they are never called artificially intelligent. So a mere performer of the sensing-action cycle in the elementary model of cognition (vide fig. 2.12) cannot be called artificially intelligent. But they can be made intelligent by replacing the standard controllers by a knowledge-based system. It should be added here that a sensing-action cycle performer too is sometimes called artificially intelligent, in case it requires some intelligent processing of the raw information. For instance, consider a robot that has to plan its trajectory from a predefined initial position of the gripper to a final position. Suppose that the robot can take images of its neighboring world by a camera and determine the 3-D surfaces around it that block its motion. There may exist a large number of possible trajectories of gripper movement and the robot has to determine the shortest path without hitting an obstacle. Such schemes obviously require much of AI tools and techniques and thus should be called artificially intelligent. But what are the tools that can make them intelligent?

The book provides an answer to this question through its next 22 chapters. A brief outline of the answer to the question, however, is presented here to hold the patience of the curious readers. A cursory view to the elementary model of cognition (vide fig. 2.12) reveals that there exist 5 mental states and 3 possible cycles that an intelligent agent can execute. The task of the agent, in the present context, is to maintain a transition of states by reasoning and learning paradigms. The reasoning schemes provide the agent new inferences, abstracted from sensory data and knowledge. It is capable of deriving inferences even in the absence of complete data and knowledge bases. The learning schemes, on the other hand, help the agent by providing him necessary actuating signals, when excited with sensory signals. The agent thus is able to maintain state-transitions with the reasoning and the learning modules. The book provides a detailed analysis of the mathematical models that can be employed to design the reasoning and learning schemes in an intelligent agent.

It must be added here that the model of cognition (vide fig. 2.12) is a generic scheme and the whole of it need not be realized in most intelligent agents. We now list some of the possible realization of the agents.

Pattern Recognition Agent: A pattern recognition agent receives sensory signals and generates a desired pattern to be used for some definitive purposes. For example, if the agent is designed for speaker recognition, one has to submit some important speech features of the speakers such as pitch, format frequencies, etc. and the system would be able to give us the speaker number. If the recognition system is to recognize objects from their visual features, then one has to submit some features such as the largest diagonal that the 2-D image can inscribe, the smallest diagonal that it can inscribe and the area of the 2-D image surface. In turn, the system can return the name of the 2-D objects such as ellipse, circle, etc. It may be mentioned here that a pattern recognition system realizes only the sensing-action cycle of the cognition.

A Path Planning Agent: A path planning agent perhaps is one of the complete agents that uses all the states in the elementary model of cognition (fig. 2.12). Such agents have ultrasonic sensors / laser range finders by which it can sense obstacles around it. It saves the sensed images in its short term memory and then extracts knowledge about the possible locations of the obstacles. This is referred to as the obstacle map of the robot's environment. For subsequent planning and the action cycle, the robot may use the obstacle map. Details of the path planning scheme of the mobile robot will be presented in chapter 24.

2.8 Summary

Cognitive science has emerged as a new discipline of knowledge that deals with the mental aspects of human beings. The chapter aims at establishing the psychological perspectives of the human cognition. It elucidated the various models of human memory and representation of imagery and cognitive maps on memory. The mechanism of understanding a problem is also presented here with special reference to representation of the problems.

Artificial Intelligence, on the other hand, is a young branch of science that rests on the theme of building intelligent machines. The chapter briefly outlined the fundamental principles of cognitive science and demonstrated the possible ways of realizing them on intelligent machines. The tools and techniques required for its possible realization have been referred to only. But their detailed mechanism will be covered throughout the book.

The special feature of the chapter is the cybernetic view to cognition. The elementary model of cognition has 5 mental states. These states can undergo transformation under appropriate reasoning and learning cycles. An intelligent agent can autonomously control the transition of states through reasoning and learning mechanisms. An agent need not always be a person. A machine that receives sensory information and acts accordingly on the environment can also play the role of an agent. Thus modern robots are ideal agents. The chapter demonstrated some applications of these agents in pattern recognition and path planning amidst obstacles.

Exercises

1. A template image of dimension of (m x m) pixels is to be searched in a digital image of dimension (n x n). Assume that mod (n / m) =0 and n >> m. If the matching of the template block with equal sized image blocks is carried out at an interleaving of (m / 2) pixels, both row and column-wise, determine the number of times the template is compared with the blocks in the image [5].

    [**Hints:** The number of comparison per row of the image = (2n / m -1).
    The number of comparison per column of the image = (2n/ m - 1).
    Total number of comparison = $(2 n / m -1)^2$.]

2. For matching a template image of dimension (m x m) pixels with a given image of dimension (n x n), where n >>m and mod (n / m) =0, one uses the above interleaving.

Further, instead of comparing pixel-wise intensity, we estimate the following 5 parameters (features) of the template with the same parameters in each block of the image [4] :

$$\text{Mean intensity } M_i = \sum_{b=0}^{L-1} b\, P(b),$$

$$\text{Variance of intensity } V_i^2 = \sum_{b=0}^{L-1} \{ (b-M_i)^2 P(b) \},$$

$$\text{Skewness of intensity } Sk_i = (1/ V_i^3) \sum_{b=0}^{L-1} \{ (b-M_i)^3 P(b) \},$$

$$\text{Kurtosis of intensity } Ku_i = (1 / V_i^4) \sum_{b=0}^{L-1} \{ (b-M_i)^4 P(b) \}, \text{ and}$$

$$\text{Energy of intensity } E_i = \sum_{b=0}^{L-1} \{P(b)\}^2,$$

where b represents the gray level of each pixel in a block i and L is the number of gray levels in the image.

The square of absolute deviation of these features of the i-th block from the corresponding features of the template is denoted by m_i^2, v_i^2, sk_i^2, ku_i^2, e_i^2 respectively.

Let d_i denote a measure of distance between the features of the i-th block to that of the template. Show (logically) that the *weakest feature match model* identifies the j-th block, by estimating

$$d_i = \text{Max} \{ m_i^2, v_i^2, sk_i^2, ku_i^2, e_i^2 \}, \qquad 1 \le \forall\ i \le (2n / m -1)^2$$

such that $d_j = \text{Min} \{ d_i \mid 1 \le \forall\ i \le (2n / m -1)^2 \}$ [4].

Also show that *the strongest feature match model* identifies the j-th block by estimating

$d_i = Min \{ m_i^2, v_i^2, sk_i^2, ku_i^2, e_i^2 \}, \quad 1 \leq \forall \ i \leq (2n / m - 1)^2$

such that $d_j = Min \{ d_i \mid 1 \leq \forall \ i \leq (2n / m - 1)^2 \}$ [4].

Further show that *the Euclidean least square model* identifies the j-th block by estimating

$d_i = \{ m_i^2 + v_i^2 + sk_i^2 + ku_i^2 + e_i^2 \}^{\frac{1}{2}}, 1 \leq \forall \ i \leq (2n / m - 1)^2$

such that $d_j = Min \{ d_i \mid 1 \leq \forall \ i \leq (2n / m - 1)^2 \}$ [4].

3. Write a program in your favorite language to match a given template with the blocks in an image, using the above guidelines. Mark the actual block in the image that you want to identify and measure the shift of the matched blocks by executing your program. Perform the experiments with different templates on the same image and then conclude which of the three measures is the best estimator of template matching.

4. Given two jugs, one 4 liters and the other 3 liters with no markings in them. Also given a water supply with a large storage. Using these 2 jugs, how can you separate 2 liters of water? Also draw the tree representing the transition of states.

    [**Hints:** Define operators such as filling up a jug, evacuating a jug, transferring water from one jug to the other etc. and construct a tree representing change of state of each jug due to application of these operators. Stop when you reach a state of 2 liters water in either of the 2 jugs. Do not use repeated states, since repeated states will result in a graph with loops. If you cannot solve it yourself, search for the solution in the rest of the book.]

5. Redraw the state-space for the water jug problem, by allowing repetition of the states. Should you call it a tree yet?

6. Show by Tulving's model which of the following information/knowledge should be kept in the episodic, semantic and procedural memories.

    a) There was a huge rain last evening.
    b) The sky was cloudy.
    c) The moon was invisible.
    d) The sea was covered with darkness.
    e) The tides in the sea had a large swinging.
    f) The boatmen could not control the direction of their boats.

g) It was a terrific day for the fishermen sailing in the sea by boats.
h) Since the sea was covered with darkness, the boatmen used a battery driven lamp to catch fish.
i) Because of the large swinging of the tides, the fishermen got a large number of fish caught in the net.
j) The net was too heavy to be brought to the seashore.
k) The fishermen used large sticks to control the motion of their boats towards the shore.

7. The Atkinson-Shiffrin's model can be represented by first ordered transfer functions, vide fig.2.18, presented below:

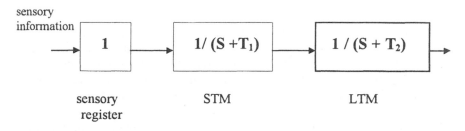

Fig.2.18: Schematic representation of the Atkinson-Shiffrin's model.

Given T_1 = 10 seconds and T_2 = 30 minutes, find the time response of the STM and the LTM, when a unit impulse is used as an excitation signal for the sensory register. Also given that the Laplace inverse of

$1 / (S + a)$ is e^{-at}.

Now suppose we replace the excitation signal by a 0.5 unit impulse. Find the response of the STM and the LTM. Can we distinguish the current responses from the last ones? If the answer is yes, what does the result imply?

References

[1] Atkinson, R. C. and Shiffrin, R. M., Human memory: A proposed system and its control process, in *The Psychology of Learning and Motivation: Advances in Research and Theory*, Spence, K. W. and Spence, J. T., Eds., vol. 2, Academic Press, New York, 1968.

[2] Baddeley, A. D., "The fractionation of human memory," *Psychological Medicine*, vol.14, pp. 259-264, 1984.

[3] Bharick, H. P., "Semantic memory content in permastore: Fifty years of memory for Spanish learned in school," *Journal of Experimental Psychology: General*, vol. 120, pp.20-33, 1984.

[4] Biswas, B., Mukherjee, A. K. and Konar, A., "Matching of digital images using fuzzy logic," *Advances in Modeling & Analysis*, B, vol. 35, no. 2, 1995.

[5] Biswas, B., Konar, A. and Mukherjee, A. K., "Fuzzy moments for digital image matching," Communicated to *Engineering Applications of Artificial Intelligence*, Elsevier/North-Holland, Amsterdam; also appeared in the *Proc. of Int. Conf. on Control, Automation, Robotics and Computer Vision*, ICARCV, '98, 1998.

[6] Chang, T. M., "Semantic memory: Facts and models," *Psychological Bulletin*, vol. 99, pp. 199-220, 1986.

[7] Downs, R. M. and Davis, S., *Cognitive Maps and Spatial Behavior: Process and Products*, Aldine Publishing Co., 1973.

[8] Duncan, E. M. and Bourg, T., "An examination of the effects of encoding and decision processes on the rate of mental rotation," *Journal of Mental Imagery*, vol. 7, pp. 33-56, 1983.

[9] Goldenberg, G., Podreka, I., Steiner, M., Suess, E., Deecke, L. and Willmes, K., Pattern of regional cerebral blood flow related to visual and motor imagery, Results of Emission Computerized Tomography, in *Cognitive and Neuropsychological Approaches to Mental Imagery*, Denis, M., Engelkamp, J., and Richardson, J. T. E., Martinus Nijhoff Publishers, Dordrecht, The Netherlands, pp. 363-373, 1988.

[10] Greeno, J. G., Process of understanding in problem solving, in *Cognitive Theory*, Castellan, N. J., Pisoni, Jr. D. B. and Potts, G. R., Hillsdale, NJ: Erlbaum, vol. 2, pp. 43-84,1977.

[11] Gross, C. G. and Zeigler, H. P., *Readings in Physiological Psychology: Learning and Memory*, Harper & Row, New York, 1969.

[12] Farah, M. J., "Is visual imagery really visual? Overlooked evidence from neuropsychology," *Psychological Review*, Vol. 95, pp. 307-317, 1988.

[13] Jolicoeur, P., "The time to name disoriented natural objects," *Memory and Cognition*, vol. 13, pp. 289-303, 1985.

[14] Kintsch, W., "The role of knowledge in discourse comprehension: A construction-integration model," *Psychological Review*, vol. 95, pp. 163-182, 1988.

[15] Kintsch, W. and Buschke, H., "Homophones and synonyms in short term memory," *Journal of Experimental Psychology*, vol. 80, pp. 403-407, 1985.

[16] Konar, A. and Pal, S., Modeling cognition with fuzzy neural nets, in *Fuzzy Logic Theory: Techniques and Applications*, Ed. Leondes, C. T., Academic Press, New York, 1999.

[17] Kosslyn, S. M., Mental imagery, In *Visual Cognition and Action: An Invitation to Cognitive Science*, Osherson, D. N. and Hollerback, J. M., Eds. , vol. 2, pp. 73-97, 1990.

[18] Kosslyn, S. M., "Aspects of cognitive neuroscience of mental imagery," *Science*, vol. 240, pp. 1621-1626, 1988.

[19] Marr, D., *Vision*, W. H. Freeman, San Francisco, 1982.

[20] Matlin, M. W., *Cognition*, Harcourt Brace Pub., Reprinted by Prism Books Pvt. Ltd., Bangalore, India, 1994.

[21] Milner, B., "Amnesia following operation on the Temporal lobe," in *Amnesia Following Operation on the Temporal Lobes*, Whitty, C. W. M. and Zangwill O. L.,Eds., Butterworth, London, pp. 109-133,1966.

[22] McNamara, T. P., Ratcliff, R. and McKoon, G., "The mental representation of knowledge acquired from maps," *Journal of Experimental Psychology: Learning, Memory and Cognition*, pp. 723-732, 1984.

[23] Moar, I. and Bower, G. H., "Inconsistency in spatial knowledge," *Memory and Cognition*, vol. 11, pp.107-113, 1983.

[24] Moyer, R. S. and Dumais, S. T., Mental comparison, in *The psychology of Learning and Motivation*, Bower, G. H., Ed., vol. 12, pp. 117-156, 1978.

[25] Murata, T., "Petri nets: Properties, analysis and applications," *Proc. of the IEEE*, vol. 77, no. 4, pp. 541-580, 1989.

[26] Paivio, A., On exploring visual knowledge, in *Visual Learning, Thinking and Communication*, Randhawa, B. S. and Coffman, W. E., Eds., Academic Press, New York, pp. 113-132, 1978.

[27] Pal, S. and Konar, A., "Cognitive reasoning with fuzzy neural nets," *IEEE Trans. on Systems, Man and Cybernetics*, Part - B, August, 1996.

[28] Peterson, M. A., Kihlstrom, J. F., Rose, P. M. and Glisky, M. L., "Mental images can be ambiguous: Reconstruals and Reference-frame reversals," *Memory and Cognition*, vol. 20, pp. 107-123, 1992.

[29] Pylyshyn, Z. W., Imagery and Artificial Intelligence, in Minnesota Studies in the Philosophy of Science, vol. 9: *Perception and Cognition Issues in the Foundations of Psychology*, University of Minnesota Press, Minneapolis, pp. 19-56, 1978.

[30] Rumelhart, D. E., McClelland, J. L. and the PDP research group, *Parallel Distributed Processing: Explorations in the Microstructure of Cognition*, vol. 1 and 2, MIT Press, Cambridge, MA, 1968.

[31] Shepard, R. N. and Cooper, L.A. (Eds.), *Mental Images and Their Transformations*, MIT Press, Cambridge, MA, 1982.

[32] Shepherd, R. N. and Metzler, Z., "Mental rotation of three dimensional objects," *Science*, vol. 171, pp. 701-703, 1971.

[33] Tulving, E., "Multiple memory systems and consciousness," *Human Neurobiology*, vol. 6, pp. 67-80, 1987.

[34] Tveter, D. R., *The Pattern Recognition Basis of Artificial Intelligence*, IEEE Computer Society Press, Los Alamitos, pp. 168-169, 1998.

[35] Washerman, P. D., *Neural Computing: Theory and Practice*, Van Nostrand Reinhold, New York, 1989.

3

Production Systems

'Production systems' is one of the oldest techniques of knowledge representation. A production system includes a knowledge base, represented by production rules, a working memory to hold the matching patterns of data that causes the rules to fire and an interpreter, also called the inference engine, that decides which rule to fire, when more than one of them are concurrently firable. On firing of a rule, either its new consequences are added to the working memory or old and unnecessary consequences of previously fired rules are dropped out from the working memory. The addition to and deletion from working memory depends on the consequent (then) part of the fired rule. Addition of new elements to the working memory is required to maintain firing of the subsequent rules. The deletion of data elements from the working memory, on the other hand, prevents a rule from firing with the same set of data. This chapter provides a detailed account of production systems, its architecture and relevance to state-space formulation for problem solving.

3.1 Introduction

Knowledge in an Expert System can be represented in various ways. Some of the well-known techniques for representation of knowledge include

Production Systems, Logical Calculus and Structured Models. This chapter is devoted to the Production System-based approach of knowledge representation. Logical Calculus-based methods for knowledge representation are covered in chapter 5, 6 and 7 while the structured models for reasoning with knowledge are presented in chapter 8. The reasoning methodologies presented in chapter 3, 5 and 6 are called *monotonic* [8] as the conclusions arrived at in any stage of reasoning do not contradict their predecessor premises derived at earlier. However, reasoning people often apply common sense, which in many circumstances results in conclusions that contradict the current or long chained premises. Such a type of reasoning is generally called *non-monotonic* [8]. A detailed account of the non-monotonic logics will be covered later in chapter 7. The reasoning methodologies covered in chapter 3- 8 do not presume any temporal and spatial variations of their problem states. The issues of spatio-temporal models for reasoning will be taken up later in chapter 11.

This chapter is an opening chapter on knowledge representation. We, therefore, discuss some elementary aspects, relevant to this chapter. Before presenting the technique for knowledge representation by Production Systems, we define the term "Knowledge", which is widely used throughout the text.

Formally, *a piece of knowledge is a function that maps a domain of clauses onto a range of clauses. The function may take algebraic or relational form depending on the type of applications.* As an example consider the production rule PR_1 , which maps a mother-child relationship between (m, c) to a Love relationship between the same pair.

PR_1: *Mother (m, c) $\rightarrow$ Loves (m, c)*

where the clause Mother (m, c) describes that "m" is a mother of child "c"; the clause Loves (m, c) denotes that "m" loves "c" and the arrow denotes the if-then condition. In brief, the rule implicates: *if "m" is a mother of child "c" then "m" loves "c".*

The production system is the simplest and one of the oldest techniques for knowledge representation. A production system consists of three items: i) a set of production rules (PR), which together forms the knowledge base, ii) One (or more) dynamic database(s), called the working memory and iii) a control structure / interpreter, which interprets the database using the set of PRs [4], [7]. The production system, which has wide applications in automata theory, formal grammars and the design of programming languages, however, entered into knowledge engineering (1978) by Buchanan and Feigenbarm [2] only a few years back. Before presenting the architecture of a production system, applied to intelligent problem solving, let us first introduce the functionaries of its modules.

3.2 Production Rules

The structure of a production rule PR_1 can be formally stated as follows:

$$PR_1: \ P_1 \ (X) \ \Lambda \ P_2 \ (Y) \ \Lambda \ .. \ P_n \ (X, \ Z) \ \rightarrow \ Q_1 \ (Y) \ V \ Q_2 \ (Z) \ V .. \ Q_m \ (Y, \ X)$$

where P_i and Q_j are predicates; x, y, z are variables; "Λ ", " V" , and " $\rightarrow$" denote the logical AND, OR and if-then operators respectively. The left-hand side of a PR is called the antecedent / conditional part and the right-hand side is called the consequent / conclusion part. Analogously, the left-side symbol P_i is called the antecedent predicate, while the right-side symbol Q_j is called the consequent predicates.

It should be pointed out that the antecedent and consequent need not be always predicates. They may equally be represented by object-attribute-value triplets. For example, (person-age-value) may be one such triplet. To represent the rules in this fashion, we consider an example, presented in PR2.

PR2 : if (person age above-21) &
     (person wife nil) &
     (person sex male)
  then (person eligible for marriage) .

It should further be noted that though object-attribute-value in PRs are often represented using variables, still the presence of constants in the triplet-form cannot be excluded. PR3, given below, is one such typical example.

PR3: if (Ram age 25) &
     (Ram wife nil) &
     (Ram sex male)
  then (Ram eligible for marriage).

In the last example person's name and age are explicit in the PR.

3.3 The Working Memory

The working memory (WM) generally holds data either in the form of clauses or object-attribute-value (OAV) triplet form. The variables in the antecedent predicates / OAV relationship of the antecedent part of PRs are matched against the data items of the WM. In case all the variable instantiation of the antecedent parts of a rule are consistent, then the rule is fired and the new consequents are added to the WM. In some production systems, the right-hand-side of the rule indicates which data are to be added to or deleted from the WM. Normally, new consequents are added to the WM and some old data of WM, which are no longer needed, are deleted from the WM to minimize the search time required for matching the antecedent parts of a rule with the data in WM. OPS5 is a production language that offers the addition / deletion features highlighted above.

3.4   The Control Unit / Interpreter

The control unit / interpreter for a production system passes through three steps, which together is called the recognize-act cycle [4].

Recognize-Act Cycle

1.   Match the variables of the antecedents of a rule, kept in a knowledge base, with the data recorded in the WM.

2.   If more than one rule, which could fire, is available then decide which rule to fire by designing a set of strategies for resolving the conflict regarding firing of the rules.

3.   After firing of a rule, add new data items to WM or delete old (and unnecessary) data, as suggested by the fired rule from the WM and go to step (1).

Generally, a start-up element is kept at the working memory at the beginning of a computation to get the recognize-act cycle going. The computation process is terminated if no rule fires or the fired rule contains an explicit command to halt.

The conflict resolution process helps the system by identifying which rule to fire. It is, however, possible to construct a rule-set where only one rule is firable at any instant of time. Such systems are called **deterministic**. Since most of the real world problems contain a **non-deterministic** set of rules, it becomes difficult for many systems to present the rule-set in a deterministic manner.

Good performance of a control unit / interpreter depends on two properties, namely, i) sensitivity and ii) stability [4]. A production system or more specifically a control unit is called **sensitive**, if the system can respond quickly to the change in environment, reflected by the new contents of the WM. **Stability**, on the other hand, means showing continuity in the line of reasoning.

3.5 Conflict Resolution Strategies

The Conflict Resolution strategies vary from system to system. However, among the various strategies, the following three are most common. In many systems a combination of two or all of the following strategies [4] are used for resolving conflict in a control unit.

1. Refractoriness

This strategy requires that the same rule should not be fired more than once when instantiated with the same set of data. The obvious way of implementing this is to discard the instantiations from the WM, which have been used up once. Another version of their strategy deletes instantiations, which were used up during the last recognition-act cycle. This actually helps the system overcome the problems of moving on loops.

2. Recency

This strategy requires that the most recent elements of the WM be used up for instantiating one of the rules. The idea is to follow the leading edge of computation, rather than doubling back to take another look at the old data. Doubling back, of course, is necessary when the reasoning in the current line of action fails.

3. Specificity

This strategy requires that the rule with more number of antecedent clauses be fired than rules handling fewer antecedent clauses. As an example, consider the following two rules, denoted as PR1 and PR2.

PR1: Bird (X) $\rightarrow$ Fly (X).

PR2: Bird (X), Not emu (X) $\rightarrow$ Fly (X).

Suppose the WM contains the data Bird (parrot) and Not emu (parrot). Then both the rules are firable. However, the second rule should be fired using the specificity strategy.

3.6 An Alternative Approach for Conflict Resolution

The MYCIN experiments [3] of Stanford University proposed another approach for resolving conflicts via metarules. Metarules too are rules, whose task is to control the direction of reasoning and not to participate in the reasoning process itself. Metarules can be either *domain-specific* or *domain-free*. A domain-specific metarule is applicable for identifying the rule to fire only in a specific domains, while domain-free metarules are of very general kinds and can be used for controlling the firing of rules in a generalized knowledge base. To illustrate this concept, we take examples from MYCIN [3].

Example 3.1: *Domain-specific metarule*

    Metarule*:*   *IF 1)   the infection is pelvic abscess and*

                     2) there are rules which   mention in their premise entero-bactoriae, and

                     3)   there are rules which mention in their premise gram- positive rods.

              THEN there exists suggestive evidence (0.4) that the former should be applied before the latter.

Example 3.2: *Domain-free rule*

    Metarule: *IF 1)   there are rules which do not mention the current goal in their premise, and*

                     2) there are rules while mention the current goal   in their premise

              THEN it is definite (1.0) that the former should be applied before the latter.

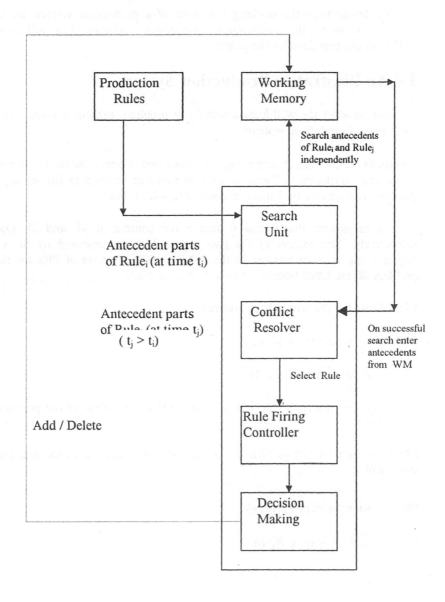

Fig. 3.1: Architecture of a typical production system.

The architecture of a production system [5] is now presented, vide fig. 3.1. The conflict resolution with two rules PRi and PRj has been demonstrated in this architecture. The other descriptions in fig. 3.1 being self-explanatory are left to the readers for interpretation.

To demonstrate the working principle of a production system, let us illustrate it using the well- known water-jug problem. The following statement can best describe the problem.

3.7 An Illustrative Production System

We now consider the well known water jug problem, presented below, as a case study of production systems.

Example 3.3: Given 2 water jugs, 4 liters and 3 liters. Neither has any measuring marks on it. There is a pump that can be used to fill the jugs. How can you get exactly 2 liters of water into 4-liter jugs?

Let us assume that u and v denote the content of 4L and 3L jugs respectively. The content of the two jugs will be represented by (u, v). Suppose, the start-up element in the WM is $(0,0)$. The set of PRs for the problem [8] are listed below.

List of PRs for the water-jug problem

PR 1. $(u, v : u < 4) \rightarrow (4, v)$

PR 2. $(u, v : v < 3) \rightarrow (u, 3)$

PR 3. $(u, v : u > 0) \rightarrow (u - D, v)$, where D is a fraction of the previous content of u.

PR 4. $(u, v : v > 0) \rightarrow (u, v - D)$, where D is a fraction of the previous content of v.

PR 5. $(u, v : u > 0) \rightarrow (0, v)$

PR 6. $(u, v : v > 0) \rightarrow (u, 0)$

PR 7. $(u, v : u + v \geq 4 \wedge v > 0) \rightarrow (4, v - (4 - u))$

PR 8. $(u, v : u + v \geq 3 \wedge u > 0) \rightarrow (u - (3 - v), 3)$

PR 9. $(u, v : u + v \leq 4 \wedge v > 0) \rightarrow (u + v, 0)$

PR 10. $(u, v : u + v \leq 3 \wedge u > 0) \rightarrow (0, u + v)$

To keep track of the reasoning process, we draw a state-space for the problem. Note that the leaves generated after firing of the rules should be stored in WM. We first consider all possibilities of the solution (i.e., without resolving the conflict). Later we would fire only one rule even though more than one are firable. The state-space without conflict resolution is given in fig. 3.2.

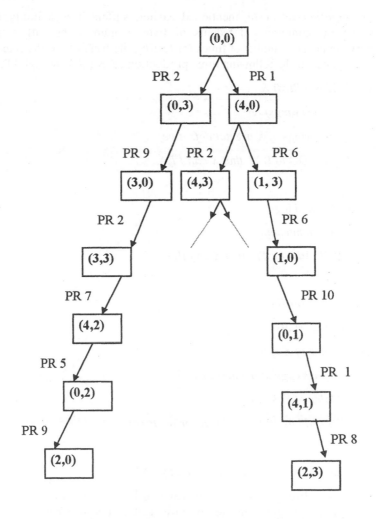

Fig. 3.2: The state-space for the water-jug problem.

To resolve conflict for this system, one can use the following strategies.

i)   *Prefer rules for firing, where u + v can be brought to 5L or 6L.*

ii)   *Avoid doubling back, whenever possible. In other words, never attempt to generate old entries.*

3.8 The RETE Match Algorithm

The *recognition-act cycle*, mentioned earlier, suffers from a limitation of matching the common antecedents of two or more rules with working memory elements a number of times for testing the friability of the rules. For instance, consider the following three production rules PR 1 through PR 3.

PR 1: *IF (X is a Bird) &*

         (X has wings) &

         (the wings of X are not defective)

         THEN (ADD to WM that X can Fly).

PR 2: *IF (X has Wings) &*

         (X is a mammal)

         THEN (Add to WM that X can fly).

PR 3: *IF (X is a Bird) &*

         (X has wings) &

         (Color of X is black) &

         (X lays eggs at the nest of Y) &

         (Color of Y is black)

         THEN (Add to WM that X is a spring Bird).

Assume that the WM contents are given by WM =

{ Cuckoo is a Bird, parrot is a Bird, Cuckoo has wings, Color of cuckoo is black, Cuckoo lays eggs at the nest of crow, Color of crow is black }.

The recognition-act cycle, in the present context, will attempt to match the antecedents of PR1 first with the data recorded in WM. Since the third antecedent clause is not available in WM, the interpreter will leave this rule and start the matching cycle for the antecedents of PR 2 with contents of WM. Since the second antecedent clause of PR 2 is not available, the interpreter would start the matching cycle for the antecedents of PR 3. So, when there exist common antecedents of a number of rules, the interpreter checks their possibility of firing one by one and thus matches the common antecedents of the rules a number of times. Such repeated matching of common antecedents can be avoided by constructing a network to keep track of these variable bindings. The word 'rete', which in Latin means net [7], refers to such a network. The RETE algorithm is illustrated below with the network of fig.3.3.

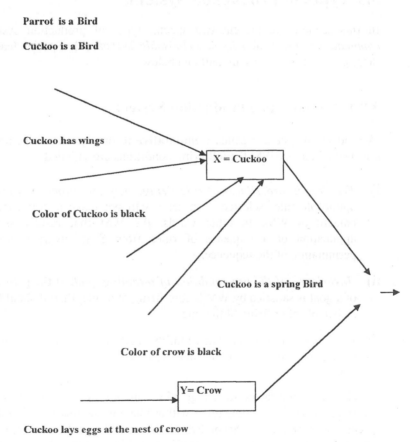

Fig. 3.3: Construction of the 'rete'.

In fig. 3.3, the antecedent clauses have been represented by circles. It may be noted that at each node there may exist more than one clause. Further, the bound values of the variables here are X = Cuckoo and Y = Crow. Thus the third rule only is selected for firing. In case more than one rule is found to be firable, then the conflict resolution strategies, described earlier, will be invoked to identify the right rule for firing.

The RETE algorithm thus constructs a network like fig. 3.3 and continues updating it as more rules are fired. It saves significant matching cycles by matching common antecedents of the rules once only.

3.9 Types of Production Systems

In this section, we present two special types of production systems: *i) commutative system* and *ii) decomposable system* [4]. Special features of these production systems are outlined below.

3.9.1 Commutative Production System

A production system is called **commutative** if for a given set of rules R and a working memory WM the following conditions are satisfied:

i) *Freedom in orderliness of rule firing:* Arbitrary order of firing of the applicable rules selected from set S will not make a difference in the content of WM. In other words, the WM that results due to an application of a sequence of rules from S is invariant under the permutation of the sequence.

ii) *Invariance of the pre-condition of attaining goal:* If the pre-condition of a goal is satisfied by WM before firing of a rule, then it should remain satisfiable after firing of the rule.

iii) *Independence of rules:* The firability condition of an yet unfired rule R_i with respect to WM remains unaltered, even after firing of the rule R_j for any j.

The most significant advantage of a commutative production system is that rules can be fired in any order without having the risk of losing the goal, in case it is attainable. Secondly, an irrevocable control strategy can be designed for such systems, as an application of a rule to WM never needs to be undone.

3.9.2 Decomposable Production System

A production system is called **decomposable** if the goal G and the working memory can be partitioned into G_i and WM_i, such that

$$G = \mathbf{AND}_i (G_i),$$

$$WM = \bigcup \{ WM_i \}$$

$$\forall i$$

and the rules are applied onto each WM_i independently or concurrently to yield G_i. The termination of search occurs when all the goals G_i for all i have been identified.

The main advantage of decomposition is the scope in concurrent access of the WM, which allows parallel firing of rules, without causing a difference in the content of the working memory WM. Decomposable production systems have been successfully used for evaluation of symbolic integration. Here a integral can be expressed as a sum of more than one integral, all of which can be executed independently.

3.10 Forward versus Backward Production Systems

Most of the common classical reasoning problems of AI can be solved by any of the following two techniques called i) forward and ii) backward reasoning. In a forward reasoning problem such as 4-puzzle games or the water-jug problem, where the goal state is known, the problem solver has to identify the states by which the goal can be reached. These class of problems are generally solved by expanding states from the known starting states with the help of a domain-specific knowledge base. The generation of states from their predecessor states may be continued until the goal is reached. On the other hand, consider the problem of system diagnosis or driving a car from an unknown place to home. Here, the problems can be easily solved by employing backward reasoning, since the neighboring states of the goal node are known better than the neighboring states of the starting states. For example, in diagnosis problems, the measurement points are known better than the cause of defects, while for the driving problem, the roads close to home are known better than the roads close to the unknown starting location of driving. It is thus clear that, whatever be the class of problems, system states from starting state to goal or vice versa are to be identified, which requires expanding one state to one or more states. If there exists no knowledge to identify the right offspring state from a given state, then many possible offspring states are generated from a known state. This enhances the search-space for the goal. When the distance (in arc length) between the starting state and goal state is long, determining the intermediate states and the optimal path (minimum arc length path) between the starting and the goal state becomes a complex problem. The issues of determining an optimal path will be taken up in detail in the next chapter.

The following example illustrates the principle of forward and backward reasoning with reference to the well-known "farmer's fox-goat-cabbage problem".

Example 3.4: The problem may be stated as follows. *A farmer wants to transfer his three belongings, a wolf, a goat and a cabbage, by a boat from the left bank of a river to its right bank. The boat can carry at most two items including the farmer. If unattended, the wolf may eat up the goat and the goat may eat up the cabbage.* How should the farmer plan to transfer the items?

The illegal states in the problem are (W,G || F,C) , (G,C || F,W), (F, W || G, C) and (F, C || W, G) where F, G, ||, W and C denote the farmer, the goat, the river, the wolf and the cabbage respectively. In the first case the wolf and the goat are at the left bank, and the farmer and the cabbage are at the right bank of the river. The second case demonstrates the presence of goat and cabbage in the left and the farmer and the wolf in the right bank. Similarly, the other illegal states can be explained easily.

A part of the knowledge base for the system is given below.

PR 1: (F, G, W, C || Nil) → (W, C || F, G)

PR 2: (W, C || F, G) → (F, W, C || G)

PR 3: (F, W, C || G) → (C || F, W, G)

PR 4: (C || F, W, G) → (F, G, C || W)

PR5: (F, G, C || W) → (G || F, W, C)

PR 6: (G || F, W, C) → (F, G || W, C)

PR 7: (F, G, || W, C) → (Nil || F,G, W, C)

PR 8 (F, W, C || G) →   (W || F, G, C)

PR 9: (W || F, G, C) → (F, G, W || C)

PR 10: (F, G, W || C) → (G | | F, W, C)

PR 11: (G| | F, W, C) → (F, G || W,C)

PR 12: (F, G || W, C) →(Nil || F, G, W, C)

Forward Reasoning: Given the starting state (F, G, W, C || Nil) and the goal state (Nil || F, G, W, C), one may expand the state-space, starting with (F,G,W,C | | Nil) by the supplied knowledge base, as follows:

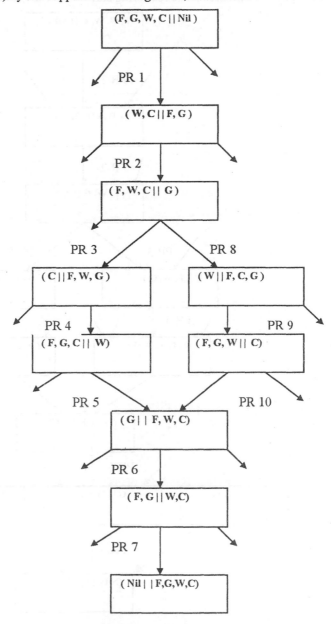

Fig. 3.4: The forward reasoning trace of the farmer's problem with a partially expanded state-space.

Backward Reasoning: The backward reasoning scheme can also be invoked for the problem. The reasoning starts with the goal and identifies a

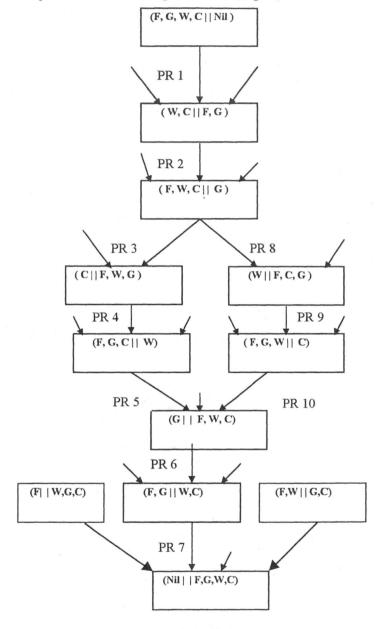

Fig. 3.5: Backward reasoning solution of the farmer's problem.

rule whose right-hand side contains the goal. It then generates the left side of the rule in a backward manner. The resulting antecedents of the rules are called sub-goals. The sub-goals are again searched among the consequent part of the rules and on a successful match the antecedent parts of the rule are generated as the new sub-goals. The process is thus continued until the starting node is obtained.

A caution about backward reasoning: Backward reasoning[1] in many circumstances does not support the logical semantics of problem solving. It may even infer wrong conclusions, when a goal or sub-goal (any intermediate state leading to the goal) has multiple causes for occurrence, and by backward reasoning we miss the right cause and select a wrong cause as its predecessor in the state-space graph. This is illustrated in the following example below with reference to a hypothetical knowledge base.

Example 3.4: Consider the following knowledge base, the starting state and the goal state for a hypothetical problem. The "," in the left-hand side of the production rules PR 1 through PR 4 denotes joint occurrence of them.

PR 1: p, q $\rightarrow$ s

PR 2: s, t $\rightarrow$ u

PR 3: p, q, r $\rightarrow$ w

PR 4: w $\rightarrow$ v

PR 5 : v, t $\rightarrow$ u

Starting state: p and q

Goal state: u.

Other facts: t.

The state-space graph for the hypothetical problem, presented in fig. 3.6, indicates that the goal can be correctly inferred by forward reasoning. However, backward reasoning may infer a wrong conclusion: p and q and r, if PR 5, PR 4 and PR 3 are used in order starting with the goal. Note that r is an extraneous premise, derived by backward reasoning. But in practice the goal is caused due to p, q and t only. Hence, backward reasoning may sometimes yield wrong inferences.

---

[1] Backward reasoning is not supported by the logic of propositions and predicates, vide chapter 5.

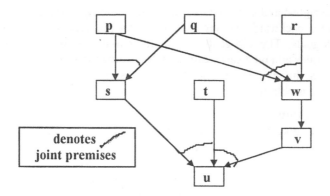

Fig. 3.6: The state-space graph of a hypothetical problem.

Bi-directional Reasoning: Instead of employing either forward or backward reasoning, both of them may be used together in automated problem solving [6]. This is required especially in situations when expanding from either direction leads to a large state-space. Fig. 3.7 (a) and (b) demonstrates the state-space created respectively by forward and backward reasoning, while fig. 3.7 (c) shows expansion of the state space from both sides together. Surely, it requires expansion of less state-space.

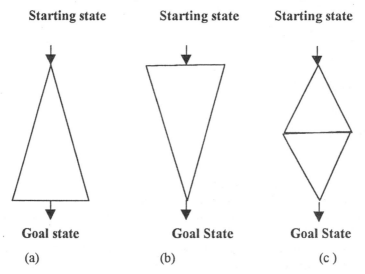

Fig. 3.7: The state-space in (a) forward, (b) backward and (c) bi-directional reasoning.

It may be added here that when instead of expanding all possible states in both the forward and backward direction, as shown in fig. 3.7 (c), a few states may be selected in both directions. This can be done by employing a heuristic search, where a heuristic function is used to select a few states among many for expansion. Heuristic search on a graph will be covered in detail in the next chapter. The resulting forward and backward reasoning state-space under this circumstance may not have an intersection, as cited in fig. 3.8. Bi-directional search in the present context is a waste of computational effort.

Starting state

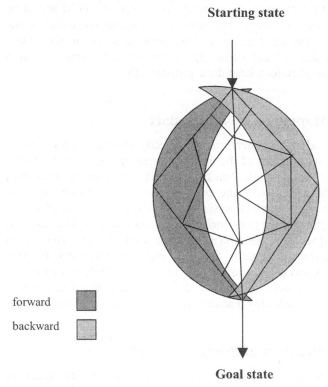

forward

backward

Goal state

Fig.3.8: Bi-directional search with minimal overlap in the state-space generated by forward and backward reasoning.

3.11 General Merits of a Production System

Production systems, as already stated, are the oldest form of knowledge representation that can maintain its traditional significance for the following reasons.

3.11.1 Isolation of Knowledge and Control Strategy

The architecture of a production system, presented in fig. 3.1, demonstrates that the knowledge base and the interpreter are realized on different modules. This has significance from the point of view of the knowledge engineer. The knowledge engineer collects the rules from the experts of her domain of interest and codes the rules in appropriate format as required for the knowledge base. Since the knowledge base is augmented with new rules and some rules are updated and sometimes deleted from the storehouse, a loose coupling between the knowledge base and the interpreter is helpful for them. Such a loose coupling protects the interpreter code from unwanted access by the knowledge engineers or users. Alternatively, updating the codes of the interpreter does not cause a change in the knowledge base due to their loose coupling. The separation of the knowledge base and the interpreter in a production system has, therefore, been done purposefully.

3.11.2 A Direct Mapping onto State-space

The modules of a production system can be directly mapped onto the state-space. For instance, the contents of the working memory represent the states, the production rules cause state transitions and the conflict resolution strategies control the selection of the promising states by firing one rule among many firable rules. Production systems thus may be compared with a problem solver that searches the goal through state-space search. The best first search algorithms that we will present in the next chapter have much similarity with a production system. The only differences between the two perhaps lies in the process of selection of the next state. While in production systems this is done by conflict resolution strategy, it is realized in the best first search algorithms by selecting a state with a minimum cost estimate[2]

3.11.3 Modular Structure of Production Rules

The production rules used in a production system generate the space of instantiation of other rules in the working memory. Thus one fired rule causes another rule to fire, thereby forming a chain of fired rule sequences. This is informally called *chaining of rules*. However, a production rule does not call other rules like function or procedure calls. Such syntactic independence of

---

[2] In a special form of the best first search algorithm like A*, the total cost of reaching a state x from the starting node (root), called $g(x)$, and the predicted cost of reaching the goal from x, called $h(x)$, is minimized. A node x is selected for expansion, if it has the minimum $g(x) + h(x)$ among many possible unexpanded nodes.

rules supports the incremental development of reasoning systems by adding, updating and deleting rules without affecting the existing rules in the knowledge base.

3.11.4 Tracing of Explanation

A production system with conflict resolution strategy selects only one rule at each recognize-act cycle for firing. Thus the fired rules are virtually time-tagged. Since the rules cause state-transition in a production system, stating the rule to the user during its firing, let the user understand the significance of the state transition. Presenting the set of the time-tagged rule in sequence thus gives the user an explanation of the sequence of the operators used to reach the goal.

3.12 Knowledge Base Optimization in a Production System

The performance of a production system depends largely on the organization of its knowledge base. The inferences derived by a production system per unit time, also called time efficiency, can be improved by reducing the matching time of the antecedents of the production rules with data in the WM. Further, if the rules are constructed in a manner so that there is no conflict in the order of rule firing, then the problem of conflict resolution too can be avoided. Another important issue of rule-base design is to select the rules so that the resulting state-space for rule firing does not contain any cycles. The last issue is to identify the concurrently firable rules that do not have conflict in their action parts. This, if realized for a rule-based system, will improve the performance to a high extent. This issue will be covered in detail in chapter 22, where the architecture of knowledge-based systems is highlighted.

For optimization of rules in a rule-based system, Zupan [9] suggested the following points.

i)   Construct by backward reasoning a state-space graph from the desired goal nodes (states) up to the nodes, which cannot be expanded further in a backward manner. Each goal node, also called fixed points, is thus reachable (has connectivity) from all possible starting states. It may be noted that some of the connectivity from the starting nodes to the goal nodes may pass through cycles. It should also be noted that the resulting state-space will not miss the shortest paths from the goal to any other state, as the predecessor states of each state are found by an exhaustive breadth first search.

ii) The common states in the graph are replaced by a single vertex and the parallel paths are identified.

iii) Do not generate an existing state.

3.13 Conclusions

Production systems are the simplest method for knowledge representation in AI. Experts, who are specialists in their respective subject-domain, need not be conversant with knowledge engineering tools for encoding knowledge in simple if-then rules. The efficiency of a production system depends mainly on the order of firing of rules and hence on the conflict resolution strategies. Selection of conflict resolution strategies, thus, is a significant issue in designing a production system. Among the other interesting properties of a production system, the two that need special mention are i) sensitivity and ii) stability. A production system with good sensitivity implies that a small change in data clauses of WM would cause a significant change in the inferences. Stability, on the other hand, means continuation of reasoning in the same line.

Exercises

1.   For the missionaries-cannibals problem, presented in Exercises of chapter 1, formulate one conflict resolution strategy. Test the advantage of using the strategy with the state-space representation of the problem.

2.   Test whether the following production systems are commutative. Justify your answer.

a) Knowledge base

   If A & B Then C.
   If C Then D.
   If A & D then E.

   Initial WM = { A, B}.

b) Knowledge base

   If A & B Then C.
   If X & Y Then C.
   If A Then E.

If B then F.

Initial WM = {A, B, X, Y}.

c) Knowledge base

If A & B Then Not (C)   [i.e. eliminate C from WM]
If C Then D.

Initial WM = {A, B, C}.

3. Which of the following two production systems is more stable ?

i) Knowledge base

If A & B Then C.
If C Then D.
If D Then E.

Initial WM = {A, B} and goal ={E}.

ii) Knowledge base

If A Then B.
If C Then E.
If A & C Then F.
If F Then A.

Initial WM = {A, C} and goal = {F}.

References

[1] Bourbakis, G. N., *Knowledge   Engineering   Shells: Systems and Techniques in Advanced Series on Artificial Intelligence*, vol. 2, World Scientific, Singapore, pp. 57-85, 1993.

[2] Buchanan, B. G. and Feigenbaum, E. A., DENDRAL and Meta DENDRAL: Their applications dimension, *Artificial Intelligence*, vol. 11 pp. 5-24, 1978.

[3]   Buchanan, B. G. and Shortliffe, E. H., Eds, *Rule-based Expert Systems*, Addison-Wesley, Reading, MA, 1984.

[4]   Jackson, P., *Introduction to Expert Systems*, Addison-Wesley, Reading, MA, 1986.

[5]   Konar, A., *Uncertainty Management in Expert Systems Using Fuzzy Petri Nets*, Ph.D. thesis, Jadavpur University, Calcutta, chapter 1, 1994.

[6]   Luger, F. G. and Stubblefield, A. W., *Artificial Intelligence: Structures and Strategies for Complex Problem Solving*, Addison-Wesley, Reading, MA, ch. 3, pp. 88-89, 1993.

[7]   Nilson, N. J., *Principles of Artificial Intelligence*, Morgan Kaufmann, San Mateo, CA, pp. 6-7, 1980.

[8]   Rich, E. and Knight, K., *Artificial Intelligence*, McGraw-Hill, New York, 1996.

[9]   Zupan, B. and Cheng, A. M. K., "Optimization of rule-based systems using state-space graphs," *IEEE Trans. on Knowledge and Data Eng.*, vol. 10, no. 2, March / April 1998.

4

Problem Solving by Intelligent Search

Problem solving requires two prime considerations: first representation of the problem by an appropriately organized state space and then testing the existence of a well-defined goal state in that space. Identification of the goal state and determination of the optimal path, leading to the goal through one or more transitions from a given starting state, will be addressed in this chapter in sufficient details. The chapter, thus, starts with some well-known search algorithms, such as the depth first and the breadth first search, with special emphasis on their results of time and space complexity. It then gradually explores the 'heuristic search' algorithms, where the order of visiting the states in a search space is supported by thumb rules, called heuristics, and demonstrates their applications in complex problem solving. It also discusses some intelligent search algorithms for game playing.

4.1 Introduction

We have already come across some of the problems that can be solved by intelligent search. For instance, the well-known water-jug problem, the number puzzle problem and the missionaries-cannibals problem are ideal examples of problems that can be solved by intelligent search. Common experience reveals that a search problem is associated with two important

93

issues: first 'what to search' and secondly 'where to search'. The first one is generally referred to as 'the key', while the second one is termed 'search space'. In AI the search space is generally referred to as a collection of states and is thus called state space. Unlike common search space, the state space in most of the problems in AI is not completely known, prior to solving the problem. So, solving a problem in AI calls for two phases: the generation of the space of states and the searching of the desired problem state in that space. Further, since the whole state space for a problem is quite large, generation of the whole space prior to search may cause a significant blockage of storage, leaving a little for the search part. To overcome this problem, the state space is expanded in steps and the desired state, called "the goal", is searched after each incremental expansion of the state space.

Depending on the methodology of expansion of the state space and consequently the order of visiting the states, search problems are differently named in AI. For example, consider the state space of a problem that takes the form of a tree. Now, if we search the goal along each breadth of the tree, starting from the root and continuing up to the largest depth, we call it *breadth first search*. On the other hand, we may sometimes search the goal along the largest depth of the tree, and move up only when further traversal along the depth is not possible. We then attempt to find alternative offspring of the parent of the node (state) last visited. If we visit the nodes of a tree using the above principles to search the goal, the traversal made is called depth first traversal and consequently the search strategy is called *depth first search*. We will shortly explore the above schemes of traversal in a search space. One important issue, however, needs mention at this stage. We may note that the order of traversal and hence search by breadth first or depth first manner is generally fixed by their algorithms. Thus once the search space, here the tree, is given, we know the order of traversal in the tree. Such types of traversal are generally called '*deterministic*'. On the other hand, there exists an alternative type of search, where we cannot definitely say which node will be traversed next without computing the details in the algorithm. Further, we may have transition to one of many possible states with equal likelihood at an instance of the execution of the search algorithm. Such a type of search, where the order of traversal in the tree is not definite, is generally termed '*non-deterministic*'[1]. Most of the search problems in AI are non-deterministic. We will explore the details of both deterministic and non-deterministic search in this chapter.

---

[1] There exists also a third variety, called stochastic (random) search, where random numbers are used to select the order of visiting the states in the search space. The execution of such search algorithms twice at a given iteration need not necessarily select the same state in the next visit.

4.2 General Problem Solving Approaches

There exist quite a large number of problem solving techniques in AI that rely on search. The simplest among them is *the generate and test* method. The algorithm for the generate and test method can be formally stated as follows:

Procedure Generate & Test

Begin

 Repeat

   Generate a new state and call it current-state;

  Until current-state = Goal;

End.

It is clear from the above algorithm that the algorithm continues the possibility of exploring a new state in each iteration of the repeat-until loop and exits only when the current state is equal to the goal. Most important part in the algorithm is to generate a new state. This is not an easy task. If generation of new states is not feasible, the algorithm should be terminated. In our simple algorithm, we, however, did not include this intentionally to keep it simplified.

But how does one generate the states of a problem? To formalize this, we define a four tuple, called state space, denoted by

$$\{ \text{ nodes, arc, goal, current } \},$$

where

  nodes represent the set of existing states in the search space;

  an **arc** denotes an operator applied to an existing state to cause transition to another state;

  goal denotes the desired state to be identified in the nodes; and

  current represents the state, now generated for matching with the goal.

The state space for most of the search problems we will cover in this chapter takes the form of a tree or graph[2]. The fig. 1.2 in chapter 1, for instance, represents the state space for a 4-puzzle problem in the form of a tree.

---

[2] The basic distinction between a tree and a graph lies in the count of parents of a node in the respective data structures. For a graph, this could be any positive integer, while for a tree it has a maximum value of one.

We will now present two typical algorithms for generating the state space for search. These are depth first search and breadth first search.

4.2.1 Breadth First Search

The breadth first search algorithm visits the nodes of the tree along its breadth, starting from the level with depth 0 to the maximum depth. It can be easily realized with a queue. For instance, consider the tree, given in fig. 4.1. Here, the nodes in the tree are traversed following their ascending ordered labels.

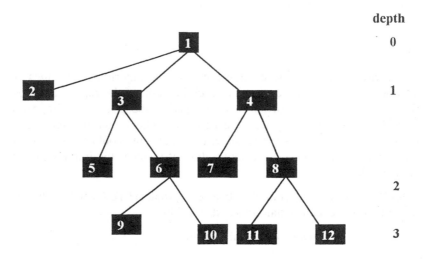

Fig. 4.1: The order of traversal in a tree of depth 3 by

breadth first manner.

The algorithm for traversal in a tree by depth first manner can be presented with a queue as follows:

Procedure Breadth-first-search

Begin

i)   Place the starting node in a queue;

ii)   **Repeat**

        Delete queue to get the front element;

        If the front element of the queue = goal,

        return success and stop;

Else do

  Begin

    insert the children of the front element,

    if exist, in any order at the rear end of

    the queue;

  End

Until the queue is empty;

End.

    The breadth first search algorithm, presented above, rests on a **simple principle**. *If the current node is not the goal add the offspring of the current in any order to the rear end of the queue and redefine the front element of the queue as the current.* The algorithm terminates, when the goal is found.

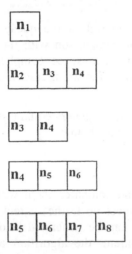

Fig. 4.2: First few steps of breadth first search on the tree of fig. 4.1.

Time Complexity

For the sake of analysis, we consider a tree of equal branching factor from each node = b and largest depth = d. Since the goal is not located within depth (d-1), the number of false search [1], [2] is given by

$$1+b+b^2+b^3+ \ldots + b^{d-1} = (b^d-1)/(b-1), \quad b \gg 1.$$

Further, the first state within the fringe nodes could be the goal. On the other hand, the goal could be the last visited node in the tree. Thus, on an average, the number of fringe nodes visited is given by

$$(1+b^d) / 2.$$

Consequently, the total number of nodes visited in an average case becomes

$$(b^d-1) / (b-1) + (1+b^d) / 2$$

$$\cong\ b^d (b+1) / 2(b-1).$$

Since the time complexity is proportional to the number of nodes visited, therefore, the above expression gives a measure of time complexity.

Space Complexity

The maximum number of nodes will be placed in the queue, when the leftmost node at depth d is inspected for comparison with the goal. The queue length under this case becomes b^d. The space complexity of the algorithm that depends on the queue length, in the worst case, thus, is of the order of b^d.

In order to reduce the space requirement, the generate and test algorithm is realized in an alternative manner, as presented below.

4.2.2   Depth First Search

The depth first search generates nodes and compares them with the goal along the largest depth of the tree and moves up to the parent of the last visited node, only when no further node can be generated below the last visited node. After moving up to the parent, the algorithm attempts to generate a new offspring of the parent node. The above principle is employed recursively to each node of a tree in a depth first search. One simple way to realize the recursion in the depth first search algorithm is to employ a stack. A stack-based realization of the depth first search algorithm is presented below.

Procedure Depth first search

Begin

1.   Push the starting node at the stack,

   pointed to by the stack-top;

2. **While** stack is not empty do

   Begin

   Pop stack to get stack-top element;

   If stack-top element = goal, return

   success and stop

   Else push the children of the stack-top

   element in any order into the stack;

   End while;

   End.

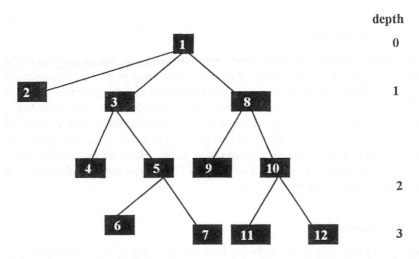

Fig. 4.3: Depth first search on a tree, where the node numbers denote

the order of visiting that node.

In the above algorithm, a starting node is placed in the stack, the top of which is pointed to by the stack-top. For examining the node, it is popped out from the stack. If it is the goal, the algorithm terminates, else its children are pushed into the stack in any order. The process is continued until the stack is empty. The ascending order of nodes in fig. 4.3 represents its traversal on the tree by depth first manner. The contents of the stack at the first few iterations are illustrated below in fig. 4.4. The arrowhead in the figure denotes the position of the stack-top.

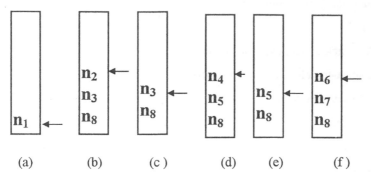

(a)   (b)   (c)   (d)   (e)   (f)

Fig. 4.4: A snapshot of the stack at the first few iterations.

Space Complexity

Maximum memory in depth first search is required, when we reach the largest depth at the first time. Assuming that each node has a branching factor b, when a node at depth d is examined, the number of nodes saved in memory are all the unexpanded nodes up to depth d plus the node being examined. Since at each level there are (b-1) unexpanded nodes, the total number of memory required = d (b -1) +1. Thus the space complexity of depth first search is a linear function of b, unlike breadth first search, where it is an exponential function of b. This, in fact, is the most useful aspect of the depth first search.

Time Complexity

If we find the goal at the leftmost position at depth d, then the number of nodes examined = (d +1). On the other hand, if we find the goal at the extreme right at depth d, then the number of nodes examined include all the nodes in the tree, which is

$$1+b+b^2 +b^3 +...+b^d = (b^{d+1} -1) / (b-1)$$

So, the total number of nodes examined in an average case

$$= (d+1) /2 + (b^{d+1} -1) / 2(b-1)$$

$$\cong b(b^d + d) / 2 (b -1)$$

This is the average case time complexity of the depth first search algorithm.

Since for large depth d, the depth first search requires quite a large runtime, an alternative way to solve the problem is by controlling the depth of the search tree. Such an algorithm, where the user mentions the initial depth cut-off at each iteration, is called an Iterative Deepening Depth First Search or simply an **Iterative deepening search**.

4.2.3 Iterative Deepening Search

When the initial depth cut-off is one, it generates only the root node and examines it. If the root node is not the goal, then depth cut-off is set to two and the tree up to depth 2 is generated using typical depth first search. Similarly, when the depth cut-off is set to m, the tree is constructed up to depth m by depth first search. One may thus wonder that in an iterative deepening search, one has to regenerate all the nodes excluding the fringe nodes at the current depth cut-off. Since the number of nodes generated by depth first search up to depth h is

$$(b^{h+1}-1) / (b-1),$$

the total number of nodes expanded in failing searches by an iterative deepening search will be

$$\{1 / (b-1) \} \sum_{h=0}^{(d-1)} (b^{h+1} -1)$$

$$\cong b(b^d - d) / (b-1)^2.$$

The last pass in the algorithm results in a successful node at depth d, the average time complexity of which by typical depth first search is given by

$$b(b^d + d) / 2 (b -1).$$

Thus the total average time complexity is given by

$$b(b^d - d) / (b-1)^2 + b(b^d + d) / 2 (b -1).$$

$$\cong (b+1) b^{d+1} / 2 (b -1)^2.$$

Consequently, the ratio of average time complexity of the iterative deepening search to depth first search is given by

$$\{(b+1) b^{d+1} / 2 (b -1)^2 \} : \{ b^{d+1} / 2 (b-1)\}$$

$= (b+1) : (b-1)$.

The iterative deepening search thus does not take much extra time, when compared to the typical depth first search. The unnecessary expansion of the entire tree by depth first search, thus, can be avoided by iterative deepening. A formal algorithm of iterative deepening is presented below.

Procedure Iterative-deepening

Begin

1. Set current depth cutoff =1;

2. Put the initial node into a stack, pointed to by stack-top;

3. **While** the stack is not empty and the depth is within the

    given depth cut-off do

    Begin

    Pop stack to get the stack-top element;

    if stack-top element = goal, return it and stop

    else push the children of the stack-top in any order

    into the stack;

    End While;

4. Increment the depth cut-off by 1 and repeat

    through step 2;

End.

The breadth first, depth first and the iterative deepening search can be equally used for Generate and Test type algorithms. However, while the breadth first search requires an exponential amount of memory, the depth first search calls for memory proportional to the largest depth of the tree. The iterative deepening, on the other hand, has the advantage of searching in a depth first manner in an environment of controlled depth of the tree.

4.2.4 Hill Climbing

The 'generate and test' type of search algorithms presented above only expands the search space and examines the existence of the goal in that space. An alternative approach to solve the search problems is to employ a function $f(x)$ that would give an estimate of the measure of distance of the goal from node x. After $f(x)$ is evaluated at the possible initial nodes x, the nodes are sorted in ascending order of their functional values and pushed into a stack in the ascending order of their 'f' values. So, the stack-top element has the least f value. It is now popped out and compared with the goal. If the stack-top element is not the goal, then it is expanded and f is measured for each of its children. They are now sorted according to their ascending order of the functional values and then pushed into the stack. If the stack-top element is the goal, the algorithm exits; otherwise the process is continued until the stack becomes empty. Pushing the sorted nodes into the stack adds a depth first flavor to the present algorithm. The hill climbing algorithm is formally presented below.

Procedure Hill-Climbing

Begin

1. Identify possible starting states and measure the distance (f) of their closeness with the goal node; Push them in a stack according to the ascending order of their f ;

2. **Repeat**

   Pop stack to get the stack-top element;

   If the stack-top element is the goal, announce it and exit

   Else push its children into the stack in the ascending order of their f values;

   Until the stack is empty;

End.

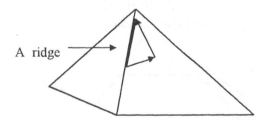

Fig.4.5: Moving along a ridge in two steps (by two successive

operators) in hill climbing.

The hill climbing algorithm too is not free from shortcomings. One common problem *is trapping at local maxima at a foothill*. When trapped at local maxima, the measure of f at all possible next legal states yield less promising values than the current state. A second drawback of the hill climbing is reaching a plateau [2]. Once a state on a plateau is reached, all legal next states will also lie on this surface, making the search ineffective. A new algorithm, called simulated annealing, discussed below could easily solve the first two problems. Besides the above, another problem that too gives us trouble is the *traversal along the ridge*. A ridge (vide fig. 4.5) on many occasions leads to a local maxima. However, moving along the ridge is not possible by a single step due to non-availability of appropriate operators. A multiple step of movement is required to solve this problem.

4.2.5   Simulated Annealing

"Annealing" is a process of metal casting, where the metal is first melted at a high temperature beyond its melting point and then is allowed to cool down, until it returns to the solid form. Thus in the physical process of annealing, the hot material gradually loses energy and finally at one point of time reaches a state of minimum energy. A common observation reveals that most physical processes have transitions from higher to lower energy states, but there still remains a small probability that it may cross the valley of energy states [2] and move up to a energy state, higher than the energy state of the valley. The concept can be verified with a rolling ball. For instance, consider a rolling ball that falls from a higher (potential) energy state to a valley and then moves up to a little higher energy state (vide fig. 4.6). The probability of such

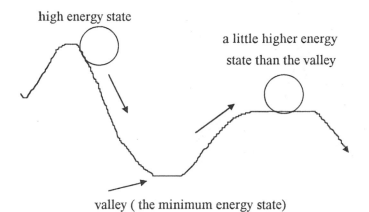

Fig. 4.6: A rolling ball passes through a valley to a higher

energy state.

transition to a higher energy state, however, is very small and is given by

$$p = \exp(-\Delta E / KT)$$

where p is the probability of transition from a lower to a higher energy state, ΔE denotes a positive change in energy, K is the Boltzman constant and T is the temperature at the current thermal state. For small ΔE, p is higher than the value of p, for large ΔE. This follows intuitively, as w.r.t the example of ball movement, the probability of transition to a slightly higher state is more than the probability of transition to a very high state.

An obvious question naturally arises: how to realize annealing in search? Readers, at this stage, would remember that the need for simulated annealing is to identify the direction of search, when the function f yields no better next states than the current state. Under this circumstance, ΔE is computed for all possible legal next states and p' is also evaluated for each such next state by the following formula:

$$p' = = \exp(-\Delta E / T)$$

A random number in the closed interval of [0,1] is then computed and p' is compared with the value of the random number. If p' is more, then it is selected for the next transition. The parameter T, also called temperature, is gradually decreased in the search program. The logic behind this is that as T decreases, p' too decreases, thereby allowing the algorithm to terminate at a stable state. The algorithm for simulated annealing is formally presented below.

Procedure Simulated Annealing

Begin

1. Identify possible starting states and measure the distance (f) of their closeness with the goal; Push them in a stack according to the ascending order of their f ;

2. **Repeat**

   Pop stack to get stack-top element;

   If the stack-top element is the goal,

   announce it and exit;

   Else do

   Begin

   a) generate children of the stack-top element N and

   compute f for each of them;

   b) If measure of f for at least one child of N is improving

        Then push those children into stack in ascending order of their f;

   c) **If** none of the children of N is better in f

      Then do

Begin

   a) select any one of them randomly, compute its p' and test whether p' exceeds a randomly generated number in the interval [0,1]; If yes, select that state as the next state; If no, generate another alternative legal next state and test in this way until one move can be selected; Replace stack-top element by the selected move (state);

   b) Reduce T slightly; If the reduced value is negative, set it to zero;

   End;

  Until the stack is empty;

End.

The algorithm is similar to hill climbing, if there always exists at least one better next state than the state, pointed to by the stack-top. If it fails, then the last begin-end bracketed part of the algorithm is invoked. This part corresponds to simulated annealing. It examines each legal next state one by one, whether the probability of occurrence of the state is higher than the random value in [0,1]. If the answer is yes, the state is selected, else the next possible state is examined. Hopefully, at least one state will be found whose probability of occurrence is larger than the randomly generated probability.

Another important point that we did not include in the algorithm is the process of computation of ΔE. It is computed by taking the difference of the value of f of the next state and that of the current (stack-top) state.

The third point to note is that T should be decreased once a new state with less promising value is selected. T is always kept non-negative. When T becomes zero, p' will be zero and thus the probability of transition to any other state will be zero.

4.3 Heuristic Search

This section is devoted to solve the search problem by a new technique, called heuristic search. The term "heuristics" stands for " thumb rules", i.e., rules which work successfully in many cases but its success is not guaranteed.

In fact, we would expand nodes by judiciously selecting the more promising nodes, where these nodes are identified by measuring their strength compared to their competitive counterparts with the help of specialized intuitive functions, called **heuristic functions**.

Heuristic search is generally employed for two distinct types of problems: i) forward reasoning and ii) backward reasoning. We have already discussed that in a forward reasoning problem we move towards the goal state from a pre-defined starting state, while in a backward reasoning problem, we move towards the starting state from the given goal. The former class of search algorithms, when realized with heuristic functions, is generally called heuristic Search for OR-graphs or the *Best First search Algorithms*. It may be noted that the best first search is a class of algorithms, and depending on the variation of the performance measuring function it is differently named. One typical member of this class is the algorithm A*. On the other hand, the heuristic backward reasoning algorithms are generally called AND-OR graph search algorithms and one ideal member of this class of algorithms is the AO* algorithm. We will start this section with the best first search algorithm.

4.3.1 Heuristic Search for OR Graphs

Most of the forward reasoning problems can be represented by an OR-graph, where a node in the graph denotes a problem state and an arc represents an application of a rule to a current state to cause transition of states. When a number of rules are applicable to a current state, we could select a better state among the children as the next state. We remember that in hill climbing, we ordered the promising initial states in a sequence and examined the state occupying the beginning of the list. If it was a goal, the algorithm was terminated. But, if it was not the goal, it was replaced by its offsprings in any order at the beginning of the list. The hill climbing algorithm thus is not free from depth first flavor. In the best first search algorithm to be devised shortly, we start with a promising state and generate all its offsprings. The performance (fitness) of each of the nodes is then examined and the most promising node, based on its fitness, is selected for expansion. The most promising node is then expanded and the fitness of all the newborn children is measured. *Now, instead of selecting only from the generated children, all the nodes having no children are examined and the most promising of these fringe nodes is selected for expansion. Thus unlike hill climbing, the best first search provides a scope of corrections, in case a wrong step has been selected earlier.* This is the prime advantage of the best first search algorithm over hill climbing. The best first search algorithm is formally presented below.

Procedure Best-First-Search

Begin

1. Identify possible starting states and measure the distance (f) of their closeness with the goal; Put them in a list L;

2. **While** L is not empty do

    Begin

    a) Identify the node n from L that has the minimum f; If there exist more than one node with minimum f, select any one of them (say, n) arbitrarily;

    b) **If** n is the goal

    Then return n along with the path from the starting node,

    and exit;

    Else remove n from L and add all the children of n to the list L,

    with their labeled paths from the starting node;

    End While;

End.

As already pointed out, the best first search algorithm is a generic algorithm and requires many more extra features for its efficient realization. For instance, how we can define f is not explicitly mentioned in the algorithm. Further, what happens if an offspring of the current node is not a fringe node. The A* algorithm to be discussed shortly is a complete realization of the best first algorithm that takes into account these issues in detail. The following definitions, however, are required for presenting the A* algorithm. These are in order.

Definition 4.1: A node is called **open** if the node has been generated and the h' (x) has been applied over it but it has not been expanded yet.

Definition 4.2: A node is called **closed** if it has been expanded for generating offsprings.

In order to measure the goodness of a node in A* algorithm, we require two cost functions: i) *heuristic cost* and ii) *generation cost*. The heuristic cost measures the distance of the current node x with respect to the goal and is denoted by h(x). The cost of generating a node x, denoted by g(x), on the other hand measures the distance of node x with respect to the starting node in the graph. The total cost function at node x, denoted by f(x), is the sum of g(x) plus h(x).

The generation cost g(x) can be measured easily as we generate node x through a few state transitions. For instance, if node x was generated from the starting node through **m** state transitions, the cost g(x) will be proportional to m (or simply m). But how does one evaluate the h(x)? It may be recollected that h(x) is the cost yet to be spent to reach the goal from the current node x. Obviously, any cost we assign as h(x) is through prediction. The predicted cost for h(x) is generally denoted by h'(x). Consequently, the predicted total cost is denoted by f'(x), where

$$f'(x) = g(x) + h'(x).$$

Now, we shall present the A* algorithm formally.

Procedure A*
Begin

1. Put a new node n to the set of ***open*** *nodes (hereafter open)*; Measure its f'(n) = g(n) + h'(n); Presume the set of closed nodes to be a null set initially;

2. **While** open is not empty do
   Begin
   If n is the goal, stop and return n and the path of n from the beginning node to n through back pointers;
   Else do
   Begin
   a) remove n from open and put it under closed;
   b) generate the children of n;
   c) **If** all of them are new (i.e., do not exist in the graph before generating them **Then** add them to open and label their f' and the path from the root node through back pointers;
   d) **If** one or more children of n already existed as open nodes in the graph before their generation **Then** those children must have multiple parents; Under this circumstance compute their f' through current path and compare it through their old paths, and keep them connected only through the shortest path from the starting node and label the back pointer from the children of n to their parent, if such pointers do not exist;
   e) If one or more children of n already existed as closed nodes before generation of them, then they too must

have multiple parents; Under this circumstance, find the shortest path from the starting node, i.e., the path (may be current or old) through which f' of n is minimum; If the current path is selected, then the nodes in the sub-tree rooted at the corresponding child of n should have revised f' as the g' for many of the nodes in that sub-tree changed; Label the back pointer from the children of n to their parent, if such pointers do not exist;

End;

End While;

End.

To illustrate the computation of f' (x) at the nodes of a given tree or graph, let us consider a forward reasoning problem, say the water-jug problem, discussed in chapter 3. The state-space for such problems is often referred to as OR graphs / trees. The production rules we would use here are identical with those in chapter 3, considered for the above problem.

Example 4.1: Let us consider the following heuristic function, where X and Y denote the content of 4-liter and 3-liter jugs respectively and x denotes an arbitrary node in the search space.

h' (x) = 2, when $0 < X < 4$ AND $0 < Y < 3$,
     = 4, when $0 < X < 4$ OR $0 < Y < 3$,
    =10, when   i) $X = 0$ AND $Y = 0$
                 OR ii) $X = 4$ AND $Y = 3$

    = 8, when   i) $X = 0$ AND $Y = 3$

                 OR ii) $X = 4$ AND $Y = 0$

Assume that g(x) at the root node = 0 and g(x) at a node x with minimum distance n, measured by counting the parent nodes of each node starting from x till the root node, is estimated to be g(x) = n. Now let us illustrate the strategy of the best first search in an informal manner using the water-jug problem, vide fig. 4.7.

In step 0, we have the node o only where g + h' = 0+10 =10. In step 1, we have two terminal nodes M and N, where (g + h' =1+8 =9) are equal. We can, therefore, choose any of these two arbitrarily for generating their offsprings. Let us select node M for generating offsprings. By expanding node M, we found nodes P and R in step 2 with g + h' =6 and 12 respectively. Now, out of these three nodes P, N and R, P has the minimum value of f'. So, we select node P for expansion. On expanding node P, we find node S, where g + h' = 3+4. Now the terminals in the tree are S, R and

N, out of which node S has the smallest f'. So, node S will be selected for expansion the next time. The process would thus continue until the goal node is reached.

Step 0 (0,0) g + h' = 0 + 10

Step 1 (0,0) g + h' = 0 + 10

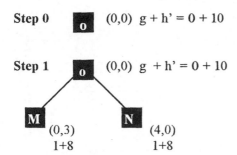

M (0,3)
1+8

N (4,0)
1+8

Step 2 (0,0) g + h' = 0 + 10

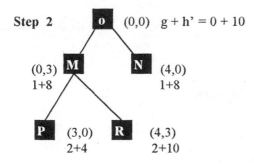

(0,3) M
1+8

N (4,0)
1+8

P (3,0)
2+4

R (4,3)
2+10

Step 3

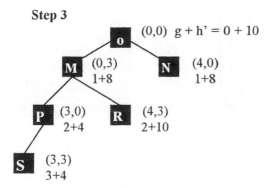

(0,0) g + h' = 0 + 10

M (0,3)
1+8

N (4,0)
1+8

P (3,0)
2+4

R (4,3)
2+10

S (3,3)
3+4

Fig. 4.7: Expanding the state-space by the A* algorithm.

Another important issue that needs to be discussed is how to select a path, when an offspring of a currently expanded node is an already existing

node. Under this circumstance, the parent that yields a lower value of g+h'
for the offspring node is chosen and the other parent is ignored, sometimes
by de-linking the corresponding arc from that parent to the offspring node.
Let us illustrate the above issue with example 4. 2.

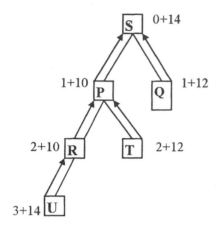

Fig. 4.8 (a): A partially expanded OR graph.

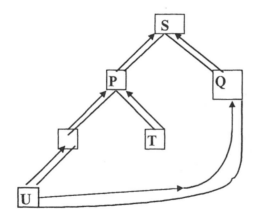

Fig. 4.8 (b): A node U having parents R and Q, the least-cost
            path being node Q to U.

Example 4.2: Consider the tree shown in fig. 4.8(a), where Q, U and T are
the free terminals (leaves). Assume that among these leaves the f' at node Q
is minimum. So, Q is selected for offspring generation. Now, suppose U is

the offspring of Q (fig. 4.8(b)) and the f' at U through Q is compared less to the f' at U through R (this in fact is obvious, since g(U) via Q is 2, while g(U) via R is 3). So, we prefer Q to R as a parent of U and, consequently, we delink the arc from node R to node U (vide fig. 4.8(b) and (c)). It may be noted that we would do the same de-linking operation, if U had offsprings too.

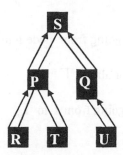

Fig. 4.8 (c): The modified tree after de-linking of the arc
from node R to node U in fig. 4.8 (b).

The third point to be discussed on the A* algorithm is to mark the arcs with back-pointers, i.e., from child to parent nodes in the search space. This helps in tracing the path from goal node to the root. Scrutinize fig. 4.8 (a)-(c) for details.

The steps of the algorithm have already been illustrated. Now, the properties of the algorithm will be presented in detail.

Properties of Heuristic Functions

The following notations will be required for understanding the properties of the heuristic functions.

Notations	Meaning
1. $C(n_i, n_j)$	cost / expenses to traverse from node n_i to node n_j
2. $K(n_i, n_j)$	cost on the cheapest path between n_i and n_j
3. γ	a goal node
4. Γ	the set of goals
5. $P_{n-\gamma}$	the path from node n to γ

6. $P_{n-\Gamma}$ the set of paths from node n to the set of goals Γ

7. $g^*(n)$ the cheapest cost of paths, going from starting (root) node s to node n,

$$g^*(n) = K(s, n)$$

8. $h^*(n)$ the cheapest cost of paths, going from node n to Γ,

$$h^*(n) = Min \quad K(n, \gamma), \text{ for all } \gamma \ \varepsilon \ \Gamma$$

9. C^* the cheapest cost of paths going from s to Γ,

$$C^* = h^*(s)$$

We now define the following properties of best first search algorithms.

a)   **Completeness:** An algorithm is   said   to be **complete**, if it terminates with a solution, when one exists.

b)   **Admissibility:** An algorithm is called **admissible** if   it is guaranteed to   return an optimal solution, whenever a solution exists.

c)   **Dominance:** An algorithm A1 is said to **dominate** A2, if every node expanded by A1 is also expanded by A2.

d)   **Optimality:** An algorithm is said to be **optimal** over a class of algorithms, if it dominates all members of the class.

We now present here some of the interesting properties of the heuristic functions.

Property I: *Any node n* on an optimal path* $P^*_{\gamma-\Gamma}$ *always satisfies equation (4.1)*

$$f^*(n^*) = C^* \qquad\qquad (4.1)$$

where C^* is the cheapest cost from s to Γ.

Proof: $f^*(n^*)$

$$= g^*(n^*) + h^*(n^*)$$

$$= K(s, n^*) + \underset{\gamma \varepsilon \Gamma}{Min} \ K(n^*, \gamma)$$

$$= \underset{\gamma \varepsilon \Gamma}{Min} \ K(s, \gamma)$$

$$= C^*. \qquad\qquad □$$

The following results directly follow from property I.

i) $f^*(s) = C^*$ and

ii) $f^*(\gamma) = C^*$.

Property II: *Any node n that does not lie on any of the optimal paths* $P^*_{s-\Gamma}$ *satisfies inequality (4.2).*

$$f^*(n) > C^*. \qquad\qquad (4.2)$$

Proof: Proof is straightforward and is therefore omitted. □

Definition 4.3: An heuristic function h is said to be **admissible** [6] if

$h(n) \leq h^*(n)$.

Property III: *At any time before A* terminates, there exists an* open node n' on $P^*_{s-\Gamma}$ with $f(n') \leq C^*$.

Proof: Consider an optimal path $P^*_{s-\gamma}$ belonging to $P^*_{s-\Gamma}$. Let $P^*_{s-\gamma}$ = s, $n_1, n_2..., n',...,\gamma$ and let n' be the shallowest (minimum depth) open node on $P'_{s-\Gamma}$. Since γ is not closed before termination, n' is an open node. Further, since all ancestors of n' are closed and since the path $s, n_1, n_2,..., n'$ is optimal, therefore, it must be that the pointers of n' are along $P^*_{s-n'}$.

Therefore, $g(n') = g^*(n')$.

Therefore, $f * (n') = g* (n') + h (n')$

$$\leq g* (n') + h* (n') \qquad \text{[by definition of admissibility]}$$

$$= f * (n')$$

$$= C*.$$

Therefore, $f (n') \leq C*$.   □

Property IV: *A* is admissible (returns optimal solution)[6]* .

Proof: Suppose A* terminates with a goal node t belonging to Γ for which $f(t) = g(t) > C*$.

However, A* tests nodes for compliance with termination criteria, only after it selects them for expansion. Hence, when t was chosen for expansion,

$\qquad f(t) \leq f (n)$, for all open n.

This means that immediately prior to termination, any node n on open satisfies:

$\qquad f (n) > C*$,

which, however, contradicts property III, which claims that there exists at least one open node n with $f(n) \leq C*$. Therefore, the terminating t must have $g (t) = C*$, which means that A* returns an optimal path.   □

Monotonicity and Consistency of Heuristics

Informally speaking, by consistency, we mean that A*   never re-opens already closed nodes. From property I and II, we find that the cheapest path constrained to pass through n cannot be less costly than the cheapest path available without this constraint, i.e.,

$\qquad g * (n) + h* (n) \geq h* (s)$, for all n

$\Rightarrow \quad K (s, n) + h* (n) \geq h* (s).$

If n' is any descendent of n, we should have

h* (n) $\leq$ K (n, n') + h* (n'),   for all (n, n').

Now, if we select h (n), the measure of h* (n) in the manner described by the last expression, we write

h (n) $\leq$ K (n, n') + h (n'),

which is the *condition for consistency* [6].

Definition 4.4: A heuristic function is said to be **monotonic / monotone** if it satisfies

h(n) $\leq$ C(n, n') + h (n') for all n, n'

such that n' is a successor of n.

Property V: *Every consistent heuristic is also admissible.*

Proof: We have

h (n) $\leq$ K (n, n') + h (n') [since h is consistent]

Replacing γ against n', we have

h (n) $\leq$ K (n, γ) + h (γ)

$\Rightarrow$ h (n) $\leq$ h* (n),

which is the *condition for admissibility.*

The following example [7] illustrates that optimal solution will never be missed, if h (n) is admissible as presented below.

Example 4.3: Consider the search-space, given in fig. 4.9(a). Note that, here h > h*, in the case of overestimation, where we made node D so bad (by making its h value too large) that we can never find the optimal path A-D-G.

On the other hand, in fig. 4.9(b), we illustrate the case of underestimation (admissibility) of h. Consider the case in fig. 4.9(b) when F, C and D are the set of expanded nodes and among these nodes C has the least value of f'. We thus expand C and fortunately reach the goal in one step. It is to be noted that we wasted some effort to generate the unnecessary nodes E and F. But, ultimately, we could correctly identify the optimal path A-C-G.

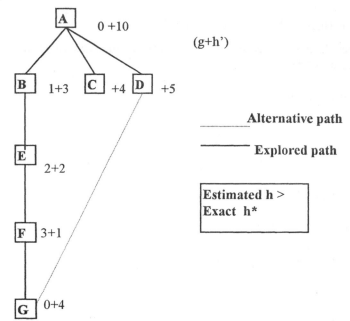

Fig. 4.9 (a): Illustration of overestimation of h in A* algorithm.

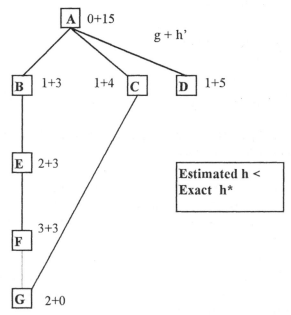

Fig. 4.9 (b): Illustration of underestimation of h in A* algorithm.

4.3.2 Iterative Deepening A* Algorithm

The iterative deepening search algorithm, discussed earlier, searches the goal node in a depth first manner at limited depth. In each pass the depth is increased by one level to test the presence of the goal node in that level. The A* algorithm, on the other hand, in each pass, selects the least cost (f) node for expansion. The iterative deepening A* (or IDA*) algorithm presented below attempts to combine the partial features of iterative deepening and A* algorithms together. Here, the heuristic measure is used to check the depth cut-off, rather than the order of the selection of nodes for expansion. The algorithm is formally presented below.

Procedure IDA*

Begin

1. Initialize the current depth cut-off $c = 1$;

2. Push a set of starting nodes into a stack; Initialize the cut-off at

    next iteration $c' = \infty$;

3. **While** the stack is not empty **do**

    Begin

    Pop stack and get the topmost element n;

    If n is the goal, **Then** report success and

    return n with the path from the starting node

    Else do

    Begin

    For each child n' of n

    If $f(n') \leq c$ **Then** push n' into the stack

    Else assign $c' := \min (c', f(n'))$;

    End For;

    End;

    End While;

4. **If** the stack is empty and $c' = \infty$ **Then** stop and exit;

5. **If** the stack is empty and $c' \neq \infty$ **Then** assign $c := c'$ and return to step 2;

End.

The above algorithm considers two depth cut-off levels. If the stack contains nodes whose children all have 'f' value lower than the cut-off value c, then these children are pushed into the stack to satisfy the depth first criteria of iterative deepening algorithms. However, when it fails, i.e., 'f' value of one or more child n' of n exceeds the cut-off level c, then the c' value of the node n is set to min (c', f(n')). The algorithm terminates when either i) the goal is identified (successful termination) or ii) the stack is empty and the cut-off value c' = $\propto$.

The main advantage of IDA* over A* lies in the memory requirement. The A* requires an exponential amount of memory because of no restriction on depth cut-off. The IDA* on the other hand expands a node n only when all its children n' have f (n') value less than the cut-off value c. Thus it saves a considerable amount of memory.

Another important point to note is that IDA* expands the same nodes expanded by A* and finds an optimal solution when the heuristic function used is optimal.

4.3.3 Heuristic Search on AND-OR Graphs

The second classical problem, where the heuristic search is applicable, is the backward reasoning problem implemented on AND-OR graphs. Before describing the technique of pruning an AND-OR graph, let us first understand the process of expanding an unconstrained graph. Consider the problem of acquiring a TV set. One has to identify the possible ways one can acquire the TV set. So, here the goal is "to acquire a TV set" and the terminals of the graph describe the possible means by which it can be achieved. The details of the possibility space are given in fig. 4.10.

For heuristic search on AND-OR graph, we use an algorithm, called an AO* algorithm. The major steps of the AO* algorithm are presented below.

1.   *Given the Goal node, hereafter called the starting state, find the possible offsprings of the starting state, such that the Goal can be derived from them by AND / OR clauses.*

2.   *Estimate the h' values at the leaves and find the leaf (leaves) with minimum h'. The cost of the parent of the leaf (leaves) is the minimum of the cost of the OR clauses plus one or the cost of the AND clauses plus the number of AND clauses. After the children with minimum h' are estimated, a pointer is attached to point from the parent node to its promising children.*

3.   *One of the unexpanded OR clauses / the set of unexpanded AND*
     clauses, where the pointer points from its parent, is now expanded
     and the h' of the newly generated children are estimated. The effect of
     this h' has to be propagated up to the root by re-calculating the f' of
     the parent or the parent of the parents of the newly created child /
     children clauses through a least cost path. Thus the pointers may be
     modified depending on the revised cost of the existing clauses.

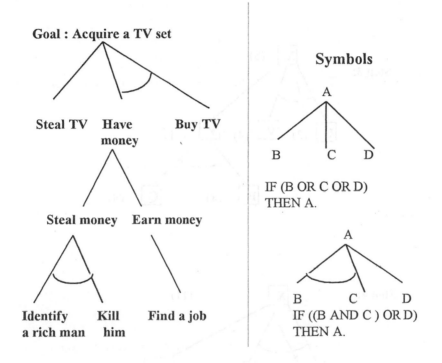

Fig. 4.10: An unconstrained AND-OR graph, where the AND, OR arcs
        are defined in side by side symbol definitions.

The few steps of the AO* algorithm are illustrated below based on the
above principle.

Step 1:   X   (7)

Step 2:
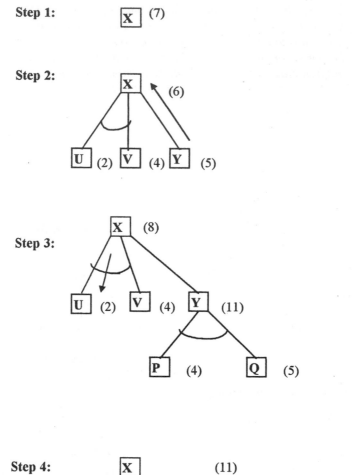

Step 3:

Step 4:
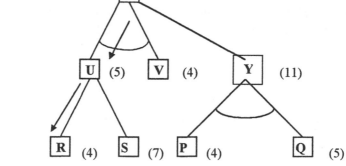

Fig. 4.11: Snapshots of the AO* algorithm.

Procedure AO*

Begin

1. Given the goal node INIT in the graph G; evaluate h' at INIT;
2. **Repeat**

> (a)Trace the marked arcs from the node INIT, if any such exists, and select one of the unexpanded nodes, named NODE, that occurs on this path, for expansion.

> (b) **If** NODE cannot be expanded, **Then** assign FUTILITY as the h' value of NODE, indicating that NODE is not solvable;

> **Else for** each such successor, called SUCCESSOR, which is not an ancestor of NODE, **do**

> **Begin**

>> (i)   Append SUCCESSOR to the graph G.

>> (ii)   **If** SUCCESSOR is a terminal node **Then** label it SOLVED and set its h' value 0.

>> (iii)   **If** SUCCESSOR is not a terminal node **Then** estimate its h' value;

> **End;**

> (c) Initialize S to NODE;

> (d) **Repeat**

>> (i) Select from S a node, none of whose descendants belong to S. Call it CURRENT and remove it from S.

>> (ii) Estimate the cost of each of the arcs, emerging from CURRENT. The cost of each arc is equal to the sum of h' value of each of the nodes at the end of the arc plus the cost of the arc itself. The new h' value of CURRENT is the minimum of the cost just computed for the arcs emerging from it.

>> (iii) Label the best path out of CURRENT by marking the arc that had the least cost as computed in the last step.

(iv) **If** all of the nodes connected to CURRENT through the new
marked arcs have been labeled SOLVED, **Then** mark the
CURRENT SOLVED.

(v) **If** CURRENT is marked SOLVED or the cost of CURRENT
was changed, **Then** propagate its new status back up the tree,
add all the ancestors of CURRENT to S.

Until S is empty.

Until INIT is labeled solved or its h' value becomes greater than a
maximum level called FUTILITY:

End.

4.4 Adversary Search

In this section we will discuss special type of search techniques required in a
game playing between two opponent players. The state space in this case is
represented by a tree or graph and includes the possible turns of both players.
Each level of the search space in the present context denotes the possible turn
of one player only. We start with a simple algorithm called MINMAX and
gradually develop more complex algorithms for game playing.

4.4.1 The MINIMAX Algorithm

The MINIMAX algorithm considers the exhaustive possibility of the state
transitions from a given state and consequently covers the entire space. The
algorithm, thus, is applicable to games having few possible state transitions
from a given trial state. One typical example that can be simulated with
MINIMAX is the NIM game. A NIM game is played between two players.
The game starts with an odd number of match sticks, normally 7 or 9, placed
on a single row, called a pile. Each player in his turn has to break a single
pile into two piles of unequal sticks, greater than zero. The game will come
to an end when either of the two players cannot give a successful move. The
player who cannot give a successful move the first time will lose the game.

According to standard convention we name the two players
MINIMIZER and MAXIMIZER. NIM is a defensive game and consequently
the opening player, here, is called the MINIMIZER. For a game such as tic-
tac-toe, where the opener always gets the benefit, the opening player is called
the MAXIMIZER. A graph space for the NIM game is presented in fig. 4.12
(a), demarcating MAXIMIZER's move from the MINIMIZER's move.

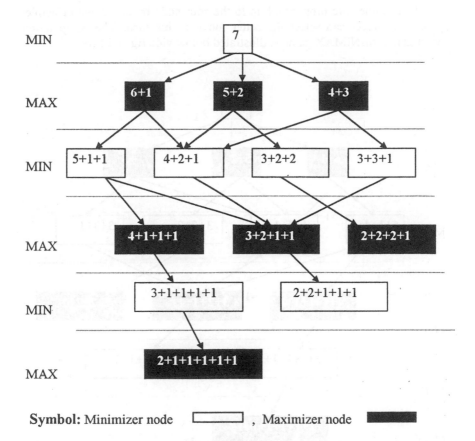

Symbol: Minimizer node [ ] , Maximizer node ▬

Fig. 4.12 (a): State Space for the NIM game.

In the MINIMAX algorithm, to be presented shortly, the following conventions will be used. The MAXIMIZER's success is denoted by +1, while the MINIMIZER's success by -1 and a draw by a 0. These values are attached with the moves of the players. A question then naturally arises: how do the players automatically learn about their success or failure until the game is over? This is realized in the MINIMAX algorithm by the following principle: *Assign a number from {+1, -1, 0} at the leaves depending on whether it is a success for the MAXIMIZER, MINIMIZER or a draw respectively. Now, propagate the values up by checking whether its parent is a MAXIMIZER or MINIMIZER node. If it is the MAXIMIZER's node then its value wiil be the maximum value possessed by its offsprings. In case it is a MINIMIZER's node then its value will presume the minimum of the values possessed by its offsprings.*

If the values are propagated up to the root node by the above principle, then each player can select the better move in his turn. The computation process in a MINIMAX game is illustrated below vide fig. 4.12 (b).

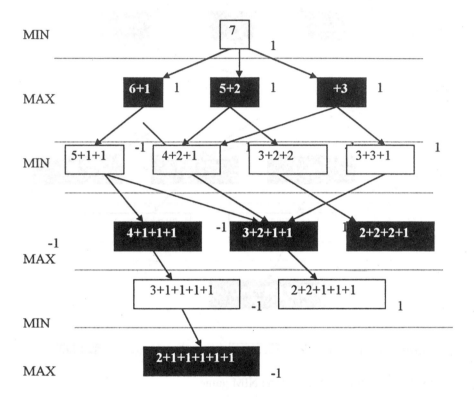

Fig. 4.12 (b): The computation in the state space for the NIM game.

The MINIMAX algorithm is formally presented below.

Procedure MINIMAX
Begin
1. Expand the entire state-space below the starting node;

2. Assign values to the terminals of the state-space from
   -1,0,+1}, depending on the success of the MINIMIZER,
   draw, or the success of the   MAXIMIZER respectively;

3. **For** each node whose all children possess values, **do**
   Begin
      if it is a MAXIMIZER node, **then** its value will be maximum

of its childrens' value; **if** it is a MINIMIZER node, **then** its
value will be the minimum of its children;
   End For;
End.

4.4.2 The Alpha-Beta Cutoff Procedure

The MINIMAX algorithm, presented above, requires expanding the entire
state-space. This is a severe limitation, especially for problems with a large
state-space. To handle this difficulty, an alternative approach is to evaluate
heuristically the status of the next ply move of the player, to select a current
move by the same player. We will demonstrate the process of computation of
the heuristic measures of a node below with the well-known tic-tac-toe game.

Consider a heuristic function e(n) [3], [5] at node n that evaluates the
difference of possible winning lines of the player and his opponent. Formally,

$$e(n) = M(n) - O(n)$$

where $M(n)$ = number of my possible winning lines
and $O(n)$ = number of opponent's winning lines.

For example, in fig. 4.13 $M(n) = 6$, $O(n) = 5$ and hence $e(n) = 1$.

Now, we will discuss a new type of algorithm, which does not require
expansion of the entire space exhaustively. This algorithm is referred to as
alpha-beta cutoff algorithm. In this algorithm, two extra ply of movements
are considered to select the current move from alternatives. Alpha and beta
denote two cutoff levels associated with MAX and MIN nodes. The alpha
value of MAX node cannot decrease, whereas the beta value of the MIN
nodes cannot increase. But how can we compute the alpha and beta values?
They are the backed up values up to the root like MINIMAX. There are a
few interesting points that may be explored at this stage. Prior to the process
of computing MAX / MIN of the backed up values of the children, the alpha-
beta cutoff algorithm estimates e(n) at all fringe nodes n. Now, the values are
estimated following the MINIMAX algorithm. Now, to prune the
unnecessary paths below a node, check whether

i)   the beta value of any MIN node below a MAX node is less than or
   equal to its alpha value. If yes, prune that path below the MIN node.

ii) the alpha value of any MAX node below a MIN node is greater than or
     equal to the beta value of the MIN node. If yes prune the nodes below
     the MAX node.

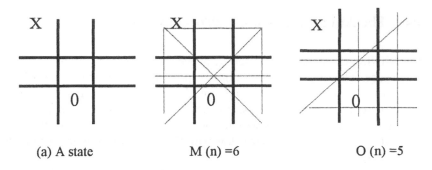

| (a) A state | M (n) =6 | O (n) =5 |

Fig. 4.13: Evaluation of e (n) at a particular state.

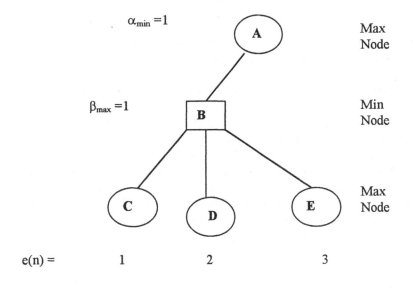

Pass - I

Fig.4.14: Two-ply move of the MAXIMIZER with computed e(n) at the
           fringe nodes: C, D, E; backed up values at node B and A and
           setting of α_{max} and β_{min} values at nodes A and B, respectively.

Based on the above discussion, we now present the main steps in the α-β search algorithm.

i)   Create a new node, if it is the beginning move, else expand the existing tree by depth first manner. To make a decision about the selection of a move at depth d, the tree should be expanded at least up to a depth (d + 2).

ii)   Compute e(n) for all leave (fringe) nodes n in the tree.

iii)   Compute α'_{min} (for max nodes) and β'_{max} values (for min nodes) at the ancestors of the fringe nodes by the following guidelines. Estimate the minimum of the values (e or α) possessed by the children of a MINIMIZER node N and assign it its β'_{max} value. Similarly, estimate the maximum of the values (e or β) possessed by the children of a MAXIMIZER node N and assign it its α'_{min} value.

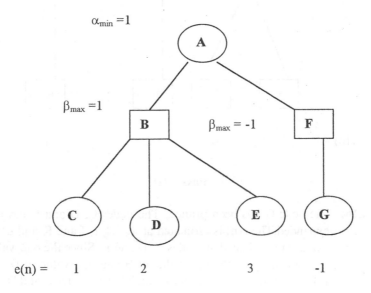

Pass - II

Fig. 4.15: A thinks of the alternative move F, and also mentally generates the next ply move G; e(G) = -1; so β_{max} at F is –1. Now, β_{max} at F is less than α_{min} of A. Thus there is no need to search below F. G may be pruned from the search space.

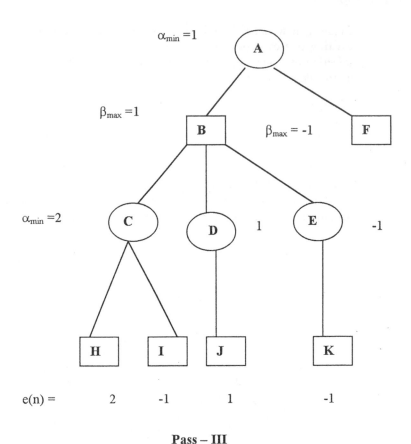

Pass – III

Fig. 4.16: The node G has been pruned; The nodes C, D and E have been expanded; The e(n) is estimated at n = H, I, J and K and the α_{min} values are evaluated at nodes C, D and E. Since the α_{min} value of C is greater than the β_{max} value of B and α_{min} value of D = β_{max} value of B, there is no need to search below nodes C and D.

iv)   If the MAXIMIZER nodes already possess α_{min} values, then their current α_{min} value = Max (α_{min} value, α'_{min}); on the other hand, if the MINIMIZER nodes already possess β_{max} values, then their current β_{max} value = Min (β_{max} value, β'_{max}).

v)   If the estimated β_{max} value of a MINIMIZER node N is less than or equal to the α_{min} value of its parent MAXIMIZER node N' then there is no need to search below the node MINIMIZER node N. Similarly, if the α_{min} value of a MAXIMIZER node N is more than the β_{max} value of its parent node N' then there is no need to search below node N.

The above steps are continued until the game is over. If we call these five steps together a pass, then the first three passes are shown in fig. 4.14-4.16. The interested reader may on his own work out the tic-tac-toe game using the definition of e (n) and the last 5 steps. We could not present it here because of the page size of this book, which cannot accommodate large trees.

4.5 Conclusions

We presented a large number of search algorithms in this chapter. We started with the breadth first and the depth first search algorithms and analysed their complexities. It is clear from the analysis that the breadth first search is not appropriate for large state space as the memory requirement is excessively high. The depth first search algorithm works well for most of the typical AI problems. Sometimes, we may not be interested to explore below a given depth. The iterative deepening search is useful under this context. Recently, the iterative deepening algorithm has been formulated in A* fashion and is called the IDA* algorithm. The IDA* algorithm has a great future, as it has been seriously studied by many researchers for realization on parallel architecture [4]. We shall take up these issues in chapter 22.

Among the heuristic search algorithms presented in the chapter, the most popular are A* and AO* algorithm. A* is used on a OR graph, while AO* is employed in an AND-OR graph. The A* is applied in problems to find the goal and its optimal path from the starting state in a state space, while the AO* determines the optimal paths to realize the goal. Heuristic search, since its inception, has remained an interesting toolbox for the researchers working in AI. Recently, Sarkar et al. [8] extended the A* algorithm for machine learning by strengthening the heuristic information at the nodes of the state-space. We, however, do not have much scope to discuss their work here.

Besides the search algorithms presented in the chapter, there exist a few more problem solving techniques we introduced in chapter 1. These are constraint satisfaction techniques, means and ends analysis and problem reductions. Constraint satisfaction techniques, being an emerging research area, will be presented in detail in chapter 19. Means and ends analysis and problem reduction techniques, on the other hand, are available in most text books [7] and we omit these for lack of space.

Exercises

1. Using the Euclidean distance of a node (x, y) from a fixed node (2, 2), i.e.,

$$h = [(x - 2)^2 + (y - 2)^2]^{1/2}$$

   solve the water-jug problem by paper and pencil by A* algorithm. Does this heuristic function return an optimal path? Consequently, can you call it an admissible heuristic?

2. The 8-puzzle problem is similar to the 4-puzzle problem we discussed in chapter 1. The only difference is that there exist 9 cells and 8 tiles instead of the 4 cells and 3 tiles of a 4-puzzle problem. Can you select a heuristic function for the 8-puzzle problem? Solve the 8-puzzle problem by the A* algorithm with your selected heuristic function.

3. Show the computation for the first 3 ply moves in a tac-tac-toe game using the α-β cut-off algorithm.

4. Consider a room whose floor space is partitioned into equal sized blocks. Suppose there is a mobile robot (MR) in one block, and we want to move to a distant block. Some of the blocks are occupied with obstacles. The robot has to plan its trajectory so that it reaches the goal position from a given initial position without touching the obstacles. Can you design a heuristic function for the problem? If yes, solve the problem using the A* algorithm on a graph paper. Assume the location of the obstacles and the starting and the goal positions.

References

[1]   Bender, E. A., *Mathematical Methods in Artificial Intelligence*, IEEE Computer Society Press, Los Alamitos, pp. 33-84, 1996.

[2]   Ginsberg, M., *Essentials of Artificial Intelligence*, Morgan Kaufmann, San Mateo, CA, pp. 49-103, 1993.

[3] Luger, G. F. and Stubblefield, W. A., *Artificial Intelligence: Structures and Strategies for Complex Problem Solving*, The Benjamin/ Cummings Publishing Co., Menlo Park, CA, pp.116-149, 1993.

[4] Mahanti, A. and Daniels, J. C., "A SIMD approach to parallel heuristic search," *Artificial Intelligence*, vol. 60, pp. 243-282, 1993.

[5] Nilson, J. N., *Principles of Artificial Intelligence*, Morgan-Kaufmann, San Mateo, CA, pp. 112-126,1980.

[6] Pearl, J., *Heuristics: Intelligent Search Strategies for Computer Problem Solving*, Addison-Wesley, Reading, MA, pp. 1-75, 1984.

[7] Rich, E. and Knight, K., *Artificial Intelligence*, McGraw-Hill, New York, pp. 29-98, 1991.

[8] Sarkar, S., Chakrabarti, P. P. and Ghose, S., " A framework for learning in search-based systems," *IEEE Trans. on Knowledge and Data Engg.*, vol. 10, no. 4, July / Aug. 1998.

5

The Logic of Propositions and Predicates

The chapter presents various tools and techniques for representation of knowledge by propositions and predicates and demonstrates the scope of reasoning under the proposed framework of knowledge representation. It begins with the syntax and semantics of the logic of propositions, and then extends them for reasoning with the logic of predicates. Both the logic of propositions and predicates require the formulation of a problem in the form of a logical theorem and aim at proving it by the syntactic and the semantic tools, available in their framework. The 'resolution principle' is the most common tool that is employed for reasoning with these logics. To prove a goal, complex sentences are first represented in 'clause forms' and the principle of resolution is employed to resolve the members of a given set, comprising of the axiomatic rules (clauses) and the negated goal. One main drawback of the predicate logic lies in its semi-decidablity that fails to disprove a statement that does not really follow from the given statements. The chapter discusses all these in detail along with the formal proofs of 'soundness' and 'completeness' of the resolution principle.

5.1 Introduction

Production systems, covered in chapter 3, has been successfully used for reasoning in many intelligent systems [1],[6]. Because of its inherent simplicity, it has been widely accepted as a fundamental tool to knowledge representation. The efficiency of production systems, however, degrades with the increase in complexity of knowledge in real world problems. For instance, a production system does not support simple rules like *if ((X is a son of Y)* *OR (X is a daughter of Y)) then (Y is a father of X)*. The logic of propositions (also called propositional logic) is an alternative form of knowledge representation, which overcomes some of the weakness of production systems. For instance, it can join simple sentences or clauses by logical connectives to represent more complex sentences. Due to the usage of logical connectives, propositional logic is sometimes called logical calculus. However, it needs mention that such logic has no relevance with Calculus, the popularly known branch of mathematics. This chapter will be devoted to representing knowledge with propositional logic. Generally, the reasoning problems in propositional logic are formulated in the form of mathematical theorems. For instance, given two facts : i) Birds fly, ii) Parrot is a bird, and one has to infer that parrot flies. This can be formally stated in the form of a theorem: given the premises *birds fly* and *parrot is a bird*, prove that *parrot flies*. We can now employ tools of propositional logic to prove (or disprove) the theorem. The chapter presents various tools and techniques for theorem proving by propositional logic.

Predicate Logic (also called first order predicate logic or simply first order logic or predicate calculus) has similar formalisms like the propositional logic. It is more versatile than the propositional counterpart for its added features. For instance, it includes two quantifiers, namely, the essential quantifier ($\forall$) and the existential quantifier ($\exists$) that are capable of handling more complex knowledge.

The chapter is organized as follows. It starts with a set of formal definitions and presents the methodology of knowledge representation by propositional logic. It then covers the semantic and syntactic methods of theorem proving by propositional logic. Next predicate logic is introduced from the first principles, and a method to represent large sentences in clause form is described. Later two fundamental properties of predicate calculus: the unification algorithm and the resolution principle, which are useful for theorem proving, are introduced. The issues of soundness and completeness are discussed briefly in the chapter.

5.2 Formal Definitions

The following definitions, which will be referred to occasionally in the rest of the book, are in order.

Definition 5.1: A **connective** is a logical operator that connects simple statements for constructing more complex statements.

The list of connectives in propositional logic and their meaning is tabulated below.

Table 5.1: Connectives in propositional logic

Operators	Notations
AND	$\wedge$
OR	$\vee$
Negation	$\neg , \sim$
If p then q	$p \rightarrow q$
If p then q and if q then p	$p \leftrightarrow q$
Implication	$\Rightarrow$
Bi-directional Implication (IFF)	$\Leftrightarrow$
Identity	$\equiv$
Logical entailment	$\models$
Derivability	$\vdash$

It should be noted that AND and OR operators are sometimes referred to as **conjunction** and **disjunction** respectively. It may further be added that the provability and implication symbols have been used in an interchangeable manner in this book. The author, however, has a strong reservation to use implication symbol in place of if-then operator and vice versa [3]. *The symbol "x $\vdash$ y " implies that y has been derived from x by following a proof procedure.* The *logical entailment relation:* "x $\models$ y " *on the other hand means that y logically follows from x.*

Definition 5.2: A **proposition** is a statement or its negation or a group of statements and/or their negations, connected by AND, OR and If-Then operators.

For instance,

> p,
> it-is-hot, the-sky-is-cloudy ,
> it-is-hot ∧ the-sky-is-cloudy,
> it-is-hot → the-sky-is-cloudy

 are all examples of propositions.

Definition 5.3: When a statement cannot be logically broken into smaller statements, we call it **atomic.**

For example, p, q, the-sky-is-cloudy are examples of atomic propositions.

Definition 5.4: A proposition can assume a **binary valuation space**, i.e., for a proposition p, its valuation space $v(p) \in \{0,1\}$.

Definition 5.5: Let r be a propositional formula, constructed by connecting atomic propositions p, q, s, etc. by operators. An **interpretation** for r is a function that maps $v(p)$, $v(q)$ and $v(s)$ into true or false values that together keep r true.

For example, given the formula: $p \wedge q$. The possible interpretation is $v(p) = $ true and $v(q) =$ true. It may be noted that for any other values of p and q the formula is false.

There may be more than one interpretation of a formula. For instance, the formula: $\neg p \vee q$ has three interpretations given below.

Interpretations:

$\{v(p) = $ true, $v(q) = $ true$\}$, $\{v(p) = $ false, $v(q) = $ false$\}$, and

$\{v(p) = $ false, $v(q) = $ true$\}$.

Definition 5.6: A propositional formula is called **satisfiable** if its value is true for some interpretation [2].

For example the propositional formula $p \vee q$ is satisfiable as it is true for some interpretations $\{v(p) = $ true, $v(q) = $ true$\}$, $\{v(p) = $ false, $v(q) = $ true$\}$ and $\{v(p) = $ true, $v(q) =$ false$\}$.

Generally, we use ⊧ p to denote that p is satisfiable.

Definition 5.7: A propositional formula is **unsatisfiable** or **contradictory** if it is not satisfiable, i.e., for no interpretation it is true.

Definition 5.8: A propositional formula is called **valid** or **tautology,** when it is true for all possible interpretations.

For example, $(p \wedge q) \wedge r \equiv p \wedge (q \wedge r)$ is a tautology, since it is true for all possible $v(p)$, $v(q)$ and $v(r) \in \{0,1\}$. Here we have 8 possible interpretations for the propositional formula, for which it is true.

The sub-sethood relationship of all formulas, and satisfiable and valid formulas is presented in Venn diagram 5.1

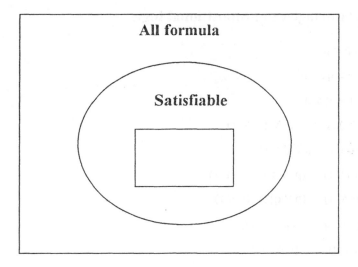

Valid $\subseteq$ Satisfiable, Satisfiable $\subseteq$ All formula,
Unsatisfiable = All formula – Satisfiable.

Fig. 5.1: Sub-sethood relationship of valid, satisfiable and all formulas.

5.3 Tautologies in Propositional Logic

The tautologies [1] may be directly used for reasoning in propositional logic. For example, consider the following statements.

p_1 = the-sky-is-cloudy, p_2 = it-will-rain, and p_3 = if (the-sky-is-cloudy) then (it-will-rain) $\equiv p_1 \rightarrow p_2$.

"p_1" and "p_2" above represent premise and conclusion respectively for the if-then clause. It is obvious from common sense that p_2 directly follows from p_1

and p_3. However to prove it automatically by a computer, one requires help of the following tautology, the proof of which is also given here.

$p_3 \equiv p_1 \rightarrow p_2$

$\equiv \neg (p_1 \wedge \neg p_2)$, since p_1 true and p_2 false cannot occur together.

$\equiv \neg p_1 \vee p_2$ (by De Morgan's law)

However, to prove p_2 from p_1 and p_3 we have to wait till example 5.1.

List of tautologies in propositional logic

1. $\neg \neg p \equiv p$

2. $p \wedge q \equiv q \wedge p$

3. $p \vee q \equiv q \vee p$

4. $(p \wedge q) \wedge r \equiv p \wedge (q \wedge r)$

5. $(p \vee q) \vee r \equiv p \vee (q \vee r)$

6. $p \wedge (q \vee r) \equiv (p \wedge q) \vee (p \wedge r)$

7. $p \vee (q \wedge r) \equiv (p \vee q) \wedge (p \vee r)$

8. $\neg (p \wedge q) \equiv \neg p \vee \neg q$

9. $\neg (p \vee q) \equiv \neg p \wedge \neg q$

10. $p \vee p \equiv p$

11. $p \wedge p \equiv p$

12. $p \wedge q \rightarrow p$

13. $p \wedge q \rightarrow q$

14. $p \rightarrow p \vee q$

15. $q \rightarrow p \vee q$

5.4 Theorem Proving by Propositional Logic

We present here two techniques for logical theorem proving in propositional logic. These are i) Semantic and ii) Syntactic methods of theorem proving.

5.4.1 Semantic Method for Theorem Proving

The following notation will be used to represent a symbolic theorem, stating that conclusion "c" follows from a set of premises $p_1, p_2, ..., p_n$

$$p_1, p_2, ..., p_n \Rightarrow c \quad \text{or} \quad p_1, p_2, ..., p_n \models c$$

In this technique, we first construct a truth table representing the relationship of p_1 through p_n with "c". Then we test the validity of the theorem by checking whether both the *forward and backward chaining methods*, to be presented shortly, hold good. The concept can be best illustrated with an example.

Example 5.1: Let us redefine $p_1 =$ the-sky-is-cloudy, $p_2 =$ it-will-rain and $p_3 \equiv p_1 \rightarrow p_2$ to be three propositions. We now form a truth table of p_1, p_2 and p_3, and then attempt to check whether forward and backward chaining holds good for the following theorem:

$$p_1, p_3 \Rightarrow p_2$$

Table 5.2: Truth Table of p_1, p_2, and $p_1 \rightarrow p_2$.

p_1	p_2	$p_3 \equiv p_1 \rightarrow p_2$ $\equiv \neg p1 \lor p2$
0	0	1
0	1	1
1	0	0
1	1	1

Forward chaining: *When all the premises are true, check whether the conclusion is true. Under this circumstance, we say that forward chaining holds good.*

In this example, when p_1 and p_3 are true, check if p_2 is true. Note that in the last row of the truth table, $p_1 = 1$, $p_3 = 1$ yield $p_2 = 1$. So, forward chaining holds good. Now, we test for backward chaining

Backward chaining: *When all the consequences are false, check whether at least one of the premises is false.*

In this example $p_2 = 0$ in the first and third row. Note that when $p_2 = 0$, then $p_1 = 0$ in the first row and $p_3 = 0$ in the third row. So, backward chaining holds   good.

As forward and backward chaining both are satisfied together, the theorem: $p_1, p_3 \Rightarrow p_2$ also holds good.

Example 5.2: Show that for example 5.1, $p_2, p_3 \ =/> \ p_1$.

It is clear from the truth table 5.2 that when $p_1=0$, then $p_2=0$ (first row) and $p_3 = 1$ (first row), backward chaining holds good. But when $p_2 = p_3 =1$, $p_1=0$ (second row), forward chaining fails. Therefore, the theorem does not hold good.

5.4.2   Syntactic Methods for Theorem Proving

Before presenting the syntactic methods for theorem proving in propositional logic, we state a few well-known theorems [4], which will be referred to in the rest of the chapter.

Standard theorems in propositional logic

Assuming p, q and r to be propositions, the list of the standard theorems is presented below.

1. $p, q \Rightarrow p \wedge q$

2. $p, p \rightarrow q \Rightarrow q$ (Modus Ponens)

3. $\neg p, \ p \vee q \Rightarrow q$

4. $\neg q, \ p \rightarrow q \Rightarrow \neg p$ (Modus Tollens)

5. $p \vee q, \ p \rightarrow r, \ q \rightarrow r \Rightarrow r$

6. $p \rightarrow q, \ q \rightarrow r \Rightarrow p \rightarrow r$ (Chaining)

7. $p, \ p \rightarrow q, \ q \rightarrow r \Rightarrow r$ (Modus Ponens & Chaining)

8. $p \vee (q \wedge \neg q) \Leftrightarrow p$

9. $p \wedge (q \vee \neg q) \Leftrightarrow p$

10. $p \rightarrow q \Leftrightarrow \neg p \vee q$

11. $\neg (p \rightarrow q) \Leftrightarrow p \wedge \neg q$

12. $p \leftrightarrow q \Leftrightarrow (p \rightarrow q) \wedge (q \rightarrow p)$

13. $p \leftrightarrow q \Leftrightarrow (p \wedge q) \vee (\neg p \wedge \neg q)$

14. $\quad p \rightarrow (q \rightarrow r) \Leftrightarrow (p \wedge q) \rightarrow r$

15. $p \rightarrow q \Leftrightarrow \neg q \rightarrow \neg p$ (contraposition theorem)

The syntactic approach for theorem proving can be done in two ways, namely, i) by the method of substitution and ii) by Wang's algorithm.

5.4.2.1 Method of Substitution

By this method, left-hand side (or right-hand side) of the statement to be proved is chosen and the standard formulas, presented above, are applied selectively to prove the other side of the statement.

Example 5. 3: Prove the contraposition theorem.

The contraposition theorem can be stated as follows. When p and q are two propositions, the theorem takes the form of $p \rightarrow q \Leftrightarrow \neg q \rightarrow \neg p$.

Now, L.H.S. $= p \rightarrow q$

$\qquad \Rightarrow \neg p \vee q \quad$ [by (10)]

$\qquad \Rightarrow q \vee \neg p$

$\qquad \Rightarrow \neg (\neg q) \vee \neg p$

$\qquad \Rightarrow \neg q \rightarrow \neg p = $ R.H.S. $\qquad \square$

Analogously, starting with the R.H.S, we can easily reach the L.H.S. Hence, the theorem bi-directionally holds good.

Example 5.4: Prove theorem (14) by method of substitution.

Proof: L.H.S. $= p \rightarrow (q \rightarrow r)$

$$\Rightarrow \quad p \to (\neg\, q \lor\, r) \quad \text{[by (10)]}$$

$$\Rightarrow \neg\, p \lor (\neg\, q \lor r) \quad \text{[by (10)]}$$

$$\Rightarrow (\neg\, p \lor \neg\, q) \lor r \quad \text{[since this is a tautology by (5)]}$$

$$\Rightarrow \quad \neg\, (p \land q) \lor r \quad \text{[by De Morgan's law]}$$

$$\Rightarrow \quad (p \land q) \to r \quad \text{[by (10)]}$$

$$= \quad \text{R.H.S.}$$

Analogously, the L.H.S. can be equally proved from the R.H.S. Hence, the theorem follows bi-directionally.   □

5.4.2.2 Theorem Proving by Using Wang's Algorithm

Any theorem of propositional logic is often represented in the following form:

$$p_1, p_2, \ \dots \ p_n \Rightarrow q_1, q_2, \ \dots, q_m$$

where p_i and q_j represent propositions. The comma in the L.H.S. represents AND operator, while that in the R.H.S. represents OR operator. Writing symbolically,

$$p_1 \land p_2 \land \ \dots \land p_n \Rightarrow q_1 \lor q_2 \lor \dots \ \lor q_m$$

This kind of theorem can be easily proved using Wang's algorithm [10]. The algorithm is formally presented below.

Wang's algorithm

Begin

Step I: *Starting condition:* Represent all sentences, involving only $\land$, $\lor$ and $\neg$ operators.

Step II: *Recursive procedure:* Repeat steps (a), (b) or (c) whichever is appropriate until the stopping condition, presented in step III, occurs.

a) Negation Removal: In case negated term is present at any side (separated by comma) bring it to the other side of implication symbol without its negation symbol.

e.g., $p, q, \neg\, r \Rightarrow s$

$\models p, q \Rightarrow r, s$

b) **AND, OR Removal:** If the L.H.S. contains $\wedge$ operator, replace it by a comma. On the other hand if R.H.S. contains $\vee$ operator, also replace it by a comma.

e.g., $\quad p \wedge r, s \Rightarrow s \vee t$

$\quad \vdash \quad p, r, s \Rightarrow s, t$

c) **Theorem splitting:** If the L.H.S. contains OR operator, then split the theorem into two sub-theorems by replacing the OR operator. Alternatively, if the R.H.S. contains AND operator, then also split the theorem into two sub-theorems.

e.g., $\quad p \vee r \Rightarrow s, t$

$\quad \vdash \quad p \Rightarrow s, t \quad \& \quad r \Rightarrow s, t \qquad$: Sub-theorems

e.g., $\quad p, r \Rightarrow s \wedge t$

$\quad \vdash p, r \Rightarrow s \quad \& \quad p, r \Rightarrow t \qquad$: Sub-theorems

Step III: *Stopping Condition:* Stop theorem proving process if either of (a) or (b), listed below, occurs.

    a) If both L.H.S. and R.H.S. contain common atomic terms, then stop.

    b) If L.H.S. and R.H.S. have been represented as a collection of atomic terms, separated by commas only and there exist no common terms on both sides, then stop.

End.

In case all the sub-theorems are stopped, satisfying condition III (a), then the theorem holds good. We would construct a tree structure to prove theorems using Wang's algorithm. The tree structure is necessary to break each theorem into sub-theorems.

Example 5. 5: Prove the chaining rule with Modus Ponens using Wang's algorithm.

Proof: The chaining rule with Modus Ponens can be described as

$$p, \ p \rightarrow q, \ q \rightarrow r \Rightarrow r$$

where p, q and r are propositions (atomic).

We now construct the tree. A node in the tree denotes one propositional expression. An arc in the tree denotes the step of Wang's algorithm, which is applied to produce the next step. The bold symbols in both the left- and right-hand side of the implication symbol describe the termination of the sub-tree by step III (a).

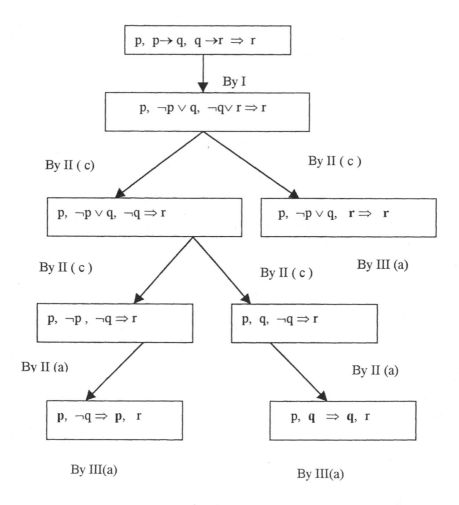

Fig. 5.2: Tree used to prove a propositional theorem by Wang's algorithm.

Since all the terminals of the tree have been stopped by using III (a), the theorem holds good. □

5.5 Resolution in Propositional Logic

The principle of resolution in propositional logic can be best described by the following theorem [7].

Resolution theorem: For any three clauses p, q and r,

$$p \vee r, \; q \vee \neg r \Rightarrow p \vee q.$$

Proof: The theorem, which can be proved by Wang's algorithm, is left as an exercise for the students. □

The resolution theorem can also be used for theorem proving and hence reasoning in propositional logic. The following steps should be carried out in sequence to employ it for theorem proving.

Resolution algorithm

Input: A set of clauses, called axioms and a goal.
Output: To test whether the goal is derivable from the axioms.

Begin
1. Construct a set S of axioms plus the negated goal.

2. Represent each element of S into conjunctive normal form (CNF) by the following steps:

   a) Replace 'if-then' operator by NEGATION and OR operation by theorem 10.

   b) Bring each modified clause into the following form and then drop AND operators connected between each square bracket. The clauses thus obtained are in conjunctive normal form (CNF). It may be noted that p_{ij} may be in negated or non-negated form.

   $[\; p_{11} \vee p_{12} \vee \ldots\ldots \vee p_{1n}] \wedge$

   $[\; p_{21} \vee p_{22} \vee \ldots\ldots\vee p_{2n}] \wedge$

   $\ldots\ldots\ldots\ldots\ldots\ldots\ldots\ldots\ldots\ldots\ldots\ldots\ldots\ldots\ldots\ldots$

   $[\; p_{m1} \vee p_{m2} \vee \quad \vee p_{mn}]$

3.   **Repeat**

   a)   Select any two clauses from S, such that one clause contains a negated literal and the other clause contains its corresponding positive (non-negated) literal.

   b)   Resolve these two clauses and call the resulting clause the resolvent. Remove the parent clauses from S.

   Until a null clause is obtained or no further progress can be made.

4.   **If** a null clause is obtained, **then** report: "goal is proved".

The following example illustrates the use of resolution theorem for reasoning with propositional logic.

Example 5.6: Consider the following knowledge base:

   1. The-humidity-is-high ∨ the-sky-is-cloudy.
   2. If the-sky-is-cloudy then it-will-rain
   3. If the-humidity-is-high then it-is-hot.
   4. it-is-not-hot

and the goal : it-will-rain.

Prove by resolution theorem that the goal is derivable from the knowledge base.

Proof: Let us first denote the above clauses by the following symbols.

   p= the-humidity-is-high,   q= the-sky-is-cloudy, r= it-will-rain, s= it-is-hot.

   The CNF form of the above clauses thus become

   1. p ∨ q
   2. ¬ q ∨ r
   3. ¬ p ∨ s
   4. ¬ s

and the negated goal = ¬ r. Set S thus includes all these 5 clauses. Now by resolution algorithm, we construct the graph of fig. 5.3. Since it terminates with a null clause, the goal is proved.

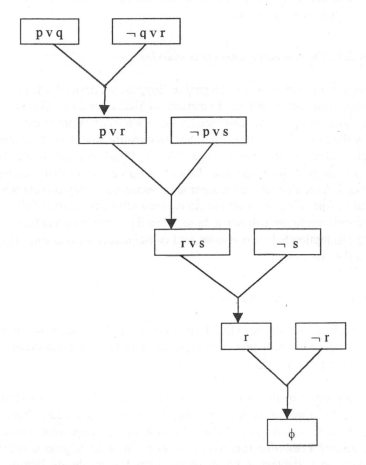

Fig. 5.3: The resolution tree to prove that it-will-rain.

5.6 Soundness and Completeness

Soundness and completeness are two major issues of the resolution algorithm. While soundness refers to the correctness of the proof procedure, completeness implicates that all the possible inferences can be derived by using the algorithm. Formal definitions of these are presented here for convenience.

Definition 5.9: A proof process is called **sound**, if any inference α has been proved from a set of axioms S by a proof procedure, i.e., $S \vdash \alpha$, it follows logically from S, i.e., $S \vDash \alpha$.

Definition 5.10: A proof process is called **complete,** if for any inference α, that follows logically from a given set of axioms S, i..e., S $\models$ α, the proof procedure can prove α, i.e., S $\vdash$ α.

Theorem 5.1: *The resolution theorem is sound.*

Proof: Given a set of clauses S and a goal α. Suppose we derived α from S by the resolution theorem. By our usual notation, we thus have S $\vdash$ α. We want to prove that the derivation is logically sound, i.e., S $\models$ α. Let us prove the theorem by the method of contradiction. So, we presume that the consequent S $\models$ α is false, which in other words means S $\models$ $\neg\alpha$. Thus $\neg\alpha$ is satisfiable. To satisfy it, we assign truth values (true / false) to all propositions that are used in α. We now claim that for such assignment, resolution of any two clauses from S will be true. Thus the resulting clause even after exhaustion of all clauses through resolution will not be false. Thus S $\vdash$ α is a contradiction. Hence, the assumption S $\models$ $\neg$ α is false, and consequently S $\models$ α is true. This is all about the proof [5]. $\square$

Theorem 5.2: *The resolution theorem is complete.*

Proof: Let α be a formula, such that from a given set of clauses S, we have S $\models$ α, i.e., α can be logically proved from S. We have to show there exists a proof procedure for α, i.e., S $\vdash$ α.

We shall prove it by the method of contradiction, i.e. let S $\vdash$ α not follow, i.e., S $\vdash$ $\neg\alpha$. In words α is not derivable by a proof procedure from S. Therefore, $S_1 = S \cup \alpha$ is unsatisfiable. We now use an important theorem, called the **ground resolution theorem,** that states "if a set of ground clauses (clauses with no variables) is unsatisfiable, then the resolution closure of those clauses contains the 'false' clause. Thus as S_1 is unsatisfiable, the resolution closure of S_1 yields the null clause, which causes a contradiction to S $\models$ α. Thus the assumption is wrong and hence S $\vdash$ α is true. $\square$

We now prove the ground resolution theorem, stated below.

Theorem 5.3: *If a set of ground clauses S is unsatisfiable, then the resolution closure T of those clauses contains the false clause.*

Proof: We prove the theorem by the method of contradiction. So, we presume that resolution closure T does not contain false clause and will terminate the proof by showing that S is satisfiable.

Let $A_S = \{A_1, A_2, \ldots, A_n\}$ be the set of atomic sentences occurring in S. Note that A_s must be finite. We now pick up an assignment (true / false) for each atomic sentence in A_S in some fixed order $\{A_1, A_2, \ldots A_K\}$ such that

i)   if a clause in T contains $\neg A_i$, with all its other literals connected through OR being false, then assign A_i to be false.

ii)   Otherwise, assign A_i to be true.

We can easily show that with this assignment, S is satisfiable, if the closure T of S does not contain false clause [9].   □

5.7 Predicate Logic

Predicate logic (also called first order predicate logic) has a similar formalism like propositional logic. However, the capability of reasoning and knowledge representation using predicate logic is higher than propositional logic. For instance, it includes two more quantifiers, namely, the essential quantifier ($\forall$) and the existential quantifier ($\exists$). To illustrate the use of the quantifiers, let us consider the following pieces of knowledge.

Knowledge 1 : *All boys like sweets.*

Using predicate logic, we can write the above statement as

$$\forall X (Boy (X) \rightarrow Likes (X , sweets))$$

Knowledge 2 : *Some boys like flying kites.*

Using predicate logic, the above statement can be represented as

$$\exists X (Boy (X) \rightarrow Likes (X, Flying\text{-}kites))$$

Before describing predicate logic (PL) or first order logic (FOL) in a formal manner, we first present the alphabets of FOL.

Alphabets of FOL

The alphabets of FOL are of the following types:

1. Constants: a, b, c

2. Variables: X, Y, Z

3. Functions: f, g, h

4. Operators: $\wedge, \vee, \neg, \rightarrow$

5. Quantifiers: $\forall, \exists$

6. Predicate: P, Q, R

Definition 5.11: A **term** is defined recursively as being a constant, variable or the result of application of a function to a term.

e.g., a, x, t(x), t(g(x)) are all terms.

To illustrate the difference between functions and predicates, we give their formal definitions with examples.

Definition 5.12: Function denotes relations defined on a domain D. They map n elements (n >0) to a single element of the domain. "father-of", "age-of" represent function symbols. An n-ary function is written as $f(t_1, t_2,.., t_n)$ where t_i s represent terms. A 0-ary function is a constant [7].

Definition 5.13: Predicate symbols denote relations or functional mappings from the elements of a domain D to the values true or false. Capital letters such as P,Q, MARRIED, EQUAL are used to represent predicates. $P(t_1, t_2, ..., t_n)$ represents an n-ary predicate where t_i are terms. A 0-ary predicate is a proposition, that is a constant predicate.

Definition 5.14: The sentences of FOL are **well-formed-formulas (WFF)**, defined as follows:

1.   If P $(t_1, t_2, ... , t_n)$ is an n-ary predicate, then P is an atomic formula.

2.   An atomic formula is a well-formed formula (WFF).

3.   If P and Q are WFF then $P \wedge Q, P \vee Q, \neg P, P \rightarrow Q$ are all WFF. Note that $\forall X$ R (X) is also an WFF.

4. If P is a WFF and X is not a quantified variable in P, then P remains a WFF even after quantification

    e.g., ∀X P   or   ∃X P are WFF.

Example 5.7: Rewrite the following sentences in FOL.

  1.   Coconut-crunchy is a biscuit.
  2.   Mary is a child who takes coconut-crunchy.
  3.   John loves children who take biscuits.
  4.   John loves Mary.

The above statements can be represented in FOL using two quantifiers X & Y.

  1.   Biscuit (coconut-crunchy)
  2.   Child (mary) ∧ Takes (mary, coconut-crunchy)
  3.   ∀X ((Child (X) ∧ ∃Y (Takes (X, Y) ∧ Biscuit (Y))) →Loves (john, X)

  4.   Loves (john, mary)

5.8 Writing a Sentence into Clause Forms

We now present a technique for representing a complex sentence into simple sentences. The technique is described below. As an example, we consider statement 3 for conversion into clause forms. The resulting expressions after application of each step are presented following the step.

Algorithm for representing a sentence into clauses

Step I: *Elimination of if-then operator*: Replace "→" operator by ¬ & ∨ operator.

By replacing 'if-then' operator by negation and OR operator, in expression (3) above, we find:

  ∀ X (¬ (Child (X) ∧ ∃Y (Takes (X, Y) ∧ Biscuit (Y))) ∨ Loves (john, X)

Step II: *Reduction of the scope of negation*: Replace ¬ sign by choosing any of the following:

$$a) ¬ (P ∨ Q) = ¬ P ∧ ¬ Q$$
$$b) ¬ (P ∧ Q) = ¬ P ∨ ¬Q$$

c) $\neg(\neg P) = P$

d) $\neg(\exists X\ P) = \forall X\ \neg P$

e) $\neg(\forall\ X\ P) = \exists X \neg P$

In the present context, we rewrite the sentence as

$\forall X\ (\neg$ Child $(X) \vee \neg\ (\exists Y$ (Takes $(X, Y) \wedge$ Biscuit $(Y))) \vee$ Loves (john, X))

$\Rightarrow \forall X\ (\neg$ Child $(X) \vee\ \forall Y\ (\neg$ Takes $(X, Y) \vee\ \ \neg$ Biscuit $(Y)) \vee$ Loves (john, X))

Step III: *Renaming the variables within the scope of quantifiers:* **Rename** $\exists X$ by $\exists Y$ when $\{\exists\ X\}$ is a subset / proper subset of $\{\forall X\}$. In the present context, since X and Y are distinct, the above operation cannot be carried out.

Step IV: *Moving of quantifiers in the front of the expression:* Bring all quantifiers at the front of the expression.

Applying this on the example yields:

$\Rightarrow \forall X \forall Y \neg$ Child $(X) \vee \neg$ Takes $(X,Y) \vee \neg$ Biscuit $(Y) \vee$ Loves (john, X)

Step V: *Replacing existential quantifier as Skolem function of essential quantifiers:* When an existential quantifier (Y) precedes an essential quantifier (X), replace Y as S (X), where S is the Skolem function [3]. In this example, since Y is not a subset of X, such a situation does not arise. Also the essential quantifier is dropped from the sentence.

Step VI: *Putting the resulting expression in conjunctive normal form (CNF):* For example, if the original expression is in the form $P \vee (Q \wedge R)$, then replace it by $(P \vee Q) \wedge (P \vee R)$.

In the present context, the resulting expression corresponding to expression (3) being in CNF, we need not do any operation at this step.

Step VII: *Writing one clause per line:* If the original expression is of the following CNF, then rewrite each clause/ line, as illustrated below.

original expression:

$(\neg P_{11} \lor \neg P_{12} ... \lor \neg P_{1n} \lor Q_{11} \lor Q_{12}.... \lor Q_{1m}) \land$

$(\neg P_{21} \lor \neg P_{22} ... \lor \neg P_{2n} \lor Q_{21} \lor Q_{22} ... \lor Q_{2m}) \land$

....

$(\neg P_{t1} \lor \neg P_{t2} ... \lor \neg P_{tn} \lor Q_{t1} \lor Q_{t2} .... \lor Q_{tm}).$

After writing one clause per line, the resulting expressions become as follows.

$P_{11}, P_{12},..., P_{1n} \rightarrow Q_{11}, Q_{12},... ,Q_{1m}$

$P_{21}, P_{22},...,P_{2n} \rightarrow Q_{21}, Q_{22},..., Q_{2m}$

...   .

$P_{t1}, P_{t2},..., P_{tn} \rightarrow Q_{t1}, Q_{t2},....., Q_{tm}$

With reference to the above, we get the following line as the final expression.

Child (X), Takes (X, Y), Biscuit (Y) $\rightarrow$ Loves (john, X).

It may be noted that the resulting expression, derived above, is not much different from the expression (3). This, however, is not the case for all complex sentences. For example, let us consider the following complex sentence and the clause forms corresponding to that.

Expression: $\forall X (\text{Loves (john, X)} \rightarrow \text{Female (X)})$

$\land \exists X (\neg \text{ Loves (X, Brother-of (X)} \land \text{ Female (X)}))$

The clause forms for the above expression are:

a)   Loves (john, X) $\rightarrow$ Female (X)

b)   Loves (s(X), Brother-of (s (X))), Female (X) $\rightarrow$ $\perp$

where the meaning of the first clause is obvious, while the second clause means that it is impossible that there exists a female X, who loves her brother. The inverted T is called a Falsum operator, which is opposite to Truam (T), meaning that the expression is true [2]. The symbol s(X) denotes a Skolem function of X, which may be read as some of X.

5.9 Unification of Predicates

Two predicates P $(t_1, t_2,..., t_n)$ and Q $(s_1, s_2,..., s_n)$ can be unified if terms t_i can be replaced by s_i or vice-versa.

Loves (mary, Y) and Loves (X, Father-of (X)) , for instance, can be unified by the substitution S ={ mary / X , Father-of (mary) / Y }.

Conditions of Unification:

i)   Both the predicates to be unified should have an equal number of terms.

ii)   Neither t_i nor s_i can be a negation operator, or predicate or functions of different variables, or if t_i = term belonging to s_i or if s_i = term belonging to t_i then unification is not possible.

The Unification Algorithm

Input: Predicates $P(t_1, t_2,..., t_n)$ and $Q(s_1, s_2,..., s_m)$
Output: Decision whether the predicates P and Q are unifiable
          and a set S that includes the possible substitution.
Procedure Unification (P, Q, S, unify)
Begin
S:= Null;
While P and Q contain a Next-symbol do
 Begin
  Symb1: = Next-symbol (P);
  Symb2: = Next-symbol (Q);
 If Symb1 ≠ Symb2 **Then do**
 Begin Case of
   Symb1 or Symb2 = Predicate: Unify: = fail;
   Symb1 = constant and symb2 = different-constant: Unify: = fail;
   Symb1 or Symb2 = ¬ : Unify: = fail;
   Symb1 and Symb2 = function: Unify: = fail;
   Symb1=variable and Symb2 =term and variable ∈ term: Unify: = fail;
   Symb2=variable and Symb1=term and variable ∈ term: Unify: = fail;
 Else If Symb1 = variable or constant and Symb2 =term **Then do**
  Begin
   S: = S ∪ {variable or constant / term};
   P: = P[variable or constant / term];
  End;
 Else If Symb2 = variable or constant and Symb1 =term **Then do**
  Begin

S: = S ∪ {variable or constant / term};
Q: = Q[variable or constant / term];
  End;
End Case of;
End while;
If P or Q contain a Next-symbol **Then** Unify: = fail
Else Unify: = Success;
End.

5.10 Robinson's Inference Rule

Consider predicates P, Q_1, Q_2 and R. Let us assume that with appropriate substitution S, Q_1 [S] = Q_2 [S] .

Then $(P \vee Q_1) \wedge (Q_2 \vee R)$ with Q_1 [S] = Q_2 [S] yields (P ∨ R) [S].

$$\text{Symbolically,} \quad \frac{P \vee Q_1, \neg Q_2 \vee R \quad Q_1 \text{ [S] } = Q_2 \text{ [S]}}{(P \vee R) \text{ [S]}}$$

The above rule is referred to as **Robinson's inference rule** [8]. It is also referred to as the **resolution principle** in predicate logic. The following example illustrates the rule.

Let P = Loves (X, father-of (X)),

   Q_1= Likes (X, mother-of (X))),

   Q_2 = Likes(john, Y),

   R = Hates (X, Y).

After unifying Q_1 and Q_2, we have

Q= Q_1 = Q_2 =Likes (john, mother-of (john))

Where the substitution S is given by

S= { john /X, mother-of (X) / Y}

  = {john / X, mother-of (john) / Y}.

The resolvent $(P \vee R)$ [s] is, thus, computed as follows.

(P ∨ R) [S]

=Loves (john, father-of (john)) ∨ hates (john, mother-of(john)).

5.10.1 Theorem Proving in FOL with Resolution Principle

Suppose, we have to prove a theorem Th from a set of axioms. We denote it by

$\{\ A_1, A_2, ..., A_n\} \models Th$

Let

A_1 = Biscuit (coconut-crunchy)

A_2 = Child (mary) $\wedge$ Takes (mary, coconut-crunchy)

$A_3 = \forall\ X\ (Child(X) \wedge \exists\ Y\ (Takes\ (X,Y) \wedge Biscuit\ (Y))) \rightarrow$
       Loves (john, X)

and Th = Loves (john, mary) = A_4 (say).

Now, to prove the above theorem, we would use Robinson's inference rule. First of all, let us express A_1 through A_4 in CNF. Expressions A_1 and A_4 are already in CNF. Expression A_2 can be converted into CNF by breaking it into two clauses:

Child (mary) and
Takes (mary, coconut-crunchy).

Further, the CNF of expression A_3 is

$\neg$Child (X) $\vee$ $\neg$Takes (X,Y) $\vee$ $\neg$Biscuit (Y) $\vee$ Loves (john, X)

It can now be easily shown that the negation of the theorem (goal) if resolved with the CNF form of expressions A_1 through A_3, the resulting expression would be a null clause for a valid theorem. To illustrate this, we will now form pairs of clauses, one of which contains a positive predicate, while the other contains the same predicate in negated form. Thus by Robinson's rule, both the negated and positive predicates will drop out and the value of the variables used for unification should be substituted in the resulting expression. The principle of resolution is illustrated below (fig. 5.4) to prove the goal that Loves (john, mary).

5.11 Different Types of Resolution

The principle of resolution can be extended to different forms. But an over-extension may cause fatal errors. This section illustrates the diversified use of the resolution principle with the necessary precautions to avoid the scope of mistakes by the beginners.

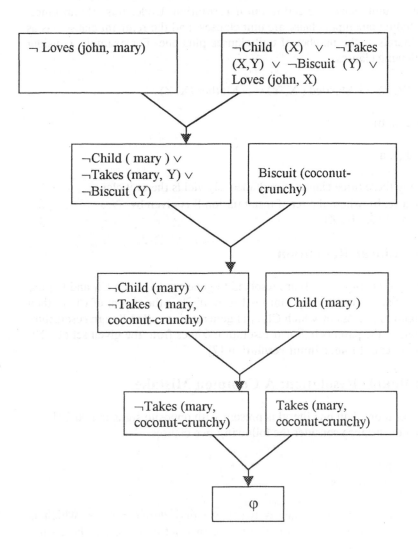

Fig 5.4: A resolution graph to prove that Loves (john, mary).

5.11.1 Unit Resulting Resolution

Typical resolutions, where two clauses of the form (p ∨ ¬ q) and (q ∨ r) are resolved to generate (p ∨ r), are called **binary resolutions**. The definition, though illustrated with propositions, is equally valid for predicates. On the other hand, when more than two clauses are resolved simultaneously to generate a unit clause, we call it a **unit resolution**. Under this circumstance, all excluding one input clause are unit clauses, and the remnant clause has as many literals as the number of unit clauses plus one. For example, consider the following clauses:

Father $(Y, Z) \lor \neg$Married $(X, Y) \lor \neg$ Mother (X, Z)

$\neg$Father (a, b).

Married (c, a)

Resolving these three clauses simultaneously yields the unit clause:
$\neg$Mother (c, b), where the set of instantiation S is given by
$S = \{a / Y, c / X, b / Z\}$.

5.11.2   Linear Resolution

Suppose two clauses Cl_1, Cl_2 are resolved to generate Cl_3, then Cl_1 and Cl_2 are called the parents of Cl_3. Now, for i = 1 to n, if Cl_i is the parent of Cl_{i+1}, then the resolution process by which Cl_{n+1} is generated is called **linear resolution.** When one of the parents in linear resolution comes from the given set of CNF clauses, we call it **linear input resolution** [7].

5.11.3 Double Resolution: A Common Mistake

Sometimes more than one literal is present with opposite signs in two CNF clauses. For instance consider the following two clauses.

$$p \lor \neg q \lor r$$
and $\neg p \lor q \lor s$.

Resolving the above clauses twice, we may derive r $\lor$ s, which is incorrect. To understand the implication of this, let us represent these rules in the following format:

$$q \rightarrow p \lor r$$
and $p \rightarrow q \lor s$.

Replacing p in the first clause by the second clause, we have

$$q \rightarrow q \lor s \lor r,$$

which implies that if q is true then either q or r or s is true, but this does not mean (q ∨ r) only is true.

A simpler but interesting example that illustrates the scope of mistakes in double resolution is given below. Let us consider the following clauses:

$$\neg p \vee q$$
and $$\neg q \vee p$$

Resolving these twice yields a null clause, which is always false. But the above system comprising of { p→ q, q→p} implicates p→p and q → q by chain rule, which in no way supports the falsehood of the resulting clause after resolution [3].

5.12 Semi-decidability

A logic is called decidable if there exists a method by which we can correctly say whether a given formula is valid or invalid. Readers may remember that validity of a formula α means satisfiability of the formula for all possible interpretations. A sound and complete proof method is able to prove the validity of a formula [3]. But if the formula is invalid, the proof procedure (by resolution principle or otherwise) will never terminate. This is called semi-decidablity. FOL is semi-decidable, as it is unable to prove the invalidity of a formula.

5.13 Soundness and Completeness

The issues of soundness and completeness of the resolution principle for propositional logic have already been discussed in a previous section. This section discusses these issues for predicate logic. To prove the completeness of the resolution theorem of predicate logic, the following definitions and theorems are presented in order.

Definition 5.15: The **Herbrand Universe (H$_S$)** for a given set of clauses S is defined as the set of all possible ground terms, constructed by replacing the variables in arguments of functions by the same or other functions or constants, so that they remain grounded (free from variables) after substitution. It is to be noted that H$_S$ is an infinite set [9].

For example, suppose that there exists a single clause in S, given by

$$Q(X, \ f (X, a)) \wedge P(X, a) \rightarrow R (X, b)$$

where {a, b} is the set of constants, {X} is a set of variables, {f} is a set of functions, {P, Q, R} is a set of predicates. Here H_S = {a, b, f(a, a), f(b, a), f(a, f(a, a)), f(a, f(a, b)) ......} is an infinite set.

Definition 5.16: Let S be a set of clauses and P be the set of ground terms. Then P (S), **the saturation of S with respect to P**, is defined [9] as the set of all ground clauses obtained by applying all possible consistent substitutions for variables in S with the ground terms in P.

For example, let P = {a, b, f (a, b)} and S = {Q (X, f (X, a)) $\wedge$ P(X, a) $\rightarrow$ R (X, b)}. Then P (S) is computed as follows.

P (S) = {Q(a, f (a, a)) $\wedge$ P(a, a) $\rightarrow$ R (a, b),
        Q(b , f (b, a)) $\wedge$ P(b, a) $\rightarrow$ R (b, b),
        Q(f (a, b), f (f(a, b), a)) $\wedge$ P(f (a, b), a) $\rightarrow$ R (f (a, b), b)}

Definition 5.17: The saturation of S, a set of clauses with respect to the Herband universe H_S, is called the **Herbrand base H_S (S).**

For example, with S = {Q (X, f (X, a)) $\wedge$ P(X, a) $\rightarrow$ R (X, b) }, H_S = {a ,b, f(a, a), f (a, b), f (f (a, a), a), f (f (a, b), a).....}, we find H_S (S) as follows.

H_S (S) = {Q(a, f (a, a)) $\wedge$ P(a, a) $\rightarrow$ R (a, b),
        Q(b , f (b,a)) $\wedge$ P(b, a) $\rightarrow$ R (b, b),
        Q(f (a, b), f (f (a, b), a)) $\wedge$ P(f (a, b),a) $\rightarrow$ R (f (a, b), b),
        ....
        ....
        Q (f (f (a, b), a), f (f (f (a, b), a), a)) $\wedge$ P(f (f (a, b), a), a) $\rightarrow$
        R (f (f (a, b), a), b)}
It may further be noted that H_S (S) too is an infinite set.

The following two theorems will be useful to prove the completeness of the resolution theorem. The proofs of these theorems are beyond the scope of this book and are thus excluded.

Herbrand's Theorem: *If a set of clauses S is unsatisfiable, then there must exist a finite subset of Herband base H_S (S) that too is unsatisfiable* [5].

Lifting Lemma: *Given that C_1 and C_2 are two clauses with no shared variables. Further given that $\underline{C_1}$ and $\underline{C_2}$ are the ground instances of C_1 and C_2 respectively. If $\underline{C}$ is a resulting clause due to resolution of $\underline{C_1}$ and $\underline{C_2}$ then there exists a clause C that satisfies the following criteria:*

i) C is a resulting clause due to resolution of C_1 and C_2, and

ii) $\underline{C}$ is a ground instance of C [9].

For example, let

$C_1 = Q\ (X, f\ (X, a)) \wedge P(\ X, c) \rightarrow R\ (X, b)$

$C_2 = W\ (\ f\ (f\ (a, b), a)\ , Z\ ) \rightarrow P\ (\ f\ (a, Y), Z\ )$

$\underline{C_1} = Q\ (\ f\ (a, b), f\ (f\ (a, b)\ , a)) \wedge P(\ f\ (a, b), c) \rightarrow R\ (\ f\ (a, b)\ , b)$

$\underline{C_2} = W\ (\ f\ (\ f\ (a, b),a)\ , c\ ) \rightarrow P\ (\ f\ (a, b\ ), c)$

Now, $C = Q\ (f\ (a, Y), f\ (X, a)) \wedge W\ (\ f\ (f\ (a, b), a)\ , Z\ )\ \rightarrow R\ (\ f\ (a, Y), b)$

and $\underline{C} = Q\ (f\ (a, b), f\ (f\ (a, b)\ , a)) \wedge W(\ f\ (f\ (a, b)\ ,a), c\ )\ \rightarrow R\ (f\ (a, b), b)$

Thus we found $\underline{C}$ as a ground instance of C.

Let us now prove the completeness theorem of predicate logic.

Theorem 5.4: *The resolution theorem of predicate logic is complete.*

Proof: Given a set of clauses S and a formula α such that $S \models \alpha$. We have to prove that $S \vdash \alpha$, i.e. there exists a logical proof of α from S. We shall prove it by the method of contradiction. Thus let $S \vdash \neg \alpha$, i.e., S is not logically provable from S. Thus $S_1 = S \cup \{\neg \alpha\}$, all expressed in clause form is unsatisfiable. So, by Herbrand's theorem, there must exist a Herbrand base $H_S(S_1)$ that is also unsatisfiable. Now, by ground resolution theorem, we find that the resolution closure of $H_S\ (S_1)$ contains the clause 'false'. Now, by lifting the lemma, if the false clause occurs in the resolution closure of $H_S\ (S_1)$ then that must also appear in the resolution closure of S_1. Now, the resolution closure of S_1 containing the false clause is a contradiction to the assumption that $S \vdash \neg \alpha$ is wrong and hence $S \vdash \alpha$ follows. □

Now, we narrate the proof of soundness of the resolution theorem in predicate logic.

Theorem 5.5: *The resolution principle of predicate logic is sound.*

Proof: To prove the soundness, we first look at the proof procedure for a particular problem that proves a formula α from a given set of clauses S, i.e., $S \vdash \alpha$. Let it be a linear resolution. It can be shown that if the soundness can be proved for linear resolution, it can be proved for other resolutions (like unit resolution) as well. To prove the soundness of the resolution theorem, we use the following three steps:

Step 1: After the proof procedure terminates, back substitute the constants by variables in the tree.

Step 2: Now instantiate these clauses with all possible constants. We thus get the Herbrand base corresponding to the clauses that participated in the proof procedure.

Step 3: The resolution theorem of propositional logic is now applied to that subset of the Herbrand base. Note that the propositional resolution theorem, employed here, is sound.

Since the elements of Herbrand base also include the clauses that participated in the resolution proof of predicate logic, the proof procedure of the resolution theorem in predicate logic is also sound [2].
□

5.14 Conclusions

The chapter presented the syntax and semantics of propositional and predicate logics and demonstrated their applications in logical theorem proving. Many AI problems, which can be represented as theorem proving problems, thus can be handled with the concept outlined in the chapter. The resolution theorem, being the fundamental theorem under the proposed framework of knowledge, its soundness and completeness have been discussed in detail. The semi-decidablity of FOL has also been covered briefly. The shortcomings of double resolution, as a common mistake, have also been pointed out. This will help the students to properly identify the use of the resolution theorem.

Exercises

1.   Prove that for the atomic propositions p, q, r and s
     a) $p, \neg q \Rightarrow r \equiv p \Rightarrow q, r$ and
     b) $p, q \Rightarrow \neg r, s \equiv p, q, r \Rightarrow s$

Could you remember the use of the above tautologies in Wang's algorithm? If yes, in which steps did you use them?

2. Verify the following theorems by Wang's algorithm.

     a) $p \lor q, p \rightarrow r, q \rightarrow r \Rightarrow r$

    b) $p \rightarrow (q \rightarrow r) \Leftrightarrow (p \wedge q) \rightarrow r$

    c) $(p \rightarrow q) \wedge (q \rightarrow p) \Leftrightarrow (p \wedge q) \vee (\neg p \wedge \neg q)$

[Note: For (b) and (c), prove the theorems first from left to right and then from right to left.]

3. Apply resolution theorem to prove the following theorem:

    $p \vee q, \; p \rightarrow r, q \rightarrow r \Rightarrow r$.

[Hints: Here, goal is r ; so resolve $\neg$ r with the CNF form of premise clauses to prove a resulting null clause.]

4.   For a triangle ABC, it is given that the sum of the interior angles: $\angle A + \angle B + \angle C = 180$ degrees. Show by resolution theorem that the exterior angle is the sum of the opposite interior angles.

    [Hints: We denote the exterior angle $\angle A$ by EXT$\angle A$. Use the following predicates:

                Equal (sum ($\angle A$, $\angle B$, $\angle C$) ,180)

                Equal (sum ($\angle A$, EXT($\angle A$), 180)

    and rules

                Equal (X, Y), Equal(Z, Y) $\rightarrow$ Equal (X, Z)

                Equal (sum (X, Y), sum (X, Z)) $\rightarrow$ Equal (Y, Z).

                Equal (Y, Z) $\rightarrow$ Equal (Z, Y).

    The rest of the proof is obvious.]

5.   Represent the following sentences into clause form:

    a)   On (X, Y) $\wedge$ (Above (Y, Z) $\vee$ On (Y, Z)) $\wedge$ On (Z, W) $\rightarrow$ On (X, W)

    b) $\forall$ X Fly (X) $\wedge \exists$ X Has-wings (X) $\rightarrow$ Bird(X) $\vee$ Kite(X)

    c) $\forall$ X Man (X) $\wedge \forall$ Y (Child (Y) $\vee$ Woman (Y)) $\rightarrow \neg$ Dislikes (X, Y)

6.   Prove that Dog (fido) follows from the following statements by the resolution theorem.

    a) $\forall$ X Barks (X) $\rightarrow$ Dog (X).

    b) $\forall$ X $\forall$ Y $\exists$ Z Has-master (X, Y) $\wedge$ Likes (X, Y) $\wedge$ Unprecedented-situation (Z) $\rightarrow$ Barks (X).

    c)   Unprecedented-situation (noise).

    d)   Likes (fido, jim).

    e)   Has-master (fido, jim).

7. Show that the following formula is valid.

$$(A(X) \vee B(Y)) \rightarrow C(Z) \Rightarrow (\neg A(X) \wedge \neg B(Y)) \vee C(Z)$$

where X, Y and Z are variables and A, B and C are predicates.

8. List all the satisfiability relations in a tabular form with four columns A, B, C and the entire formula (say, Y) for the last formula.

9. Given $X \varepsilon \{a1, a2\}$, $Y \varepsilon \{b1, b2\}$ and $Z \varepsilon \{c1, c2\}$, and $S = \{A(X) \vee B(Y)) \rightarrow C(Z) \Rightarrow (\neg A(X) \wedge \neg B(Y)) \vee C(Z)\}$, find the Herbrand universe and the Herbrand base.

10. Illustrate the lifting lemma with the following parameters.

$$C_1 = P(X, f(X)) \wedge Q(Y, c) \rightarrow R(X, b)$$

$$C_2 = W(f(Y), Z) \rightarrow Q(Y, Z)$$

$$\underline{C}_1 = P(a, f(a)) \wedge Q(b, c) \rightarrow R(a, b)$$

$$\underline{C}_2 = W(f(b), c) \rightarrow Q(b, c)$$

References

[1] Anderson, D. and Ortiz, C., "AALPS: A knowledge-based system for aircraft loading," *IEEE Expert*, pp. 71-79, Winter 1987.

[2] Ben-Ari, M., *Mathematical Logic for Computer Science*, Prentice-Hall, Englewood Cliffs, NJ, pp. 11-87, 1993.

[3] Bender, E. A., *Mathematical Methods in Artificial Intelligence*, IEEE Computer Society Press, Los Alamitos, chapter 1, pp. 26, 1996.

[4] Dougherty, E. R. and Giardina, C. R., *Mathematical Methods for Artificial Intelligence and Autonomous Systems*, Prentice-Hall, Englewood Cliffs, NJ, 1988.

[5] Herbrand, J., *Researches sur la Theorie de la Demonstration*, Ph.D. Thesis, University of Paris, 1930.

[6]   Leinweber, D., "Knowledge-based system for financial applications," *IEEE Expert*, pp. 18-31, Fall 1988.

[7]   Patterson, D, W., *Introduction to Artificial Intelligence and Expert Systems*, Prentice Hall, Englewood-Cliffs, pp. 345-347, 1990.

[8]   Robinson, J. A., "A machine oriented logic based on the resolution principle," *Journal of the ACM*, vol. 12. no.1, pp. 23-41.

[9]   Russel, S. and Norvig, P., *Artificial Intelligence: A Modern Approach*, Prentice-Hall, Englewood Cliffs, pp. 286-294, 1995.

[10]   Wang, H., "Toward mechanical mathematics," *IBM Journal of Research and Development*, vol. 4, pp. 2-22, 1960.

6

Principles in Logic Programming

*This chapter is an extension of chapter 5 to study in detail a specialized class of Predicate Logic based reasoning programs, called Logic Programs. PROLOG, which is an acronym of **PRO**gramming in **LOG**ic, is commonly used for handling this class of reasoning problems. Various issues of Logic Programming have been narrated in this chapter with special emphasis on the syntax and semantics of PROLOG programming language. It may be mentioned here that besides PROLOG, another well-known programming language, called LISP (**LIS**t **P**rocessing language), is also used for programming in Artificial Intelligence. However, LISP is well suited for handling lists, whereas PROLOG is designed for Logic Programming.*

6.1 Introduction to PROLOG Programming

To introduce readers to the syntax and semantics of logic programming, we first take a look at a few examples. Let us, for instance, consider the problem of a 'classroom scene' interpretation We assume that the scene has been passed through various stages of low and medium level image processing [7] and the objects in the scene thus have been recognized [8] and labeled by a computing machine. The data (clauses) received from the scene and the knowledge used to analyze it are presented in both English and Predicate Logic below.

Database:

Object (board)
Writes-on (john, board, classhour)
Sits-on (mita, bench, classhour)
Sits-on (rita, bench, classhour)
Person (john)
Person (rita)
Person (mita)

The above are called **clauses** in PROLOG.

Fig. 6.1: A classroom scene, where John, the teacher, writes on
the board and Mita and Rita, the students, sit on a bench.

Knowledge base:

1. A board (blackboard) is an object, where teachers used to write in classhours.

In Predicate Logic, the above sentence can be written as

Object (board) ∧ Writes-on (X, board, Time) ∧ Teacher (X)
→Equal (Time, classhour).

2. A Teacher is a person who writes on a board during classhour.

In Predicate Logic, the above sentence is written as

∀ X (Person(X) ∧ Writes-on (X, board, classhour) → Teacher (X))

3. A student is a person who sits on a bench during classhour.

In Predicate Logic, the above sentence can be written as

∀Y (Person (Y) ∧ Sits-on (Y, bench, Time) ∧ Equal (Time, classhour) →
Student (Y))

*4. If at least one person sits on a bench in classhour and a second person
writes on the board at the same time then time = classhour.*

In Predicate Logic, the above expression can be described by

∃ Y ((Person (Y) ∧ Sits-on (Y, bench, classhour)) ∧
∃ X (Person (X) ∧ Writes-on (X, board, Time)) → Equal (Time, classhour))

The above 4 statements can be written in PROLOG language as follows:

1. Equal (Time, classhour) :-
    Object (board),
    Writes-on (X, board, Time),
    Teacher (X).

2. Teacher (X) :-
    Person (X),
    Writes-on (X, board, classhour).

3. Student (Y) :-
    Person (Y),
    Sits-on (Y, bench, Time),
    Equal (Time, classhour).

4. Equal (Time, classhour) :-
    Person (Y),

Sits-on (Y, bench, classhour),
Person (X),
Writes-on (X, board, Time).

It may be added that the above pieces of knowledge are also called **clauses** in PROLOG.

6.2 Logic Programs - A Formal Definition

We are now in a position to formally define **Logic Programs**. We first define a **Horn clause**, the constituents of a logic program

Definition 6.1: A clause consists of two parts: the head and the body. One side of the clause to which the arrowhead (if-then operator) points to is called the head and the other side of it is called the body. A **Horn clause** contains at most one literal (proposition / predicate) at the head of the clause [6].

Example 6.1: The following are two examples of Horn clauses.

$$\text{i) P (X), Q (Y)} \rightarrow \text{W (X,Y)}$$

$$\text{ii) R (X, Y), Q (Y)} \rightarrow \text{?}$$

In (i) W (X, Y) is the head and P (X), Q (Y) is the body. In (ii) R(X, Y), Q(Y) is the body and the head comprises of a null clause (sub-clause). In fact (ii) represents a query, asking whether R (X, Y), Q (Y) is true, or what are the set of instantiations for X and Y, which makes R (X, Y) $\land$ Q (Y) true.

Definition 6.2: A **Logic Program** is a program, comprising of Horn clauses. The following example illustrates a logic program.

Example 6.2: Consider the following two sentences in Predicate Logic with one head clause in each sentence.

Father (X, Y) ← Child (Y, X), Male (X).
Son (Y, X) ← Child (Y, X), Male (Y).

The above two clauses being horn clauses are constituents of a Logic Program. Let us now assume that along with the above knowledge, we have the following three data clauses:

Child (ram, dasaratha).

Male (ram).

Male (dasaratha).

Suppose that the above set of clauses is properly structured in PROLOG and the program is then compiled and executed. If we now submit the following queries (goals), the system responds correctly as follows.

1.   *Goal:* Father (X, Y)?

     Response: Father (dasaratha, ram).

2.   *Goal:* Son (Y, X)?

     Response: Son (ram, dasaratha).

But how does the system respond to our queries? We shall discuss it shortly. Before that let us learn a little bit of syntax of PROLOG programs. We take up the scene interpretation problem as an example to learn the syntax of PROLOG programming.

6.3 A Scene Interpretation Program

/* PROLOG PROGRAM FOR SCENE INTERPRETATION */

Domains
     Time, X, Y, Z, W, board, classhour, bench = **symbol**

Predicates

     Teacher (X)
     Writes-on (X, board, Time)
     Equal (Time, classhour)
     Person (X) Person (Y) Person (Z)
     Sits-on (Y, bench, Time) Sits-on (Z, bench, Time)
     Student (Y)
     Student (Z)
     Object (W)

Clauses

     Object (board).   1

     Writes-on (john, board, classhour).   2

Sits-on (mita, bench, classhour).   3

Sits-on (rita, bench, classhour).   4

Person (john).   5

Person (mita).   6

Person (rita).   7

Equal (Time, classhour):-   8
    Object (board),
    Writes-on (X, board, Time),
    Teacher (X).

Equal (Time, classhour):-   9
    Person (Y),
    Sits-on (Y, bench, classhour),
    Person (X),
    Writes-on (X, board, Time).

Teacher (X):-   10
    Person (X),
    Writes-on (X, board, classhour).

Student (Y) :-   11
    Person (Y),
    Sits-on (Y, bench, Time),
    Equal (Time, classhour).

This is all about the program. Readers may note that we mentioned no procedure to solve the problem of scene interpretation; rather we stated the facts only in the program. Here lies the significance of a logic program.

Now, suppose, the user makes a query:

Goal: Teacher (X)?

System prompts: Teacher (john).

Further, if the user asks:

Goal: Equal (Time, classhour) ?

System prompts: Yes.

6.4 Illustrating Backtracking by Flow of Satisfaction Diagrams

To explain how the system answers these queries, we have to learn a very useful phenomenon, called **backtracking**.

Let us now concentrate on the query: Teacher (X)?
Since

> Teacher (X) ←
> Person (X),
> Writes-on (X, board, classhour).   (10)

to satisfy the Goal: Teacher (X), one has to satisfy the sub-goals: Person (X), Writes-on (X, board, classhour). Now, PROLOG searches a sub-goal Person( ) for the predicate Person (X). At clause 5, it finds a match and X is instantiated to john (fig. 6.2 (a)). PROLOG puts a marker at clause 5. Now, it continues searching Writes-on (john, board, classhour) in the remaining clauses. But it fails to find so, since Writes-on (john, board, classhour) is at 2nd position in the list of clauses (fig. 6.2 (b)). So, it has to trace back above the marker place (fig. 6.2(c)) and then ultimately it finds Writes-on (john, board, classhour) (fig. 6.2(d)). Since the sub-goals are succeeded, the goal also succeeds, yielding a solution: Teacher (john). The concept of backtracking is illustrated below with the help of **flow of satisfaction diagrams** [2] fig. 6.2(a) to (d)).

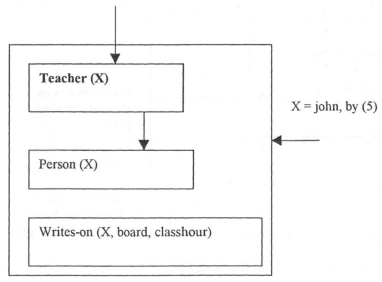

Fig. 6.2 (a): Unification of the first sub-goal.

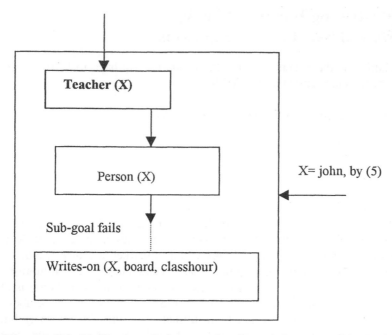

Fig. 6.2 (b): Unification of the second sub-goal is not possible in the clauses following the marked unified clauses.

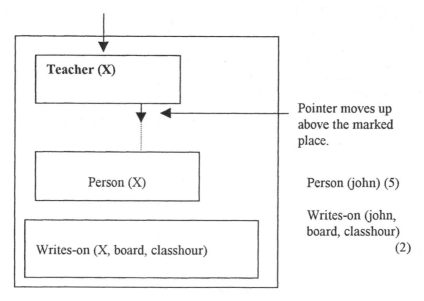

Fig. 6.2 (c): Back-tracking in the set of clauses.

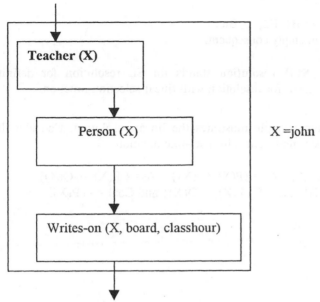

Fig.6. 2(d): Successful goal: Teacher (john).

For answering the query Equal (Time, classhour), a number of backtracking is required. We omit this for space constraints and ask the reader to study it herself. The next important issue that we will learn is SLD (Select Linear Definite clauses) resolution.

6.5 The SLD Resolution

We start this section with a few definitions and then illustrate the SLD resolution with examples.

Definition 6.3: A **definite program clause** [1] is a clause of the form

$$A \leftarrow B_1, B_2,..., B_n$$

which contains precisely one atom (viz. A) in its consequent (head) and a null, one or more literals in its body (viz. B_1 or B_2 or ... or B_n).

Definition 6.4: A **definite program** is a finite set of definite program clauses.

Definition 6.5: A **definite goal** is a clause of the form

$$\leftarrow B1, B2, ..., Bn$$

i.e., a clause with an empty consequent.

Definition 6.6: **SLD resolution** stands for SL resolution for definite clauses, where SL stands for resolution with linear selection function.

Example 6.3: This example illustrates the linear resolution. Consider the following OR clauses, represented by a set-like notation.

Let $S = \{A_1, A_2, A_3, A_4\}$, $A_1 = \{P(X), Q(X)\}$, $A_2 = \{P(X), \neg Q(X)\}$, $A_3 = \{\neg P(X), Q(X)\}$, $A_4 = \{\neg P(X), \neg Q(X)\}$ and Goal $= \neg P(X)$.

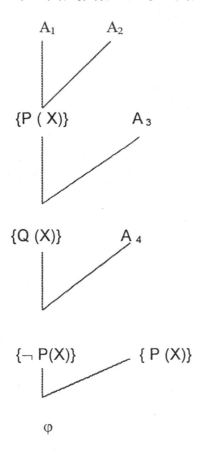

Fig. 6.3: The linear selection of clauses in the resolution tree.

The resolution tree for linear selection is presented in fig. 6.3. It is clear that two clauses from the set $S_1 = S \cup \{ \neg \text{ Goal} \}$ are first used for resolution and the resolvent is next used for resolution with a third clause from the same set S_1. The process is continued until a null clause is generated. In the linear selection process, one clause, however, can be used more than once for resolution.

An alternative way to represent the resolution process in a tree with linear selection is presented below. Such trees are generally referred to as **SLD trees**. Let us now consider the following Logic program and draw the SLD tree for the program.

Example 6.4: The Logic program built with definite clauses and the goal are presented below.

1. P (X, Z) ← Q (X, Y) , P (Y, Z)

2. P (X, X) ←

3. Q (a, b) ←

Goal :   ← P (X, b)

For the construction of the SLD tree, we would match the head of a clause with the same consequent clause in another's body, during the process of resolution of two clauses. This, however, in no way imposes restriction in the general notion of resolution. Rather, it helps the beginners to mechanically realize the resolution process. The SLD tree of a Logic Program becomes infinite, if one uses the same rule many times. For instance using rule 1 many times, we find an infinite SLD tree like fig. 6.4 for the current example.

6.6 Controlling Backtracking by CUT

Since a PROLOG compiler searches for the solution using a "depth first search" strategy, the leftmost sub-tree in fig. 6.4 being infinite does not yield any solution of the system. Such a problem, however, can be taken care of by appropriately positioning "CUT" statements in the Logic Program. Since depth first search in PROLOG programs are realized by stack, an infinitely large SLD tree results in a **stack overflow**.

Example 6.5, presented below, illustrates how unnecessary search in SLD trees can be avoided by placing CUT statements in the Logic Program.

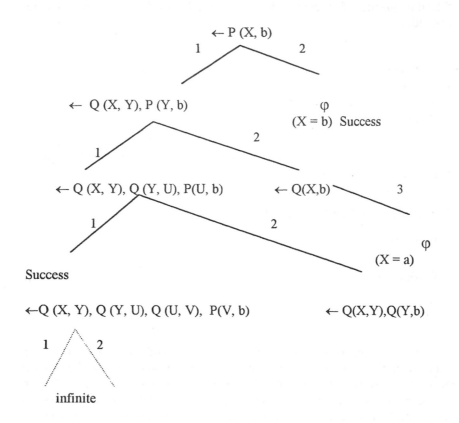

Fig. 6.4: An infinite SLD tree.

Example 6.5: Consider the Logic Program, where "!" denotes the CUT predicate.

   1)   A ← B, C

   ...

   4) B ← D, !, E

   ...

   7) D ←

and Goal: ← A

The SLD tree for the above Logic Program is presented in fig. 6.5. Let us now explain the importance of the CUT predicate in fig. 6.5. For all predicates preceding CUT, if unifiable with other clauses, then CUT is automatically satisfied. Further, if any of the predicates preceding CUT are not unifiable, then backtracking occurs to the parent of that clause for finding alternative paths. However, suppose all predicates preceding CUT are unifiable and any of the predicates following CUT in the clause are not unifiable. Under this circumstance, backtracking occurs to the root of the SLD tree and the control attempts to find alternative solutions from the root.

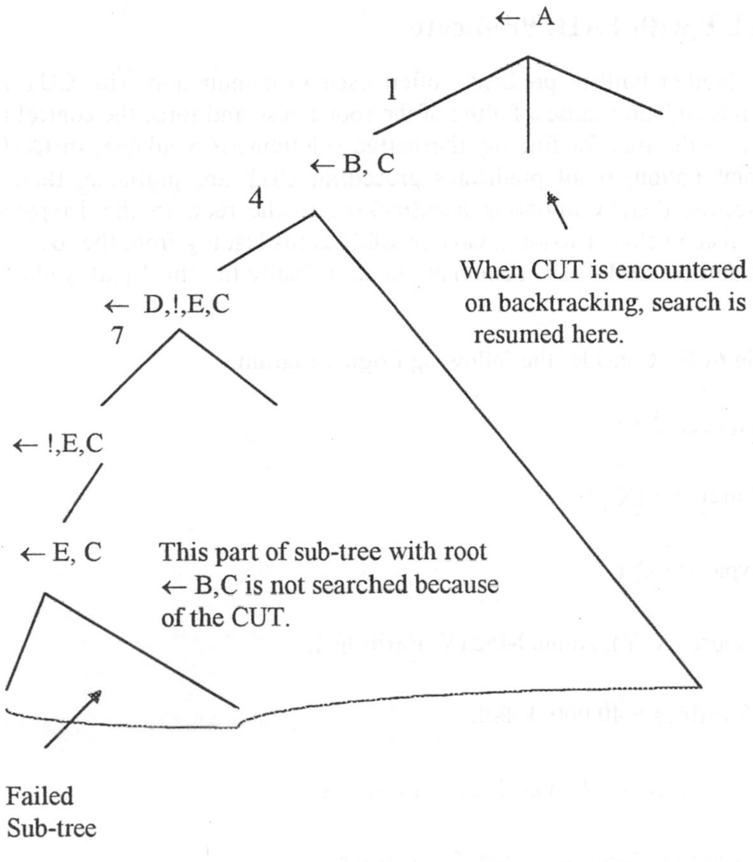
← A
1
← B, C
4
When CUT is encountered on backtracking, search is resumed here.
← D,!,E,C
7
← !,E,C
← E, C
This part of sub-tree with root ← B,C is not searched because of the CUT.
Failed Sub-tree

Literals preceding CUT are unifiable with the same literals in the head of other clauses. So, ! is automatically satisfied. Since ← E, C cannot be resolved with any more clauses, the control returns to the root of the tree ← A for generating alternative solution.

Fig. 6.5: Controlling backtracking by using CUT.

6.6.1 Risk of Using CUT

It is to be noted that while expanding a node in the SLD tree, the PROLOG compiler attempts to unify the clause, representing the node with the subsequently labeled clauses in order. Thus, if we have an additional clause, say clause number 10, given by B ← D, it will not be used to unify with ←B,C. So, due to failure of the sub-tree with root ← E, C (fig. 6.5) the control returns to the second alternative of the sub-tree with root ← A, thereby keeping the option to lose a possible solution. Controlling backtracking in an SLD tree, thus, undoubtedly saves computational time, but an improper use of it is risky for it may lose solutions.

6.6.2 CUT with FAIL Predicate

FAIL is another built-in predicate, often used in conjunction with CUT in order to intentionally cause a failure of the root clause and force the control to backtrack to the root for finding alternative solutions, if available. In CUT-FAIL combination, if all predicates preceding CUT are unifiable, then a failure occurs, thereby initiating backtracking to the root. In the Taxpayer problem, listed below, two successive possible paths leading from the root are forced to fail by CUT-FAIL combinations and finally the third path yields a solution.

Example 6. 6: Consider the following Logic Program.

1.   Taxpayer (X) ←

         Foreigner (X), !, fail.

2.   Taxpayer (X) ←

         Spouse (X, Y), Annual-Inc (Y, Earnings),

         Earnings > 40,000, !, fail.

3.   Taxpayer (X) ← Annual-Inc (X, Earnings),

         30000 < Earnings, 50000 > Earnings.

4.   Foreigner (ram) ←

5.   Spouse (ram, mita) ←

6.   Annual-Inc (mita, Earnings) ←

7.   Earnings = 45,000 ←

8.   Annual -Inc (lakshman, 35,000) ←

Query : ← Taxpayer (X)

The SLD tree for the above Logic Program is presented in fig. 6.6.

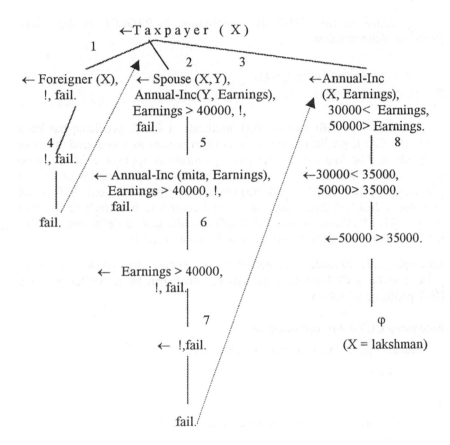

Fig. 6.6: CUT-FAIL combination forces the control to backtrack to the root
from the left two sub-trees and generate alternative paths (rightmost
sub-tree) for solutions.

In the SLD tree for the Taxpayer problem, the control first attempts to find solutions from the leftmost sub-tree. On getting the fail predicate following CUT, it backtracks to the root node and attempts to generate the second alternative solution. Then expanding the second sub-tree, it finds FAIL following CUT predicate and returns to the root. Finally, the third sub-tree yields the solution: X = lakshman.

6.7 The NOT Predicate

An alternative of the CUT-FAIL combination in PROLOG is the NOT predicate, defined below:

1. NOT (P) ← CALL (P), !, FAIL.
2. NOT (P) ←

In the above definition of NOT predicate, a CALL predicate has been used. The CALL predicate simply treats its argument as a goal and attempts to satisfy it. The first rule of the NOT predicate is applicable, if P can be shown and the second rule is applicable otherwise. As a matter of fact, if PROLOG satisfies CALL (P), it abandons satisfying NOT goal. If P is not provable, CALL (P) fails, thus forcing the control to backtrack to the root from CALL (P). Consequently, PROLOG uses the second definition of NOT predicate and it succeeds, signifying that P is not provable.

Example 6.7: Consider the definition of income through a pension of persons using a CUT-FAIL combination, which can be re-written using a NOT predicate as follows:

Rule using CUT-FAIL combination

  Annual-Inc (X, Y) ← Receives-Pension (X, P),

  P < 30,000,

  !, fail.

  Annual-Inc (X,Y) ← Receives-Pension (X, P), Y = P.

The same rule using NOT predicate

  Annual-Inc (X, Y) ← Receives-Pension (X, P),

  NOT (P < 30,000),

  Y= P.

6.8 Negation as a Failure in Extended Logic Programs

It is evident from our background in Predicate Logic that negated clauses have a significant role in representing knowledge. Unfortunately, however, the HORN-clause based programs do not allow negated clauses in their body. To facilitate the users with more freedom of knowledge representation, recently, Logic programs have been extended to include negated atomic clauses in the body of a *non-Horn clause*, presented below:

$$p \leftarrow q, r, \neg s, \neg t$$

where p, q, r, s and t are atomic propositions or predicates.

The *principle* of **negation as failure** [9] states: *For a given formula P, if one cannot prove P, then it is reasonable to deduce that* $\neg P$ *is true.*

For illustrating the principle consider the following extended logic program:

1. Subset (A, B) ← ¬ Non-subset (A, B).

2. Non-subset (A, B) ← Member (X | A), ¬ member (X | B).

Goal: Subset ((2, 4, nil), (1, 2, 3, 4, nil)) →

To prove the above goal, we resolve the goal clause with clause (1) and get

     ← ¬ Non-subset ((2,4, nil), (1, 2, 3, 4, nil)).

Now, by negation as a failure, we want to satisfy the sub-goal

     ← Non-subset ((2,4, nil), (1,2,3,4, nil)).

Now, by (2) we find that

Member (X | 2 ,4,nil) , ¬ Member (X | 1,2,3,4, nil)

fails, which consequently proves that non-subset ((2,4, nil), (1,2,3,4,nil)) fails and thus Subset ((2, 4, nil), (1, 2, 3, 4, nil)) is a valid inference.

6.9 Fixed Points in Non-Horn Clause Based Programs

A non-Horn clause based program can have many **minimal (least) fixed points** [3], i.e., many possible models / interpretations exist for a given program. For instance, for the following logic program:

$$P \leftarrow \neg\, Q.$$

there exist two interpretations (P is true, Q is false) and (P is false and Q is true). For determining minimal fixed points, one generally has to assign values to a minimum set of literals so that the given clauses are consistent. However, here none of the above models are smaller than the other and thus determining the fixed points for such non-Horn programs is difficult.

One approach to evaluate fixed points for such programs is to write each clause in a **stratified** [4-5] manner as follows.

$$P \leftarrow$$
$$\neg\, Q.$$

The independent ground literals in the body of clauses are then assigned Boolean values so that the head is true. The process is applied recursively to all clauses until the set of clauses is exhausted. The interpretation, thus obtained, will be minimal and the Boolean values of the minimal set of literals together form fixed points. By this method, the fixed point for the above clause is (P is true).

6.10 Constraint Logic Programming

Logic programming has recently been employed in many typical problems, like game scheduling in a tournament, examination scheduling, flowshop scheduling etc., where the goal is to find a solution by satisfying a set of logical constraints. Such logic programs include a set of constraints in the body of clauses and are called **constraint logic programs (CLP).** The structure of a typical Horn clause based CLP is presented below:

$$P\,(t) \leftarrow Q_1\,(t),\ Q_2(t),\dots..,Q_m(t),C_1(t),C_2\,(t),\dots\dots,C_n(t),$$

where P, Q_i are predicate symbols, C_j are constraints and t denotes a list of terms, which need not be the same for all literals. In a constraint logic program, all constraints are equally useful for finding viable solutions for the problem. However, there exist situations, when no solution is found that satisfies all the constraints. For handling these type of problems, the strengths S_i of the constraints C_i are attached with them in the program clauses as presented below:

$$P(t) \leftarrow Q_1(t), Q_2(t), \ldots, Q_m(t), S_1C_1(t), S_2C_2(t), \ldots, S_nC_n(t).$$

Logic programs built with such type of clauses are called **Hierarchical Constraint Logic Programs (HCLP)** [11].

We now illustrate the formulation of an HCLP and the approach to its solution for the *problem of editing a table on a computer screen* in order to keep it visually appealing.

Here, the spacing between two successive lines is a variable, whose minimum value, for obvious reason, is greater than zero. However, we want it to be less than 10. We also prefer that the table fit on a single page of 30 lines. Further, there could be a default space of 5 lines, i.e., if other constraints are satisfiable, then one may attempt to satisfy the default constraint.

Let us define the strength of the constraints into following 4 levels,

> essential,
> strongly preferred,
> preferred,
> default.

The logic program for the above problem is presented below.

Table (page-length, type-size, no-of-lines, space) ←
essential (space + type-size) * no-of-lines = page-length,
essential space >0,
strongly preferred space < 10,
preferred page-length <= 30,
default space =5.

For solving HCLP, one has to first find solutions satisfying the essential constraints and then filter those solutions that satisfy the next labeled constraints. The process is thus continued recursively until all the constraints are satisfied or the solution converges to a single set of values for the variables, whichever occurs earlier.

6.11 Conclusions

The main advantage of logic programming lies in the declarative formulation of the problem without specifying an algorithm for its solution. Non-Horn clause based logic programs are currently gaining significance for their inherent advantage of representing negated antecedents. These programs have many interpretations and determination of the least interpretation requires stratification of the clauses. For the real world applications, the constraints of a specific problem are also added with the classical definition of the problem. These programs are referred to as CLP. When the constraints have unequal weights, we call the logic programs HCLP. In HCLP, if solutions satisfying all the constraints are not available, then the solutions that satisfy some of the hierarchically ordered constraints are filtered. Logic programming has been successfully used in many commercial applications including prediction of the share market fluctuations [5] and scheduling teams of players in tournaments [5].

Exercises

1. Draw the flow of satisfaction diagram to show that the clause: Equal (Time, classhour) is true for the Scene interpretation program.

2. Design a logic program to verify the truth table of the CMOS Inverter, presented below (fig. 6.7) with the given properties of channel conduction with gate triggering.

    Properties of the MOSFETS: When the Gate (G) is high for n-channel MOSFET, the Drain (D) and the Source (S) will be shorted, else they are open (no connection). Conversely, when the gate of a p-channel MOSFET is low, its drain and source will be shorted, else they are open.

    [Hints: Inverter (In, Out) ←
    Pwr (P),
    Gnd (Q),
    Ptrans (P, In, Out),
    Ntrans (Out, In, Q).

    Ntrans (X, 1, X) ← .

    Ntrans (X, 0, Y) ← .

Ptrans (X, 0, X) ←.

Ptrans (X, 1, Y) ←.

Goal: Inverter (1, Out) →
Inverter (0, Out) →]

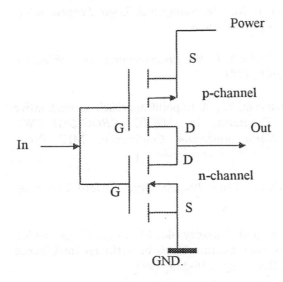

Fig. 6.7: The CMOS Inverter.

3. Show how a stack helps backtracking in PROLOG with respect to the Taxpayer program.

[**Hints:** As you go down the SLD tree, go on pushing the address of the parent node in the stack. On failure or success, when the control has to move up the tree, pop the stack.]

4. Design an HCLP to get change for a US $500. Preferred at least two currency of US $50. Strongly preferred one currency of US $10.

[**Hints:** US$ $(X_{100}, X_{50}, X_{20}, X_{10}, X_5, X_2, X_1)$←

essential $100 * X_{100} + 50 * X_{50} + \ldots + 1 * X_1 = 500$,
strongly preferred $X_{10} >= 1$,

preferred $X_{50} >= 2$,
default $X_{100} <= 4$,

where X_j denotes the number of currency of value j.]

References

[1]   Alferes, J. J. and Pereira, L. M., *Reasoning with Logic Programming*, Springer-Verlag, Berlin, pp. 1-28, 1996.

[2]   Clocksin, W. F. and Mellish, C. S., *Programming in PROLOG*, Springer-Verlag, New York, 1981.

[3]   Dung, P.M. and Kanchanasut, K., A fixpoint approach to declarative semantics of Logic Programs in *LOGIC PROGRAMMING: Proceedings of the North American Conference*, MIT Press, Cambridge, MA, 1989.

[4]   Hogger, J. C., *Essentials of Logic Programming*, Oxford University Press, Oxford, 1990.

[5]   Marek, V. W., Nerode, A. and Truszczynski, M., *Logic Programming and Nonmonotonic Reasoning*, Lecture Notes in Artificial Intelligence series, Springer-Verlag, Berlin, pp. 236-239, 1991.

[6]   Nerode, A. and Shore, R. A., *Logic for Applications*, Springer-Verlag, Berlin, 1993.

[7]   Pratt, W. K., *Digital Image Processing*, Wiley-Interscience Pub., John Wiley and Sons, New York, pp. 570-587, 1978.

[8]   Patterson, D. W., *Introduction to Artificial Intelligence and Expert Systems*, Prentice-Hall, Englewood Cliffs, NJ, pp. 285-326, 1990.

[9]   Spivey, M., *An Introduction to Logic Programming through Prolog*, Prentice-Hall Inter. Series, Englewood Cliffs, NJ, 1996.

[10]   Townsend, C., *PROLOG*, BPB Publications, Indian Reprint, New Delhi, 1990.

[11]   Winston, M. and Borning, A., Extending hierarchical constraint logic programming: Non-monotonicity and inter hierarchy comparison in *LOGIC PROGRAMMING: Proceedings of the North American Conference*, MIT Press, Cambridge, MA, 1989.

7

Default and Non-Monotonic Reasoning

This chapter briefly outlines the scope of automated reasoning in the absence of complete knowledge about the world. The incompleteness of facts or knowledge may appear in a reasoning system in different forms. There is no general notion to treat the different types of incompleteness of data and knowledge (clauses) by a unified approach. In this chapter, we cover 4 distinct techniques, all of which, however, may not be applicable to a common problem. The techniques are popularly known as default logic, non-monotonic logic, circumscription and auto-epistemic logic. Informally, default logic infers a consequent, when its pre-conditions are consistent with the world knowledge. The non-monotonic logic works with two modal operators and can continue reasoning in the presence of contradictory evidences. Circumscription is an alternative form of non-monotonic reasoning, which attempts to minimize the interpretations of some predicates and thus continues reasoning even in the presence of contradictions. The auto-epistemic logic that works on the belief of the reasoning system, occasionally changes inferences, as new evidences refute the existing belief.

7.1 Introduction

Predicate Logic is monotonic [3] in the sense that a derived theorem never contradicts the axioms or already proved theorems [4], from which the former theorem is derived. Formally, if a theorem T_1 is deduced from a set of axioms $A = \{a_1, a_2, a_3,...,a_n\}$, i.e., $A \vdash T_1$, then a theorem T_2, derived from $(A \cup T_1)$ (i.e., $(A \cup T_1) \vdash T_2$), never contradicts T_1. In other words T_1 and T_2 are consistent. Most of the mathematical theorem provers work on the above principle of monotonicity. The real world, however, is too complex and there exist situations, where T_2 contradicts T_1. This called for the development of non-monotonic logic.

7.2 Monotonic versus Non-Monotonic Logic

The monotonicity of Predicate Logic can be best described by the following theorem.

Theorem 7.1: *Given two axiomatic theory T and S, which includes a set of axioms (ground clauses) and a set of first order logical relationships like Modus Ponens, Modus Tollens, etc. If T is a subset (or proper subset of) S then Th(T) is also a subset (or proper subset) of Th(S), where Th(T) (or TH(S)) means the theorems derivable from T (or S) [4].*

An interpretation of the above theorem is that adding new axioms to an axiomatic theory preserves all theorems of the theory. In other words, any theorem of the initial theory is a theorem of the enlarged theory as well. Because of default assumptions in reasoning problems, the monotonicity property does not hold good in many real world problems. The following example is used to illustrate this phenomenon.

Example 7.1: Let us consider the following information about birds in an axiomatic theory T :

    Bird (tweety)
    Ostrich (clide)
    $\forall$ X Ostrich (X) $\rightarrow$ Bird (X) $\land \neg$ Fly (X).

Further, we consider the default assumption R, which can be stated as

R: $\left\{ \begin{array}{l} (T \mathbin{\vert\!\sim} \text{Bird} (X)) , \\[2ex] (T \mathbin{\vert\!\not\sim} \neg \ \text{Fly}(X)) \rightarrow (T \mathbin{\vert\!\sim} \text{Fly} (X)) \end{array} \right.$

where $\mid\sim$ denotes a relation of formal consequence in extended Predicate Logic and $\mid\sim/$ denotes its contradiction. The assumptions stated under R can be presented in language as follows: R: *"If it cannot be proved that the bird under consideration cannot fly, then infer that it can fly."*

Now, the non-monotonic reasoning is started in the following manner.

1. Bird (tweety)

2. T $\mid\sim/$ $\neg$ Fly (tweety)

3. Monotonicity fails since T $\cup\{$ Fly (tweety)$\}$ $\mid$-/ falsum

4. From default assumption R and statement (2) above, it follows that T $\mid\sim$ Fly (tweety).

A question then naturally arises: can $\mid\sim$ be a first order provability relation $\mid$- ? The answer, of course, is in the negative, as discussed below.

The First Order Logic being monotonic, we have

$\{$ T $\subseteq$ S $\rightarrow$ Th (T) $\subseteq$ Th (S)$\}$ (Theorem 7.1)

Let

   T $\mid$- Fly (tweety)   (7.1)

and S = T U $\{$ $\neg$ Fly (tweety)$\}$.   (7.2)

Now, since Fly (tweety) is a member of T, from (7.1) and (7.2) above, we find

   S $\mid$- Fly (tweety).   (7.3)

Again, by definition (vide expression (7.2)),

   S $\mid$- $\neg$ Fly (tweety).   (7.4)

Expression (7.3) and (7.4) shows a clear mark of contradiction, and thus it fails to satisfy Th (T) $\subseteq$ Th (S). This proves all about the impossibility to replace $\mid\sim$ by first order relation $\mid$-.

7.3 Non-Monotonic Reasoning Using NML I

In this section, we will discuss the formalisms of a new representational language for dealing with non-monotonicity. McDermott and Doyle [8]

proposed this logic and called it non-monotonic logic I (NML I). The logic, which is an extension of Predicate Logic, includes a new operator, called the consistency operator, symbolized by ◆. To illustrate the importance of the operator, let us consider the following statement.

$\forall$ X Bird (X) $\land$ ◆ Fly (X) $\rightarrow$ Fly (X)

which means that if X is a bird and if it is consistent that X can fly then infer that X will fly.

It is to be noted that NML I has the capability to handle default assumptions. Thus the Default Logic of Reiter [11], to be presented shortly, is a special case of NML I.

Let us now attempt to represent the notion of the non-monotonic infer-encing mechanism by the consistency operator. We will consider that

T $\not\hspace{-0.3em}\sim$ $\neg$ A

$\Rightarrow$ T $\hspace{-0.3em}\sim$ ◆ A,

which means that if $\neg$ A is not non-monotonically derivable from the axiomatic theory T, then infer that A is consistent with any of the theorems provable from T.

7.4 Fixed Points in Non-Monotonic Reasoning

To understand the concept on fixed points (also called fixpoints), let us consider an example, following McDermott and Doyle [8].

Example 7.2: Let T be an axiomatic theory, which includes

◆ P $\rightarrow$ $\neg$ Q and ◆ Q $\rightarrow$ $\neg$ P.

Formally, T = { ◆ P $\rightarrow$ $\neg$ Q , ◆ Q $\rightarrow$ $\neg$ P },

which means that if P is consistent with the statements of the axiomatic theory T then $\neg$Q and if Q is consistent with the statements in T then $\neg$ P. McDermott and Doyle called the system having two fixed points (P, $\neg$ Q) and ($\neg$P, Q). On the other hand, if T = { ◆ P$\rightarrow$ $\neg$ P}, then there is no fixed points in the system.

Davis [4], however, explained the above phenomena as follows.

$$\{ \blacklozenge P \rightarrow \neg Q , \blacklozenge Q \rightarrow \neg P \} \hspace{0.2em}\mid\!\sim (\neg P \vee \neg Q)$$

and $\{ \blacklozenge P \rightarrow \neg P \} \hspace{0.2em}\mid\!\sim$ falsum.

Problems encountered in NML I

McDermott and Doyle identified two basic problems in connection with reasoning with NML I. These are

i) $\blacklozenge$ A cannot be inferred from $\blacklozenge (A \wedge B)$

ii) $T = \{ \blacklozenge P \rightarrow Q, \neg Q \} \hspace{0.2em}\mid\!\sim$ falsum,

which means that the axiomatic theory T is inconsistent in NML I.

7.5 Non-Monotonic Reasoning Using NML II

In order to overcome the above difficulties, McDermott and Doyle recast non-monotonic logic with the help of another modal [4] operator like consistency, called the necessitation operator and symbolized by $\square$. This operator is related to modal operator $\blacklozenge$ by the following manner:

$$\square P \equiv \neg \blacklozenge \neg P$$

or $\blacklozenge P \equiv \neg \square \neg P$

where the former notation denotes that P is necessary could be described alternatively as negation of P is not consistent. The second definition implicates that P is consistent could be written alternatively as the negation of P is not necessary.

The significance of the necessitation operator can be understood from the following example.

Example 7.3: Given that $T \mid\!\sim \blacklozenge$ A, i.e., A is consistent with the derived consequences from T. Then it can be inferred that

$T \mid\!\sim \neg \square \neg$ A,

which means that it is not necessary to assume that $\neg$ A is derivable from T. This is all about the example.

We now present a few modal properties, as presented in NML II.

1. ◆ P = ¬ □ ¬ P   (by definition)   (property 1)

2. □ (P → Q) ⇒ (□P → □Q),   (property 2)

which means, if it is necessary that P → Q, then infer that if P is necessary then Q is necessary.

3. □ P ⇒ P,   (property 3)
which may be read as if P is necessary, then infer P.

4. □ P⇒ □□ P,   (property 4)

which is to be read as if P is necessary then imply that it is necessary that P is necessary.

How the basic two limitations of NML I could have been overcome in NML II are presented next.

Example 7.4: Show that◆ (A ∧ B) ⇒ ◆ A

Proof: The proof consists of the following 6 steps.

Step 1: By Predicate Logic

$$\neg\ A \Rightarrow (\neg A \vee \neg B) \Rightarrow \neg\ (A \wedge B) \tag{7.5}$$

Step 2: Preceding with necessitation operator in (7.5), we have

$$\square\ (\neg A \Rightarrow \neg\ (A \wedge B)) \tag{7.6}$$

Step 3: Now by identity (2), we have

$$\square\ (\neg\ A \Rightarrow \neg(A \wedge B)) \Rightarrow (\square\ \neg A \Rightarrow \square\ \neg(A \wedge B)) \tag{7.7}$$

Step 4: Now, from (7.6) and (7.7) by Modus Ponens, we find

$$\square\neg\ A \Rightarrow \square\neg\ (A \wedge B) \tag{7.8}$$

Step 5: Now by applying the contraposition theorem of Propositional Logic, we find expression (7.9).

$$\neg\ \square\neg\ (A \wedge B) \Rightarrow \neg\ \square\ \neg\ A \tag{7.9}$$

Step 6: Now, replacing $\neg \Box \neg P$ by $\blacklozenge P$ in expression (7.9), we get the desired expression (7.10)

$$\blacklozenge (A \wedge B) \Rightarrow \blacklozenge A \qquad\qquad (7.10)$$

The second difficulty of NML I can also be overcome in NML II, as demonstrated by the following example.

Example 7.5: Given an axiomatic theory T, where T = { $\blacklozenge P \Rightarrow Q, \neg Q$ }. Prove that $\neg P$ follows from T using NML II.

Proof: Given the expressions (7.11) and (7.12)

Step 1: $\neg Q$ $\qquad\qquad\qquad\qquad\qquad\qquad$ (7.11)

Step 2: $\blacklozenge P \Rightarrow Q$ $\qquad\qquad\qquad\qquad\qquad$ (7.12)

Step 3: $\neg \Box \neg P \Rightarrow Q$, by property 1 $\qquad\qquad$ (7.13)

Step 4: $\neg Q \Rightarrow \Box \neg P$, by contraposition theorem $\qquad$ (7.14)

Step 5: $\Box \neg P$, from (7.11) and (7.14) by Modus Ponens $\qquad$ (7.15)

Step 6: $\Box \neg P \Rightarrow \neg P$, by property 3 $\qquad\qquad$ (7.16)

Step 7: $\neg P$, from (7.15) and (7.16) by Modus Ponens $\qquad$ (7.17)

This is all about the proof. $\qquad\qquad\qquad\qquad\qquad\qquad$ □

In the next section, we discuss about a practical system for non-monotonic reasoning.

7.6 Truth Maintenance System

Doyle's truth maintenance system (TMS) [5] is one of the most practical systems for non-monotonic reasoning. The TMS works with an expert or decision support system and helps the reasoning system to maintain the consistency among the inferences generated by it. The TMS itself, however, does not generate any inferences. The functional architecture of an expert reasoning system that employs a TMS for maintaining the consistency among the generated inferences is presented in fig. 7.1.

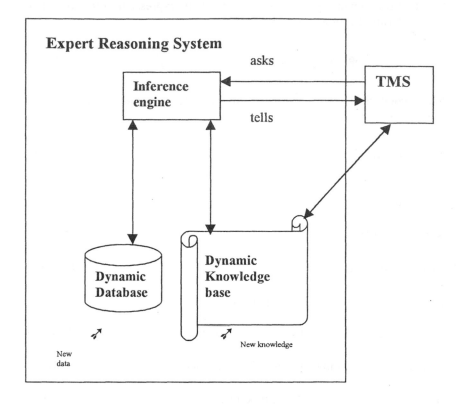

Fig. 7.1: Architecture of a reasoning system that includes a TMS.

In fig. 7.1, the inference engine (IE) in an expert reasoning system interprets the database based on the available pieces of knowledge, stored in the knowledge base. Since both the database and knowledge base are dynamic, the process of reasoning would be continued as long as new data or knowledge are entered into the system from the external world. The TMS asks the inference engine about the current inferences that it derives and attempts to resolve the inconsistency between the old and current inferences, after the inference engine delivers the derived inferences to the TMS. The TMS then groups the set of consistent information and reports the same to the IE [10].

The current status of all inferences/information is labeled by the TMS by IN or OUT nodes. IN nodes mean that the nodes (information) are active, while OUT nodes signify that the nodes have to be retracted from the current reasoning space. It needs emphasizing that the dependency of the information in the reasoning space is represented by a dependency-directed graph, where the nodes in the graph denote information, while the arcs stand for cause-effect dependency relation among the nodes. Justification records maintained for the nodes are also attached with the nodes in the graph (network). There are two types of justification, namely *support list (SL)* and *conditional proofs* (CP). The data structure for these two justifications is discussed below:

(SL <in-nodes> <out-nodes>)
(CP <consequent> <in-hypotheses> <out-hypotheses>)

To illustrate the reasoning process carried out by the TMS, we take the following example.

Example 7.6: Mita likes to accompany her husband when buying some household items in the market place, located at Gariahat in Calcutta. Further, she prefers to visit the market next Monday evening, since the crowd is normally less on Monday evenings. However, on Saturday it was learnt that the market offers a special rebate on a few items, starting next Monday. So, Mita revised her decision not to go to the market this Monday and selects the next Monday for her visit. Later it was found that her husband had an official emergency meeting on the next Monday evening and accordingly Mita had to postpone her visit on that evening. Later, the time of the emergency meeting was shifted to an early date and Mita happily agreed to visit the market with her husband the next Monday evening.

Let us attempt to resolve this kind of non-monotonicity by TMS. The knowledge base for this system consists of the following production rules (PR).

PR1: Crowd-is (large, at-market, at-time-T) →
  ¬Visits (Wife, market, at-time-T).

PR2: Visits (Wife, market-at-time-T) →
  Accompanies (Husband, Wife, at-time-T).

PR3: Offers (market, rebate, at-time -T) →
  Crowd-is (large, at-market, at-time-T).

PR4: Has-Official-meeting (Husband, at-time-T) →
  Unable-to-accompany (Husband, Wife, at-time-T).

PR5: ¬Visits (Wife, market, Monday-evening) →
       Visits (Wife, market, next-Monday-evening).

Further, the database includes the following data:

1. Offers (market, rebate, Monday-evening)
2. Has-Official-meeting (Husband, next-Monday-evening)
3. Meeting-shifted (from, next-Monday-evening)

The TMS, now, resolves inconsistency in the following manner.

Node	Status	Meaning	SL / CP
n1	IN	Offers (mar, reb, mon-eve)	(SL ( ) ( ))
n2	IN	Crowd-is (lar, at-mar, mon-eve)	(SL(n1) ( ))
n3	IN	¬Visits (mita, mar, mon-eve)	(SL (n2) ( ))
n4	IN	Visits (mita, mar, next-mon-eve)	(SL (n3) ( ))
n5	IN	Has-meeting (ram, next-mon-eve)	(SL ( ) ( ))
n6	IN	Unable-to-Acc. (ram, mita, next-mon-eve)	(SL (n5) ( ))
n7	IN	Accompanies (ram, mita, next-mon-eve)	(SL (n4) (n6))
n8	IN	contradiction	(SL (n6, n7) ( ))
n9	IN	no-good n7	(CP n8 (n6, n7) ( ))

Now with a new data item 3

Node	Status	Meaning	SL / CP
n10	IN	Meeting-shifted (from, next-mon-eve)	(SL ( ) (n5))
n11	IN	contradiction	(SL (n5, n10) ( ))
n12	IN	no-good n5	(CP n11 (n5, n10) ( ))
n5	OUT		(SL (n12) ( ))

-

The dependency-directed backtracking scheme, invoked by the TMS for the above problem, is illustrated below.

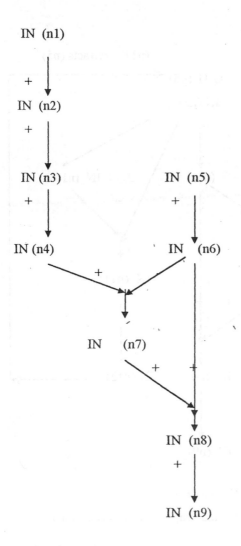

IN (nj) denotes node nj is IN. Sign against an arc denotes positive / negative consequences of input antecedents.

Fig 7.2: A section of dependency graph, demonstrating the working principles of the TMS.

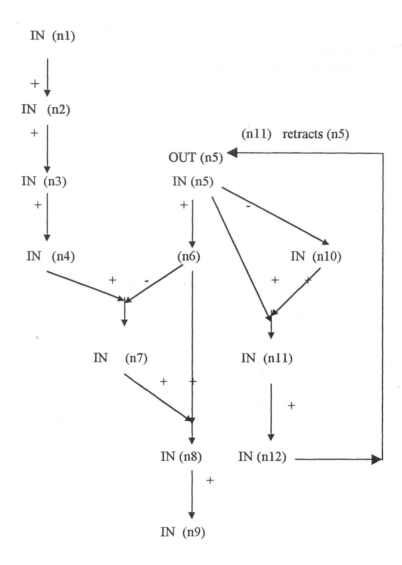

OUT (nj) denotes node nj is out.

Fig.7.3: The dependency directed graph, after receiving data 3.

Fig. 7.2 represents a dependency graph, where the nodes denote the events and the arcs denote the causal relationship. The abbreviation used in the figures is self-explanatory, and thus requires no explanation. A positive arc from node Ni to node Nj represents that Ni causes Nj, whereas a negative arc from Ni to Nj represents that Ni refutes Nj. The reasoning in fig. 7.2 is based on the first two data: Offers (market, rebate, Monday-evening), Has-Official-meeting (Husband, next-Monday-evening). When the third data: Meeting-shifted (from, next-Monday-evening) is added to the reasoning system, it starts backtracking and finally changes the status of node n5 (fig. 7.3) from IN-node to OUT-node. The TMS thus always maintains consistency among the inferences derived by the inference engine.

7.7 Default Reasoning

Reiter's default logic [11] is an alternative form of non-monotonic logic, which includes a few First Order statements along with one or more default assumptions. Formally, let Del be a set of statements, including a set of axioms W and a set of Default statements D, i.e.,

$$Del = \{D, W\}$$

where the elements of D are of the form:

P: $Q_1, Q_2, ..., Q_n$
_____

R

which may be read as: Given P is true, as long as Q_1 through Q_n are consistent with P, one may infer that R holds. Further, when P is absent from the above statement, i.e.,

: $Q_1, Q_2, ..., Q_n$

_____

R

it means that as long as Q_1 through Q_n are consistent with known beliefs, it may be inferred that R holds.

Another special case of a Default statement is of the form

P:

_____

R

which means that " if P is true, one may infer that R is true". The last statement is different from P $\rightarrow$ R, which means "if P is true then R is true", i.e, when P holds R must hold. But in the previous case "when P holds, R may hold (and not must hold). An example, illustrating the concept of default logic, is presented below.

Example 7.7: Let Del = { D, W} where

W = { Bird (parrot), Bird (penguin),

Bird (Penguin) $\rightarrow \neg$ has-aeronautical-prowess (penguin)}

$$D = \frac{\text{Bird (X) : has-aeronautical-prowess (X)}}{\text{Fly (X)}}$$

The inference derived by the system is: Fly (parrot). Since has-aeronautical-prowess (Penguin) is inconsistent with Bird (penguin), the default rule D blocks Fly (Penguin) to be an inference.

Types of Default Theories: There exist varieties of Default theories. Among them, i) Normal, ii) Semi-normal, iii) Ordered and iv) Coherent theories need special mention.

Definition 7.1: Any Default knowledge of the form

$$\text{Del}_1 = \{ \ \frac{A: \quad B}{B} \ \} \ , \text{Del}_1 \text{ being a subset of Del,}$$

is called **normal**. Even when "A" is absent from the statement Del_1, the default knowledge remains normal.

Definition 7.2: Any default knowledge of the form

$$\text{Del}_1 = \{ \ \frac{A: \quad B \wedge C}{B} \ \}, \ \text{Del}_1 \text{ being a subset of Del,}$$

is said to be **Semi-normal.**

The concept of **Ordered Default Theory** requires a detailed formalization of "precedence relation" << and <<=, where the first and second relation are referred to as **strong and weak precedence relations**. For example, in the default knowledge:

$$Del_1 = \{\frac{A: \quad B \wedge \neg C}{B}\}$$

B has strong precedence over C, denoted by C<< B.

Definition 7.3: A semi-normal Default theory is said to **be ordered** if there is no literal y, such that y << y. For example, in the Default knowledge base D, given by

$$D = \{\frac{: A \wedge \neg B}{A}, \frac{: B \wedge \neg C}{B}, \frac{: C \wedge \neg A}{C}\}$$

B << A, C << B and A << C and consequently A << A and thus the default theory is not ordered.

Definition 7.4: A Default theory that has at least one extension, i.e., that can infer at least one inference, is called **Coherent.**

It is to be noted that an ordered semi-normal default theory has at least one extension and thus is coherent.

Stability of Default Theory: The notion of stability in Default theory is a significant issue for it helps partitioning the knowledge base under Del into stable (time-invariant) fragments, if possible. For understanding the concept of stability informally, let us consider the following example.

Example 7.8: Let Del = { D, W } where

$$D = \{ Del_1 = \frac{A: B \wedge \neg C}{B}, \quad Del_2 = \frac{A: C \wedge \neg B}{C}\}$$

and W = { A }.

It is clear from common sense that both {Del_1, W} and {Del_2, W} are consistent and **stable extensions** of the given default theory Del. The resulting stable consequences that follow from the above extensions are {A, B} and {A, C} respectively. The Default theory need not always yield stable consequences. The following example illustrates this concept.

Example 7.9: Consider the Default theory Del = { D, W} where

$$D = \{ \frac{: \neg A}{A} \} \text{ and } W = \{\ \}.$$

The default statement under Del means as long as ¬ A is consistent with the known beliefs, infer that A holds good. However, from common sense reasoning it follows that "A" cannot be inferred when ¬ A is consistent with known beliefs. Thus the Default theory itself is incoherent and the derived consequence {A} is not stable (unstable).

7.8 The Closed World Assumption

The readers by this time can understand that non-monotonicity mainly creeps into a reasoning system because of incompleteness of information. Thus for reasoning in a non-monotonic domain, one has to presume either a **closed world assumption** or add new facts and knowledge to continue reasoning. The phrase "closed world assumption" means that the ground predicates not given to be true are assumed to be false. For instance, consider the following clauses.

Bird (X) ∧ ¬ Abnormal (X) → Fly (X)

Bird (parrot).

Suppose we want to derive whether Fly (parrot) is true. However, we cannot derive any inference about Fly (parrot), unless we know ¬ Abnormal (parrot).

The closed world assumption (CWA), in the present context, is the assumption of the negated predicate: ¬ Abnormal (parrot). In fact, the closed world assumption requires the ground instances of the predicate to be false, unless they are found to be true from the supplied knowledge base.

Thus the CWA of the following clauses:
Bird (X) ∧ ¬ Abnormal (X) → Fly (X)
Bird (parrot)
¬Abnormal (parrot)

infers: Fly (parrot).

The following two points are noteworthy in connection with the closed world assumption:

i)   The CWA makes a reasoning system complete. For instance, given the set of clauses {Bird (parrot), Bird (penguin), Bird(X) → Fly (X)}, the Modus Ponens is not complete as neither Fly (penguin) or ¬ Fly (penguin) can be derived by Modus Ponens. But we can make the Modus Ponens complete by adding ¬ Fly (penguin) in the set.

ii)   The augmented knowledge base, constructed by CWA, is inconsistent. For instance, consider the following clause: Bird (penguin) ∨ Ostrich (penguin). Since none of the ground literal of the clause is derivable, we add:

$$\neg Bird\ (penguin)\ \ and$$

$$\neg\ Ostrich\ (penguin)$$

to the reasoning space by CWA. But the reasoning system, comprising of {Bird (penguin) ∨ Ostrich (penguin), ¬Bird (penguin), ¬ Ostrich (penguin)} is inconsistent, as there exists no interpretation for which Bird (penguin) or Ostrich (penguin) is true. It may, however, be noted that if the knowledge base comprises of Horn clauses and is consistent, then closed world extension is consistent [10].

7.9 Circumscription

Circumscription [7] is the third main type of non-monotonic reasoning following NMLs and default logic. Developed by John McCarthy, circumscription attempts to minimize the interpretation of specific predicates, thus reducing the scope of non-monotonicity. For instance, suppose that a child knows only two kinds of birds: parrot and sparrow. Formally, we can write this as

$$Bird\ (parrot) \lor Bird\ (sparrow).$$

So, he defines a rule:

$$\forall\ X\ Bird\ (X) \to Bird\ (parrot) \lor Bird\ (sparrow).$$

The expression of the form: X = parrot ∨ X = sparrow is called a **predicate expression.** It is so named, because it seems to replace the predicate Bird (X).

In this example, we call ψ = Bird (parrot) $\vee$ Bird (sparrow) a **formula** and X is a distinguished variable [2]. We can now formalize circumscription as follows.

CIR (KB: Bird)

$= KB \wedge [\forall \psi \, KB (\psi) \wedge (\forall X \, \psi (X) \rightarrow Bird (X))]$

$\rightarrow \forall X (Bird (X) \rightarrow \psi (X))$

where KB denotes the set of the given pieces of knowledge, connected by AND operator; KB (ψ) denotes a knowledge base, where every occurrence of Bird is replaced by ψ [10].

It is to be noted that ψ has been quantified in the circumscription schema above, thereby making it a second order formula.

The semantic interpretation that circumscription constructs a minimal follows directly from the last example. We, for instance, found that

$\forall X \, Bird (X) \rightarrow (X = parrot \vee X = sparrow),$

which is equivalent to

$\forall X ((\neg X = parrot \vee X = sparrow) \rightarrow \neg \, Bird (X)).$

The last expression implicates there exists no interpretation of Bird(X), where X is not a parrot or sparrow.

7.10 Auto-epistemic Logic

The word 'auto-epistemic' means reflection upon self-knowledge [9]. Developed by Moore in the early 80's this logic attempts to derive inferences based on the belief of the speaker. For instance, consider the sentence: "Are the Olympics going to be held in Japan next year?" The answer to this: No, otherwise I must have heard about it. Now, suppose, the next morning I found in the newspaper that the Olympics will be held in Japan next year. Now, my answer to the above question will be: yes. But it is to be noted that my long-term knowledge that 'if something is going to be held in my neighborhood, then I will know about it" is still valid.

Generally, auto-epistemic expressions are denoted by L. For instance, to represent that 'I know that a tiger is a ferocious animal,' we would write it in auto-epistemic logic by

$\forall$ X L (Tiger (X) $\rightarrow$Ferocious (X))

where L denotes the modal operator, meaning 'I know'.

Some of the valid auto-epistemic formula are:

L (p), L (L (p)), L ($\neg$ L (q)) , ($\neg$ L ($\neg$p $\vee$ L (r))), where p, q and r are predicates or propositions. But what do they mean? L (L (p)), for example, means that I know that I know p. The other explanations can be made analogously.

One important aspect of auto-epistemic logic is stability. A deductively closed set E of auto-epistemic formula is called **stable**, iff

i) $\varphi \in E \Rightarrow L \varphi \in E$
ii) $\varphi \notin E \Rightarrow \neg L \varphi \in E$

The concept of stability reflects introspection of the auto-epistemic reasoning. For instance, if A is in my knowledge, then I know A, or I know that I know A. In other words, if I do not know B then I know that I do not know B, i.e.,

$\neg$ B $\Rightarrow$ L ($\neg$ B) $\Rightarrow$ $\neg$ L (B).

Auto-epistemic formula are defined as the smallest set that satisfy the following:

i)   All closed FOL formula is an auto-epistemic formula.
ii)   If φ is an auto-epistemic formula, then L φ will also remain as an auto-epistemic formula.
iii)   If φ and ψ are auto-epistemic formula, then $\neg\varphi$, ($\varphi \vee \psi$), ($\varphi \vee \psi$) ($\varphi \rightarrow \psi$) are all auto-epistemic formula.

An auto-epistemic theory is a set of auto-epistemic formula.

7.11 Conclusions

The chapter presented four different types of non-monotonic reasoning and illustrated each of the reasoning methodologies by examples. The most important factor in non-monotonic reasoning is 'stability'. Much importance has been given to determine stability of many non-monotonic programs. The closed world assumption seems to be a powerful tool for non-monotonic reasoning. Circumscription and auto-epistemic theory are two different extensions of the closed world assumption. There exists ample scope to unify the theories by a more general purpose theory. We hope for the future when the general purpose theory for non-monotonic reasoning will take shape. Further, in the short run Logic Programming and Non-monotonic Logics will merge together to provide users with a uniform theory for automated reasoning [1], [6].

Exercises

1. Represent the following sentences by default logic. Also mention the sets D and W.

   a)   Typically molluscs are shell-bearers [2].
   b)   Cephalopods are molluscs.
   c)   Cephalopods are not shell-bearers.

   [**Hints:** W = {Cephalopod (X) → Molluscs(X)} and

$$D = \{ \frac{Molluscs\ (X):\quad Shell\text{-}bearer(X)}{Shell\text{-}bearer(X)} \quad \}]$$

2. Represent the following sentences by default logic.

   a)   John is a high school leaver.
   b)   Generally, high school leavers are adults.
   c)   Generally, adults are employed.

Now, determine the extension of the default theory. Do you think that the extension is logically rational?

3. Replace rule (c) in the last problem by the following rule.

$$\frac{Adult(X):\quad Employed(X) \wedge \neg\ School\text{-}leaver\ (X)}{Employed(X)}$$

What do you gain by doing so? Can you hence identify the limitations of 'normal defaults'?

4. Test whether the following default theory is ordered:

$$\frac{f:\ \neg P \wedge q}{q}, \frac{f:\ \neg q \wedge r}{r}, \frac{f:\ \neg r\ \wedge}{p}.$$

5. Test whether the following default theory is ordered.

$$\frac{:\neg\, Q \wedge P\,(a)}{P(a)}, \frac{:\neg\, R \wedge Q}{Q}, \frac{:\neg\, P(b) \wedge R}{R}.$$

[Hints: Q < P(a), R< Q, P(b) < Q; No circularity and hence ordered.]

6. Find the stable extensions for the following default theory.
   Given Del ={W, D} where

W = {a, b} and

$$D = \{\frac{a:\ c \wedge b}{c}, \frac{b:\ b \wedge \neg c}{b}\ \}$$

[Ans. {a, b, c} and {a, b}]

7. Show that ◆ P → Q, ¬ Q ⇒ ¬ P by using the following axiomatic theory
   [4].

   a) (◆ P → Q) ⇒ (P → (◆ P → Q)),

   b) P → (◆ P → Q) ⇒ (P → ◆ P) → (P → Q) and

   c) P→ ◆ P.

   [Hints: With given ◆ P → Q and (a) by Modus Ponens get (P → (◆ P → Q)). Next with this result and (b) by Modus Ponens derive (P → ◆ P) → (P → Q), which with (c) by Modus Ponens yields P → Q. Now given ¬ Q and P → Q, we derive ¬ P by Modus Tollens.]

8. Show by definition that the ground instances of the following statements generate an auto-epistemic theory.

a) Boy(X) $\wedge \neg$ L $\neg$ Likes (X, chocolates) $\rightarrow$Likes (X, chocolates)
b) Boy (ram)
c) Boy (john)

[**Hints:** Get 4 formulas: Boy (ram), Boy (john) and the ground instances of (a) with X = ram and X= john. Since these are closed first order formula, these must be auto-epistemic formula. Then the collection of these formula is the auto-epistemic theory (see text for the definition).]

9. Add the following fact and knowledge with Mita's marketing visit problem and show with the dependence graph: how the TMS works.

Fact: Carries-Influenza (mita, next-Monday-evening).

Knowledge: Carries-Influenza (X, Day) $\rightarrow \neg$ Visits (X, market, Day).

References

[1]   Alferes, J. J. and Pereira, L. M., *Reasoning with Logic Programming*, Springer-Verlag, Berlin, pp.1-35, 1996.

[2]   Antoniou, G., *Nonmonotonic Reasoning*, The MIT Press, Cambridge, MA, pp. 21-160, 1997.

[3]   Bender, E. A., *Mathematical Methods in Artificial Intelligence*, IEEE Computer Society Press, Los Alamitos, pp. 199-254, 1996.

[4]   Besnard, P., *An Introduction to Default Logic*, Springer-Verlag, Berlin,   pp. 27-35, pp.163-177, 1989.

[5]   Doyle, J., A Truth Maintenance System, *Artificial Intelligence*, vol. 12, pp. 231-272, 1979.

[6]   Marek, V. W., Nerode, A. and Truszczynski, M., *Logic Programming and Nonmonotonic Reasoning*, Lecture Notes in Artificial Intelligence series, Springer-Verlag, Berlin, pp. 388-415, 1995.

[7]   McCarthy, J., "Circumscription- a form of non-monotonic reasoning," *Artificial Intelligence*, vol. 13, pp. 27-39, 1980.

[8]   McDermott, D. and Doyle J., "Nonmonotonic Logic I," *Artificial Intelligence*, vol. 13 (1-2), 1980.

[9]   Moore, R. C., Autoepistemic Logic, in *Non-Standard   Logics for Automated Reasoning,*   Smets, Ph., Mamdani, E. H., Dubois, D., and Prade, H., Eds., Academic Press, London.

[10]   Patterson, D. W., *Introduction to Artificial Intelligence and Expert Systems*, Prentice Hall, Englewood Cliffs, NJ, pp. 80-93, 1990.

[11]   Reiter, R., A logic for default reasoning, *Artificial Intelligence*, vol. 13, pp. 81-132, 1980.

8

Structured Approach to Knowledge Representation

The chapter addresses various structured models for knowledge representation and reasoning. It starts with semantic nets and provides in detail its scope in reasoning with monotonic, non-monotonic and default knowledge bases. The principles of defeasible reasoning by semantic nets have been introduced. The concept of partitioned semantic nets to represent complex knowledge including quantifiers has been illustrated with examples. Among the other structured approaches special emphasis has been given to frames. The principles of single and multiple inheritance have been elucidated for frame structures. The rest of the chapter includes discussions on conceptual dependency graphs, scripts and Petri nets.

8.1 Introduction

This chapter provides an alternative approach for knowledge representation and reasoning by structured models. Such models have immense significance over the non-structured models by the following counts. Firstly, the knowledge base can be easily represented by modular fashion, thus ensuring a compartmentalization of the related chunks of knowledge for efficient access. Secondly, explanation tracing in knowledge based systems becomes easier with structured models. Thirdly, a structured model offers provision for representing monotonic, non-monotonic and default logic on a common platform. Fourthly, the fragments of a piece of knowledge can be accessed concurrently by many modules, thereby providing the reasoning systems a scope for sharing of resources. Such sharing of resources call for minimal structure and consequently less hardware / software devices for realization of the knowledge bases. Finally, many modules of the structural models can be made active concurrently, thus providing a scope for massive parallelism in the reasoning process. The time efficiency of the inference engines thus can be improved with structured models.

In this chapter, a number of structured models for knowledge representation will be covered. Among these structures, some include very rigid dependence relationships among the events. These structured models are called strong filler and slot methods. Conceptual dependencies and scripts are ideal examples of such structured models. The other type of structural models does not impose much restrictions on the dependence relationships among events and is hence called weak filler and slot methods. This includes semantic nets and frames. The next section will explore the reasoning mechanism with semantic nets.

8.2. Semantic Nets

Semantic nets at the time of its origin were used mainly in understanding natural language, where the semantics (meaning) of the associated words in a sentence was extracted by employing such nets. Gradually, semantic nets found a wider application in reasoning of knowledge based systems. A semantic net consists of two elementary tuples: **events** denoted by nodes and **relationship** between events, denoted by links / arcs. Generally, a linguistic label is attached to each link to represent the association between the events. A simple semantic net that describes a binary predicate: Likes (X, Y) is presented in fig 8.1.

Fig.8.1: A semantic net corresponding to the predicate Likes(X, Y).

Readers should not confuse from the last figure that a semantic net can represent only a binary predicate, i.e., a predicate having two arguments. It can represent non-binary predicates as well. For instance, consider the unary predicate boy (jim). This can be represented by a binary predicate Instance-of (jim, Boy) and consequently can be represented by fig 8.2.

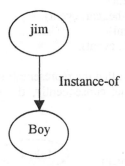

Fig 8.2: Representation of instance-of (jim, Boy).

Further, predicates having more than two arguments can also be represented by semantic nets [6]. For example, consider the ternary predicate: Gave (john, the-beggar, 10$). This can be represented by semantic nets, vide fig 8.3.

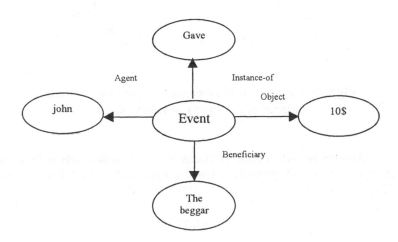

Fig. 8.3: A semantic net, representing a ternary predicate.

It is thus clear from fig 8.3 that a predicate of arity > 2   can be represented by a number of predicates, each of arity 2, and then the resulting predicates can be represented by a semantic net. For instance, the ternary predicate Gave(john, the-beggar, 10$) has been represented in fig 8.3. as a collection of 4 predicates given by

>Agent (john, event)
>Beneficiary (the-beggar, event)
>Object (10$, event)
>Instance-of (give, event).

It may be added here that the representation of a higher-arity predicate in binary form is not unique; consequently, the semantic net of predicates of arity > 2 is also not unique.

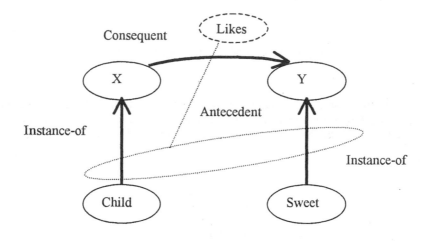

Fig 8.4: A semantic net to represent Instance-of (X, Child) ∧
Instance-of (Y, sweet) → Likes (X, Y) .

Another important issue of discussion on semantic nets is its capability of representation of quantifiers. For example, suppose we like to represent FOL based sentence:

>∀X ∃Y (Child (X) ∧ Sweet(Y) → Likes(X,Y)).

This can be transformed into the following forms of binary predicates:

>∀X ∃Y (Instance-of (X, child) ∧ Instance-of (Y, sweet)
>→ Likes (X, Y)).

Consequently, we represent the above clause by a semantic net (vide fig 8.4).

The semantic net shown in fig 8.4 describes a mapping from relationships Instance-of (X, child) and Instance-of (sweet, Y) to relationships Likes (X, Y). To represent the given clause by an alternative form of semantic nets, we, for example, first express it in CNF. The CNF of $\forall X \ \exists Y$ (Instance-of (X, child) $\wedge$ Instance-of (Y, sweet)$\rightarrow$ Likes (X, Y)) is $\forall X \ \exists Y$ ($\neg$Instance-of (X, child) $\vee$ $\neg$Instance-of (Y, sweet) $\vee$ likes(X, Y)).

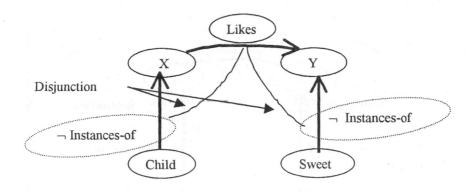

Fig . 8.5: A representation of $\neg$ Instance -of (X, Child) V
$\neg$ Instance- of (Y, Sweet) V Likes (X , Y).

Semantic nets can be partitioned into modules, where one or more modules are subsets of the others. Such modular structure helps reasoning in a hierarchical fashion and thus is more time efficient. For instance, consider the semantic net of fig. 8.6, where node 'g' corresponds to the assertion given in the above problem, whereas GS is a general statement. In other words, g is a special instance of GS. The form of node g, and $\forall X$, are also shown clearly in fig. 8.6. It is to be further noted that the entire semantic net is now partitioned into two modules, one called SI, while the entire space is called SA. The spaces in a **partitioned semantic net** [3] are associated with each other by a subsethood hierarchical relation. For instance, space SI in fig. 8.6 is included in space SA. The search in a partitioned semantic net thus can be done from inner to outer space; searching variables or arcs in an opposite direction (i.e., from outer to inner space) is not allowed [9].

8.3 Inheritance in Semantic Nets

A binary relation x < y is a **partial order** if i) x </ x (i.e., x < x fails
unconditionally) and ii) when x < y and y <z, we have x < z, i.e., transitive
relationship holds good. If the concept in a semantic net has a partial order,
we call the net an **inheritance system**. While drawing the net, we, however,
always omit the arc representing **transitive inheritance**, i.e., if there is a
directed edge from node u to v and v to w, the edge u to w is obvious and thus
omitted.

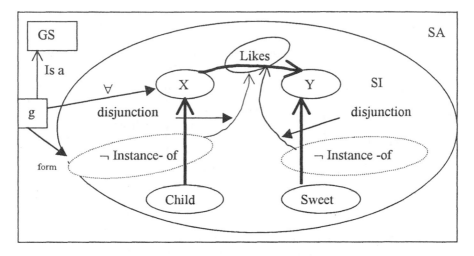

Fig. 8.6: Induction of ∀X in the semantic net of fig 8.5.

Example 8.1: In this example, we demonstrate an inheritance relationship
among various biological species connected through Is-a relationship. It is
clear from fig 8.7 that since bacteria is-a protozoa, and protozoa is an
invertebrate, therefore, bacteria is an invertebrate. By reasoning in the same
line, it further can be shown that bacteria is a biological mass.

8.4 Manipulating Monotonic and Default Inheritance in Semantic Nets

In this section we present a uniform notion to handle both the FOL and default
inheritance in semantic nets. The following nomenclatures will be used to
discuss these issues.

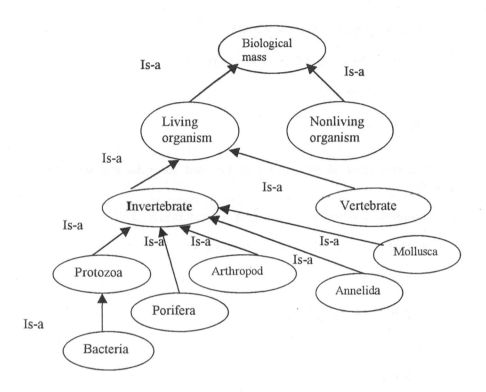

Fig. 8.7: The Is-a hierarchy of biological species.

i) The monotonic / absolute links will be denoted by '→' , whereas default links will be denoted by ' ------→' .

ii) Nodes in the graph denote either constants / predicates. No links should point towards a constant. The monotonic edges and their negations may point away from a constant.

iii) A link p→q denotes the predicate Q(p), when p is constant; otherwise it denotes

$$\forall X \ (p(X) \rightarrow q(X)) \ ,$$

when p and q are predicates.

iv) Consequently, a link $p /\rightarrow q$ denotes $\neg Q(p)$ when p is a constant and

$\forall X \, (p(X) /\rightarrow q(X))$,

when p and q are predicates. Further,

$\forall X \, (p(X) /\rightarrow q(X))$, $\forall X \, (p(X) \rightarrow \neg q(X))$ and

$\forall X \, (\neg p(X) \rightarrow \neg q(X))$

are **equivalent**.

The following rules of inference may be used for reasoning with '$\rightarrow$' operator.

i) **Symmetry:** If $p/\rightarrow q$, where p is not a constant then $q /\rightarrow p$.

The proof of the above is simple as follows.

$p /\rightarrow q$

$\models p \rightarrow \neg q$ (by definition)

$\models \neg p \vee \neg q$ (by rule of implication)

$\models \neg q \vee \neg p$

$\models q \rightarrow \neg p$

$\models q /\rightarrow p$

ii) **Positive domains:** If $p \rightarrow q \rightarrow r \rightarrow s$ then $p \rightarrow s$.

The proof follows directly from the elementary definition of inheritance.

iii) **Negative link:** If $p_1 \rightarrow p_2 \rightarrow p_3 \rightarrow p_4 \text{ ---- } \rightarrow p_k$ and $q_1 \rightarrow q_2 \rightarrow q_3 \text{ --- } \rightarrow q_m$

and $p_k /\rightarrow q_m$ then $p_1 / \rightarrow q_1$, provided q_1 is not a constant.

Example 8.2: This example illustrates that the /→ links have definitely some significance. The knowledge base of fig 8.8 represented in FOL is given by

  Green-bird (parrot)
  Singer-bird (cuckoo)
  Bird (Green-bird)
  Bird(Singer-bird)
  Avis (Bird)
  ¬Singer-bird (parrot)
  ¬Green-bird (cuckoo)

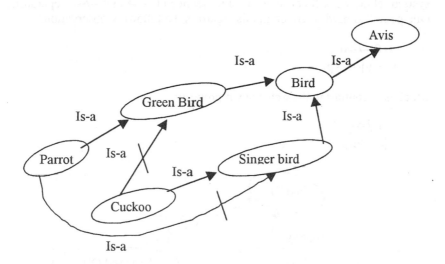

Fig. 8.8: A semantic net used to illustrate '→'and '/→' operation.

The validity of the system inferential rules (i) – (iii) is illustrated here in this example. For instance, since parrot /→ singer-bird is existent in fig 8.8, by inferential rule (i) we may add Singer-bird /→ parrot as well.

Secondly, since

  Parrot → green-bird → bird → avis,

we may add a link parrot → avis by inferential rule (ii).
Lastly, we have in fig. 8.8,

Parrot → green-bird → bird → avis
cuckoo → singer-bird → bird → avis ,
parrot /→ singer-bird and
cuckoo /→ green-bird.

Now,   suppose we add one extra rule: green-bird /→ singer-bird. Let p_2 = green-bird, q_2 = singer-bird, p_1 = parrot and q_1 = cuckoo; thus, we have

p_1 → p_2, q_1 → q_2, p_2 / → q_2.

So, by inference rule (iii), we derive: p_1 /→ q_1, i.e., parrot /→ cuckoo.

So far in our discussion, we have considered only **monotonic inheritance system**. Now, we will consider the significance of ----> and --/--> operators. One point that needs mention in this regard is the following observation:

P --/->q
$\models$ ¬ (q--/-> p)

which is in contrast to the well-known property

p /→q
$\models$ q /→p

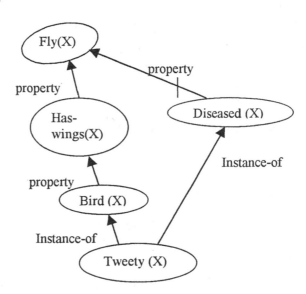

Fig. 8.9: A semantic net showing contradictions: Tweety(X) → Fly (X)
        and Tweety (X) /→ Fly (X).

A common question that may be raised is why to use --> and /--> operators, when there exist monotonic → and /→ operators. The answer to this is that the --> and -/--> operators can **avoid contradiction**, which is not possible with → and /→ .

The following example illustrates how the contradiction that arises due to use of → and /→ only can be avoided by employing ---> and -/-->.

Example 8.3: Consider the symantic net shown in fig 8.9. Here, by inheritance we have

Tweety(X) → Bird(X)→ Has-wings(X) → Fly(X) and thus

Tweety(X) → Fly(X).

Next, we have

Tweety(X) → Diseased (X) /→Fly(X)

i.e., Tweety(X) /→Fly(X) .

So, we find a contradiction:

Tweety(X)→ Fly(X) and

Tweety(X) /→Fly(X) .

This contradiction, however, can be avoided if we use ---> operator as shown in fig 8.10. Here, Has-wings(X) --> Fly(X) denotes that it is a default property that anyone having wings should fly. But the specialized property states that

Diseased(X) /→ Fly(X)

or,   Diseased(X) → ¬Fly(X).

Since specialized property should be given importance, we infer

Tweety(X) → ¬Fly(X).

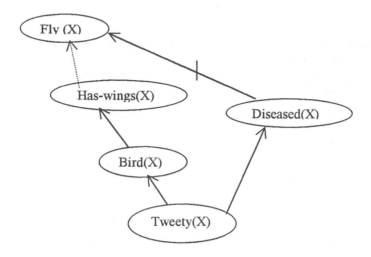

Fig .8.10: A modified semantic net that includes Has- wings(X) --->
Fly(X) instead of Has-wings(X) → Fly(X).

8.5 Defeasible Reasoning in Semantic Nets

The last example demonstrates whenever we have inconsistent inferences like
Fly(X) and ¬Fly(X), we resolve it by giving favour to the "more restrictive"
rules. But how to determine which rule is more restrictive? This section
presents a principle, by which one will be able to select a better inference,
even when the reasoning is constrained with contradictions.

Rules employing --> type of implications are called **defeasible rules**
[7]-[8]. Rules using → are called **absolute / monotonic**. Rules using -/--> are
called **'defeaters'** [2]. For reasoning with defeasible and defeater rules we
now present the following principle. We define three possible kinds of
derivations [1], 1) monotonic / absolute derivation, 2) defeasible derivation
and 3) defeating.

1) **Monotonic / absolute derivation:** If p→q is in the knowledge and p
    is a conjunction of literals $p_1, p_2, p_3, \cdots, p_n$, all of which are either
    available or absolutely derivable, then q is absolutely derivable.

Formally, $p_1, p_2, \cdots, p_n \rightarrow q$
             $P_1 \leftarrow$
             $P_2 \leftarrow$
             -
             -
             $p_n \leftarrow$

$\therefore$ q is true.

2) **Defeasible derivation**: It has already been mentioned that defeasible rules employ --> type of implication.

Given that, the knowledge base includes p -->q, where p is a conjunction of literals p_1, p_2, ---,p_n. Now **q is defeasibly derivable** if

a) q is absolutely derivable by another absolute rule r$\rightarrow$q, where r is either available in the knowledge base or absolutely derivable.

or b) p_1, p_2, ---, p_n all are **defeasibly derivable**, if p $\rightarrow$ q is not defeated, i.e., p /-->$\neg$ q is guaranteed.

or c) p --->q is not defeated, when p_1, p_2, -- ,p_n all are defeasibly derivable literals.

3) **Defeating:** An absolute rule p$\rightarrow$q can be defeated by an absolute contradictory rule or fact. A defeasible rule p--> q can also be defeated by a defeasible rule r---> $\neg$q or r -/--> q, if p is not 'more restrictive' than r. The definition of 'more restrictiveness' will be presented under c(ii).

A rule p$\rightarrow$q is defeated by (a) and (b) below, while a rule p---> q is defeated by (a), (b) and (c) below.

(a) $\neg$q is available in the knowledge base.

(b) r$\rightarrow$$\neg$q is present in the knowledge base, where r is a conjunction of defeasibly derivable literals [1].

(c) r is a conjunction of defeasibly derivable literals r_j , so that

   i)   r --> $\neg$ q or r --/--> q   is in the knowledge base.

   ii) One or more r_j is not absolutely derivable from p_1, p_2, .., p_n and the absolutely derivable rules in the knowledge base.

The principles stated above are illustrated with the following example.

Example 8. 4: Consider the following knowledge base:

Rule1: Person(X) ---> Mortal(X)
Rule2: Poet(X) -/--> Mortal(X)
Rule3: Person (tagore).
Rule4: Poet (tagore).

In the example, we illustrate the concept of defeating by c(i) and c(ii), presented above. Here rule 1 is defeated by rule 2 as c(i) and c(ii) are both satisfied. c(i) is satisfied by rule 2 of the example and c(ii) is satisfied as Poet (tagore) (= r_1, say) is not absolutely derivable from Person (tagore) (= p_1, say) from the rules in the knowledge base.

It is to be noted from this example that both the p_1, p_2, -- ,p_n and the rules in the knowledge base should be used to prove that one of the r_js is not absolutely derivable.

8.6 Frames

A frame [4] is defined as a structure that contains a number of slots, where the attributes of the frame are sorted. Usually the slots in a frame are filled with values, but it may contain a frame as well. Two important issues in a frame are i) containment and ii) specialization.

Containment: It means that a slot in the frame contains another frame. For example, the 'seat arrangement' slot of frame B contains the default frame C and the 'preferential arrangement' slot of frame C contains the default frame D (fig. 8.11). It is to be noted that default containment is denoted by ---> . A solid arrow ($\rightarrow$) , on the other hand, corresponds to specialization of frame, presented below.

Specialization: The frame B, describing an examination hall of Jadavpur University, is a specialization of the frame A, representing a generic examination hall   (fig. 8.11). Here, the frame B has some specialized properties, where it may override the elementary properties of frame A, but otherwise it will maintain the property of frame A. For instance, the slot of question paper is inherited in frame B from A, while the slot of answer scripts is specialized in B and is thus different from A.

8.7 Inheritance in Tangled Frames

In fig 8.12, we presented a hierarchical frame, where a node 'Tweety' has more than one parent. Such a directed acyclic structure is called a **tangled hierarchy**. The most important issue for tangled hierarchy is how to inherit the features of parents. For inference, should Tweety inherit the features of a Bird or 'Bird having lost aeronautical powers'?

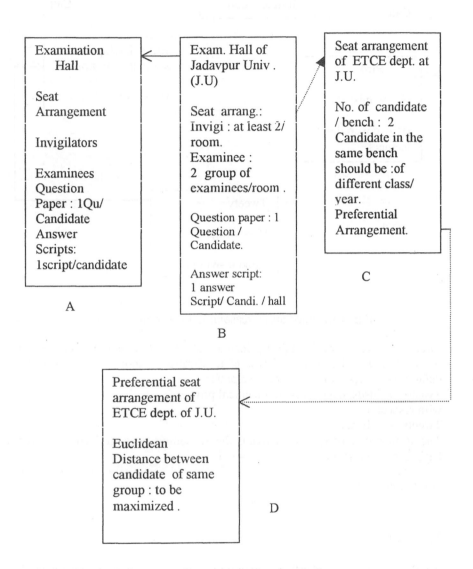

Fig. 8.11: A frame describing an examination hall.

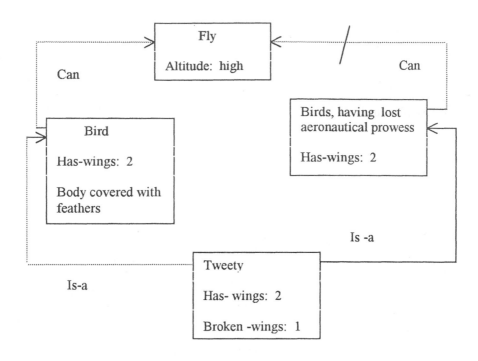

Fig. 8.12: Illustrating reasoning in a frame.

This is resolved here through a principle called "*follow minimal inferential distance path*" [11]. Since '→' type of implication is more specific than the default '--->' type of implication , we prefer
Tweety → Birds, having lost aeronautical prowess
with respect to
Tweety ---> Birds.
The former rule being predominant, the reasoning goes in favor of Birds having lost aeronautical power ----/--> Fly.

So, finally we have
        Tweety --/--> Fly as the conclusion.

8.8 Petri Nets

Another structured model that is gaining popularity currently is Petri nets. A Petri net is a directed bipartite graph consisting of places and transitions. It is getting popular for its capability of reasoning in a parallel and distributed manner. In this section we limit our discussion on FOL based reasoning with

Petri nets [5]. It, however, does not mean that Petri nets can be used as an automated tool for FOL based reasoning only. In fact, Petri nets can be used for reasoning with a rule based system or (and) can be extended to handle uncertainties modeled with stochastic and fuzzy techniques. A separate chapter on fuzzy Petri nets thus has been covered in the book, keeping in mind its increasing demand in knowledge engineering. Let us now represent the following FOL clauses by a Petri net.

FOL clauses

Rule1: Father (X,Y)→ Son (Y,X) ∨ Daughter (Y,X).
Rule2: Daughter (Y,X) ∧ Female (X) → Mother (X,Y).

The Petri net of 8.13 describes the above rules. It is to be noted that the variables are recorded against the arc. The argument of the valid predicate recorded against the arc is called an **arc function**. The input arc functions of a transition are positive, while the output arc functions are negative.

Reasoning: For reasoning, one has to assign the atomic clauses in the Petri net. For instance let us assume that we are given the following clauses:

Rule3: Father (d, r) ←
Rule4: ¬Son (r, d) ←
Rule5: ¬Mother (k, l) ←
Rule6: Female (k) ←

The argument of the above clauses along with the sign of the predicates is assigned as tokens (inside places) in the Petri net (vide fig. 8.14). The following rules of transition firing are then used to derive the resulting instances.

1. A transition is **enabled in the forward direction**, if all its input places and all but one of its output places possess properly signed tokens with **'consistent variable bindings'**. Consistent variable bindings are checked by the following procedure:

   i) The value of signed variables in the arc function is bound with the signed tokens of the associated places.

   ii) The above process is repeated for all the arc functions associated with the transition.

iii) The common value of each (sign-free) variable is then identified. The value of the set of variables, thus obtained, for each transition is called **consistent bindings.**

2. A transition is enabled in the backward direction, if all its output places and all but one of its input places possess consistent bindings.

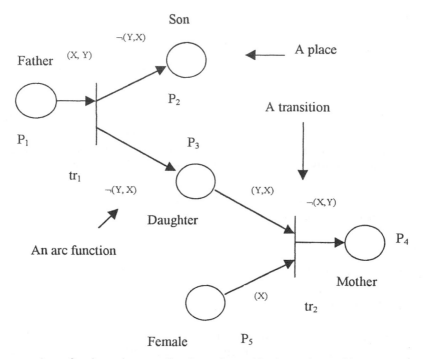

An arc function at the output (input) arc of a transition is negative (positive).

Fig. 8.13: A Petri net representing FOL clauses.

An enabled transition fires by generating tokens following the arc function but with opposite sign of the arc functions and saves it at appropriate places, associated with the transition. However, a multiple copy of the same token is not kept at a given place.

With reference to fig. 8.14, we find for transition tr_1 two set of bindings

$$X = d, Y = r \qquad \text{(see place } p_1 \text{ and the associated arc function)}$$

and $\quad \neg Y = \neg r, \neg X = \neg d \quad$ (see place p_2 and the associated arc function).

The resulting consistent binding for tr_1 is thus $X = d \ \& \ Y = r$.

Since p_1 and p_2 contain tokens and the variable bindings for transition tr_1 are consistent, tr_1 is enabled and thus fires, resulting in a new token $<r, d>$ in place p_3, following the opposite sign of the arc function $\neg(Y,X)$.

Analogously, for transition tr_2, the bound value of variables is

$\neg X= \neg k$, $\neg Y= \neg l$ (see token at place p_4 and the associated arc function)

$X= k$ (see token at place p_5 and the associated arc function)

$\therefore$ the consistent bindings are $X= k$, $Y= l$.

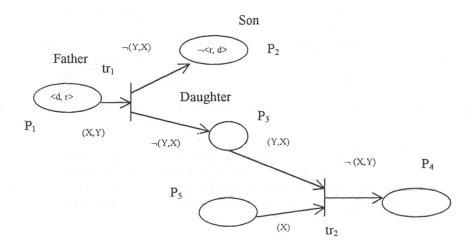

Fig. 8.14: A Petri net of fig. 8.13 with tokens.

Since the output place p_4 and one of the two input places p_5 contains consistent bindings, transition tr_2 fires, resulting in a new token at place p_3. The value of the token is $\neg<l, k>$, following the opposite sign of the arc function (Y, X). So, at place p_3 we have two resulting tokens $\neg<l, k>$ and $<r, d>$. It may be noted that these two tokens can be generated **concurrently**. Chapter 22 covers in detail the concurrent realization of the Petri net models.

8.9 Conceptual Dependency

Conceptual dependency, abbreviated as CD, is a specialized structure that describes sentences of natural languages in a symbolic manner. One significant feature of CD is its structural independence on the languages in which it is expressed. The basic difference between semantic nets and CD lies in the naming of the connecting links. In semantic nets, one may name the connecting links between events according to its relevance to the context and consequently the name differs for different users. A CD, on the other hand, requires a standard assignment of a dependence relationship and is, therefore, independent of the users. An English sentence and its corresponding CD representation is presented below to visualize the issues discussed above. Consider the sentence 'She gave me a flower'. This is represented below in fig. 8.15.

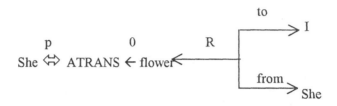

Fig 8.15: Representation of 'She gave me a flower' in CD .

In the last figure, R denotes a recipient case relation, o denotes an object case relation, p denotes past tense, ⇐ denotes a two way link between the actor and action and ATRANS stands for transfer of possession. It is called a primitive action. The set of primitive actions in CD, proposed by Schank and Abelson [10], is presented below.

Primitive actions

PTRANS: Transfer of physical location by an object (like move)

ATRANS: Transfer of abstract relationship (like give)

MOVE: Movement of a pair of one's body (like stretch)

PROPEL: Application of force to an object (like pull)

GRASP: Catching hold of an object (like clutch)

INGEST: Ingestion of food by animals

EXPEL: Expulsion of material from the physique of an animal (like cry)

MBUILD: Building information from existing information (like decide)

MTRANS: Transfer of mental ideas (like say)

ATTEND: Activating sensory organ toward stimulus (like listen)

SPEAK: Generation of sounds (like say)

For building dependency structures, we also require 4 primitive conceptual categories.

Conceptual categories:

Conceptual categories	Meaning
ACTs	Actions
PPs	Picture
PAs	Picture aiders
AAs	Action aiders

Schank defined a set of conceptual tenses also, as presented below [9].

Conceptual Tenses	Meaning
p	past
f	future
t	transition
t_s	start transition
t_f	finish transition
k	continuing
?	interrogative
/	negative
nil	present
c	conditional
delta	timeless

Schank listed 14 typical dependencies, the details of which are explained in [9]. We here illustrate a few to make the readers aware of the concept.

In the dependencies, presented below, 'o' stands for object. The rest of the notations in the above dependencies being obvious are not discussed further.

Like any representations CDs too have merits and demerits from the point of view of reasoning. The following points support knowledge representation by CD [9].

i) A few inferential rules are required, when knowledge is represented by CD. This, however, is not feasible when other forms of knowledge representation are used.

ii) Inferences, in many circumstances, are directly available in the representation itself.

iii) The initial structure of CD that corresponds to one sentence must have many holes (gap), which will be used as an attention focusser in the program that recognizes the subsequent sentences.

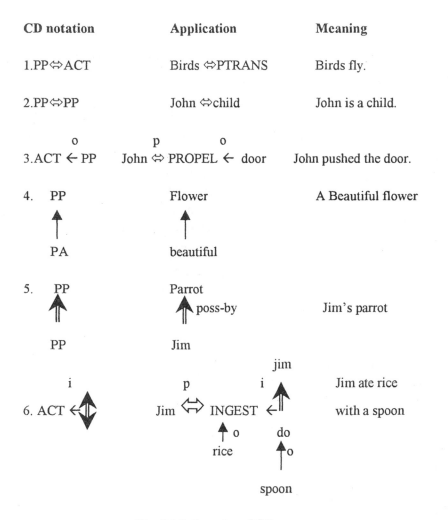

CD notation	Application	Meaning
1.PP⇔ACT	Birds ⇔PTRANS	Birds fly.
2.PP⇔PP	John ⇔child	John is a child.

Fig. 8.16: Samples of CD.

The most significant drawback of CDs is that it can merely represent the events but there exists other information in complex programs, which CDs fail to represent.

8.10 Scripts

Scripts represent stereotyped sequence of events in a particular context. For instance, it can represent scenes in a restaurant, marketplace or examination hall. A script of a restaurant includes the scenes of ordering the waiter, to bring the menu card, then ordering him to bring desired food, then taking up the food items, paying the bill and leaving the restaurant. In each scene there exists a sequence of operations. For example, when the waiter is ordered, he keeps the ordered item in his memory; then moves to the cook; the cook serves him items; he brings the items to the table, where from the order was placed and finally places the dishes on the table.

A script consists of a number of components, as defined below:

i) **Entry conditions:** The conditions that must be satisfied before the events described in the script occur.

ii) **Scenes:** The sequences of events that occur. The events are described following the formalisms of CDs.

iii) **Roles:** These are slots representing people / actors involved in the events.

iv) **Props:** Slots used for objects involved in the events of a script.

v) **Track:** It stands for 'specific variations on a more general pattern that is represented by a particular script [9]'.

vi) **Results:** The consequences after the events in the script have occurred.

A script of an examination hall is represented in fig. 8.17. These are 8 scenes altogether; each is clearly defined. The components of the scripts are illustrated in the left side margin.

Script: Example Track: Hall no.1 Props: Seats (S) Questions (Q) Answer Scripts (AS) Desk Stapler Roles: C= Candidates H= Helping Staff I = Invigilator	Scene1: Candidates entering the hall C PTRANS C into hall C ATTENDS his seat C PTRANS C towards his seat C MOVES his body part to sit
	Scene2: Invigilator enters the hall I PTRANS I into hall I ATTENDS his seat I PTRANS I towards his seat I MOVES his body part to sit
Entry Conditions: C has the seat no. in the particular hall. I has duty in the particular hall. Results: AS filled with Writing and returned. Q exhausted.	Scene 3: Answer scripts distributed I GRASPS AS I PTRANS I to each seat I MOVES hand down to place the AS to seat
	Scene 4: Question papers distributed I GRASPS Q I PTRANS I to each seat I MOVES hand down to place Q to seat
	Scene 5: Invigilator verifies candidature I PTRANS to each seat I ATTENDS AS
	Scene 6: Candidates writing and submitting scripts after writing C ATRANS AS , C GRASPS AS , C PTRANS AS to I
	Scene 7: Candidates leaving out of the hall C PTRANS C out of the hall
	Scene 8: Invigilator leaving out of the hall I PTRANS I out of the hall

Fig. 8.17: A script of an examination hall.

Once a script structure for a given context is desired, answering queries for a particular incident that can be mapped to the given script is possible.

For instance, suppose we have the examination script of fig 8.17. We are now told a story as follows:

" Jim entered the examination hall and came out with a question paper ."

Now, if we ask, "Did Jim appear at the examination?"

We immediately answer "Yes".

A program that realizes a script can also answer such queries. One important issue that we do not discuss here is that the capacity of inheritance of the slots in a script. A script is a collection of slots. So, it also supports the capability of inheritance like a frame.

8.11 Conclusions

The chapter covered a wide range of structured models, including semantic nets, frames, Petri nets, scripts and conceptual dependency graphs (CDs). Semantic nets are useful for their application in monotonic, non-monotonic and defeasible reasoning. CDs are more powerful tools for knowledge representation, but have limited use in monotonic systems. Scripts are mainly useful to represent complex scenes, which by other means are too difficult to be realizable. The work of Schank and his group in building a Script Applier Mechanism (SAM) system at Yale University, in this regard, needs special mention. The above system reads a text and reasons with it to understand stories.

Among the recently developed structured models, Petri nets are the most popular for their parallel and distributed architecture. They can also handle the inexactness of data and knowledge by fuzzy or stochastic tools. However, if someone wants to use binary logic only, then semantic nets and their inheritance in the presence of defeasible reasoning should be adopted. There exists an ample scope of work on defeasible reasoning and its realization on different structured models. A unified model that can handle all typical kinds of reasoning has yet to be developed.

Exercises

1. Represent each of the the following pieces of knowledge by a semantic net. (a) Loves (mary, john), (b) Loves (mary, john) $\land$ Hates (john, mita), (c) Loves (mary, john) $\rightarrow$Hates (mita, john).

2. Draw a partitioned semantic net to represent the knowledge: $\forall X$ Adult (X) $\rightarrow$Loves (X, children).

3. Draw a script to represent a restaurant, explaining the entry at the restaurant, ordering of items, waiting for the items, serving the items, enjoying the meals, collecting the bills for payment and exiting from the site.

4. Represent the following statements by a Petri net: (a) Graduate-Student(X) $\rightarrow$Married (X), (b) Employed (X) $\rightarrow$Married (X), (c) Married(X) $\rightarrow$ Has-son (X) $\vee$ Has-daughter (X).

5. Adding the data clauses (a) $\neg$Has-daughter (john) $\leftarrow$, (b) Graduate-student (john) $\leftarrow$, (c) $\neg$Has-son (john) $\leftarrow$ to the previous clauses, can we derive $\neg$Employed (john) and $\neg$Married (john)?– justify. Clearly show the forward and / or the backward firing of the transitions. Does more than one transition fire concurrently here?

6. Construct examples to illustrate the cases when a given rule is defeated.

References

[1]   Bender, E. A., *Mathematical Methods in Artificial Intelligence*, IEEE Computer Society Press, Los Alamitos, pp. 234-250,1996.

[2]   Cerro, L. F. del. and Penttonen, M. (Ed.), *Intensional Logics for Programming*, Clarendon Press, Oxford, 1992.

[3]   Hendrix, G. G., "Expanding the utility of semantic networks through partitioning," *Proc. of Int. Joint Conf. on Artificial Intelligence*, Tbilisi, Georgia, pp. 115-121, 1977.

[4]   Minisky, M., A framework for representing knowledge, in *The Psychology of Computer Vision*, Winston, P., Ed., McGraw-Hill, New York, 1975.

[5]   Murata, T. and Yamaguchi, H., " A Petri net with negative tokens and its application to automated reasoning," *Proc. of the 33rd Midwest Symp. on Circuits and Systems*, Calgary, Canada, pp. 762-765, 1990.

[6]   Nilsson, N. J., *Principles of Artificial Intelligence*, Springer-Verlag, Berlin, pp. 361-414, 1990.

[7]   Nute, D., *Basic Defeasible Logic*, in [2], pp. 125-154.

[8]   Pollock, J. L., "How to reason defeasibly," *Artificial Intelligence*, vol. 57, pp. 1-42, 1992.

[9]   Rich, E. and Knight, K., *Artificial Intelligence*, McGraw-Hill, New York, pp. 251-303, 1993.

[10]   Schank, R. C. and Abelson, R. P., *Scripts, Plans, Goals and Understanding*, Erlbaum, Hillsdale, NJ, 1977.

[11]   Touretzky, D., *The Mathematics of Inheritance Systems*, Morgan Kaufmann, Palo Alto, CA, 1986.

9

Dealing with Imprecision and Uncertainty

In the various methods of knowledge representation, discussed in the last few chapters, it has been presumed that the database consists of precise data elements and the knowledge base contains no uncertainty among its constituents. Such an assumption restricts the scope of application of the proposed techniques. In fact, for many real world problems, imprecision of data and uncertainty of knowledge are, by nature, part of the problem itself and continuing reasoning in their presence without proper modeling tools may lead to generating inaccurate inferences. This chapter discusses the tools and techniques required for handling the different forms of inexactness of data and knowledge. We here cover the stochastic techniques including Pearl's evidential reasoning and the Dempster-Shafer theory, the certainty factor based schemes and the fuzzy relational algebra for modeling imprecision and uncertainty of data and knowledge respectively. Examples have been given to illustrate the principles of reasoning by these techniques.

9.1 Introduction

This chapter covers various tools for reasoning in the presence of imprecision of facts and uncertainty of knowledge. Before formally presenting the techniques, we first introduce the notion of imprecision and uncertainty and their possible sources with examples. Consider a problem of medical diagnosis. Here, the pieces of knowledge in the knowledge base describe a mapping from symptom space to disease space. For example, one such piece of knowledge, represented by production rules, could be

Rule: *IF   Has-fever (Patient) AND*
Has-rash (Patient) AND
Has-high-body-ache (Patient)
THEN Bears-Typhoid (Patient).

Now, knowing that the patient has fever, rash and high body pain, if the diagnostic system infers that the patient is suffering from typhoid, then the diagnosis may be the correct one (if not less) in every hundred cases. A question then naturally arises: is the knowledge base incomplete? If so, why don't we make it complete? In fact, the above piece of knowledge suffers from the two most common forms of incompleteness: firstly, there is a scope of many diseases with the same symptoms, and secondly, the degree or level of the symptoms is absent from the knowledge. To overcome the first problem, the knowledge engineer should design the knowledge base with more specific rules (i.e., rules with maximum number of antecedent clauses, as far as practicable; see the specificity property in chapter 3). For rules with identical symptoms (antecedent clauses), some sort of measures of coupling between the antecedent and the consequent clauses are to be devised. This measure may represent the likelihood of the disease, as depicted in the rule among its competitive disease space. Selection of the criteria for this coupling may include many issues. For example, if the diseases are seasonal, then the disease associated with the most appropriate season may be given a higher weightage, which, in some ways should be reflected in the measure of coupling. The second problem, however, is more complex because the setting of the threshold level at the symptoms to represent their strength is difficult even for expert doctors. In fact, the doctors generally diagnose a disease from the relative strength of the symptoms but quantification of the relative levels remains a far cry to date. Besides the above two problems of incompleteness / inexactness of knowledge, the database too suffers from the following kinds of problems. Firstly, due to inappropriate reporting of facts (data) the inferences may be erroneous. The inappropriate reporting includes omission of facts, inclusion of non-happened fictitious data, and co-existence of inconsistent data, collected from multiple sources. Secondly, the level or strength of the facts submitted may not conform to their actual strength of happening, either due to media noise of the communicating sources of data or incapability of the sources to judge the correct level/ strength of the facts. Further, the observed data in many circumstances do not tally with the antecedent clauses of the knowledge base. For example, consider the following piece of knowledge and the observed data. The degree of the data is quantified by the adjective "very". How can we design an inference engine that will match such data with clauses of the knowledge to infer: IS-Very-Ripe (banana)?

Rule: *IF IS-Yellow (banana)*

  THEN IS-Ripe(banana).

Data: *IS-Very-Yellow (banana).*

Inference: *IS-Very-Ripe (banana).*

The last problem discussed is an important issue, which will be analyzed in detail in this chapter.

For reasoning in an expert system in the presence of the above forms of inexactness of data and knowledge, the following methodologies are presented in order.

1)   **Probabilistic techniques**, which have been developed by extending the classical Bayes' theorem for application to a specialized network model of knowledge, representing the cause-effect relationship [4] among the evidences. An alternative form of probabilistic reasoning using the Dempster-Shafer theory is also included in this section.

2)   **Certainty factor-based reasoning** is one of the oldest techniques for reasoning, used in MYCIN experiments of Stanford University. This is a semi-probabilistic approach, where the formalisms are defined in a more or less ad hoc basis, which do not conform to the notion of probability.

3)   **Fuzzy techniques**, which are comparatively new techniques, based on the definition of Fuzzy sets and Logic, proposed by Zadeh. Both Probabilistic and Fuzzy techniques are stochastic in the sense that each variable in the system has a finite valuation space and for each value of the variable, we attach a probability or fuzzy membership value (to be discussed shortly).

9.2   Probabilistic Reasoning

Probabilistic techniques, generally, are capable of managing imprecision of data and occasionally uncertainty of knowledge. To illustrate how the degree of precision of data and certainty of knowledge can be modeled with the notion of probability, let us consider example 9.1.

Example 9.1: Consider the following production rule PR$_i$.

PR$_i$:   *{ IF   (the-observed-evidences-is-rash)   (x),*
            AND   (the-observed-evidences-is-fever)   (y),
            AND   (the-observed-evidences-is-high-pain) (z),
        THEN (the-patient-bears-German-Measles)}   (CF)

where x, y, z denote the degree of precision / beliefs / conditional probabilities that the patient bears a symptom, assuming he has German Measles. On the other hand, CF represents the degree of certainty of the rule or certainty factor/ probability that the patient bears German Measles, assuming the prior occurrence of the antecedent clauses. It is, now, clear that the same problem can be modeled by many of the existing techniques; of course, for the purpose of reasoning by a technique we need a particular kind of parameter.

9.2.1 Bayesian Reasoning

Under this context, we are supposed to compute P(H$_i$ / E$_j$) or P (H$_i$ / E$_1$, E$_2$, ... E$_m$) where H$_i$ represent a hypothesis and E$_j$ represents observed evidences where $1 \leq j \leq m$. With respect to a medical diagnosis problem, let H$_i$ and E$_j$ denote the i-th disease and j-th symptoms respectively. It is to be noted that under Bayesian reasoning we have to compute the inverse probability [P (H$_i$ / E$_j$)], rather than the original probability, P(E$_j$ /H$_i$). Before describing the Bayesian reasoning, let us revise our knowledge on conditional probabilities.

Definition 9.1: **Conditional probability** [1] P (H / E) is given by

$$P(H / E) = P(H \cap E) / P(E) = P(H \& E) / P(E),$$

where H and E are two events and P (H $\cap$ E) denotes the joint occurrence of the events H & E. Analogously,

$$P(E / H) = P(E \& H) / P(H) .$$

Now, since P(E & H) = P(H & E), we find

$$P(E \& H) = P(E / H) . P (H) = P(H / E) .P (E)$$

$$\text{So, } P(H/E) = \frac{P(E/H) . P(H)}{P(E)} \qquad (9.1)$$

which is known as *Bayes' theorem*.

Example 9.2: Consider the following problem to illustrate the concept of joint probability of mutually independent events and dependent events.

A box contains 10 screws, out of which 3 are defective. Two screws are drawn at random. Find the probability that none of the two screws are defective using a) sampling with replacement and b) sampling without replacement.

Let A: First drawn screw is non-defective.

B: Second drawn screw is non-defective.

a) Sampling with replacement

P(A) = 7/10

P(B) = 7/10

P(A ∩ B)= P(A).P(B) = 0.49

b) Sampling without replacement

P(A) = 7/10

P(B/A) = 6 / 9 =2 / 3

P (A ∩ B) = P(A). P (B/A) = 14 / 30 = 0.47 ☐.

Now, for reasoning under uncertainty, let us concentrate on the well-known Bayes' theorem. With reference to our medical diagnosis problem,

$$P(D/S) = P(S/D).P(D)/P(S) \qquad (9.2)$$

where D and S stand for disease and symptoms respectively. Normally, P(S) is unknown. So to overcome the difficulty the following formalism is used in practice.

We compute: $P(\neg D / S) = P(S / \neg D) . P(\neg D) / P(S)$. (9.3)

Now from expressions (9.2) and (9.3) we find,

$$\frac{P(D/S)}{P(\neg D /S)} = \frac{P(S/D)}{P(S/\neg D)} \times \frac{P(D)}{P(\neg D)}$$

i.e., $O(D/S) = L(S/D) \times O(D)$ (9.4)

where O and L denotes the **odds of an event** and the **likelihood ratio** [1].

However, in some circumstances, where P (S) is known, we can directly use the Bayes' theorem.

In our formal notations, consider the set H and E to be partitioned into subsets, vide fig. 9.1.

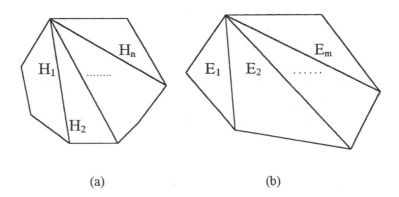

(a)   (b)

Fig. 9.1: (a) Partitioned H_j, $1 \le j \le n$ and (b) Partitioned E_i, $1 \le i \le m$.

The expression (9.1) can now be extended to take into account the joint occurrence of E_i with all H_j for $1 \le j \le n$.

$P(E_i)$
$= P(E_i \cap H_1) + P(E_i \cap H_2) + ... + P(E_i \cap H_n)$
$= P(E_i / H_1) \times P(H_1) + P(E_i / H_2) \times P(H_2) + ... + P(E_i / H_n) \times P(H_n)$

(9.5)

It may be noted that some of the probability of joint occurrences [$P (E_i \& H_j)$] may be zero. It follows from expression (9.5) and expression (9.1) that

$$P (H_j / E_i) = \frac{P (E_i / H_j) \times P (H_j)}{\sum\limits_{1 \leq k \leq n} P (E_i / H_k) \times P (H_k)} \qquad (9.6)$$

In the real world, however, the hypothesis H_j depends on a number of E_i s. Thus,

$P (H_i / E_1, E_2, ..., E_m)$

$$= \frac{P \{ (E_1, E_2,..., E_m) / H_j\} \times P (H_j)}{\sum\limits_{1 \leq k \leq n} P \{ (E_1, E_2, ..., E_m) / H_k\} \times P (H_k)} \qquad (9.7)$$

However, the conditional probability of the joint occurrences of E_1, E_2,...., E_n when H_j has happened, in many real life problems, is unknown. So E_i is considered to be statistically independent, which unfortunately is never true in practice.

When E_j s are independent, we can write

$P (E_1, E_2,..., E_m / H_j)$

$= P (E_1/ H_j) \times P (E_2 / H_j) \times ... \times P (E_m / H_j)$

$$= \prod_{i=1}^{m} P (E_i / H_j) \qquad (9.8)$$

Substituting the right-hand side of expression (9.8) in (9.7) we find,

$$P (H_j / E_1, E_2,..., E_m) = \frac{\{ \prod\limits_{i=1}^{m} P (E_j / H_i) \} \times P (H_i)}{\sum\limits_{k=1}^{n} \{ \prod\limits_{i=1}^{m} P (E_i / H_k) \} \times P (H_k)} \qquad (9.9)$$

To illustrate the reasoning process using expression (9.9), let us consider the following example.

Example 9.3: Consider the hypothesis space and the evidence space for a medical diagnosis problem. The rules, representing the cause-effect relationship between the evidence space and the hypothesis space, are also given along with the conditional probabilities.

Determine the probable disease that the patient bears.

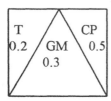 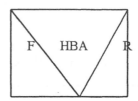

T = Typhoid   F = Fever

GM = German Measles   R = Rash

CP = Chicken Pox   HBA = High Body Ache

Hypothesis space   *Evidence Space*

Fig. 9.2: The hypothesis and the evidence space for example 9.3.

Here,

P (T) = 0.2,

P (CP) = 0.5,

P (GM) = 0.3

where P (X) denotes the probability that a patient has disease X.

Set of Rules

Rule 1: *IF symptoms are*
         F   (P (F / T) = 0.9),
         HBA (P (HBA / T) = 0.6)
       THEN the Patient hopefully bears T.

Rule 2: *IF symptoms are*
         F (P(F/ GM) = 0.8),
         R (P (R / GM) = 0.7),
         HBA (P (HBA / GM) = 0.8)
       THEN the patient hopefully bears GM.

Rule 3: *IF symptoms are*
$$F (\ P \ (F / CP) = 0.6),$$
$$R \ (P \ (R/ CP) = 0.9),$$
$$HBA \ (P \ (HBA / CP) = 0.8)$$
THEN the patient hopefully bears CP.

We compute P(CP/ R, F, HBR), P(T / F, HBA) and P(GM /R, F, HBA) and then find the highest among them and hence draw a conclusion in favor of the disease with the highest probability. Now,

$$P \ (CP /R, F, HBA) = \frac{P \ (CP) \ P \ (R/ CP) \ P \ (F / CP) \ P \ (HBA / CP)}{\underset{x \, \varepsilon \, \{CP,GM, T\}}{\sum} \ P \ (F /x) \ P \ (HBA /x) \ P \ (x) \ \{P \ (R/GM) + P(R/CP)\}}$$

(9.10)

Analogously, we can compute P(GM / R, F, HBA) and P(T / F, HBA). It is to be noted that the denominator of (9.10) is common to the three conditional probabilities. So, evaluation of their numerators only is adequate for comparison.

Now, P (CP) .P (R/CP). P (F/CP). P(HBA/CP)
  = 0.5 x 0.9 x 0.6 x 0.7 = 0.189

P (GM). P (R/GM). P (F/GM). P(HBA/GM) = 0.134

and   P (T). P (F/T) . P (HBA/T) = 0.108

From the above results it is clear that the patient bears CP. The system would respond to the user in the following manner. "*It is highly probable that the patient bears CP for P (CP/ conditional events) is the highest among the competitive conditional probabilities for other diseases.*"

Limitation of Bayesian reasoning

1. Since $\sum\limits_{j=1}^{n} P \ (H_j) = 1$, if a new hypothesis is discovered,

    $P(H_j)$ for all $1 \le j \ \le n+1$ then have to be redefined by an expert team.

2.   It might happen that none of the H_i has happened. This, in general, does not occur in a conventional Bayesian system, unless we intentionally put a H_k in the Hypothesis set with a given probability.

3.   That observed evidences, which are considered to be statistically independent, do not behave so in realistic situations.

9.2.2   Pearl's Scheme for Evidential Reasoning

A Bayesian belief network [2]-[3] is represented by a directed acyclic graph or tree, where the nodes denote the events and the arcs denote the cause-effect relationship between the parent and the child nodes. Each node, here, may assume a number of possible values. For instance, a node A may have n number of possible values, denoted by $A_1, A_2, ..., A_n$. For any two nodes, A and B, when there exists a dependence A→B, we assign a conditional probability matrix [P (B/A)] to the directed arc from node A to B. The element at the jth row and i^{th} column of P(B/A), denoted by $P(B_j /A_i)$, represents the conditional probability of B_j assuming the prior occurrence of A_i. This is described in fig. 9.3.

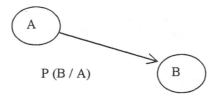

Fig. 9.3: Assigning a conditional probability matrix in the directed arc connected from A to B.

Given the probability distribution of A, denoted by $[P(A_1)\ P(A_2)\ .....P(A_n)]^T$, we can compute the probability distribution of event B by using the following expression:

$$P(B) = [P(B_1)\ P(B_2)\ ....P(B_m)]^T_{m \times 1}$$

$$= [P(B/A)]_{m \times n}\ [P(A_1)\ P(A_2)\ ....P(A_n)]^T_{n \times 1}$$

$$= [P(B/A)]_{m \times n}\ \times\ [P(A)]_{n \times 1}.$$

We now illustrate the computation of P(B) with an example.

Example 9.4: Consider a Bayesian belief tree describing the possible causes of a defective car.

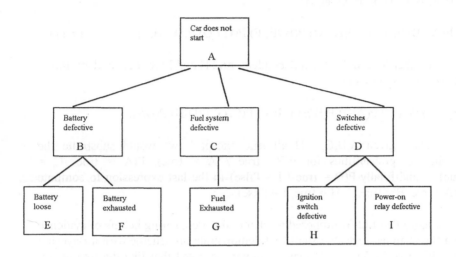

Fig. 9.4: A diagnostic tree for a car.

Here, each event in the tree (fig. 9.4) can have two possible values: true or false. Thus the matrices associated with the arcs will have dimensions (2 x 2). Now, given $P(A) = [P(A = true) \ P(A = false)]^T$, we can easily compute P(B), P(C), P(D), P(E), ...,P(I) provided we know the transition probability matrices connected with the links. As an illustrative example, we compute P(B) with P(B/A) and P(A).

$$Let \ P(A) = [P(A = true) \ P(A = false)]^T$$

$$= [\quad 0.7 \quad\quad 0.3 \quad]^T$$

$$
\begin{array}{c|cc}
 & \multicolumn{2}{c}{B_j \rightarrow} \\
A & B = true & B = false \\
\hline
P(B/A) = A = true & 0.8 & 0.2 \\
A = false & 0.4 & 0.6 \\
\end{array}
$$

So, $\quad P(B) = \quad P(B/A) . P(A) = [\ 0.62 \quad 0.46]^T$

One interesting property of Bayesian network is that we can compute the probability of the joint occurrence easily with the help of the topology. For instance, the probability of joint occurrence of A, B, C, D, E, F, G, H, I (see fig. 9.4) is given by

P(A, B, C, D, E, F, G, H, I)

= P(A / B).P(A / C).P(A / D).P(B /E, F).P(C / G).P(D / H, I)   (9.11)

Further, if E and F are independent, and H and I are independent, the above result reduces to

P(B/A).P(A/C).P(A/D).P(B/E).P(B/F).P(C/G).P(D/H).P(D/I).

Thus, given A,B,C,…,H all true except I, we would substitute the conditional probabilities for P(B= true / A = true), P(A = true /C = true)…..and finally P(D = true / I = false) in the last expression to compute P(A= true, B = true,….H = true, I = false).

Judea Pearl [2-3] proposed a scheme for propagating beliefs of evidence in a Bayesian network. We shall first demonstrate his scheme with a Bayesian tree like that in fig. 9.4. It may, however, be noted that like the tree of fig. 9.4, each variable, say A,B,…., need not have only two possible values. For example, if a node in a tree denotes German Measles (GM), it could have three possible values like severe-GM, little-GM, moderate-GM.

In Pearl's scheme for evidential reasoning, he considered both the causal effect and the diagnostic effect to compute the **belief function** at a given node in the Bayesian belief tree. For computing belief at a node, say V, he partitioned the tree into two parts: i) the subtree rooted at V and ii) the rest of the tree. Let us denote the subset of the evidence, residing at the subtree of V by e_v^- and the subset of the evidence from the rest of the tree by e_v^+. We denote the belief function of the node V by Bel(V), where it is defined as

$$Bel\ (V) = P\ (V/e_v^+, e_v^-)$$

$$= P\ (e_v^-/V).P(V/e_v^+)\ /\alpha$$

$$= \lambda(V)\ \Pi\ (V)/\alpha \qquad\qquad (9.12)$$

where, $\lambda\ (V) = P(e_v^-/V),$

$\Pi(V) = P(V/e_v^+),$ $\left.\begin{array}{c} \\ \\ \end{array}\right\}$ (9.13) and α is a normalizing constant, determined by

$$\alpha = \sum_{v \in (\text{true, false})} P(e_v^-/V) \cdot P(V/e_v^+) \qquad (9.14)$$

It seems from the last expression that v could assume only two values: true and false. It is just an illustrative notation. In fact, v can have a number of possible values.

Let node V have n offsprings, vide fig. 9.5. For computing $\lambda(V)$, we divide e_v^- into n disjoint subsets $e_{Z\,i}$, $1 \le i \le n$, where Zi is a child of V.

So, $\lambda(V) = P(e_v^-/V)$.

$$= P(e_{Z1}^-, e_{Z2}^-, ..., e_{Zn}^- / V)$$

$$= P(e_{Z1}^-/V) \cdot P(e_{Z2}^-/V) .... P(e_{Zn}^-/V).$$

$$= \Pi^n_{i=1} \lambda_{Zi} \cdot (V) \qquad (9.15)$$

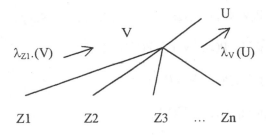

Fig. 9.5: Propagation of λs from the children to the parent in an illustrative tree.

We now compute $\Pi(V)$ using the message $\Pi_v(U) = P(U|e_v^+)$ from the parent U of V.

$$\Pi(V) = P(U|e_v^+)$$

$$= \sum_{u \in (\text{true, false})} P(V | e_v^+, U = u) \; P(U = u | e_v^+)$$

$$= \sum_{u \in (\text{true, false})} P(V | U = u) \cdot P(U = u | e_v^+)$$

$$= \sum_{u \in (\text{true, false})} P(V | U = u) \cdot \Pi_v(U = u)$$

$$= [P(V|U)]^T_{2 \times 2} \; \times \; [\; \Pi_v(0) \quad \Pi_v(1)]^T_{2 \times 1} \qquad (9.16)$$

We now compute the messages that node V sends to its parents U and each of its children Z_1, Z_2, ...,Z_n to update their values. Each of these two messages is a conditional probability, given that the condition holds and the probability given that it does not.

Now, the message from V to parent U, denoted by λ_v (U), is computed as

$$\lambda_v(U) = \Sigma_{v \,\in(\text{true, false})}\ P(e_v^-|U, V = v)\ P(V = v \mid U)$$

$$= \Sigma_{v\in(\text{true,false})}\ P(e_v^-|,V= v)\ P\ (V = v\ |U)$$

$$= \Sigma_{v\,\in(\text{true,false})}\ P(V = v \mid U)\ \lambda\ (V = v)$$

$$= [P(V \mid U)]_{2\times 2}\ \text{x}\ [\lambda(0) \qquad \lambda(1)]^T_{2\times 1} \qquad (9.17)$$

Lastly, the message from V to its child Z_j is given by

$$\Pi_{Zj}\,(V)$$
$$= P(V \mid e_{Zi}^+)$$

$$= P(V \mid e_v^+, e_{Z1}^-, e_{Z2}^-, ..., e_{Zi\text{-}i}^-, e_{Zi+i}^-, ..., e_{Zn}^-)$$

$$= \beta\ \Pi_{j\neq i}\ P(e_{Zi}^- \mid V, e_v^+)\ P(V \mid e_v^+)$$

$$= \beta\ \Pi_{j\neq i}P(e_{Zi}^- \mid V)\ P(V \mid e_v^+)$$

$$=\beta(\ \Pi_{j\neq i}\lambda_{Zi}\,(V))\ \Pi(V)$$

$$=\beta(\lambda(V)/\ \lambda_{Zj}(V))\ \Pi(V)$$

$$=\beta\ \text{Bel}\ (V)/\ \lambda_{Zj}(V) \qquad\qquad (9.18)$$

where β is a normalizing constant computed similarly as α.

The belief updating process at a given node B (in fig. 9.4) has been illustrated based on the above expressions for computing the λ and Π messages. We here assumed that at each node and link of the tree (fig. 9.4) we have one processor [7]. We call these node and link processor respectively. The functions of the node and the link processors are described in fig. 9.6.

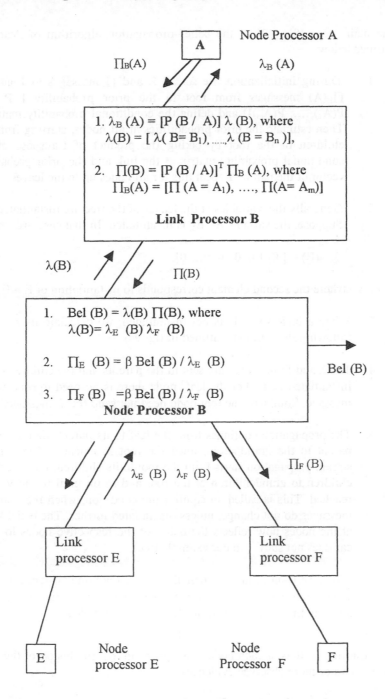

Fig. 9.6: The computation and propagation of λ and Π messages from and to node B of fig. 9.4.

The main steps [8], [12] of the belief–propagation algorithm of Pearl are outlined below.

1. During initialization, we set all λ and Π messages to 1 and set $\Pi_B(A)$ messages from root to the prior probability $[\ P\ (A_1)\ P(A_2),...., P(A_m)]^T$ and define the conditional probability matrices. Then estimate the prior probabilities at all nodes, starting from the children of the root by taking the product of transpose of the conditional probability matrix at the link and the prior probability vector of the parent. Repeat this for all nodes up to the leaves.

2. Generally the variables at the leaves of the tree are instantiated. Suppose, the variable $E = E_2$ is instantiated. In that case, we set [7]

$$\lambda_E\ (B) = [\ 0\ \ 1\ 0\ 0\ \ 0\ 0...\ 0],$$

where the second element corresponds to instantiation of $E = E_2$.

3. When a node variable is not instantiated, we calculate its λ values following the formula, outlined in fig. 9.6.

4. The λ and Π messages are sent to the parents and the children of the instantiated node. For the leaf node there is no need to send the Π message. Similarly, the root node need not send the λ message.

5. The propagation continues from the leaf to its parent, then from the parent to the grandparent, until the root is reached. Then down stream propagation starts from the root to its children, then from the children to grandchildren of the root and so on until the leaves are reached. This is called an equilibrium condition, when the λ and Π messages do not change, unless instantiated further. The belief value at the nodes now reflects the belief of the respective nodes for 'the car does not start' (in our example tree).

6. When we want to fuse the beliefs of more than one evidence, we can submit the corresponding λ messages at the respective leaves one after another, and repeat from step 3, otherwise stop.

The resulting beliefs at each node now appear to be the fusion of the joint effect of two or more observed evidences.

We presented Pearl's scheme for evidential reasoning for a tree structure only. However, the belief propagation scheme of Pearl can also be extended to **polytrees**, i.e., graphs where the nodes can have more than one parent, but there must be a single arc between each parent to a child and the graph should not have any cycles [12]. We do not derive the formula for belief propagation here, but only state it and illustrate with an example.

9.2.3 Pearl's Belief Propagation Scheme on a Polytree

Let U and V be predecessors of node X, and Y and Z are the successors of node X, as shown in fig. 9.7. Here, we denote the value of a variable, say V, by lower case notations, say v. Let P(x/ u, v) be the fixed conditional probability matrix that relates the variable to its parents u and v. Let Π_X (u) be the current strength of the causal support, contributed by U to X. Let $\lambda_Y(x)$ be the current strength of the diagnostic support contributed by Y to X. Causal support represents evidence propagating forward from parents to children, while diagnostic support represents feedback from children to their parents.

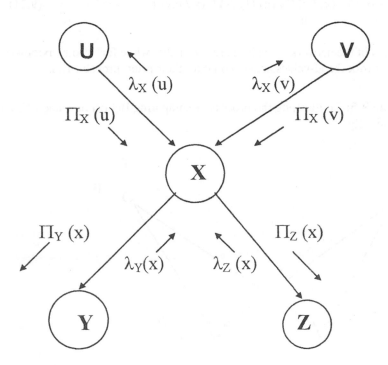

Fig. 9.7: Propagation of belief through a piece of belief network.

Updating a node X thus involves updating not only its belief function (Bel (x)) but also its λ and Π functions. Belief updating is carried out by the following formula.

$$\text{Bel}(x) = \alpha\ \lambda_Y(x)\ \lambda_Z(x)\ \sum_{u,v} P(x/u,v)\ \Pi_X(u)\ \Pi_X(v) \tag{9.19}$$

where α is a normalizing constant that makes $\sum_{\forall x} \text{Bel}(x) = 1$.

The process of λ and Π updating is now presented below with reference to fig. 9.7.

$$\lambda_X(u) = \alpha\ \sum_v [\Pi_X(v)\ \sum_x [\lambda_Y(x)\ \lambda_Z(x)\ P(x/u,v)\,]\,] \tag{9.20}$$

$$\Pi_Y(x) = \alpha\ \lambda_Z(x)\ [\ \sum_{u\ v} \Pi_X(u)\ \Pi_X(v)\ P(x/u,v)\,] \tag{9.21}$$

For leaves in the network, λ-values are set to one, while for roots (axioms) in the network the Π-values are set equal to their prior probabilities.

Example 9.5: To illustrate the process of computing Bel (x) at a node X, let us consider fig. 9.8.

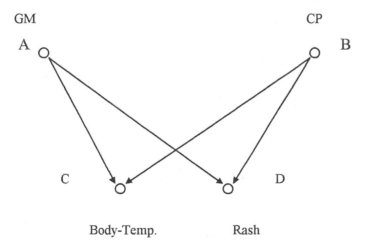

GM   CP

A   B

C   Body-Temp.   Rash   D

Fig. 9.8: A causal network, representing hypothesis (disease) and evidence (symptom) relationship.

Let the possible value of the hypothesis and evidences be as follows.

GM = {high-GM, low-GM}

CP = {high-CP, little-CP}

Body-Temp = { BT <= 98°F, BT > = 100° F}
Rash = {round, oval-shaped}

The matrices that are associated with the links (arcs) of fig. 9.7 are presented below.

Let M_{AC} =

	i → BT≤ 98° F	BT≥ 100° F	
↓ j			
high-GM	0.2	0.8	
low-GM	0.7	0.3	= P (C /A)

M_{BC} =

	i→ BT ≤ 98° F	BT ≥ 100° F	
j ↓			
High-CP	0.3	0.7	
Low-CP	0.6	0.4	= P(C/B)

M_{AD} =

	i → round	oval-shaped	
j ↓			
high-GM	0.9	0.1	
low-GM	0.8	0.2	= P (D/A)

and M_{BD} =

i→	round	oval-shaped
j ↓		
high-CP	0.3	0.7
low-CP	0.4	0.6

$$= P(D/B)$$

Suppose, we are interested to compute:

Bel (BT ≤ 98° F)

and Bel (BT ≥100° F).

Now, with reference to Pearl's nomenclature, we thus assume the following items in fig. 9.8:

Π_{BT} (high-GM) = 0.6

Π_{BT} (low-GM) = 0.1

Π_{BT} (high-CP) = 0.25 and

Π_{BT} (low-CP) = 0.05

Here, λ_Y (body-temp) =1.0 and

λ_Z (body-temp) = 1.0

Since body-temp (C) is a terminal node in fig, 9.8. the λ values incoming to it should be unity. Thus by expression (9.19), we find

Unnormalized Bel (BT ≤ 98° F)

= Π_{BT} (high-GM) x Π_{BT} (high-CP) x P (BT ≤ 98° F/ high-GM, high-CP)
+ Π_{BT} (high-GM) x Π_{BT} (low-CP) x P (BT ≤98° F / high-GM, low-CP)
+ Π_{BT} (low-GM) x Π_{BT} (high-CP) x P (BT ≤ 98° F / low-GM, high-CP)
+Π_{BT} (low-GM) x Π_{BT} (low-CP) x P (BT ≤ 98° F / low-GM, low-CP)

= (0.6x 0.25x 0.2x 0.3) + (0.6x 0.05x 0.2x 0.6) + (0.1x 0.25x 0.7x 0.3) + (0.1x 0.05x 0.7x 0.6)

= 0.01995

Suppose analogously, we find Bel (BT≥100° F) = β (say).

Then α = 1 / (0.01995 + β).

So, normalized Bel (BT ≤ 98° F) = α x 0.01995

and normalized Bel (BT ≥ 100° F) = α x β

The λ and Π messages can also be calculated by the formulas supplied. According to Pearl [2], the belief computation in the polytree is done in an asynchronous manner, and at some point of time, the beliefs at all nodes do not change. We call it an equilibrium condition. The belief of the nodes in the polytree at this condition is consistent with the theory of probability.

9.2.4   Dempster-Shafer Theory for Uncertainty Management

The Bayesian formalism assigns a positive belief to a proposition, but it does not take into account of the disbelief of the propositions. Dempster-Shafer (DS) theory, on the other hand, allows information integration by considering both their belief and disbelief. To illustrate this point, let us consider an example. Suppose that one of the three terrorist groups: A, B and C planted a bomb in an office building in a country. Further, suppose, we have adequate evidence to believe that group C is the guilty one with a measure of belief P(C) = 0.8, say. On the other hand, without any additional knowledge / fact, we do not like to say that P(B)+ P(A) = 0.2. Unfortunately, we are forced to say so using conventional probability theory as it presumes P(¬C) = 1- P(C) =P(B) + P(A). This prompted Dempster and his follower Shafer to develop a new theory, well known as the DS theory in the AI community.

In the DS theory, we often use a term, **frame of discernment** (FOD) θ.To illustrate this, let us consider an example of rolling a die. In rolling a die, the set of outcomes could be described by a statement of the form: "the-number-showing-is-i" for $1 \leq i \leq 6$. The frame of discernment in the die example is given by

FOD θ = { 1, 2, 3, 4, 5, 6 }.

Formally, the set of all possible outcomes in a random experiment is called the frame of discernment. Let $n = |\theta|$, the cardinality of θ. Then all the 2^n subsets of theta are called the propositions in the present context. In the die example, the proposition, "the-no-showing-i-is-even" is given by $\{2, 4, 6\}$.

In the DS theory, the probability masses are assigned to subsets of θ, unlike Bayesian theory, where probability mass can be assigned to individual elements (singleton subsets). When a knowledge-source of evidence assigns probability masses to the propositions, represented by subsets of θ, the resulting function is called a **basic probability assignment** (BPA).

Formally, a BPA is m

where $m : 2^\theta \rightarrow [0,1]$

where $0 \leq m(.) \leq 1.0, \; m(\phi) = 0$

and $\sum_{x \subseteq \theta} m(x) = 1.0$ (9.22)

Definition 9.2: Subsets of θ, which are assigned nonzero probability masses are called **focal elements** of θ.

Definition 9.3: A **belief function** [5-6] Bel (x), over θ, is defined by

$$Bel(x) = \sum_{Y \subseteq X} m(Y)$$ (9.23)

For example, if the frame of discernment θ contains mutually exclusive subsets A, C and D, then

Bel $(\{ A, C, D \})$

$= m(\{A,C,D\}) + m(\{A, C\}) + m(\{A,D\}) + m(\{C,D\}) + m(\{a\}) + m(\{c\}) + m(\{d\})$.

In DS model, belief in a proposition is represented by the belief interval. This is the unit interval $[0,1]$, further demarcated by two points j and k, $k \geq j$. Suppose that the belief interval describes proposition A. Then the sub-interval $[o, j)$ is called Belief (A) and the subinterval $(k,1]$ is called the disbelief (A) and the remainder $[j, k]$ is called Uncertainty (A). **Belief (A)** is the degree to which the current evidence supports A, the **Disbelief (A)** is the degree to which the current evidence supports ¬ A and **Uncertainty (A)** is the degree to which we believe nothing one way or the other about proposition A. As new evidences are collected, the remaining uncertainty will decrease [5], and each piece of length that it loses will be given to Belief (A), or Disbelief (A). The concept can be best described by fig. 9.9. We denote Belief (A), Disbelief (A), Plausibility (A) and Uncertainty (A) by Bel (A), Disbel (A), Pl (A) and U (A) respectively.

Further, Pl (A) = Bel (A) + U (A)
and D (A) = Disbel (A) + U (A).

It can be easily shown that

i)   Pl (A) ≥ Bel (A)
ii) Pl (A) + Pl (¬ A) ≥ 1
iii) Bel (A) + Bel (¬ A) ≤1.

Further, for A being a subset of B,

Bel (A) ≤ Bel (B) and
 Pl (A) ≤ Pl (B).

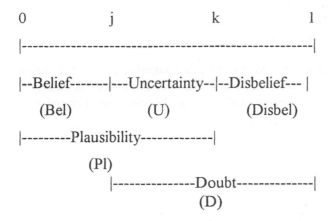

Fig. 9.9: The belief interval.

The orthogonal summation of belief functions

Assume that two knowledge sources KB1 and KB2 submit two frames of discerrnments θ_1 and θ_2 respectively. Let m_1 (.) and m_2 (.) be the BPA at the subsets of θ_1 and θ_2 respectively. The new BPA, m (.) can be computed based on m_1 (.) and m_2 (.) by using

$$m(X) = K \sum_{X = X_i \cap X_j} m_1(X_i) \cdot m_2(X_j) \qquad (9.24)$$

and $\quad K = 1 - \sum_{X_i \cap X_j = \phi} m_1(X_i) \cdot m_2(X_j)$

where X_i and X_j are focal elements of θ_1 and θ_2 respectively. We denote the **orthogonal summation operation**, referred to above, by $\quad m = m_1 \oplus m_2$.

To illustrate the orthogonal summation process, let us consider the BPAs that are assigned by two knowledge sources through a image recognition process.

Let us assume that knowledge source 1 (KS1) claims that an unknown object in a scene could be

a chair with $m_1(\{C\}) = 0.3$,
a table with $m_1(\{T\}) = 0.1$,
a desk with $m_1(\{D\}) = 0.1$,
a window with $m_1(\{w\}) = 0.15$,
a person with $m_1(\{P\}) = 0.05$,
and the frame θ, with $m_1(\{\theta\}) = 0.3$.

The assignment of BPA = 0.3 to θ means that knowledge source 1 knows that something in θ has occurred, but it does not know what it exactly is. Analogously, knowledge source 2 (KS2) claims the same object in the scene to be

a chair with $m_2(\{C\}) = 0.2$,
a table with $m_2(\{T\}) = 0.05$,
a desk with $m_2(\{D\}) = 0.25$,
a window with $m_2(\{W\}) = 0.1$,
a person with $m_2(\{P\}) = 0.2$,
and the frame θ with $m_2(\{\theta\}) = 0.2$

Now, suppose, we are interested to compute "What is the composite belief of the object to be a chair ?"

To compute this we construct the following table.

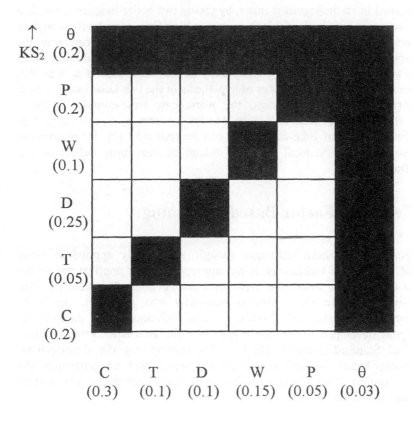

$$KS_1 \rightarrow$$

Fig. 9.10: Illustrating the principle of orthogonal summation.

Now, m_{12} ({C})

$$= \frac{m_1(\{C\}).\, m_2(\{C\}) + m_1(\{\theta\}).\, m_2(\{C\}) + m_1(\{C\}).\, m_2(\{\theta\})}{\text{Sum of the area of the shaded blocks}}$$

$$= \frac{0.2 \times 0.3\ + 0.2 \times 0.3 + 0.2 \times\ 0.3}{0.555}$$

$= 0.32$

i.e., $Bel_1(\,C) + Bel_2\,(C\,) = 0.32$.

The orthogonal summation operations of more than two belief functions can be computed in an analogous manner, by taking two belief functions, one at a time. The major drawback of this technique is high time-complexity, which in the worst case may be as high as $p_1 \times p_2$, where p_1 and p_2 represent the hypothesis space of the two sources of evidences. Thus for combining belief from n sources, the overall time-complexity in the worst case is $p_1 \times p_2 \times .... p_n$, where p_i represents the number of hypothesis in the i-th knowledge source. Summarizing, the above concept, the worst case time-complexity for n composition of beliefs from n sources is $O(p^n)$, where $p_1 = p_2 = ... p_n = p$, say. This exponential time-complexity can be reduced [7], by performing belief combinations on local families, instead of combining beliefs on the entire frames.

9.3 Certainty Factor Based Reasoning

The Bayesian reasoning technique, though successfully applied in many areas of science and technology, is not appropriate for applications in the domain of problems, where the hypotheses are not mutually exclusive. An alternative technique for evidential reasoning was, therefore, needed to meet the crisis. In the early 1970's, a new technique based on certainty factors was developed under the aegis of the Heuristic Programming Project of Stanford University [9], [11]. The context was the development of computer-based medical consultation systems and in particular the MYCIN project [94] which was concerned with replicating a consultant in the anti-microbial therapy.

'Certainty factor' (CF) in the treatises [9-10] was considered to be associated with a given priori hypothesis. This factor ranges from -1, representing the statement 'believed to be wholly untrue', to +1, representing the statement 'believed to be wholly true'. Further, there is no assumption like $\sum CF(i) = 1$ for i= 1 to n, where n is the number of the hypotheses. Thus the method is not in any sense probabilistic in its origin or basis. The CF itself is computed as the difference between two measures: the current measure of belief (MB) and the current measure of disbelief (MD):

$$CF(H:E) = MB(H:E) - MD(H:E)$$

for each hypothesis H, given evidence E.

The belief and disbelief measures both range from 0 to 1. The belief updating of a hypothesis supported by evidences E1 and E2, as reported in the literature [95] is given by

MB (H : E1,E2) = MB (H : E1) + [MB (H:E2) * {1 - MB (H :E1)}]

= MB (H: E1) + MB (H:E2) -MB (H:E1)* MB (H:E2).

This formula has a number of pragmatic attractions:

i)   It is symmetric with respect to the accrual of evidence from different sources. It does not matter whether we discover evidence E1 or E2 first.

ii)   It is a cumulative measure of beliefs for different evidences which confirm the hypothesis and, therefore, accords both with intuition and information theory.

We do not discuss much of certainty factor based reasoning as it is obsolete nowadays. Interested readers may get it in any textbook [12], [1] or in Shortliffe's original works [9].

9.4 Fuzzy Reasoning

Fuzzy sets and logic is a relatively new discipline that has proved itself successful in automated reasoning of expert systems. It is a vast area of modern research in Artificial Intelligence. In this section, we briefly outline this discipline and illustrate its application in reasoning with inexact data and incomplete knowledge.

9.4.1 Fuzzy Sets

In conventional set theory an element (object) of a universal set U may (or may not) belong to a given set S. In other words, the degree of membership of an object in set S is either zero or one. As an example, let us consider the set S of positive integers, formally defined as

S = { s : s = positive integer}.

Since the definition of positive integer is very clear, there exists no doubt to identify which elements of the universal set of numbers U belong to this set S. In case the universal set contains only positive and negative integers, then we can definitely say that elements like -1, -2,....up to - ∞, all belong to set S with a membership value zero, while the elements 0, +1, +2,...to + ∞ belong to set S with membership value one. We can express this as follows.

$$S = \{ \; 0 \; / \; 1.0, \; +1 \; / \; 1.0, \; +2 \; / \; 1.0, \; +3 \; / \; 1.0, \; ..., \; + \infty \; / \; 1.0,$$
$$-1 \; / \; 0.0, \; -2 \; / \; 0.0, \; -3 \; / \; 0.0, \; ..., \; - \infty \; / \; 0.0 \; \}$$

where b in (a / b) form in set S represents the degree of membership of "a". Unlike such sets where membership values could be either zero or one, **fuzzy sets** represent sets, whose elements can possess degree of membership lying in the closed interval of [0,1]. As an example, let us consider a set named AGE which has a range from (0 - 120) years. Now, suppose one assumes the age of a person by observation, since he (she) does not have a proof of age. We may classify the person under subset: Young with certain degree of membership, Old with other membership, Very-Old with a third membership value. For example, if the age of the person seems to be between (20-22), say, then he (she) is called young with a degree of membership = 0.9, say, old with a degree of membership = 0.4, say, and very-old with a degree of membership = 0.2, say. It is to be noted that the sum of these three membership values need not be one. Now, assume that in the universal set U we have only four ages: 10, 20, 30, 40. Under this circumstance, subsets Young, Old and Very-Old might take the following form.

Young = { 10 / 0.1, 20 / 0.9, 30/ 0.5 , 40 / 0.3}

Old = { 10 / 0.01, 20 / 0.3, 30 / 0.9, 40 / 0.95 }

Very-Old = { 10 / 0.01, 20 / 0.1, 30 / 0.7, 40/ 0.9}

A question may be raised as to how to get the membership value of the persons. To compute these from their respective ages, one can use the **membership distribution curves** [14], generated intuitively from the commonsense knowledge (fig. 9.11).

To represent the membership value of an object u in set (subset) A, we use the notation: μ_A (u). As an example, the membership value of a person having age = 80 to belong to subset very-old = 0.7 can be represented as

μ Very-old (age = 80) = 0.7.

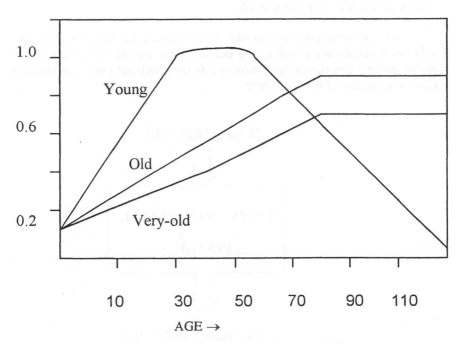

Fig. 9.11: Membership distribution for subsets of AGE.

9.4.2 Fuzzy Relations

For modeling a physical system, whose variations of output parameters with input parameters are known, one can use fuzzy relations. To illustrate this point, let us consider a relational system that relates Fast-Runners with Young by the following production rule PR1.

PR1: *IF X-is Young*

THEN X-is-a Fast-Runner.

Suppose the membership distribution of subset Young is given in the following form:

Young = {10 / 0.1, 20 / 0.6, 30 / 0.8, 40 / 0.6},

where a / b in subset Young represents Age / membership value of having that age. Further, let us consider the subset Fast-Runner, which can be stated as:

Fast-Runner = {5 / 0.1, 8 / 0.2, 10 / 0.4, 12 / 0.9}

where a / b in fast-runner subset represents [(speed of persons in meters /sec) /(membership of having that speed)].

Now, to represent a system (fig. 9.12) whose input is the membership value of a person being young, and output is the membership value of the person being a fast-runner, we construct a fuzzy relational matrix by taking a Cartesian product of the two subsets.

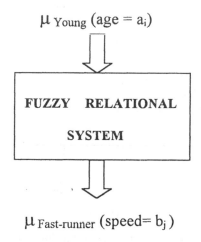

$\mu_{Young} (age = a_i)$

FUZZY   RELATIONAL

SYSTEM

$\mu_{Fast\text{-}runner} (speed = b_j)$

Fig. 9.12: A fuzzy system which relates $\mu_{Fast\text{-}runner} (speed = b_j)$
to $\mu_{Young} (age = a_i)$.

The relation obtained through a Cartesian product can be denoted as $\mu_R(age, speed)$. To illustrate the Cartesian product operation we first consider this example by forming the relational matrix [15].

u_R (age, speed) =

		Age →			
		10	20	30	40
Speed ↓					
	5	0.1*0.1	0.6*0.1	0.8*0.1	0.6*0.1
	8	0.1*0.2	0.6*0.2	0.8*0.2	0.6*0.2
	10	0.1*0.4	0.6*0.4	0.8*0.4	0.6*0.4
	12	0.1*0.9	0.6*0.9	0.8*0.9	0.6*0.9

The "*" operation in the above relational matrix could be realized by different implication functions. For example, Zadeh used fuzzy AND (MIN) operation [14] to represent the implication function. When "*" denotes fuzzy MIN operator, we can formally write it as follows:

$a_i * b_j = a_i$, if $a_i < b_j$ and
$\quad\quad = b_j$ if $b_j \leq a_i$.

Many researchers prefer a decimal multiplication operation to describe the "*" operation [15]. Polish logician Lukasiewicz described the "*" operator as follows:

$$a_i * b_j = Min\ [\ 1,\ (1- a_i + b_j\ )].$$

If we consider the "*" to be Zadeh's fuzzy AND operator, then the above relational matrix reduces to

$$
\mu_R\ (age,\ speed) =
\begin{array}{c}
\\
speed \downarrow
\end{array}
\begin{array}{c}
Age \rightarrow \\
\begin{array}{cccc}
10 & 20 & 30 & 40
\end{array}
\end{array}
$$

speed	10	20	30	40
5	0.1	0.1	0.1	0.1
8	0.1	0.2	0.2	0.2
10	0.1	0.4	0.4	0.4
12	0.1	0.6	0.8	0.6

Now, suppose that the measured distribution of a young person is as follows:
Young = { 10 / 0.01, 20 / 0.8, 30 / 0.7, 40 / 0.6}

which means that the same person has an age 10 with a membership value 0.01, age 20 with a membership value 0.8 and so on. Now, the fuzzy membership distribution of the person being a Fast-Runner can be estimated by post-multiplying the relational matrix by the Young vector. Thus,

$\mu_{Fast\text{-}runner}\ (speed = b_j)\ =\ \mu_R\ (age,\ speed)\ o\ \mu_{Young}\ (age = a_i)$

where the "o" denotes a fuzzy AND-OR composition operator, which is executed in the same way, while computing product in matrix algebra, with the replacement of addition and multiplication operators by fuzzy OR (Maximum) and AND (Minimum) operations respectively.

The estimated membership distribution in the present context becomes

$$\mu_{\text{Fast-runner}} \text{ (speed} = b_j \text{)} = \{\ 5\ /\ 0.1,\ \ 8\ /\ 0.2,\ 10\ /\ 0.4,\ 12\ /\ 0.7\}.$$

Thus the person is a fast-runner having an estimated speed of 5 m / s with a membership value of 0.1, 8 m / S with a membership value of 0.2 and so on.

9.4.3 Continuous Fuzzy Relational Systems

Consider the following set of rules, where x and y represent two variables and A_i and B_i denote fuzzy sets.

Rules:

IF x-is A_1 Then y-is B_1.

IF x-is A_2 Then y-is B_2
    ...
    ...

IF x-is A_n Then y-is B_n.

In more specific cases, x and y, for example, could be age and speed, while A_i could be fuzzy subsets like Young, Old, Very-Old, etc. and B_i could be fuzzy sets like Slow-Runner, Medium-Fast-Runner, Fast-Runner, etc. It may be noted that all rules, as stated above, are applicable to measure the membership distribution of y-is B_i, given the distributions of x-is A_i. Suppose the measured distribution of x-is A_i and y-is B_i for $1 \le i \le n$ is known. With these known distributions we can design the fuzzy relations. Then if we know the observed distribution of x-is A' we would be able to infer the distributions for y-is B_i' for $1 \le i \le n$. Then to arrive at a final decision about the distribution of y-is B', we would OR the resulting distributions of y-is B_i' for $1 \le i \le n$.

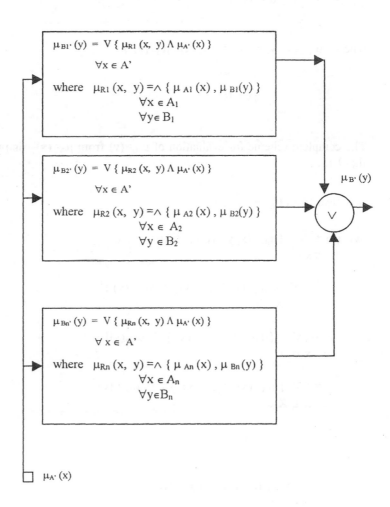

Fig. 9.13: Evaluation of $\mu_{B'}(y)$ from $\mu_{A'}(x)$.

This may be formally written as

$$\mu_{Bi'}(y) = V \quad \{ \mu_{Ri}(x, y) \wedge \mu_{A'}(x) \}$$
$$\forall x \, \epsilon \, A'$$

$$\mu_{B'}(y) = V \{ \mu_{Bi'}(y) \}$$
$$\forall i$$

where $\mu_{Ri}(x, y) = \wedge \{ \mu_{Ai}(x), \mu_{Bi}(y) \}$
$$\forall x \, \epsilon \, A_i$$
$$\forall y \, \epsilon \, B_i$$

The complete scheme for evaluation of $\mu_{B'}(y)$ from $\mu_{A'}(x)$ is presented in fig. 9.13.

It is clear from fig. 9.13 that

$$\mu_{Bi'}(y) = V \quad \{ \mu_{Ri}(x, y) \wedge \mu_{Ai}(x) \}$$
$$\forall x \in X$$

$$= V \quad [\{ \mu_{Ai}(x) \wedge \mu_{Bi}(y) \} \wedge \mu_{Ai'}(x)]$$
$$x \in X$$

$$= V \quad [\{ \mu_{Ai}(x) \wedge \mu_{Ai'}(x) \} \wedge \mu_{Bi}(y)]$$
$$x \in X$$

$$= [V \{ \mu_{Ai}(x) \wedge \mu_{Ai'}(x) \}] \wedge \mu_{Bi}(y)$$
$$x \in X$$

$$= \alpha_i \wedge \mu_{Bi}(y)$$

where $\alpha_i = V \{ \mu_{Ai}(x) \wedge \mu_{Ai'}(x) \}$.
$$x \in X$$

Finally, $\mu_{B'}(y) = V \quad \{ \mu_{Bi'}(y) \}$.
$$1 \leq i \leq n$$

The computation of $\mu_{B'}$ (y) by the above expressions was proposed by Togai and Watanabe [13], who first realized a fuzzy inference engine on a VLSI chip. The architecture of their inference engine will be discussed now.

9.4.4 Realization of Fuzzy Inference Engine on VLSI Architecture

Togai and Watanabe [13] considered a set of fuzzy production rules of the following form.

Rule: IF x-is A_i Then y-is B_i

where A_i and B_i each can assume 16 possible subsets, i.e. $1 \leq i \leq 16$. Moreover, the membership distributions μ_{Ai} (x) and μ_{Bi} (y) can assume membership values from the following set:

Membership set = { 0, 1/15, 2/15, 3/15,...., 14/15, 1},

the elements of which, if multiplied by a scale factor of 15, can be converted to 4-bit binary numbers. Thus to represent $[\mu_{A1}$ (x) μ_{A2} (x) ..... μ_{A16} (x)] they used 64 bit registers, 4-bit each for one field depicting μ_{Ai} (x), as presented in fig. 9.14.

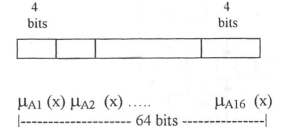

$$\mu_{A1} (x) \mu_{A2} (x) ..... \qquad \mu_{A16} (x)$$
|-------------------- 64 bits ---------------|

Fig. 9.14: The 64-bit register to hold μ_{Ai} (x).

For μ_{Bi} (y) s, $\mu_{A'}$ (x) and $u_{B'}$ (y) **s,** we also use 3 more registers, each of 64 bits. For computing α_i s, Togai and Watanabe used the computational scheme (vide fig. 9.15).The fuzzy MIN boxes in fig. 9.15 determine the minimum of the two input signals applied to them. The 4-bit shift register holds the cumulative maximum of the two successive outputs of the MIN box. The α_i s thus produced are ANDed with μ_{Bi} (y) s to yield $\mu_{Bi'}$ (y) s, which finally are ORed to produce the $\mu_{B'}$ (y) distribution. The 16-input OR function is carried out with the help of an OR tree, shown in fig. 9.16. The system implemented by Togai and Watanabe [13] has an execution speed of 80,000 fuzzy logical inferences per second, when a 20.8 M-Hz crystal is used as a basic timing unit.

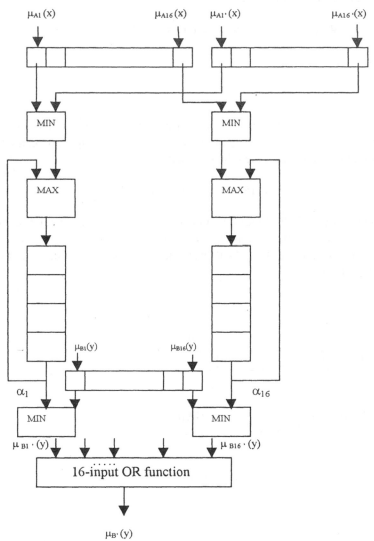

Fig. 9.15: The logic architecture of the fuzzy inference engine.

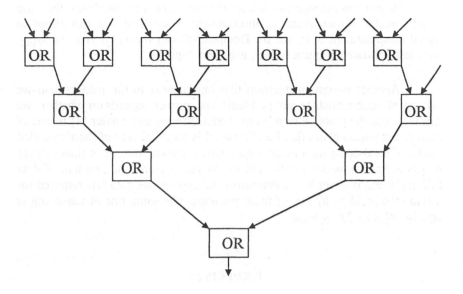

Fig. 9.16: The OR-tree realizing the 16-input OR function in the logic
architecture of fig. 9.15.

9.5 Comparison of the Proposed Models

We covered the Bayesian, Dempster-Shafer and fuzzy reasoning model based
on Zadeh's implication function and also briefly outlined the certainty factor
based model for reasoning with incomplete data and knowledge. A common
question naturally arises: which model to use when? If we prefer a
mathematically sound model then one should prefer Pearl's model or the
Dempster-Shafer theory. But it is to be noted that Pearl's model requires a
number of conditional probabilities, which are difficult to extract for most of
the real world problems. The Dempster-Shafer theory, on the other hand,
requires only the Basic Probability Assignments to the propositions, which
can be done easily by a knowledge engineer for the problems of her domain.
But due to its exponential computational costs, it cannot be used on a large
frame of discernment. The certainty factor based reasoning has no formal
mathematical basis, and thus it should not be used nowadays, when other
models are available. The fuzzy models require defining the membership
functions only, which can be done from the intuitive knowledge of the
problem domains. Further, fuzzy computation is not at all costly. So, it has a
good potential in the next generation expert systems. So, what is the rule of
selecting the models? A simple thumb rule is presented below.

"If you can manage to evaluate the conditional probabilities, then use Pearl's model. If your frame of discernment is small or you can afford to spend computational time, use the Dempster-Shafer theory. If you are happy with less accurate but quick results, use fuzzy logic."

Another important question that may appear to the readers: can we use any of these models to any problem? The answer depends on whether we can represent the problem by a given model. Shenoy and Shafer [7] in one of their recent papers claim that Pearl's model is a special case of their extended model of local computations on a qualitative Markov tree. It is thus evident that both Pearl's model and the DS theory can perform the problem of data fusion [2]. But the user has to determine the appropriate data sets required for solving the problem by any of these methods. The same line of reasoning is equally valid for fuzzy logic.

Exercises

1. List the set of parameters (conditional probabilities by name) and the inputs (a priori probabilities and λs at the instantiated leaf node) required to apply the belief propagation scheme, when node E (in fig. 9.4) is instantiated at $E = E_2$.

2. After instantiation of node E (in fig. 9.4), compute manually the Bel values, λ and Π messages at the equilibrium state, based on your initial assignments. Can you construct a program to realize this?

3. Suppose a crime was committed by either of three suspects A, B, or C. There are two knowledge sources. The knowledge source 1 submits the following data:

    $m_1\{(A)\} = 0.1$, $m_1\{(B)\}= 0.3$, $m_1\{(C)\} = 0.3$ and $m_1 \{(\theta)\} = 0.3$.

    The knowledge source 2 submits the following data:

    $m_2\{(A)\}= 0.4$, $m_2\{(B)\} = 0.2$, $m_2\{(C)\} = 0.2$ and $m_2\{(\theta)\} =0.2$.

    Find $m_{12}\{(A)\}$, $m_{12}\{(B)\}$ and $m_{12}\{(C)\}$ and hence comment on the culprit.

4. Find the relational matrices for the following systems:

a) when x, y∈{10, 20, 30} and x = y;

b) when x, y∈{10, 20, 30} and membership = 0.6, if x > y,
$$= 0.3, \text{ if } x < y \text{ and}$$
$$= 1.0, \text{ if } x = y.$$

5. Given $\mu_A (x) = [\, 0.1 \quad 0.5 \quad 0.3]$ and $\mu_B (y) = [\, 0.7 \quad 0.8 \quad 0.4]$. Compute

$R (x, y) = [\mu_A (x)]^T \circ [\mu_B (y)]$.

Now, with R (x, y) and a given $\mu_A{'} (x) = [0.3 \quad 0.4 \quad 0.7]$, find $\mu_B{'} (y)$.

6. Prove the closure of A→ B ⇔ ¬ B → ¬ A by the Lukasiewicz implication function f $(a_i, b_j) = \text{Min } [1, (1 - a_i + b_j)\,]$ for the implication rule A→ B.

[Hints: For $a_i \to b_j$, f = Min $[1, (1 - a_i + b_j)\,]$. Now, for ¬ $b_j \to ¬ \, a_i$, f = Min $[\, 1, (1-(1-b_j) + (1-a_i))] = \text{Min } [1, (1 - a_i + b_j)]$. Hence the result follows.]

References

[1] Patterson, D. W., *Introduction to Artificial Intelligence and Expert Systems*, Prentice-Hall, Englewood Cliffs, NJ, pp. 107-119, 1990.

[2] Pearl, J., "Fusion, propagation and structuring in belief networks," *Artificial Intelligence*, vol. 29, pp. 241-288, 1986.

[3] Pearl, J., "Distributed revision of composite beliefs," *Artificial Intelligence*, vol. 33, pp. 173-213, 1987.

[4] Peng, Y. and Reggia, J. A., "A probabilistic causal model for diagnostic problem solving," *IEEE Trans. on Systems, Man and Cybernetics*, SMC- 17, no. 3, pp. 395-408, May-June 1987.

[5] Shafer, G., *A Mathematical Theory of Evidence*, Princeton University Press, Princeton, NJ, 1976.

[6] Shafer, G. and Logan, R., "Implementing Dempster's rule for hierarchical evidence," *Artificial Intelligence*, vol. 33, pp. 271-298, 1987.

[7]   Shenoy, P. P. and Shafer, G., "Propagating belief functions with local computations," *IEEE Expert*, pp. 43-52, Fall 1986.

[8]   Shoham, Y., *Artificial Intelligence Techniques in PROLOG*, Morgan Kaufmann, San Mateo, CA, pp. 183-185, 1994.

[9]   Shortliffe, E. H., *Computer based medical consultations: MYCIN*, American Elsevier, New York, 1976.

[10]   Shortliffe, E. H. and Buchanan, B. G., "A model of inexact reasoning," *Mathematical Biosciences*, vol. 23, pp. 351-379, 1975.

[11]   Shortliffe, E. H., Buchanan, B. G. and Feigenbaum, E. A., "Knowledge engineering for medical decision-making: A review of computer based clinical decision aids," *Proc. IEEE*, vol. 67, pp. 1207-1224, 1979.

[12]   Stefik, M., *Introduction to Knowledge Systems*, Morgan Kaufmann, San Mateo, CA, pp. 493-499, 1995.

[13]   Togai, M. and Watanabe, H., "Expert system on a chip: An engine for real time approximate reasoning," *IEEE Expert*, pp. 55-62, Fall 1986.

[14]   Zadeh, L. A., "Outline of a new approach to the analysis of complex systems and decision processes," *IEEE Trans. Systems, Man and Cybernetics*, vol. 3, pp. 28-45, 1973.

[15]   Zimmerman, H. J., *Fuzzy Set Theory and Its Applications*, Kluwer Academic, Dordrecht, The Netherlands, pp. 131-162,1996.

10

Structured Approach to Fuzzy Reasoning

Reasoning in expert systems in the presence of imprecision and inconsistency of the database and uncertainty of a knowledge base, is itself a complex problem, and becomes intractable when the knowledge base has an explicit self-reference to itself. A self-referential knowledge base, if instantiated with appropriate data elements, gives rise to formation of cycles in the reasoning space. The chapter presents a structured approach to fuzzy reasoning using Petri nets to handle all the above problems of inexactness of data and knowledge by a unified approach.

A structural analysis of the model has been undertaken to examine its properties with respect to 'reachability of places' in the network for subsequent applications in the detection of cycles. A behavioral analysis of the model, with special reference to stability, reciprocity and duality, has also been presented. The results of the above analysis have been applied to fuzzy reasoning of a diverse nature including Modus Ponens, Modus Tollens, Abduction and non-monotonism.

10.1 Introduction

Databases associated with the real world problems are invariably contaminated with imprecision and inconsistency of data. Besides, the knowledge base is found to contain pieces of knowledge, with doubtful certainty. A number of techniques have, of late, been developed for the management of imprecision [12],[38],[2] and inconsistency [25],[4] of data and uncertainty of knowledge[2], [6], [14],[39], [28] to facilitate reasoning in expert systems (ES). However, none of them are adequate if one or more of the above forms of incompleteness of data and knowledge coexist in a given situation.

More recently, fuzzy logic has been successfully applied to a specialized structure of knowledge, called **Fuzzy Petri nets** (FPN), [1],[3]-[9], [19], [23], [37] for handling one or more of the above problems. The concept of the management of imprecision of data with FPN was pioneered by Looney [24], who considered an acyclic model of FPN, for estimating the degree of truth of a proposition with a foreknowledge of its predecessors in the network. Chen et al. [7] presented an alternative model and an interactive algorithm for reasoning in the presence of both imprecision and uncertainty. Bugarin and Barro [1] refined the underlying concept of the model in [9] and extended it in the light of classical fuzzy logic [42]. The most challenging part of their work was reasoning under incomplete specification of knowledge. Yu improved the concept of structural mapping of knowledge onto FPN [40] and presented a new formalism [41] for reasoning with a knowledge base, comprising of fuzzy predicates [42], instead of fuzzy propositions [7]. Scarpelli et al. presented new algorithms for forward [36] and backward [35] reasoning on FPN which is of much interest. A completely different type of model of FPN using fuzzy t and s norms [13] was proposed by Pedrycz [29] for applications in supervised learning problems. There exists an extensive literature on FPN models [11], [1], [3], [8], which cannot be discussed here for lack of space. However, to the best of the author's knowledge, none of the existing models of FPN can handle the complexities in a reasoning system created by the coexistence of imprecision and inconsistency of data and uncertainty of knowledge. The complexity of the reasoning system is further complicated, when the knowledge base has an explicit self-reference to itself. The chapter presents new models of FPN [20], pivoted around the work of Looney, for dealing with the above problems by a unified approach.

In this chapter a FPN has been constructed first from a set of rules and data, represented by predicates and clauses respectively. The FPN, so formed, with a self-referential knowledge base under the instantiation space of appropriate clauses, may contain cycles, which, when subjected to reasoning, may result in sustained oscillation in the level of precision of the inferences [16]. The detection of cycles in a FPN, if any, and the analysis of stability of the model, which is of paramount importance, are discussed. Among other significant issues discussed in the chapter are 'backward reasoning' and 'reciprocity under bi-directional IFF' type reasoning. Backward reasoning has been carried out with the help of a new definition of inverse fuzzy relational matrix, Q, which when pre- or post-composed with a given relational matrix, R, yields a matrix which is closest to the identity matrix, I, in a global sense. The condition of reciprocity, which ensures regaining of fuzzy tokens [26] at all places of the FPN after n-forward steps followed by n-backward steps of reasoning (and vice versa), has been derived. Since the condition of reciprocity imposes relationships between the structure of the FPN and its relational matrices, determination of the matrices for a given network topology, therefore, is a design problem. Networks whose relational matrices support reciprocity conditions can generate tokens at all places consistently, when the tokens of only a few terminal or non-terminal places are given. Such networks may ideally be used for diagnostic problems, where the tokens of the terminal places, representing measurement points, are known and the tokens of the independent starting places, representing defects, are to be evaluated. Another problem of interest, considered in the chapter, is the transformation of a given primal FPN into its dual form, using the classical modus tollens property of predicate logic. The dual FPN is useful for estimation of the degree of precision of the negated predicates, when the degree of precision of one or more negated predicates is known. Lastly, the principle of management of contradiction of data and knowledge, hereafter called non-monotonic reasoning, has been presented briefly in the chapter.

In section 10.2 of the chapter, an algorithm for formation of FPN is presented along with an algorithm for detection of cycles in a FPN with the help of reachability analysis. Section 10.3 is devoted to the state-space formulation of the model and its stability analysis. Section 10.4 includes an algorithm for forward reasoning. In section 10.5, the formulation of the backward reasoning problem along with its solution with inverse fuzzy relational matrix is presented. Reciprocity analysis under bi-directional IFF type reasoning is presented in section 10.6. Details of primal to dual transformation and its application are covered in section 10.7. The principles of non-monotonic reasoning are presented in section 10.8. The conclusions are summarized in section 10.9.

10.2 Structural Model of FPN and Reachability Analysis

In this section, an algorithm for formation of FPN from a set of database and knowledge base is presented. An analysis of reachability with special reference to detection of cycles in a FPN is also included in this section.

Definition 10.1: A FPN is a directed bipartite graph with 9 tuples, formally denoted by FPN = { P, D, N, Tr, t, th, I, O, R_i} where P ={p_1 , ,p_2 ,...., p_n } is a finite set of places, D= {d_1, d_2 ..., d_n } is a finite set of predicates, each d_i having a correspondence to each p_i for $1 \le i \le n$, N = {n_1,, n_2 ,...,n_n } is a finite set of discrete fuzzy membership distributions, called belief distributions, each distribution n_i having correspondence to each predicate d_i. Tr = {tr_1,, tr_2, ..., tr_m } is a finite set of transitions; $P \cap Tr \cap D = \varnothing$. ' t' and 'th' represent respectively sets of fuzzy truth token (FTT) distribution and thresholds, associated with each transition. I and O: Tr $\rightarrow$ P represent mapping from transitions tr_i to their input and output places. R_i , associated with each transition tr_i, represents the certainty factor (CF) of a rule: I(tr_i) $\rightarrow$O (tr_i) and is represented by a fuzzy relational matrix.

Example 10.1: To bring out the implications of the above definitions, let us consider the following production rules and database.

Production Rules (PR)

PR1: Tall(x), Stout (x) →Fast-runner (x)
PR2: Fast-runner (x) →Has-nominal-pulse-rate (x), Stout (x).

Database: *Tall (ram), Stout (ram).*

In the PR above, Tall (x), Stout (x), etc. denote predicates and the comma in the left and right hand sides of the implication sign (→) denote AND and OR operations respectively. Given the measured membership distribution of Tall(x), Stout (x) and Fast-runner (x) in PR1, one can easily construct a relational matrix, representing the CF of the rule for each possible membership values of the antecedent and consequent predicates under the rule. For example, let us consider the membership distribution of Tall (x), Stout (x) and Fast-runner (x) as shown in fig. 10.1. The relational matrix for the rule can be constructed first by ANDing the distribution of Tall (X) and Stout (x) and then by using an implication function [30-32] over the derived distribution and the distribution of Fast-runner (x).

$\mu_{tall}(x) \wedge \mu_{stout} (x)$

$= [0.2\ 0.4\ 0.6\ 0.8]^T \wedge [0.1\ 0.2\ 0.9\ 0.2\ ]^T = [\ 0.1\ 0.2\ 0.6\ 0.2]^T.$

Here T denotes the transposition operator and the '$\wedge$' operation between two vectors has been computed by taking component-wise minimum of the two vectors.

$R_1 = [\ \mu_{tall}(x) \wedge \mu_{stout}(x)]\ o\ [\mu_{fast-runner}(x)]^T$
$\quad = [0.1\ 0.2\ 0.6\ 0.2\ ]^T\ o\ [0.1\ 0.2\ 0.6\ 0.9]$

$$
= \quad
\begin{array}{r}
\\
5'\wedge 40\ \text{kg} \\
6'\wedge 50\ \text{kg} \\
7'\wedge 60\ \text{kg} \\
8'\wedge 80\ \text{kg}
\end{array}
\begin{array}{cccc}
5 & 6 & 8 & 10 \ \ \text{Speed (in m/S)} \\
0.1 & 0.1 & 0.1 & 0.1 \\
0.1 & 0.2 & 0.2 & 0.2 \\
0.1 & 0.2 & 0.6 & 0.6 \\
0.1 & 0.2 & 0.2 & 0.2
\end{array}
$$

where 'o' denotes the fuzzy AND-OR composition operator [31]. This, however, is not a unique method for estimation of R_1. In fact, R_1 can be constructed by several ways [21] by substituting appropriate operators in place of the composition operator. A set of relational matrices is thus formed, each corresponding to one rule.

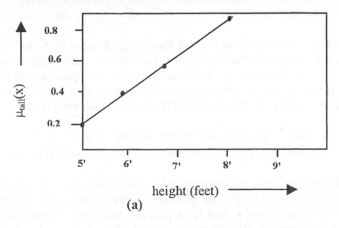

(a)

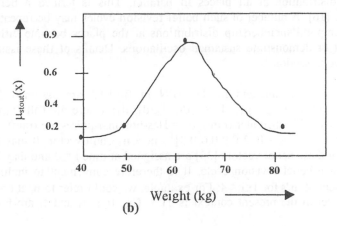

(b)

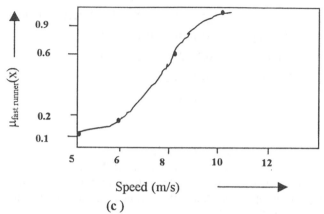

Speed (m/s)

(c)

Fig. 10.1: Membership distribution of (a) tall(x), (b) stout (x) and (c) fast runner (x).

With the given set of database and knowledge base, a Petri net (vide fig. 10.2) is now constructed. The algorithm for formation of the Petri net will be presented shortly. However, before describing the algorithm, let us first explain the problem under consideration. Assuming that the observed membership distributions for Tall (ram) and Stout (ram) respectively are

$$\mu_{tall} (\text{ram}) = [0.6/5'\ \ 0.8/6'\ \ 0.9/7'\ \ 0.4/8']^T = \mathbf{n_1}$$
$$\mu_{stout} (\text{ram}) = [0.2/40kg\ \ 0.9/50kg\ \ 0.6/60kg\ \ 0.3/80kg]^T = \mathbf{n_2}$$

and the distribution of all other predicates to be null vectors, one can estimate steady-state distribution [17] for all predicates. Such estimation requires updating of FTT distributions t_1 and t_2 in parallel, followed by updating of membership distribution at all places in parallel. This is termed a belief revision cycle [16]. A number of such belief revision cycles may be repeated until fuzzy temporal membership distributions at the places become either time-invariant or demonstrate sustained oscillations. Details of these issues will be covered in section 10.3.

The following parameters in the FPN in fig. 10.2 are assumed for illustration. $P = \{p_1, p_2, p_3, p_4\}$, $D = \{d_1, d_2, d_3, d_4\}$ where d_1=Tall (ram), d_2=Stout (ram), d_3 = Fast-runner (ram), d_4 = Has-nominal-pulse-rate (ram). $\mathbf{n_1}$ = $[0.6\ 0.8\ 0.9\ 0.4]^T$, $\mathbf{n_2}$ = $[0.2\ 0.9\ 0.6\ 0.3]^T$, $\mathbf{n_3}$ = $\mathbf{n_4}$= null vector. It may be added here that these belief vectors [33] are assigned at time t = 0 and may be updated in each belief revision cycle. It is therefore convenient to include time as argument of n_i's for 1≤i ≤ 4. For example, we could refer to $\mathbf{n_i}$ at t = 0 by $\mathbf{n_i}$ (0). Tr set in the present context is Tr = $\{tr_1, tr_2\}$; t_1 and t_2 are FTT vectors associated in the tr_1 and tr_2 respectively. Like n_i's, t_j's are also time varying quantities and are denoted by $t_j(t)$. R_1 and R_2 are relational matrices associated with tr_1 and tr_2 respectively. $I(tr_1) = \{p_1, p_2\}$, $I(tr_2) = \{p_3\}$, $O(tr_1) = \{p_3\}$, $O(tr_2) = \{p_2, p_4\}$.

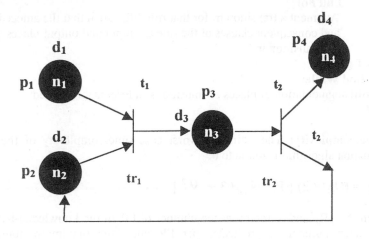

d_1=Tall (Ram), d_2=Stout (Ram), d_3=Fast runner (Ram),
d_4= Has-nominal-pulse-rate (Ram)

Fig. 10.2: An illustrative FPN.

10.2.1 Formation of FPN

Given the database in the form of clauses (Predicates with constant arguments) and knowledge base in the form of if-then rules, comprising of Predicates with variable arguments, one can easily construct a FPN by satisfying the knowledge base with the data clauses.

Procedure FPN-Formation (DB-file, KB-file, FPN)
Begin
 Repeat
   While not EOF of KB-file **do Begin**   // KB file consists of Rules //
    Pick up a production rule;
    If the Predicates in the antecedent part of the rule are
       unifiable with clauses in the DB-file and instantiation of
       all the variables is consistent (i.e., a variable in all the
       predicates of the antecedent part of the rule assumes same value)
    Then do Begin
     Substitute the value of the variables in the predicates

present in the consequent part;
Augment the derived consequent predicates(clauses) in the DB-file;
For each of the antecedent and consequent clauses of the rule
  If the place representing that clause is absent from the FPN
  Then augment the place in the FPN;
End For;
Augment a transition tr$_i$ for that rule PR$_i$, such that the antecedent
and consequent clauses of the rule are input and output places of
the transition tr$_i$;
End If;
End While;
Until augmentation of places or transitions in FPN is terminated;
End.

Time-complexity: The estimated worst case time-complexity of the FPN-formation algorithm is found to be

$$T_{FPN} = (N_{pr}{}^2 / 2) [P_p . N_{pr} / 3 + V^2]$$

where N_{pr}, P_p and V represent the number of PR in the knowledge-base, the maximum number of predicates per PR and the maximum number of variables per PR respectively.

10.2.2 Reachability Analysis and Cycle Identification

While analyzing FPNs, reachability of places [7], [1] and reachability of markings [28] are commonly used. In this chapter, the concept of reachability of places, as defined below, is used for identifying cycles in the FPN.

Definition 10.2: If $p_i \in$ I(tr$_a$) and $p_j \in$ O(tr$_a$) then p_j is **immediately reachable** from p_i. Again, if p_j is immediately reachable from p_i and p_k is immediately reachable from p_j , then p_k is **reachable** from p_i. The reachability property is the reflexive, transitive closure of the immediate reachability property [9]. We would use IRS (p_i) and RS (p_i) operators to denote the set of places immediately reachable and reachable from the place p_i respectively.

Moreover, if $p_j \in$ [IRS{IRS(IRS... k-times (p_i)}], denoted by IRS^k (p_i), then p_j is reachable from p_i with a **degree of reachability k**. For reachability analysis two connectivity matrices [12] are defined.

Definition10.3: A **place to transition connectivity (PTC) matrix Q** is a binary matrix whose elements $q_{jk} = 1$ if $p_k \in$ I (tr$_j$), otherwise $q_{jk} = 0$. If the FPN has n places and m transitions, then the **Q** matrix is of   (m × n) dimension.

Definition 10.4: A **transition to place connectivity (TPC) matrix P** is a binary matrix whose element $p_{ij} = 1$ if $p_i \in O$ (tr_j), otherwise $p_{ij} = 0$. With n places and m transitions in the FPN, the P matrix is of $(n \times m)$ dimension.

Since the binary product (AND-OR composition) $(P \; o \; Q)$ represents mapping from places to their immediately reachable places, therefore, the presence of a 'one' in the matrix $M_1 = (P \; o \; Q)$ at position (j, i) represents that $p_j \in IRS(p_i)$. Analogously, a 'one' at position (j, i) in matrix $M_r = (P \; o \; Q)^r$ for positive integer r represents $p_j \in IRS^r$ (p_i), i.e., p_j is reachable from p_i with a degree of reachability r.

Definition 10.5: If an element m_{ij} of matrix $M_k = (P \; o \; Q)^k$ is unity for positive integer k, then p_i and p_j are called **associative places with respect to m_{ij}**.

Theorem 10.1: *If the diagonal elements m_{ii} of the matrix $M_k = (P \; o \; Q)^k$ are unity, then the associative places p_i for all i lie on cycles through k transitions in each cycle.*

Proof: The proof is presented in Appendix C.

Corollary 1: *In a purely cyclic FPN [34], where all transitions and places lie on a cycle, $M_k = (P \; o \; Q)^k = I$, where k is the number of transitions (places) in the FPN.*

For identifying cycles in a FPN, the matrix $M_k = (P \; o \; Q)^k$ for k = 1 to m is to be computed, where m is the number of transitions in the network. Then by theorem 10.1, the associative places corresponding to the diagonal elements of M_k will lie on a cycle with k transitions on each cycle. However, if more than k number of diagonal elements are unity, then places lying on a cycle are to be identified by finding immediate reachability of places on the cycle using $M_1 = (P \; o \; Q)$ matrix.

The algorithm for **cycle-detection** consists of several procedures. Procedure **Find-places-on-cycle** determines the set of places S_k that lie on cycles with k transitions. Procedure **Find-IRS-places** saves in L_k the connectivity between pairs of immediately reachable set of places, lying on cycles with k transitions. Procedure **Find-connected-places-on-cycles** determines the list of places ordered according to their immediate reachability on cycles with k transitions and saves them in **Newlist$_k$**. Procedure **Put-transitions** positions appropriate transitions in the list of places in **Newlist$_k$**, so that places preceding and following a transition in the modified list **Finallist$_k$** are its input and output places on a cycle with k-transitions. The variable **'cycles'** in procedure Cycle-detection denotes the list of cycles.

Procedure Cycle-detection (P, Q, m, cycles)
Begin
  cycles := $\varnothing$;
  For k := 1 to m **do Begin** // m = no. of transitions in the FPN //
    Find-places-on-cycles(**P, Q**, k, S_k);
    If $S_k = \varnothing$ **Then** $L_k := \varnothing$;
    Find-IRS-places (S_k, L_k);
    Find-connected-places-on-cycles (L_k, Newlist$_k$);
    Put-transitions (Newlist$_k$, Finallist$_k$);
    cycles := cycles $\cup$ Finallist$_k$;
  End For;
End.

Procedure Find-places-on-cycles (P, Q, k, S_k)
Begin
$S_k := \varnothing$
$M_k := (P \; o \; Q)^k$;
For i:= 1 to n // n = no. of places in the FPN //
  For j := 1 to n
    If m_{ii} in $M_k = 1$
    Then $S_k := S_k \cup \{p_i\}$;
  End For;
  End For;
End.

Procedure Find-IRS-places (S_k, L_k)
Begin
  If $S_k \neq \varnothing$ **Then do Begin**
    $L_k := \varnothing$; M_1 : = **P o Q** ; **End** ;
    For all places $p_i, p_j \in S_k$
      If $m_{pi,pj}$ in $M_1 = 1$
        // p_i and p_j are row & column indices in M_1 //
      Then $L_k := L_k$ U $\{p_j \rightarrow p_i\}$;
    End For;
  End ;
End.

Procedure Find-connected-places-on-cycle (L_k, Newlist$_k$)
Begin
  Newlist$_k$:= $\varnothing$;
  While $L_k \neq \varnothing$ **do Begin**
    For all j, h **do Begin**
    If $\{p_j \rightarrow p_h\} \in L_k$ **Then do Begin**
    List$_k$:= $\{p_j \rightarrow p_h\}$; Stringfirst:= p_j;
    End ;
      If for some j, h $\{p_j \rightarrow p_h\}$ is a terminal string in List$_k$
      // e.g., In $P_u \rightarrow p_v \rightarrow p_r \rightarrow p_j \rightarrow p_h$, $p_j \rightarrow p_h$ is the terminal string//
      Then do Begin
      Temp := $\{h\}$; S := $\varnothing$ R:= $\varnothing$;
        For w $\in$ Temp **do Begin**
        For i:= 1 to n **do Begin**
            augmentation :=0;
          If $\{p_w \rightarrow p_i\} \in L_k$ **Then do Begin**
            Stringlast :=p_i; List$_k$:= List$_k$ ∇ $\{\rightarrow p_i\}$;
                    // ∇ is an augmentation operator//
              While (augmentation $\leq$(k-1)) **do Begin**
                  If (Stringfirst = Stringlast) **Then do Begin**
                    If (augmentation =(k-1)) **Then do Begin**
                    Newlist$_k$:= Newlist$_k$ $\cup$List$_k$;
                    Save the string in Newlist$_k$ starting with p_r,
                    where r is the minimum index among the
                    indices of places in the string;
                    End
                    Else List$_k$:= List$_k$ Δ $\{ \rightarrow p_i \}$; //Δ is the de-
                                      augmentation operator//
                  Else do
                  Begin S:= $\{i\}$; R:= R $\cup$S; Temp:= R;
                  End;
                End While;
          End ;
          End For;
      End For;
    End ; L_k : = L_k -$\{p_j \rightarrow p_h\}$;
  End For;
  End While;
End.

Procedure Put-transition (Newlist$_k$, Finallist$_k$)
Begin
  If Newlist$_k$ = $\varnothing$ **Then** Finallist$_k$:= $\varnothing$;
  For all i, j

If $\{p_i \rightarrow p_j\}$ is in Newlist$_k$
Then for all tr$_v$ in **Q**
   If Q (tr$_v$, p$_i$) = **P** (p$_j$, tr $_v$) = 1
   // p$_i$, p$_j$, tr$_v$ are indices of **P** and **Q** //
   Then replace $\{p_i \rightarrow p_j\}$ of Newlist$_k$ by $\{p_i \rightarrow tr_v \rightarrow p_j\}$
   and re-write the entire string as many times as the
   number of tr$_v$'s in Finallist$_k$;
End For;
End.

The merits of the algorithm cycle detection stems from the Porcedure Find-places-on-cycles. In fact *a priori* identification of the set of places on cycles significantly reduces the search time-complexity for detecting the cycles. Quantitatively, with no foreknowledge about the places on cycle, the worst case search complexity for identifying cycles with k transitions = $k.n^2$. Consequently, the overall complexity for $1 \leq k \leq m$ becomes

$$\Sigma \, k \, n^2 \approx O\,(\, m.\, n\,)^2.$$

However, when z number of places on cycles are identified, the overall search complexity is reduced to $O\,(m.\, z\,)^2 \approx O\,(m^2\,)$, since in most realistic cases z $<\,<$ m.

Example 10.2: The trace of the algorithm for cycle-detection is illustrated with reference to the FPN of fig. 10.3. The **P** and **Q** matrices for the FPN are first identified from fig. 10.3 and $M_k = (\mathbf{P\;o\;Q})^k$ for k = 1 to m (= 6, here) are evaluated. The diagonal entries of M_k for k = 1 to 6 are presented in Table-10.1 for quick reference. $M_1 = (\mathbf{P\;o\;Q})$ matrix is also presented below for ready reference in connection with estimation of L_k in Procedure Find-IRS-places.

To places	From Trans. tr$_1$	tr$_2$	tr$_3$	tr$_4$	tr$_5$	tr$_6$
p$_1$	0	1	0	1	1	0
p$_2$	1	0	0	0	0	0
P = p$_3$	0	0	0	0	0	1
p$_4$	0	0	0	1	0	0
p$_5$	0	0	1	0	0	0

To Trans.	From places p$_1$	p$_2$	p$_3$	p$_4$	p$_5$
tr$_1$	1	0	0	0	0
tr$_2$	0	1	0	0	0
Q = tr$_3$	0	1	1	0	0
tr$_4$	0	0	0	0	1
tr$_5$	0	0	0	1	0
tr$_6$	0	0	0	1	0

Table 10.1: The diagonal entries of matrix $\mathbf{M_k}$.

Iteration k	$M_k(1, 1)$	$M_k(2, 2)$	$M_k(3, 3)$	$M_k(4, 4)$	$M_k(5, 5)$
1	0	0	0	0	0
2	1	1	0	0	0
3	1	1	1	1	1
4	1	1	0	1	1
5	1	1	0	0	1
6	1	1	1	1	1

$$
M_1 = \begin{array}{c c} & \begin{array}{c} \text{From places} \\ p_1 \; p_2 \; p_3 \; p_4 \; p_5 \end{array} \\ \begin{array}{c} p_1 \\ p_2 \\ p_3 \\ p_4 \\ p_5 \end{array} & \left| \begin{array}{c c c c c} 0 & 1 & 0 & 1 & 1 \\ 1 & 0 & 0 & 0 & 0 \\ 0 & 0 & 0 & 1 & 0 \\ 0 & 0 & 0 & 0 & 1 \\ 0 & 1 & 1 & 0 & 0 \end{array} \right. \end{array}
$$

To places / From places

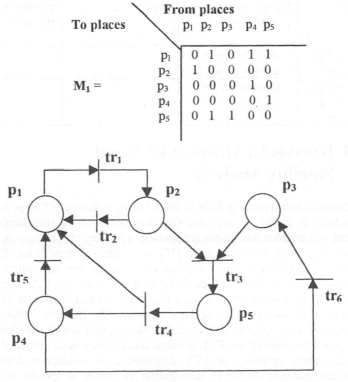

Fig. 10.3: A FPN for illustrating the cycle detection.

The trace of the algorithm cycle- detection is presented in table 10.2, where cycles = $\cup$ {finallist$_k$}.
   $1 \le k \le 6$

Table 10.2: Trace of the algorithm cycle-detection.

k	S_k	L_k	Newlist$_k$	finallist$_k$
1	$\varnothing$	$\varnothing$	$\varnothing$	$\varnothing$
2	$\{p_1, p_2\}$	$\{p_1 \rightarrow p_2, p_2 \rightarrow p_1\}$	$\{p_1 \rightarrow p_2 \rightarrow p_1\}$	$\{p_1 \rightarrow tr_1 \rightarrow p_2 \rightarrow tr_2 \rightarrow p_1\}$
3	$\{p_1, p_2, p_3, p_4, p_5\}$	$\{p_1 \rightarrow p_2, p2 \rightarrow p1, p_2 \rightarrow p_5, p_3 \rightarrow p_5, p_4 \rightarrow p_1, p_4 \rightarrow p_3, p_5 \rightarrow p_1, p_5 \rightarrow p_4\}$	$\{p_1 \rightarrow p_2 \rightarrow p_5 \rightarrow p_1, p_3 \rightarrow p_5 \rightarrow p_4 \rightarrow p_3\}$	$\{p_1 \rightarrow tr_1 \rightarrow p_2 \rightarrow tr_3 \rightarrow p_5 \rightarrow tr_4 \rightarrow p_1, p_3 \rightarrow tr_3 \rightarrow p_5 \rightarrow tr_4 \rightarrow p_4 \rightarrow tr_6 \rightarrow p_3\}$
4	$\{p_1, p_2, p_4, p_5\}$	$\{p_1 \rightarrow p_2, p_2 \rightarrow p_1, p_2 \rightarrow p_5, p_4 \rightarrow p_1, p_5 \rightarrow p_1, p_5 \rightarrow p_4\}$	$\{p_1 \rightarrow p_2 \rightarrow p_5 \rightarrow p_4 \rightarrow p_1\}$	$\{p_1 \rightarrow tr_1 \rightarrow p_2 \rightarrow tr_3 \rightarrow p_5 \rightarrow tr_4 \rightarrow p_4 \rightarrow tr_5 \rightarrow p_1\}$
5	$\{p_1, p_2, p_5\}$	$\{p_1 \rightarrow p_2, p_2 \rightarrow p_1, p_2 \rightarrow p_5, p_5 \rightarrow p_1\}$	$\varnothing$	$\varnothing$
6	$\{p_1, p_2, p_3, p_4, p_5\}$	same as for k=3	$\varnothing$	$\varnothing$

10.3 Behavioral Model of FPN and Stability Analysis

The dynamic behavior of a FPN is modeled by updating FTTs at transitions and beliefs at places. In fact the enabling condition of transitions is first checked. All enabled transitions are fireable; on firing of a transition, the FTT distribution at its outgoing arcs [16] is estimated based on the belief distribution of its input places and the relational matrix associated with the transition. It may be noted that on firing of a transition, the belief distribution of its input places are not destroyed like conventional Petri nets [26]. After the FTTs at all transitions are updated concurrently, the belief distribution at the places is also updated concurrently. The revised belief distribution at a place p_j is a function of the FTT of those transitions whose output place is p_j. The concurrent updating of FTT distribution at transitions followed by concurrent updating of belief distribution at places is termed as a **belief revision cycle**.

10.3.1 The Behavioral Model of FPN

Let us consider a transition tr_i, where $I(tr_i) = \{p_k, p_m\}$ and $O(tr_1) = \{p_u, p_v\}$. Assume that th_i is the threshold vector, associated with the transition. The transition tr_i is enabled if

$$R_i \; o \; (n_k \wedge n_m) \geq th_i.$$

An enabled transition fires, resulting in a change in the FTT vectors at its output arcs. It is to be noted that the FTT vectors at all its output arcs are equal. In case the transition tr_i is not enabled, the FTT distribution at its output arcs is set to null vector. The model of FPN, designed after Looney [24] and based on the above considerations, is now formally presented.

$$t_i(t+1) = t_i(t) \wedge [\, R_i \; o \; (n_k(t) \wedge n_m(t))] \wedge$$
$$U[R_i \; o \; (n_k(t) \wedge n_m(t)) - th_i \,] \qquad (10.1)$$

In expression (10.1), U denotes a unit step vector, each component of which becomes one when its corresponding argument ≥ 0 and becomes zero, otherwise. In fact, the enabling condition of the transition tr_i is checked by this vector. Moreover, the $\wedge$ operation between two vectors is done component-wise like column vector addition in conventional matrix algebra. It may be noted that if tr_i has m input places $p_1, p_2, \ldots p_m$ and k output places $p_{m+1}, p_{m+2}, \ldots p_{m+k}$ (fig. 10.4) then expression (10.1) can be modified with the replacement of

$$n_k(t) \wedge n_m(t) \quad \text{by} \quad \overset{m}{\underset{w=1}{\wedge}} n_w(t).$$

After the FTT distribution at all the transitions in the FPN are updated concurrently, the belief distribution at all places can be updated in parallel following expression (10.2). Let us consider a place p_j such that $p_j \in [O(tr_1) \cap O(tr_2) \cap \ldots \cap O(tr_s)]$ (fig.10.5). The updating of belief distribution n_j at place p_j is given by

$$n_j(t+1)$$

$$= n_j(t) \vee [t_1(t) \vee t_2(2) \vee \ldots \vee t_s(t)]$$

$$= n_j(t) \vee (\overset{s}{\underset{r=1}{\vee}} t_r(t)). \qquad (10.2)$$

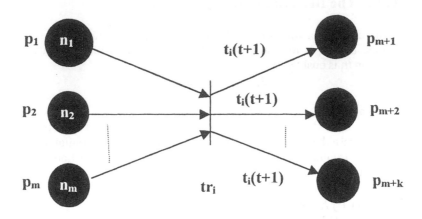

Fig. 10.4: A transition tr_I with m input and k output places.

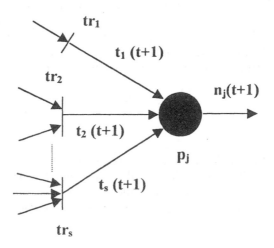

Fig. 10.5: A place p_j that belongs to output places of s transitions.

10.3.2 State Space Formulation of the Model

For convenience of analysis and simplicity of realization of FPN by arrays, instead of linked list structures, the dynamic behavior of the entire FPN is represented by a single vector-matrix (state-space) equation. For the formulation of the state space model, we first construct a belief vector N (t) of dimension (z. n) × 1, such that

$$N(t) = [\ n_1\ (t)\ n_2\ (t)\ ....n_n\ (t)\ ]^T,$$

where each belief vector n_i (t) has z components. Analogously, we construct a FTT vector **T** and threshold vector **Th** of dimension (z. m) ×1, such that

$$T(t) = [\ t_1\ (t)\ t_2\ (t)\ ...t_m\ (t)\ ]\quad \text{and}\quad Th = [\ th_1\ th_2,\ ...\ ,th_m\ ]$$

where each t_j (t) and th_i have z components. We also form a relational matrix **R**, given by

$$
R \ = \
\begin{bmatrix}
R_1 & \phi & \phi & -- & \phi \\
\phi & R_2 & \phi & -- & \phi \\
\phi & \phi & R_3 & & \phi \\
- & - & - & & -- \\
\phi & \phi & \phi & -- & R_m
\end{bmatrix}
$$

where R_i for $1 \le i \le m$ is the relational matrix associated with transition tr_i and Φ denotes a null matrix of dimension equal to that of R_i's, with all elements = 0. It may be noted that position of a given R_i on the diagonal in matrix **R** is fixed. Moreover, extended **P** and **Q** matrices denoted by **P'** and **Q'** respectively may be formed by replacing each unity and zero element in **P** and **Q** by square identity and null matrices respectively of dimensions equal to the number of components of t_i and n_j respectively. Now consider a FPN with n places and m transitions. Omitting the **U** vector for brevity, the FTT updating equation at a given transition tr_i may now be described by expression (10.3)

$$t_i\,(t+1) = t_i\,(t)\ \Lambda[\ R_i\ o\ (\ \overset{n}{\underset{\exists\ w=1}{\Lambda}}\ n_w\,(t)\ )\ ] \tag{10.3}$$

$$= t_i\,(t)\ \Lambda[\ R_i\ o\ \{\ \overset{n}{\underset{\exists\ w=1}{V}}\ n^c{}_w\,(t)\}^c\ ]$$

$$= t_i\,(t)\ \Lambda\ [\ R_i\ o\ \{\ \overset{n}{\underset{\forall\ w=1}{V}}\ q_w{}'\ \{\ V\ n^c{}_w\,(t)\}^c{}_\}\ ]$$

where the elements $q_w{}' \in \{0,\ 1\}$ and c over a vector denote the one's complements of its corresponding elements.

Combining the FTT updating equations for m transitions, we have

$$T(t + 1) = T (t) \wedge [R o (Q' \ o N^c (t))^c].$$ (10.4)

Similarly the belief updating equations for n places can now be combined using expression (10.2) as follows:

$$N (t + 1) = N (t) V [P' \ o T(t + 1)].$$ (10.5)

Combining expressions (10.4) and (10.5) yields

$$N(t+1) = N(t) V [P' o \{T(t) \wedge \{R o (Q' o N^c (t))^c\}\}].$$ (10.6)

Including the **U** vector in the expression (10.6), we have

$$N(t +1) = N (t) V P' o [\{T(t) \wedge \{R o (Q'o N^c (t))^c \}\}$$
$$\wedge U\{R o (Q 'o N^c (t))^c - Th\}].$$ (10.7)

Estimation of N(r) for r>1 from N(o) can be carried out by updating N(r) iteratively r times using expression (10.7) .

10.3.3 Special Cases of the Model

In this section, two special cases of the above model, obtained by eliminating state feedback [21] from FTT and beliefs, are considered. This renders the belief that any given place is influenced by not only its parents [20], as in the general model, but its global predecessors also [27].

Case I: In expression (10.1), the current value of FTT distribution is used for estimation of its next value. Consequently, the FTT distribution t_i (t+1) at a given time (t+1) depends on the initial value of t_i (0). Since t_i (0) for any arbitrary transition tr_i in the FPN is not always available, it is reasonable to keep t_i (t+1) free from t_i (t). If t_i (t) is dropped from expression (10.1), the modified state space equation can be obtained by setting all components of T(t) = 1 in the expressions (10.6) and (10.7). The revised form of expression (10.6), which will be referred to frequently, is rewritten as expression (10.8).

$$N(t+1) = N(t) V [P' o \{R o (Q' o N^c (t))^c \} \wedge$$
$$\{U \{ R o (Q ' o N^c (t))^c - Th \}\}].$$ (10.8)

Case II: A second alternative is to keep both t_i (t+1) and n_j (t+1), $\forall i,j$ independent of their last values. However when n_j (t) is dropped from expression (10.2), places with no input arcs, called axioms [35], cannot restore their belief distribution, since t_r (t) for r=1 to s in expression (10.2) are zero. In order to set n_j (t+1) = n_j (t) for axioms, we consider self-loop around each axiom through a virtual transition tr_k, such that the R_k and th_k are set to identity matrix and null vector respectively. The P', Q' and R matrices are thus modified and denoted by P'_m , Q'_m and R_m respectively. The state space model for case II without U is thus given by

$$N(t+1) = P'_m \text{ o } R_m \text{ o } (Q'_m \text{ o } N^c (t))^c. \qquad (10.9)$$

Example 10.3: In this example, the formation of P', Q' and R matrices (vide fig. 10.6) and P'_m , Q'_m , and R_m matrices (vide fig.10.7) are demonstrated for the models represented by expression (10.7) and (10.9) respectively.

The P and Q matrices for the FPN of fig. 10.6 are given by

	From		
To	tr_1	tr_2	tr_3
p_1	0	0	1
P = p_2	0	0	0
p_3	1	1	0

	From		
To	p_1	p_2	p_3
tr_1	0	1	0
Q = tr_2	1	1	0
tr_3	0	0	1

Assuming the n_j and t_i vectors of dimension (3x1) we construct the P' and Q' matrix

$$P' = \begin{pmatrix} \Phi & \Phi & I \\ \Phi & \Phi & \Phi \\ I & I & \Phi \end{pmatrix}, \quad Q' = \begin{pmatrix} \Phi & I & \Phi \\ I & I & \Phi \\ \Phi & \Phi & I \end{pmatrix}$$

where Φ and I denote null and identity matrices each of dimension (3 x 3).

The relational matrix R in the present context is given by

$$R = \begin{pmatrix} R_1 & \Phi & \Phi \\ \Phi & R_2 & \Phi \\ \Phi & \Phi & R_3 \end{pmatrix}$$

Further, $N = [n_1 \; n_2 \; n_3]^T$ $T = [t_1 \; t_2 \; t_3]^T$ $Th = [th_1 \; th_2 \; th_3]^T$. Expression (10.7) can be used for updating N with the above parameters. For updating N with expression (10.9), we , however, redraw the FPN with a virtual self loop around place p_2 [vide fig. 10. 7] and reconstruct P, Q and consequently P'_m , Q'_m and R_m matrices. It may be noted that the virtual transitions around place p_j should be named tr_j (j= 2 , here) for satisfying equation (10.9) and other transitions should be renamed distinctively.

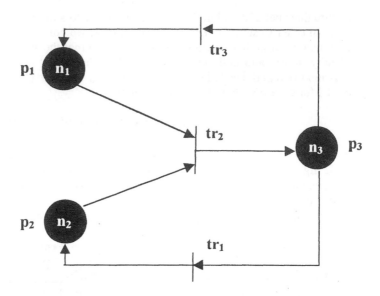

Fig. 10.6: A FPN for illustrating the formation of **P'** , **Q'**
and **R** matrices.

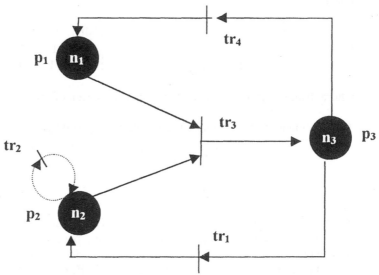

Fig. 10.7: The modified form of fig. 10.6 with self-loop around
place p_2 and renamed transitions.

10.3.4 Stability analysis

In this section, the analysis of the dynamic behavior of the proposed model will be presented. A few definitions, which are used to understand the analysis, are in order.

Definition 10.6: A FPN is said to have reached an equilibrium state (steady-state) when $N(t^* +1) = N(t^*)$ for some time $t= t^*$, where t^* is the minimum time when the equality of the vectors is first attained. The t^* is called the equilibrium time.

Definition 10.7: A FPN is said to have **limit cycles** if the fuzzy beliefs n_j of at least one place p_j in the network exhibits periodic oscillations, described by $n_j (t + k) = n_j (t)$ for some positive integer $k>1$ and sufficiently large t, numerically greater than the number of transitions in the FPN.

The results of stability analysis of the proposed models are presented in the Theorems 10.2 through 10.4.

Theorem 10.2: *The model represented by expression (10.1) is unconditionally stable and the steady state for the model is attained only after one belief revision step in the network.*

Proof: Proof of the theorem is given in Appendix C.

Theorem 10.3: *The model represented by expression (10.8) is unconditionally stable and the non-zero steady state belief vector N * satisfies the inequality (10.10).*

$$N^* \geq P' \text{ o } \{ R \text{ o } (Q' \text{ o } N^{*\text{ c}})^{\text{ c}}\},$$

when $R \text{ o } (Q \text{ o } N^c (t))^c \geq Th$, $\forall t \geq 0$. (10.10)

Proof: Proof is given in Appendix C.

The following definitions will facilitate the analysis of the model represented by expression (10.9) .

Definition 10.8: An arc $tr_i \times p_j$ is called **dominant** at time τ if for $p_j \varepsilon$ $(\exists k \cap O(tr_k))$, $t_i (\tau) > t_k (\tau)$; alternatively, an arc $p_x \times tr_v$ at time τ is **dominant** if $\forall w$, $p_w \varepsilon I (tr_v)$, $n_x(\tau) < n_w (\tau)$, provided $R_v \text{ o } (\forall w, \wedge n_w) > Th_v$.

Definition 10.9: An arc is called **permanently dominant** if after becoming dominant at time $t = \tau$, it remains so for all time $t > \tau$.

The limit cycle behavior of the model, represented by expression (10.9), is stated in Theorem (10.4).

Theorem 10.4: If *all the n number of arcs on any of the cycles of a FPN remains dominant from r_1 -th to r_2- th belief revision step, by using the model represented by expression (10.9), then each component of the fuzzy belief distribution at each place on the cycle would exhibit*

> *i) at least 'a' number of periodic oscillations, where a= integer part of $\{(r_2 - r_1)/n\}$ and*

> *ii) limit cycles with $r_2 \to \infty$.*

Proof: Proof is available in [16] and omitted for space limitation.

The model represented by expression (10.9) also yields an equilibrium condition, if none of the cycles have all their arcs permanently dominant. The number of belief revision steps required to reach the equilibrium condition for this model is estimated below.

Let l_1 = the worst number of belief revision steps required on the FPN for transfer of fuzzy belief distribution from the axioms to all the places on the cycles, which are directly connected to the axioms through arcs lying outside the cycle,

l_2 = the worst number of belief revision steps required for the transfer of fuzzy belief distribution from the places on the cycles to the terminal places in the network,

n = number of transitions on the largest cycle,

l_3 = the worst number of belief revision steps required for the transfer of fuzzy belief distribution from the axioms to all the terminal places through the paths, which do not touch the cycles.

Theorem 10.5: *In case steady state is reached in a FPN, by using the model, represented by the expression (10.9), then the total number of belief revision steps required in the worst case to reach steady state is given by*

$$T_{worst} = Max\{ l_3, (l_1 + l_2 + n - 1)\}. \tag{10.11}$$

Proof: Proof is available in [16] and hence omitted for space limitations.

It may be added that the number of belief revision steps required for the model, represented by (10.8), is the same as computed above.

10.4 Forward Reasoning in FPN

Forward reasoning is generally carried out in fuzzy logic by extending the principle of generalized modus ponens (GMP) [36]. For illustration, consider the following rule having fuzzy quantifiers and the observed antecedent.

Rule: if x-is-A AND y-is-B Then z-is-C
Observed antecedent: x-is-A' AND y-is-B'

Conclusion: z-is-C'

The conclusion z-is-C ' is inferred by the reasoning system based on the observed level of quantifiers A' and B'. While representing the above problem using FPN, we consider that two discrete membership distributions are mapped at places p_1 and p_2 with proposition d_1 = x-is-A and d_2= y-is-B respectively. Further, let p_1 ,p_2 $\in I(tr_i)$, then p_3 which corresponds to d_3 = z-is-C is an element of $O(tr_i)$. Here, the membership distribution of z-is-C' may be estimated using the distribution of x-is-A' and y-is-B'.

Further, for representing chained modus ponens, Petri net is an ideal tool. For example, consider the second rule z-is-C$\rightarrow$ w-is-D and the observed antecedent z-is-C'. We subsequently infer w-is-D'. This too can be realized by adding one transition tr_j and a place p_4 such that $p_3 \in I(tr_j)$ and $p_4 \in O(tr_j)$.

The most complex and yet unsolved problem of forward reasoning, perhaps, is reasoning under self-reference. This problem too can be easily modeled and solved by using FPN.

We, now, present an algorithm for forward reasoning that is applicable to all the above kinds of problems independent of their structures of the FPNs. Procedure forward reasoning is described below based on the state space equation (10.8) , which is always stable.

Procedure forward-reasoning (FPN,R,P',Q',N(0), Th)
Begin
  N(t): = N(0) ;
  While N(t+1) $\neq$ N(t)
    Temp : = R o (Q ' o N^c (t))^c ;
    N(t+1):= N(t) V P ' o [Temp $\wedge$ U (Temp - Th))];
      N(t) := N(t+1);
  End while;
End.

The procedure forward reasoning is used to compute the steady state belief of **N** (**t**) from its initial value **N(0)**. In application, like criminal investigation [20], these steady state values of the predicates are used to identify the culprit from a given set of suspects. After the culprit, described by a terminal place of the FPN, is identified, procedure reducenet, presented below, is invoked to find the useful part of the network for generating an evidential explanation for the culprit.

Procedure reducenet (FPN, axioms, goal, parents);
  Begin
     nonaxioms:= goal;
    Repeat
      Find-parents (nonaxioms); //Find parents of non-axioms.//
      Mark the generated parent place, hereafter called parents and
      the transitions connected between parents and nonaxioms;
      nonaxioms:=parents - axioms; //nonaxiom parents detection.//
    Until parents $\in$ axioms;
      trace the marked places and transitions;
  End.

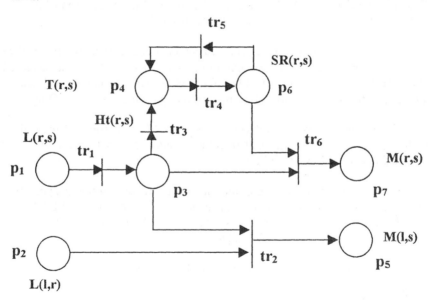

L = Loves, T = Tortures, SR = Strained-relationships-between ,
M = Murders, r= Ram, s= Sita, l= Lata

Fig.10.8: A FPN representing a murder history of a housewife 's' where the husband 'r' and the girl friend 'l' of 'r' are the suspects.

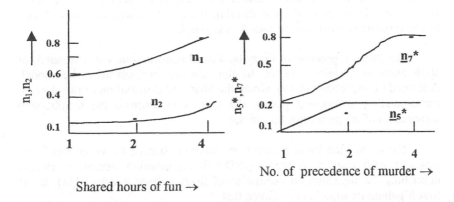

Fig. 10.9(a): Initial belief distribution   **Fig. 10.9(b):** Steady-state distribution
of n_1 and n_2.   of n_5 and n_7, denoted by
   n_5^*, n_7^*.

The worst case time complexity of the Procedure Forward reasoning and Procedure Reducenet are O (m) and O (a . n) respectively, where 'm' , 'n' and 'a' denote the number of transitions, number of places before reduction of the network and number of axioms respectively.

Example 10.4: In the FPN (fig. 10.8) the fuzzy belief distribution corresponding to places p_1 and p_2 is shown in fig. 10.9(a). The initial belief distribution of all other places are null vectors. Further, we assumed **R = I**. The steady-state belief distribution at all places in the entire FPN is obtained after 5 iterations using forward reasoning algorithm, and their distributions at places p_5 and p_7 are shown in fig. 10.9(b). Since for all components, n_7 is larger than n_5, p_7 is marked as the concluding place and then Procedure reducenet is invoked for tracing explanation for the problem.

10.5 Backward Reasoning in FPN

'Backward reasoning ' [33] in fuzzy logic is concerned with inferring the membership distribution of the antecedent clauses, when the if-then rule and the observed distribution of the consequents are available. For example, given the rule and the observed consequent clause, the inference follows.

Rule: *If x-is-A AND y-is-B THEN z-is-C*
Observed evidence: *z-is-C'*

Inferred: *x-is-A' AND y-is-B'*

In the above example, A, B, and C are three fuzzy quantifiers. C' is an observed quantifier of z and A' and B' are the inferred quantifiers of x and y respectively. Here, given the membership (belief) distribution of z-is-C', one has to estimate the distribution x-is-A' and y-is-B'.

The classical problem of fuzzy backward reasoning may be extended for application in cycle-free FPNs. In this chapter, we consider the model, described by expression (10.9). Given the observed distribution of the clauses, corresponding to terminal places, the task is to estimate the membership distribution of the predicates for axioms.

For solving the above problem, we have to estimate the inverse of fuzzy matrices with respect to fuzzy AND-OR composition operators. Before describing the algorithm for estimation of fuzzy inverse matrices [34], let us first highlight its significance. Given that

$$N(t+1) = P'_{fm} [R_{fm} \; o \; (Q'_{fm} \; o \; N^c(t))^c] \qquad (10.9a)$$

where the suffix 'f' represents that the model corresponds to forward reasoning.

Pre-multiplying both sides of the above equation by the fuzzy inverse(pre-inverse to be specific [34]) of P'_{fm}, denoted by $P'_{fm}{}^{-1}$, we find

$$P'_{fm}{}^{-1} \; o \; N(t+1) = R_{fm} \; o \; (Q'_{fm} \; o \; N^c(t))^c.$$

After some elementary fuzzy algebra, we get

$$N(t) = [Q'_{fm}{}^{-1} \; o \; \{R_{fm}{}^{-1} \; o \; (P'_{fm}{}^{-1} \; o \; N(t+1))\}^c]^c. \qquad (10.12)$$

For estimation of the belief distribution of the axiom predicates, from the known belief distribution of the concluding predicates, the following steps are to be carried out in sequence.

i)   All the concluding predicates at terminal places should have self-loops through virtual transitions. This would help maintain the initial belief distribution at these places. The threshold and the relational matrices for the virtual transitions should be set to null vector and identity matrices respectively to satisfy the above requirement.

ii)   The $N(t+1)$ vector in expression (10.12) is to be initialized by assigning non-zero vectors at the concluding places and null vectors at all other places. Call this belief vector N_{ini}.

iii) The expression (10.12) should be updated recursively in backward time until $N(t) = N(t+1)$.

The algorithm for backward reasoning may be formally stated as follows:

Procedure backward-reasoning (FPN, P'_{fm}, R_{fm}, Q'_{fm}, N_{ini})
Begin
    $t := m$ // m= no. of transitions in the FPN //;
    $N(t+1) := N_{ini}$;
    While $N(t) \neq N(t+1)$ **do Begin**
        $N(t) := [Q'_{fm}{}^{-1} \text{ o } \{ R_{fm}{}^{-1} (P'_{fm}{}^{-1} \text{ o } N(t+1)) \}^c]^c$;
        $N(t+1) := N(t)$;
    End while;
End.

The worst case time complexity of the procedure backward-reasoning is proportional to $O(m)$, where m denotes the number of transitions in the FPN.

It is apparent from the above procedure that for estimating $N(t)$ from $N(t+1)$, the inverse of the fuzzy / binary matrices with respect to fuzzy composition operators is to be computed. Now, we present an algorithm for estimation of pre-inverse of a fuzzy matrix.

Definition 10.10: A matrix **Q** is called the **pre-inverse** of a fuzzy / binary matrix **R** if $Q \text{ o } R = I' \rightarrow I$, where **I** is the identity matrix and $I' \rightarrow I$ means I' is close enough to **I** in the sense of Euclidean distance.

Definition 10.11: Q_{best} is called the **best pre-inverse** of R if $\|(Q_{best} \text{ o } R) - I\| \leq \|(Q \text{ o } R) - I\|$ for all real matrix Q, with elements $0 \leq q_{ij} \leq 1$, where $\| \delta \|$ means sum of square of the elements of matrix δ.

It may be added here that **Q** is called the **post-inverse** of matrix **R**, if $R \text{ o } Q = I' \rightarrow I$. Analogously, Q_{best} is called the **best post-inverse** of R if $\|(R \text{ o } Q_{best}) - I\| \leq \|(R \text{ o } Q) - I\|$ for all real matrix Q, where $0 \leq q_{ij} \leq 1$.

For estimation of **Q**, the pre-inverse matrices of **R** of dimension (n x n) let us consider the k-th row and i-th column of (Q o R), given by

$$(Q \text{ o } R)_{k,i} = \bigvee_{j=1}^{n} (q_{kj} \wedge r_{j,i}).$$

For obtaining **Q**, one has to satisfy $Q \text{ o } R = I'$, sufficiently close to **I**, which requires

$$\bigvee_{j=1}^{n} (q_{kj \wedge} r_{jk}) \text{ to be close to 1 (criterion 1)}$$

$$\text{and} \quad \bigvee_{j=1, i \neq k}^{n} (q_{kj} \wedge r_{ji}) \text{ to be close to 0 (criterion 2)}$$

Criterion 1 may be satisfied, without much constraint, by choosing each of the individual terms $(q_{k1} \wedge r_{1k})$, $(q_{k2} \wedge r_{2k})$,..,$(q_{kn} \wedge r_{nk})$ close to 1 [34]. Similarly, the second criterion may be satisfied if each individual term $(q_{k1} \wedge r_{1i})$, $(q_{k2} \wedge r_{2i})$,..,$(q_{kn} \wedge r_{ni})$ [i $\neq$ k] is close to zero. Further, it can be proved, without any loss of generality, that the choice of q_{kj} may be confined to the set $\{r_{j1}, r_{j2}, ..., r_{jk}, ... , r_{jn}\}$ (vide a theorem in [38]) instead of the wide interval [0, 1] which will result in significant saving in computational time. The choice of q_{kj}, $\forall k$, j from the set $\{r_{j1}, r_{j2}, ..., r_{jk}, ..., r_{jn}\}$ is, therefore, governed by the following two criteria

i) $(q_{kj} \wedge r_{jk})$ is to be maximized.

ii) $(q_{kj} \wedge r_{ji})$ is to be minimized for $1 \leq \forall i \leq n$ but $i \neq k$.

The above two criteria can be combined to a single criterion as depicted below

$$(q_{kj} \wedge r_{jk}) - \bigvee_{i=1, i \neq k}^{n} (q_{kj} \wedge r_{ji}) \text{ is to be maximized, where } q_{kj} \epsilon \{r_{j1}, r_{j2}, ..., r_{jn}\}.^{[1]}$$

Procedure pre-inverse is designed based on the last criterion.

Procedure Pre-inverse (Q, R);
Begin
  For k: =1 to n
  For j: =1 to n
      For w: =1 to n
          compute α_w: = $(r_{jw} \wedge r_{jk}) - \bigvee_{i=1, i \neq k}^{n} (r_{jw} \wedge r_{ji})$
  End For;
      sort (α_w, β_w) ‖ this procedure sorts the elements of the array α_w
                    and saves them in β_w in descending order ‖

---

[1] For post -inversion of R

$$(r_{jk} \wedge q_{kj}) - \bigvee_{i=1, i \neq j}^{n} (r_{ik} \wedge q_{kj}) \text{ is to be maximized, where } q_{kj} \epsilon \{r_{1k}, r_{2k}, .., r_{nk}\}.$$

```
    For w:= 1 to n-1
       if β₁ = β_{w+1}
          q_{kj} : = r_{jw} ;
          print q_{kj};
       End For;
    End For;
  End For;
End.
```

It may be added that for identifying Q_{best} among all the Q_k's, one has to estimate $\Sigma\Sigma\delta_{ij}^2$ for all j, for all I, where δ_{ij} is the (i, j)th element of $((Q_k \text{ o } R) - I)$ for each Q_k. The Q_k with the smallest $((Q_k \text{ o } R) - I)$ is declared as the Q_{best} [34]. In case more than one Q_k yields the smallest value of $((Q_k \text{ o } R) - I)$, any one of them may be picked up as the best.

Example 10.5: Consider a relational matrix **R** ,

$$R = \begin{pmatrix} 0.3 & 0.5 & 0.6 \\ 0.4 & 0.6 & 0.9 \\ 0.8 & 0.7 & 0.2 \end{pmatrix}$$

By using procedure pre-inverse (Q, R), we find eight inverse matrices, namely Q_1, Q_2 through Q_8 for **R**. The best pre-inverse matrix, Q_{best}, is then computed by the method described above. Here,

$$Q_{best} = \begin{pmatrix} 0.3 & 0.4 & 0.8 \\ 0.3 & 0.4 & 0.2 \\ 0.6 & 0.9 & 0.2 \end{pmatrix}$$

The algorithm for backward reasoning is applied to a diagnostic reasoning problem in example 10.6.

Example 10.6: Consider the problem of diagnosis of a 2-diode full wave rectifier circuit. The expected rectifier output voltage is 12 volts, when the system operates properly. Under defective condition, the output could be close to 0 volts or 10 volts depending on the number of defective diodes. The knowledge base of the diagnosis problem is extended into a FPN (vide fig. 10.10). The task here is to identify the possible defects: defective (transformer) or defective(rectifier). Given the belief distribution of the predicate close-to (rectifier-out, 0 v) and more-or-less (rectifier-out, 10 v) (vide fig. 10. 11 and 10.12), one has to estimate the belief distribution of the predicate: defective (transformer) and defective (rectifier). However, for this estimation, one should have knowledge of the relational matrices corresponding to input-output place pairs of each transition and the thresholds.

Let us assume for the sake of simplicity that the thresholds are zero and the relational matrices for each input-output pair of transition tr_1 through tr_7 are equal. So, for $1 \le i \le 7$ let

$$R_i = \begin{pmatrix} 0.3 & 0.5 & 0.6 \\ 0.4 & 0.6 & 0.9 \\ 0.8 & 0.7 & 0.2 \end{pmatrix}$$

be the same matrix, as chosen in the example 10. 5. The R_i for $8 \le i \le 9$ will be the identity matrix. Thus R_{fm} in the present context will be a (27 x 27) matrix, whose diagonal blocks will be occupied by R_i. Further, since all the non-diagonal block matrices are null matrix, the R_f^{-1} can be constructed by substituting R_i in R_{fm} by R_i^{-1}. The P'_{fm} and Q'_{fm} in the present context are also (27 x 27) matrices. N_{ini} is a (27 x 1) vector, given by

$$N_{ini} = [\; 000\; 000\; 000\; 000\; 000\; 000\; 000\; 0.2\; 0.1\; 0.0\; 0.0\; 0.4\; 0.5\; 0.6\;]^T_{27x1}$$

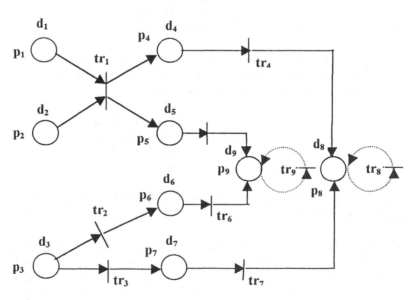

d_1= defective (transformer), d_2 =close-to (primary, 230), d_3=defective (rectifier), d_4= Close-to (trans-out, 0V), d_5 = Open (one half-of-secondary-coil), d_6 = Defective (one-diode), d_7 = Defective (two-diodes), d_8 =Close-tp (rectifier-out, 0V), d_9 =More-or-less (rectifier-out, 0V)

Fig. 10.10: A FPN representing diagnostic knowledge of a 2- diode full wave rectifier.

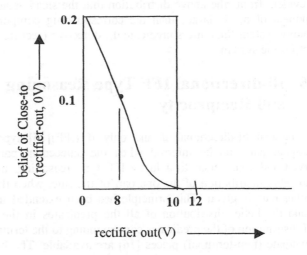

Fig. 10.11: Belief distribution of Close-to (rectifier-out, 0V).

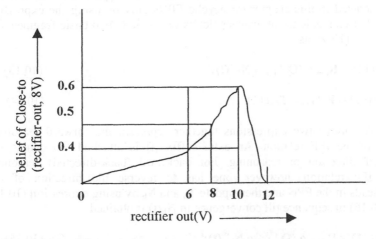

Fig. 10.12: Belief distribution of More-or-Less (rectifier-out, 10V).

The $P'_{f m}{}^{-1}$ and $Q'_{f m}{}^{-1}$ are now estimated and the algorithm for backward reasoning is then invoked. The steady-state belief distribution obtained after 3 iterations is given by

$N_{s.s}$ =

$[.5 .4 .6 \quad .5 .4 .6 \quad .2 .2 .2 \quad .2 .2 .2 .6 .4 .5 \quad .6 .4 .5 \quad .2 .2 .2 \quad .2 .1 0.0 . 4 .5 .6]^{T}$

It is evident from the above distribution that the steady-state values of each component of n_1 is larger than the corresponding components of n_3. The reasoning system thus infers predicate d_1: defective (rectifier) as the possible defect for the system.

10.6 Bi-directional IFF Type Reasoning and Reciprocity

In a classical bi-directional if and only if (IFF)[15] type reasoning, the consequent part can be inferred when the antecedent part of the rule is observed and vice versa. In a fuzzy IFF type reasoning, one can infer the membership distribution of the one part of the rule, when the distribution of the other part is given. This principle has been extended in this section to estimate the belief distribution of all the predicates in the FPN, when the belief distribution of the predicates corresponding to the terminal places or the intermediate (non-terminal) places [16] are available. The bi-directional IFF type reasoning in FPN has, therefore, a pragmatic significance.

Like backward reasoning, bi-directional IFF type reasoning too has been modeled in this chapter for acyclic FPNs only by using the expression (10.9). For convenience of analysis, let us reformulate two basic fragments of expression (10.9), as

$$T_f(t+1) = R_{fm} \, o \, (Q\,'_{fm} \, o \, N_f{}^c(t))^c \qquad\qquad (10.13)$$

$$N_f(t+1) = P\,'_{fm} \, o \, T_f(t+1). \qquad\qquad (10.14)$$

The above two expressions together represent the forward reasoning model of the IFF relation. An extra suffix (f) is attached to denote the 'forward' direction of reasoning. For backward (back-directed) reasoning using IFF relation, however, one has to reverse the direction of the arrowheads in the FPN and then update T_b and N_b by using expression (10.15) and (10.16) in sequence till convergence in $N_b(t)$ is attained.

$$T_b(t+1) = R_{bm} \, o \, (Q\,'_{bm} \, o \, N_b{}^c(t))^c \qquad\qquad (10.15)$$

$$N_b(t+1) = P\,'_{bm} \, o \, T_b(t+1). \qquad\qquad (10.16)$$

The suffix 'b' in expression (10.15) and (10.16) stands for the backward direction of reasoning. It may be noted that, once $P\,'_{fm}$ and $Q\,'_{fm}$ are known, $P\,'_{bm}$ and $Q\,'_{bm}$ may be obtained using Theorem (10.6).

Theorem 10.6: $P\,'_{bm} = (Q\,'_{fm})^T$ *and* $Q\,'_{bm} = (P\,'_{fm})^T$.

Proof: Proof is presented in Appendix A.   ☐

When the belief distribution of the axiom predicates are given, one has to use the forward reasoning model. On the other hand, when the belief distribution of the predicates for the concluding places is known, one should use the back-directed reasoning model of the IFF relation. Moreover, when the belief distributions of the predicates at the non-terminal places are available, one has to use both forward and back-directed reasoning models to estimate the belief distribution of the predicates corresponding to respective predecessors and successors of the given non-terminal places. However, under this case, the estimated beliefs of the predicates may not be consistent. In other words, after obtaining steady-state beliefs at all places, if one re-computes beliefs of the non-axiom predicates with the known beliefs of the axiom predicates, the computed beliefs may not tally with their initial values. In order to overcome this problem, one requires a special relationship, called reciprocity [26]. It may be noted that in a FPN that holds (perfect) reciprocity property, n successive steps of forward (backward) reasoning followed by n successive steps of backward (forward) reasoning restores the value of the belief vector $N(t)$.

Definition 10.12: A FPN is said to hold **reciprocity property** if updating FTT (belief) vector in the forward direction followed by updating of FTT (belief) vector in the backward direction restores the value of the FTT (belief) vector.
Formally, we estimate $T_f(t+1)$ from given $N_f(t)$ and $N_f(t)$ from $T_f(t+1)$ in succession,

i.e., $T_f(t+1) = R_{fm} \, o \, (Q'_{fm} \, o \, N_f{}^c(t))^c$ (10.17)

and $N_f(t) = P'_{bm} \, o \, T_f(t+1).$ (10.18)

Combining equations (10.17) and (10.18), we have

$N_f(t) = P'_{bm} \, o \, R_{fm} \, o \, (Q'_{fm} \, o \, N_f{}^c(t))^c$

       $= (Q'_{fm})^T \, o \, R_{fm} \, o \, (Q'_{fm} \, o \, N_f{}^c(t))^c.$ [by theorem 10.5]   (10.19)

Further, from the definition 10.14, one may first estimate $N_f(t+1)$ from $T_f(t+1)$ and then $T_f(t+1)$ from $N_f(t+1)$. Formally

$N_f(t+1) = P'_{fm} \, o \, T_f(t+1)$ (10.20)

and $T_f(t+1) = R_{bm} \, o \, (Q'_{bm} \, o \, N_f{}^c(t+1)).$ (10.21)

Combining (10.20) and (10.21) we have

$T_f(t+1) = R_{bm} \, o \, (Q'_{bm} \, o \, N_f{}^c(t+1))^c$

$$= R_{bm} \, o \, ((P'_{fm})^T \, o \, N_f^c(t+1))^c \quad [\text{ by theorem } 10.5]$$

$$= R_{bm} \, o \, [(P'_{fm})^T \, o \, (P'_{fm} \, o \, T_f(t+1))^c]^c. \tag{10.22}$$

Expression (10.19) and (10.22), which are identities of N_f and T_f respectively, taken together is called a reciprocity relation. For testing reciprocity conditions, one, however, has to use the results of Theorem 10.7.

Theorem 10.7: *The condition of reciprocity in a FPN is given by*

$$(Q'_{fm})^T \, o \, R_{fm} \, o \, (Q'_{fm} \, o \, I^c)^c \; = \; I \tag{10.23 (a)}$$

and $\quad R_{bm} \, o \, [(P'_{fm})^T \, o \, (P'_{fm})^c]^c \; = \; I \tag{10.23(b)}$

Proof: Proof is presented in Appendix C.   □

Example 10.7: Consider the FPN given in fig. 10.13. Given $R_{fm} = I$ and $R_{bm} = I$, we want to test the reciprocity property of the FPN.

Here, $\quad P'_{fm} = \begin{pmatrix} \phi & \phi & I \\ I & \phi & \phi \\ \phi & I & \phi \end{pmatrix} \quad$ and $\quad Q'_{fm} = \begin{pmatrix} I & \phi & \phi \\ \phi & I & \phi \\ \phi & \phi & I \end{pmatrix}$

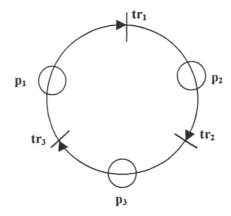

Fig. 10.13: A FPN used to illustrate the reciprocity property.

where ϕ and **I** denote null and identity matrices of dimension (3 x 3) respectively. With these values of **P'**$_{fm}$ and **Q'**$_{fm}$, we found that the reciprocity conditions 10.23(a) and (b) hold good.

It is clear from expressions 10.23(a) and (b) that the condition of reciprocity depends on both the structure of the FPN and the relational matrices associated with it. Thus for a given structure of an FPN, identification of the relational matrices (**R**$_{fm}$, **R**$_{bm}$) satisfying the reciprocity conditions is a design problem. In fact, rearranging expression 10.23 (a) and (b) ,we find **R**$_{fm}$ and **R**$_{bm}$ as follows

$$\mathbf{R}_{fm} = [(\mathbf{Q'}_{fm})^{T}]^{-1}_{pre} \; 0 \; [(\mathbf{Q'}_{fm} \, 0 \, \mathbf{I}^{c})^{c}]^{-1}_{post} \qquad (10.24)$$

$$\mathbf{R}_{bm} = [\{(\mathbf{P'}_{fm})^{T} \, 0 \, (\mathbf{P'}_{fm})^{c}\}^{c}]^{-1}_{post} \qquad (10.25)$$

where the suffix 'pre' and 'post' denote pre-inverse and post-inverse of the matrices.

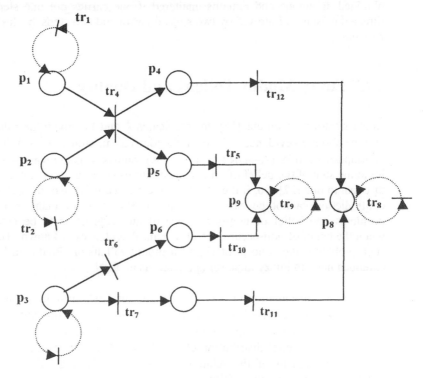

Fig. 10.14: The FPN of fig. 10.10 with self-loop around axioms (and renamed transitions) that supports reciprocity theorem.

Since such choice of $\mathbf{R_{fm}}$ and $\mathbf{R_{bm}}$ satisfy the reciprocity condition, it is expected that the belief distribution at a given place of the FPN would retrieve its original value after n-forward steps followed by n-backward steps of reasoning in the network. Consequently the steady-state belief distribution at all places in the FPN will be consistent independent of the order of forward and backward computation. This, in fact, is useful when the initial belief distribution of the intermediate [12] places only in the FPN is known.

Example 10.8: Consider the diagnosis problem, cited in example 10.7. We assume that the bi-directional IFF relationship exists between the predicates corresponding to input-output place pairs of the transitions in the network of fig. 10.14. We also assume that the belief distribution at places p_4 and p_5 only is known and one has to estimate the consistent beliefs at all places in the network. In the present context, we first estimate $\mathbf{R_{f\ m}}$ and $\mathbf{R_{bm}}$ by using expressions (10.24) and (10.25) and then carry out one step forward reasoning followed by two steps back-directed reasoning using expression (10.13) through (10.16). It has been checked that the steady-state belief vector, thus obtained, is unique and remains unaltered if one carries out one step back-directed reasoning followed by two steps forward and two steps back-directed reasoning.

10.7 Fuzzy Modus Tollens and Duality

In classical modus tollens [15], for predicates A and B, given the rule A→B and the observed evidence ¬ B, then the derived inference is ¬A. Thus the contrapositive rule: (A→B)⇔ (¬ B→ ¬ A) follows. It is known that in fuzzy logic the sum of the belief of an evidence and its contradiction is greater than or equal to one [22]. So, if the belief of an evidence is known, the belief of its contradiction cannot be easily ascertained. However, in many real world problems, the belief of non-occurrence of an evidence is to be estimated, when the belief of non-occurrence of its causal evidences is known. To tackle such problems, the concept of classical modus tollens of Predicate logic is extended here to Fuzzy logic for applications in FPN.

Before formulation of the problem, let us first show that implication relations (A→B) and (¬B→ ¬A) are identical in the fuzzy domain, under the closure of Lukasiewciz implication function. Formally let a_i , $1\le i \le n$ and b_j , $1\le j \le m$ be the belief distribution of predicates A and B respectively. Then the (i, j)th element of the relational matrix $\mathbf{R_1}$ for the rule A→B by Lukasiewciz implication function is given by

$$\mathbf{R_1}\ (\ i\ ,j\ )\ = \text{Min}\ \{\ 1,\ (\ 1\text{-}a_i + b_j\ )\ \} \qquad (10.26)$$

Again, the (i, j) th element of the relational matrix R_2 for the rule $\neg B \rightarrow \neg A$ using Lukasiewciz implication function is given by

$$R_2(i,j) = Min[1, \{1-(1-b_j) + (1-a_i)\}] = Min\{1, (1-a_i + b_j)\} \quad (10.27)$$

Thus it is clear from expressions (10.26) and (10.27) that the two relational matrices R_1 and R_2 are equal. So, classical modus tollens can be extended to fuzzy logic under Lukasiewciz implication relation. Let us consider a FPN (vide fig. 10.15(a)), referred to as the primal net, that is framed with the following knowledge base

> *rule 1:* $d_1, d_2 \rightarrow d_3$
> *rule 2:* $d_2 \rightarrow d_4$
> *rule 3:* $d_3, d_4 \rightarrow d_1$

The dual of this net can be constructed by reformulating the above knowledge base using the contrapositive rules as follows:

> *rule 1:* $\neg d_3 \rightarrow \neg d_1, \neg d_2$
> *rule 2:* $\neg d_4 \rightarrow \neg d_2$
> *rule 3:* $\neg d_1 \rightarrow \neg d_3, \neg d_4$

Here the comma in the R.H.S. of the if-then operator in the above rules represent OR operation. It is evident from the reformulated knowledge base that the dual FPN can be easily constructed by replacing each predicate's d_i by its negation and reversing the directivity in the network. The dual FPN of fig. 10.15(a) is given in fig. 10.15(b).

Reasoning in the primal model of FPN may be carried out by invoking the procedure: forward reasoning. Let $R = R_p$, $P' = P_p$ and $Q' = Q_p$ denote the matrices for the primal model in expression (10.8). Then for forward reasoning in the dual FPN, one should initiate $P' = (Q_p)^T$ and $Q' = (P_p)^T$ (vide theorem 10.5), and $R = R_p$ prior to invoking the procedure forward reasoning. If the belief distributions at the concluding places are available, the belief distribution at other places of the dual FPN may be estimated by invoking the procedure backward-reasoning with prior assignment of $Q'_{fm} = (P_p)^T$, $P'_{fm} = (Q_p)^T$ and $R_{fm} = R_p$.

Example 10.9: Consider the FPN of fig. 10.15(a), where $d_1 \equiv$ Loves (ram, sita), $d_2 \equiv$ Girl-friend (sita, ram), $d_3 \equiv$ Marries (ram, sita) and $d_4 \equiv$ Loves(sita, ram). Suppose that the belief distribution of $\neg$loves (ram, sita) and $\neg$Loves(sita, ram) are given as in fig. 10.16(a) and (b) respectively. We are interested to estimate the belief distribution of $\neg$girl-friend (sita, ram). For the sake of simplicity in calculation, let us assume that Th = 0 and R = I and estimate the steady-state belief distribution of the predicates in the network by using forward reasoning. The steady-state belief vector is obtained only after one step of belief revision in the network with the steady-state value of the predicates $\neg d_2$ equals to [0.85 0.9 0.95]^T

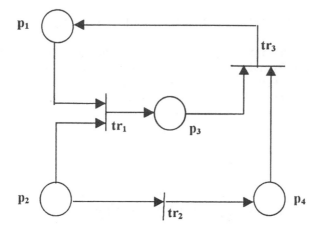

Fig. 10.15(a): The primal fuzzy Petri net.

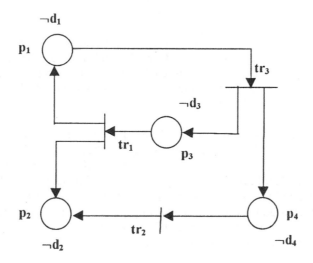

Fig. 10.15(b): The dual fuzzy Petri net corresponding to the primal of fig. 10.15(a).

10.8 Non-Monotonic Reasoning in a FPN

Inconsistent information often enter into a FPN because of i) the occurrence of inconsistent data in the database, ii) presence of inconsistent rules in the knowledge base and iii) imprecision of data in the database [16]. Inconsistencies are first detected by the reasoning system through consultation with a list that contains pairs of inconsistent information. Once the inconsistent pair of information are detected, the reasoning system attempts to resolve inconsistency by eliminating one information from each contradictory pair through voting. The voting in the present context is implemented by opening the output arcs of the contradictory pairs, such that these information cannot take part in the voting process and then continuing belief revision in the entire network by using the model represented by expression (10.8) until steady state is reached. The steady-state beliefs of the contradictory pairs are compared. The one with higher steady-state belief is selected out of each pair (for subsequent reasoning), while the belief of its contradiction is permanently set to zero, so that it cannot influence the reasoning process. In case the steady-state belief of both the contradictory pairs is equal, both of these are discarded from the reasoning space, by setting their beliefs permanently to zero. The arcs opened earlier are then re-established and the original beliefs of each place are re-assigned to all places, excluding the places whose beliefs are set to zero permanently [18].

In our prototype ES CRIMINVES [20], designed for criminal investigation, we implemented the above scheme for non-monotonic reasoning [16]. It may be mentioned here that reasoning of all possible types, covered in the last few sections, can be similarly carried out in the presence of contradictory evidences

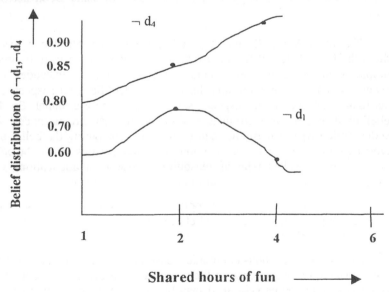

Fig. 10.16: Belief distribution of ¬Loves (ram, sita) and ¬Loves (sita; ram).

10.9 Conclusions

The chapter presented a new methodology for reasoning in ES, which is contaminated with imprecision and inconsistency of database and uncertainty of knowledge base. Most of the common types of first order logic tools, such as modus ponens, modus tollens, and abduction, which are applicable to monotonic systems only, have been extended in this chapter for inexact and non-monotonic systems using FPN. Moreover, self reference in the knowledge base, which was intractable so far, can be handled following the techniques presented here.

The reasoning methodologies presented in this chapter have applications to a wide range of real world problems. For example, fuzzy modus ponens type of reasoning has been used for identifying criminals from imprecise and inconsistent word description [20] of criminal history. The fuzzy abductive reasoning scheme can be applied to diagnostic problems. For instance, in medical diagnosis problems, the (steady-state) fuzzy belief of the possible diseases, mapped at the axioms, can be computed if the initial fuzzy beliefs of the observed evidences, mapped at the concluding places, are supplied. The predicates corresponding to the axioms with the highest steady state belief may be inferred as the possible disease for the given symptoms. Since the fuzzy belief of an axiom predicate (a hypothesis) is computed from the fuzzy belief of a number of concluding predicates (observed evidence), the proposed scheme has analogy with data fusion [38] using Dempster-Shafer theory.

The fuzzy Modus Tollens type of reasoning is applicable in systems, where the fuzzy belief of the nonexistence of one or more facts is used for computing the belief of non-existence of the other facts embedded in the system. Suppose that in a criminal history one identifies four suspects and it was found that one of the suspects, say A, cannot be a criminal. So, fuzzy belief of A not being a criminal is known. In such circumstances, fuzzy Modus Tollen type of reasoning can be used for computing the belief of not being a criminal for any one of the remaining three persons. It may be pointed out that Modus Tollen type of reasoning is useful in applications where sensitivity analysis [15] of a system in fuzzy domain is required.

The reciprocity property has applications in estimating consistent fuzzy beliefs at all places from known beliefs of one or more predicates located at the intermediate places.

The scheme for non-monotonic reasoning is used in an FPN for deriving stable fuzzy inference in the presence of inconsistent / contradictory evidences in a reasoning system. It may be noted that steady-state vector N^* for such a system contains a stable fuzzy belief of each predicate with respect to all others in the network, even in the presence of inconsistent predicates like p and negation of p. The results of the reasoning process considered in this chapter differ from that of McDermott's logic [25] on the following considerations. In McDermott's logic, there exists two stable points in a system represented by $p \rightarrow \neg q$, and $q \rightarrow \neg p$, whereas the present method leads to only one stable point, involving either p or q, depending on the initial fuzzy beliefs of p, q and their supporting evidences.

Exercises

1. For the FPN given in fig. 10.3, identify the $\mathbf{P}$, $\mathbf{Q}$, $\mathbf{P'_{fm}}$, $\mathbf{Q'_{fm}}$ matrices. Assuming that the relational matrices associated with the transitions to be the identity matrix and an arbitrary belief vector $\mathbf{N(0)}$, compute $\mathbf{N(2)}$ by an appropriate forward reasoning model. What guideline should you suggest to identify the appropriate reasoning model for a given FPN?

2. Identify the cycle in the FPN of fig. 10.6 by using the algorithm for cycle detection.

3. From the given belief vectors n_5 and n_7 in the FPN of fig. 10.8, determine the belief vectors $\mathbf{n_1}$ and $\mathbf{n_2}$ by using the backward reasoning algorithm. Assume that the relational matrices are $\mathbf{I}$.

4. Prove that for a purely cyclic net $(\mathbf{P} \ o \ \mathbf{Q})^k = \mathbf{I}$ when k = number of transitions in the cycle.

5. Given that pre- and post-inverse of matrix $\mathbf{I}$ is $\mathbf{I}$. Hence show that reciprocity relations hold perfectly for a purely cyclic net. Also show that $\mathbf{R_{fm}}$ and $\mathbf{R_{bm}}$ for such net = $\mathbf{I}$.

6. Prove logically that a dual net can always be constructed by reversing the arrowheads in a primal net.

7. Can you devise an alternative formulation of the fuzzy inversion of matrices? [*open ended problem*]

8. Does the algorithm for computing fuzzy inverse apply to binary matrices? If yes, can you use it for diagnostic applications in switching circuits? [*open ended problem*] [32]-[33].

References

[1]   Bugarin, A. J. and Barro, S., "Fuzzy reasoning supported by Petri nets", *IEEE Trans. on Fuzzy Systems,* vol. 2, no.2, pp 135-150,1994.

[2]   Buchanan, B. G., and Shortliffe E. H., *Rule Based Expert Systems: The MYCIN Experiment of the Stanford University,* Addison-Wesley, Reading, MA, 1984.

[3]   Cao, T. and Sanderson, A. C., "A fuzzy Petri net approach to reasoning   about uncertainty in robotic systems," *in Proc. IEEE Int. Conf. Robotics and Automation,* Atlanta, GA, pp. 317-322, May 1993.

[4]   Cao, T., "Variable reasoning and analysis about uncertainty with fuzzy Petri nets," *Lecture Notes in Computer Science,* vol. 691, Marson, M. A., Ed., Springer-Verlag, New York, pp. 126-145, 1993.

[5]   Cao, T. and Sanderson, A. C., "Task sequence planing using fuzzy Petri nets," *IEEE Trans. on Systems, Man and Cybernetics,* vol. 25, no.5, pp. 755-769, May 1995.

[6]   Cardoso, J., Valette, R., and Dubois, D., "Petri nets with uncertain markings", in *Advances in Petri nets,* Lecture Notes in Computer Science, Rozenberg, G., Ed., vol.483, Springer-Verlag, New York, pp. 65-78, 1990.

[7]   Chen, S. M., Ke, J. S. and Chang, J. F., "Knowledge representation using fuzzy Petri nets," *IEEE Trans. on Knowledge and Data Engineering,* vol. 2 , no. 3, pp. 311-319, Sept. 1990.

[8]   Chen, S. M., "A new approach to inexact reasoning for rule-based systems," *Cybernetic Systems,* vol. 23, pp. 561-582, 1992.

[9]   Daltrini, A., "Modeling and knowledge processing based on the extended fuzzy Petri nets," *M. Sc. degree thesis,* UNICAMP-FEE0DCA, May 1993.

[10]   Doyle, J., "Truth maintenance systems," *Artificial Intelligence,* vol. 12, 1979.

[11]   Garg, M. L., Ashon, S. I., and Gupta, P. V., "A fuzzy Petri net for knowledge representation and reasoning", *Information Processing Letters,* vol. 39, pp.165-171,1991.

[12] Graham, I. and Jones, P. L., *Expert Systems: Knowledge, Uncertainty and Decision*, Chapman and Hall, London, 1988.

[13] Hirota, K. and Pedrycz, W., " OR-AND neuron in modeling fuzzy set connectives," *IEEE Trans. on Fuzzy systems*, vol. 2 , no. 2 , May 1994.

[14] Hutchinson, S. A. and Kak, A. C., "Planning sensing strategies in a robot workcell with multisensor capabilities," *IEEE Trans. Robotics and Automation*, vol. 5, no. 6, pp.765-783, 1989.

[15] Jackson, P., *Introduction to Expert Systems*, Addison-Wesley, Reading, MA, 1988.

[16] Konar, A. and Mandal, A. K., "Uncertainty management in expert systems using fuzzy Petri nets ," *IEEE Trans. on Knowledge and Data Engineering*, vol. 8, no. 1, pp. 96-105, February 1996.

[17] Konar, A. and Mandal, A. K., "Stability analysis of a non-monotonic Petri net for diagnostic systems using fuzzy logic," *Proc. of 33rd Midwest Symp. on Circuits, and Systems*, Canada, 1991.

[18] Konar, A. and Mandal, A. K., "Non-monotonic reasoning in expert systems using fuzzy Petri nets," Advances in Modeling & Analysis, B, *AMSE Press*, vol. 23, no. 1, pp. 51-63, 1992.

[19] Konar, S., Konar, A. and Mandal, A. K., "Analysis of fuzzy Petri net models for reasoning with inexact data and knowledge-base," *Proc. of Int. Conf. on Control, Automation, Robotics and Computer Vision*, Singapore, 1994.

[20]   Konar, A., "*Uncertainty Management in Expert System using Fuzzy Petri Nets*," Ph. D. dissertation , Jadavpur University, India, 1994.

[21] Konar, A. and Pal, S., Modeling cognition with fuzzy neural nets, in *Fuzzy Theory Systems: Techniques and Applications*, Leondes, C. T., Ed., Academic Press, New York, 1999.

[22] Kosko, B., *Neural Networks and Fuzzy Systems*, Prentice-Hall, Englewood Cliffs, NJ, 1994.

[23] Lipp, H. P. and Gunther, G., "A fuzzy Petri net concept for complex decision making process in production control," in *Proc. First European congress on fuzzy and intelligent technology (EUFIT '93)*, Aachen, Germany, vol. I, pp. 290 – 294, 1993.

[24] Looney, C. G., "Fuzzy Petri nets for rule-based decision making," *IEEE Trans. on Systems, Man, and Cybernetics*, vol. 18, no.1, pp.178-183, 1988.

[25] McDermott, V. and Doyle, J., "Non-monotonic logic I," *Artificial Intelligence*, vol. 13 (1-2), pp. 41-72, 1980.

[26] Murata, T., "Petri nets: properties, analysis and applications", *Proceedings of the IEEE*, vol. 77, no.4, pp. 541-580,1989.

[27] Pal, S. and Konar, A., "Cognitive reasoning using fuzzy neural nets," *IEEE Trans. on Systems , Man and Cybernetics*, August, 1996.

[28] Peral, J., "Distributed revision of composite beliefs," *Artificial Intelligence*, vol. 33, 1987.

[29] Pedrycz, W. and Gomide, F., "A generalized fuzzy Petri net model," *IEEE Trans. on Fuzzy systems*, vol . 2, no.4, pp 295-301, Nov 1994.

[30] Pedrycz, W, *Fuzzy Sets Engineering*, CRC Press, Boca Raton, FL, 1995.

[31] Pedrycz, W., "Fuzzy relational equations with generalized connectives and their applications," *Fuzzy Sets and Systems*, vol. 10, pp. 185-201, 1983.

[32] Pedrycz, W. and Gomide, F., *An Introduction to Fuzzy Sets: Analysis and Design*, MIT Press, Cambridge, MA, pp. 85-126, 1998.

[33] Saha, P. and Konar, A.,"Backward reasoning with inverse fuzzy relational matrices," *Proc. of Int. Conf. on Control, Automation, Robotics and Computer vision* , Singapore, 1996.

[34] Saha, P. and Konar, A., "A heuristic approach to computing inverse fuzzy relation," Communicated to *Information Processing Letters*.

[35] Scarpelli, H. and Gomide, F., "High level fuzzy Petri nets and backward reasoning," in *Fuzzy Logic and Soft Computing*, Bouchon-Meunier, Yager and Zadeh, Eds., World Scientific, Singapore, 1995.

[36] Scarpelli, H., Gomide, F. and Yager, R.., "A reasoning algorithm for high level fuzzy Petri nets," *IEEE Trans. on Fuzzy Systems*, vol. 4, no. 3 , pp. 282-295, Aug. 1996.

[37]   Scarpelli, H. and Gomide, F., "Fuzzy reasoning and fuzzy Petri nets in manufacturing systems modeling," *Journal of Intelligent Fuzzy Systems*, vol. 1, no. 3, pp. 225-241, 1993.

[38]   Shafer, G., *A Mathematical Theory of Evidence*, Princeton University Press, Princeton, NJ, 1976.

[39]   Shenoy, P. P. and Shafer, G., "Propagating belief functions with local computations," *IEEE Expert*, pp. 43-51, Fall 1986.

[40]   Yu, S. K.,"Comments on 'Knowledge Representation Using Fuzzy Petri nets'," *IEEE Trans. on Knowledge and Data Engineering*, vol. 7, no.1, pp. 190-191, Feb 1995.

[41]   Yu, S. K., " Knowledge representation and reasoning using fuzzy Pr\ T net-systems," *Fuzzy Sets and Systems* , vol. 75, pp. 33-45, 1995.

[42]   Zadeh, L. A. "The role of fuzzy logic in the management of uncertainty in expert system," *Fuzzy Sets and Systems*, vol .11, pp. 199-227,1983.

11

Reasoning with Space and Time

The chapter presents the models for reasoning with space and time. It begins with spatial axioms and illustrates their applications in automated reasoning with first order logic. Much emphasis has been given on the formalization of the spatial relationships among the segmented objects in a scene. Fuzzy spatial relationship among 2-D objects has been briefly outlined. The application of spatial reasoning in navigational planning of mobile robots has also been highlighted. The second half of the chapter deals with temporal reasoning. The principles of temporal reasoning have been introduced from the first principles by situation calculus and first order temporal logic. The need for reasoning with both space and time concurrently in dynamic scene interpretation is also outlined at the end of the chapter.

11.1 Introduction

The reasoning problems we came across till now did not involve space and time. However, there exist many real world problems, where the importance of space and time cannot be ignored. For instance, consider the problem of navigational planning of a mobile robot in a given workspace. The robot has to plan its trajectory from a pre-defined starting point to a given goal point. If the robot knows its world map, it can easily plan its path so that it does not touch the obstacles in its world map. Now, assume that the robot has no prior knowledge about its world. In that case, it has to solely rely on the data it receives by its sonar and laser sensors or the images it grabs by a camera and processes these on-line. Representation of the space by some formalism and developing an efficient search algorithm for matching of the spatial data, thus, are prime considerations. Now, let us assume that the obstacles in the robot's world are dynamic. Under this circumstance, we require information about both space and time. For example, we must know the velocity and displacements of the obstacles at the last instance to determine the current speed and direction of the robot. Thus there is a need for both spatial and temporal representation of information. This is a relatively growing topic in AI and we have to wait a few more years to get a composite representation of both space and time.

Spatial reasoning problems can be handled by many of the known AI techniques. For instance, if we can represent the navigational planning problem of a robot by a set of spatial constraints, we can solve it by a logic program or the constraint satisfaction techniques presented in chapter 19. Alternatively, if we can represent the spatial reasoning problem by predicate logic, we may employ the resolution theorem to solve it. But how can we represent a spatial reasoning problem? One way of doing this is to define a set of spatial axioms by predicates and then describe a spatial reasoning problem as clauses of the spatial axioms. In this book we used this approach for reasoning with spatial constraints.

The FOL based representation of a spatial reasoning problem sometimes is ambiguous and, as a consequence, the ambiguity propagates through the reasoning process as well. For example, suppose an object X *is not very close* to object Y in a scene. Can we represent this in FOL? If we try to do so then for each specific distance between two objects, we require one predicate. But how simple is the representation in fuzzy logic! We need to define a membership function of '*Not-very-close*' versus distance, and can easily obtain the membership value of *Not-very-close* (X, Y) with known distance between X and Y. The membership values may later be used in fuzzy reasoning. A section on fuzzy reasoning is thus introduced for spatial reasoning problems.

Reasoning with time is equally useful like reasoning in space. How can one represent that an occurrence of an event A at time t, and another event B at time t+1, causes the event C to occur at time t+2? We shall extend the First order logic to two alternative forms to reason with this kind of problem. First one is called the **situation calculus**, after John McCarthy, the father of AI.

The other one is an extension by new temporal operators; we call it **propositional temporal logic**.

Section 11.2 describes the principles of spatial reasoning by using a set of spatial axioms. The spatial relationship among components of an object is covered in section 11.3. Fuzzy spatial representation of objects is presented in section 11.4. Temporal reasoning by situation calculus and by propositional temporal logic is covered in section 11.5 and 11.6 respectively. The formalisms of interval temporal logic is presented in section 11.7. The significance of the spatial and temporal reasoning together in a system is illustrated in section 11.8.

11.2 Spatial Reasoning

Spatial reasoning deals with the problems of reasoning with space. Currently, to the best of the author's knowledge, there exist no well-organized formalisms for such reasoning. So we consider a few elementary axioms based on which such reasoning can be carried out. These axioms for spatial reasoning we present here, however, are not complete and may be extended for specific applications.

Axioms of Spatial Reasoning

Axiom 1: Consider the problems of two non-elastic objects O_i, O_j. Let the objects be infinitesimally small having 2D co-ordinates (x_i, y_i) and (x_j, y_j) respectively. From commonsense reasoning, we can easily state that

$$\forall O_i, O_j, \ x_i \neq x_j \ \text{and} \ y_i \neq y_j.$$

Formally,

$$\forall O_i, O_j \ \text{Different} \ (O_i, O_j) \rightarrow \neg(\ Eq(x_i, x_j) \wedge Eq \ (y_i, y_j) \ ).$$

An extension of the above principle is that no two non-elastic objects, whatever may be their size, cannot occupy a common space. If S_i and S_j are the spaces occupied by O_i and O_j respectively,

then $S_i \cap S_j = \phi$,

$\Rightarrow \neg (Si \cap Sj) = \text{true}$

$\Rightarrow \neg S_i \cup \neg S_j \ \text{is true}.$

Formally,

$$\forall O_i, O_j \quad S_i(O_i) \wedge S_j(O_j) \wedge \neg Eq(O_i, O_j) \rightarrow \neg S_i(O_i) \neg \vee S_j(O_j).$$

In the above representation, the AND ($\wedge$) and OR ($\vee$) operators stand for intersection and union of surfaces or their negations (complements).

Further, $\forall O_i, O_j$ means $O_i, O_j \in S$, where S is the entire space that contains O_i, and O_j, vide fig. 11.1.

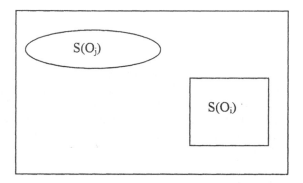

Fig. 11.1: Space S containing object O_i and O_j having 2-D
surfaces $S(O_i)$ and $S(O_j)$.

In our formulation, we considered two dimensional spaces S , $S(O_i)$ and $S(O_j)$. However, we can easily extend the principle to three dimensions.

Axiom 2: When an object O_i enters the space S , $S \cap S(O_i) \neq \phi$, which implies
$$S \wedge S(O_i) \text{ is true}.$$

Formally ,
$$\forall O_i \ S(O_i) \wedge Enter(O_i, S) \rightarrow S \wedge S(O_i).$$

Similarly, when an object O_i leaves a space S, $S \cap S(O_i) = \phi$,

Or, $\neg S \vee \neg S(O_i)$ is true.

Formally, $\forall O_i \ S(O_i) \wedge Leaves(O_i, S) \rightarrow \neg S \vee \neg S(O_i)$.

Axiom 3: When the intersection of the surface boundary of two objects is a non-null set, it means either one is partially or fully embedded within the other, or they touch each other. Further, when a two dimensional surface touches another, the common points must form a 2-D line or a point. Similarly, when a 3-dimensional surface touches another, the common points must be a 3-D /2-D surface or a 3-D / 2-D line or a point. It is thus evident that two objects touch each other, when their intersection of surface forms a surface of at most their dimension. Formally,

$\forall O_i$, $\forall$ O_j Less-than-or-Equal-to (dim($S(O_i) \wedge S(O_j)$)), dim ($S(O_l)$)) $\wedge$
Less-than-or-Equal-to (dim($S(O_i) \wedge S(O_j)$)), dim ($S(O_j)$)) $\rightarrow$ Touch (O_i, O_j)

Where 'dim' is a function that returns the dimension of the surface of its argument and dim ($S(O_i) \wedge S(O_j)$)) represents the dimension of the two intersecting surfaces: O_i and O_j. The $\wedge$-operator between the predicates Less-than-or-Equal-to denotes logical AND operation.

Axiom 4: Now, for two scenes if d_{ij1} and d_{ij2} denote the shortest distance between the objects O_i and O_j in scene 1 and 2 respectively, then if $d_{ij2} < d_{ij1}$, we can say the objects O_i and O_j are closer in scene 2 compared to that in scene 1. Formally,

$\forall$ O_i , O_j Exists (O_i , O_j , in-scene1) $\wedge$ Shortest -distance ($d_{ij\,1}$, Oi , Oj , in-scene1) $\wedge$ Exists (O_i , O_j, in -scene2) $\wedge$ Shortest -distance ($d_{ij}2$, O_i , O_j , in-scene2) $\wedge$ smaller ($d_{ij\,2}$, $d_{ij\,1}$) $\rightarrow$ Closer (O_i , O_j , in-scene2 , wrt-scene =1);

where the predicate Exists (O_i, O_j ,in-scene k) means O_i and O_j exists in scene k; Shortest distance (d_{ijk} , O_i , O_j , in-scene k , wrt-scene =1) denotes that d_{ijk} is the shortest distance between O_i , and O_j in scene k with respect to scene 1.

The axioms of spatial reasoning presented above can be employed in many applications. One typical application is the path planning of a mobile robot. Consider, for example, the space S, where a triangular shaped mobile robot has to move from a given starting to goal point, without touching the obstacle O1, O2, O3, O4, ... , O7.

We can construct a **constraint logic program** (CLP) to solve this problem. We assume that the robot R can sense the obstacles from a distance by ultrasonic sensors, located around the boundary of it. The CLP of this problem is presented below.

```
Move(R, Starting -position , goal -position ) : -
Move S( R) in S,
not Touch(S(R ), S (O_i)) ∀ O_i.

Move(R, goal-position, goal-position).
```

The above program allows the robot R to wander around its environment, until it reaches the goal-position. The program ensures that during the robot's journey it does not hit an obstacle. Now, suppose, we want to include that the robot should move through a shortest path. To realize this in the CLP we define the following nomenclature.

1. Next-position(R): It is a function that gives the next-position of a robot R.

2. S (next-position (R)): It is a function, representing the space to be occupied by the robot at the next-position of R.

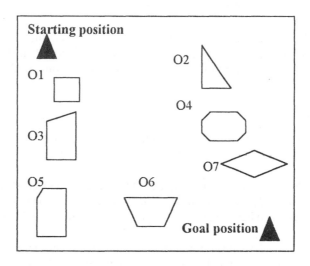

Space S
Fig 11.2: Path planning of a robot R in space S.

It is to be noted that the robot should select arbitrary next position from its current position and then would test whether the next-position touches any object. If yes, it drops that next-position and selects an alternative one until a next-position is found, where the robot does not touch any obstacle. If more than one next-position is found, it would select that position, such that the sum of the distances between the current and the next-position, and between the next position and the goal, is minimum.

The CLP for the above problem will be presented next. A pseudo Pascal algorithm is presented below for simplicity.

Procedure Move-optimal (R , Starting-position, goal-position)
Begin
 Current-position (R) := Starting-position (R);
 While goal not reached **do**
 Begin
   Repeat
     Find-next-position (R) ;
     j:=1;
     If S (next-position(R)) does not touch $S(O_i)$ $\forall i$;
     Then do
      Begin
      Save next-position (R) in A[j] ;
       j=j +1;
      End ;
   Until all possible next positions are explored;
   $\forall j$ Find the next-position that has the minimum distance from the
     current position of R and the goal; Call it A[k].
  current-position(R) := A [k] ;
 End while
End

We now present the CLP that takes care of the two optimizing constraints: i)
movement of the robot without touching the obstacles, and ii) traversal of an
optimal (near- optimal) path.

Move-optimal (R, Starting-position, goal-position):-
    Move S(R) in S,
    Not Touch(S(R), S (O_i)) $\forall$ i,
    Distance (next-position (R) , current-position (R)) +
    Distance (next-position (R) , goal-position)
    is minimum $\forall$ feasible next-position(R),
    current-position (R) $\leftarrow$ next-position (R),
    Move-optimal (R, current-position, goal-position).

Move-optimal (R, goal-position, goal-position).

It is to be noted that here we need not explicitly define Touch (S (R), S (O_i))

as it is presumed to be available in the system as a standard predicate,
following axiom 3. Further, we can re-define the distance constraint in the last
program by axiom 4 as follows:

Closer (next-position (R), current-position (R), in-scene k, w.r.t scene $\neq$ k),

Closer (next-position (R), goal-position, in-scene k, w.r.t scene ≠ k).

The significance of the spatial axioms now is clear. It helps in declaring the problem specifications in simpler terms, rather than formulating the problem from the grass-root level.

11.3 Spatial Relationships among Components of an Object

Many physical and geometric objects can be recognized from the spatial relationship among its components. For instance, let us define a chair as an object consisting of two planes abcf and cdef having an angle θ between them, where $\theta \leq 90^0+\alpha$ and where $0 \leq \alpha \leq 45^0$. Further, one of its plane is perpendicular to at least 3 legs (the 4th one being hidden in the image). So, we define:

Object(chair):-

        Angle-between (plane1, plane2, 90+α) ,
        Greater-than(α, 0),
        Less-than (α, 45) ,
        Parallel (line1, line2, line3) ,
        perpendicular (line1, plane1),!

For actual realization of the small program presented above, one has to define equation of lines and planes; then one has to check the criteria listed in the logic program. It may be noted here that finding equation of a line in an image is not simple. One approach to handle this problem is to employ a stochastic filter, such as Kalman filtering [1] . We shall discuss this issue once again in chapter 17 on visual perception. However, for the convenience of interested readers, we say a few words on the practical issues.

A skeleton of a chair, which can be obtained after many elementary steps of image processings is presented in fig. 11.3. Now, the equation of the line segments is evaluated approximately from the set of 2-dimensional image points lying on the lines. This is done by employing a Kalman filter. It may be noted that the more the number of points presented to the filter, the better would be accuracy of the equation of the 2-dimensional lines. These 2-D lines are then transformed to 3-D lines by another stage of Kalman filtering. Now, given the equation of the 3-D lines, one can easily evaluate the equation of the planes framed by the lines by using analytical geometry. Lastly, the constraints like the angles between the planes, etc. are checked by a logic program, as described above. The graduate students of the ETCE department at Jadavpur University verified this method of recognizing a 3-D planer object from its skeletal model.

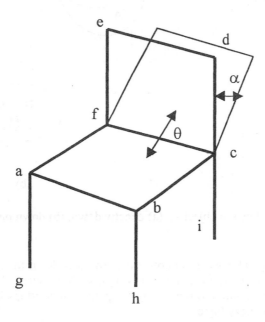

Fig. 11.3: Spatial relations among components of a skeleton chair.

11.4 Fuzzy Spatial Relationships among Objects

Consider the objects A and B in fig. 11.4 (a) and (b). We would say that B is left to A. It , however, is to be noted that B and A have some overlap in (a) but there is no overlap in (b).

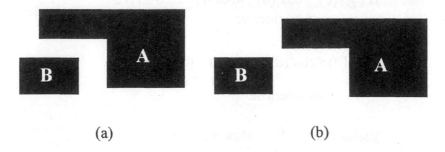

(a)   (b)

Fig. 11.4: Object B is left to object A: (a) with overlap, (b) without overlap.

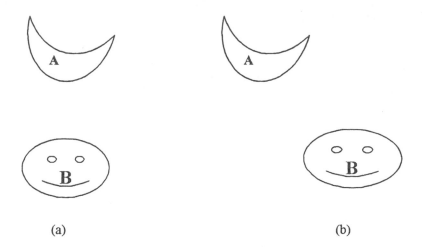

(a)   (b)

Fig. 11.5: Object B is down to object A: (a) exactly down, (b) down but right
         shifted.

Now consider fig. 11.5 where in both (a) & (b) B is down to A; but in
(a) B is exactly down to A, whereas in (b) it is right shifted a little. To define
these formally, we are required to represent the spatial relationships between
the objects A and B by fuzzy logic.

Let us first define spatial relations between points A and B. We
consider four types of relations: right, left, above and below. Here following
Miyajima and Ralescu [4], we define the membership function as a square of
sine or cosine angles θ (vide fig. 11.6), where θ denotes the angle between
the positive X axis passing through point A and the line joining A and B. The
membership functions for the primitive spatial relations are now given as
follows:

$$\mu_{right}(\theta) = \cos^2(\theta) \text{ , when } -\Pi/2 \leq \theta \leq \Pi/2 \text{ ,}$$
$$= 0 \text{, otherwise.}$$

$$\mu_{left}(\theta) = \cos^2(\theta) \text{ , when } -\Pi < \theta < -\Pi/2 \text{ ,}$$
$$\text{and } \Pi/2 \leq \theta \leq \Pi$$
$$= 0 \text{, otherwise.}$$

$$\mu_{below}(\theta) = \sin^2\theta \text{ , when } 0 \leq \theta \leq \Pi \text{ ,}$$
$$= 0 \text{, otherwise.}$$

$$\mu_{above}(\theta) = \sin^2\theta \quad , \text{ when } -\prod \le \theta \le 0 ,$$
$$= 0, \quad \text{otherwise.}$$

A common question that now arises is why we select such functions. As an example, we consider the 'below membership function'. Let us compute $\mu_{below}(\theta)$ at a regular interval of $\theta = \prod/4$, in the graph $0 \le \theta \le \prod$. Fig. 11.5 presents the membership values for different θ. It is clear from the figure that when B is exactly below A (fig. 11.6(c)) $\mu_{below}(\theta = \prod/2) = 1$, which is logically appealing. Again when $\theta = \prod/4$ or $\theta = 3\prod/4$ (fig. 11.6 (b) & (d)), the membership value of $\mu_{below}(\theta) = 1/2$; that too is logically meaningful. When $\theta = 0$ (fig. 11.6(a)) or $\prod$, $\mu_{below}(\theta) = 0$, signifying that B is not below A. The explanation of other membership functions like $\mu_{right}(\theta)$, $\mu_{left}(\theta)$, $\mu_{above}(\theta)$ can be given analogously.

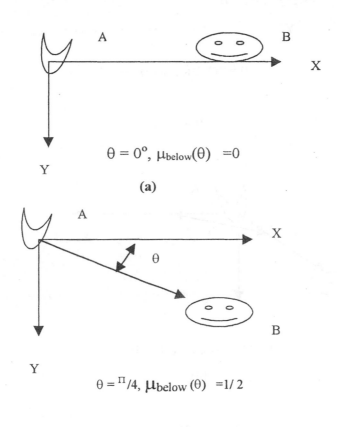

$$\theta = 0°, \mu_{below}(\theta) = 0$$

(a)

$$\theta = {}^\prod/4, \mu_{below}(\theta) = 1/2$$

(b)

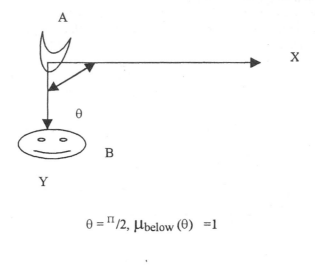

$$\theta = {}^{\Pi}/2, \; \mu_{below} \, (\theta) \; = 1$$

(c)

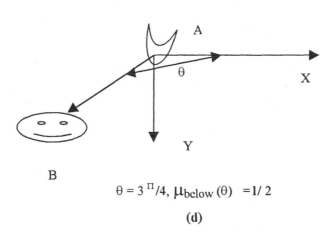

$$\theta = 3 \, {}^{\Pi}/4, \; \mu_{below} \, (\theta) \; = 1/ \, 2$$

(d)

Fig. 11.7: Illustrating significance of the $\mu_{below} \, (\theta)$ function.

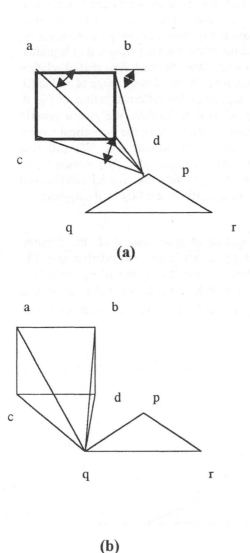

(a)

(b)

Here in (a) and (b) computation of angles (w.r.t the horizontal axis) of the lines joining the vertices of the rectangle to the vertices p and q of the triangle have been illustrated. Similar computations have to be performed for the line joining the vertices of the rectangle to the vertex r of the triangle. All 8 angles have not been shown in the figure for clarity.

Fig. 11.7: Demonstrating f (θ) / (m .n) computation.

So far we discussed spatial relationship between two points by fuzzy measure. Now, we shall discuss the spatial relationships between two objects. Let A and B be two objects and $\{a_i, 1 \le i \le n\}$, $\{b_j, 1 \le j \le m\}$ be the set of points on the boundary A and B respectively. We first compute the angle θ_{ij} between each two points a_I and b_j. Since there are n a_I points and m b_j points, the total occurrence of θ_{ij} will be (m x n). Now, for each type spatial relation like b_j below a_{Ii}, we estimate $\mu_{below}(\theta_{ij})$. Since θ_{ij} has a large range of value [0, Π], we may find equal value of $\mu_{below}(\theta_{ij})$ for different values of θ_{ij}. A frequency count of $\mu_{below}(\theta_{ij})$ versus θ_{ij} is thus feasible. We give a generic name f (θ) to the frequency count. Since f (θ) can have the theoretical largest value (n. m), we divide f (θ) by (m. n) to normalize it. We call that normalized frequency $\underline{f}(\theta) = $ f (θ) /(m .n). We now plot $\underline{f}(\theta)$ versus θ and find where it has the largest value. Now to find the spatial relationship between A and B, put the values of θ in $\mu_{below}(\theta)$ where $\underline{f}(\theta)$ is the highest.

In fig. 11.7 we illustrate the method of measurement of the possible θ_{ij}s. Since abcd is a rectangle and pqr is a triangle, considering only the vertices, m .n =3. 4 =12. We thus have 12 possible values of θ_{ij}. So $\underline{f}(\theta) = $ f(θ)/12. It is appearing clear that $\underline{f}(\theta)$ will have the largest value at around 45 degrees (fig. 11.8); consequently $\mu_{below}(\theta=45^\circ)$ gives the membership of pqr being below abcd.

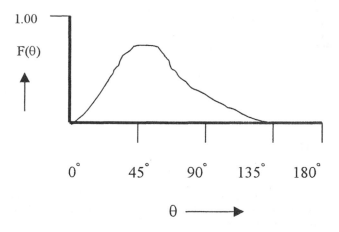

Fig. 11.8: Theoretical $\underline{f}(\theta)$ versus θ for example cited in fig. 11.7.

11.5 Temporal Reasoning by Situation Calculus

'Temporal reasoning', as evident from its name, stands for reasoning with time. The problems THE real world contain many activities that occur at a definite sequence of time. Further there are situations, when depending upon the result of occurrence of one element at a time t, a second event occurs at some time greater than t. One simple way to model this is to employ 'situation calculus' devised by John McCarthy [3].

The reasoning methodology of situation calculus is similar with first order predicate logic. To understand the power of reasoning of situation calculus, we are required to learn the following terminologies.

Definition 11.1: An **event** stands for an activity that occurs at some time.

Definition 11.2: A **fluent** is a fact that is valid at some given time frame but becomes invalid at other time frames.

Definition 11.3: A **situation** is an interval of time during which there is no change in events.

The following example illustrates the above definitions.

Example 11.1: Consider the following facts:

1. It was raining one hour ago.
2. People were moving with umbrellas in the street.
3. The rain has now ceased.
4. It is now noon.
5. The sky is now clear.
6. The sun is now shining brightly.
7. Nobody now keeps an umbrella open.

Statement 1, 2, 3, 6 and 7 in this example stand for events, while all statements 1-7 are fluent. Further, it is to be noted that we have two situations here; one when it was raining and the other when the rain ceased.

11.5.1 Knowledge Representation and Reasoning in Situation Calculus

To represent the statements 1-7 in situation calculus, we use a new predicate 'Holds'. The predicate Holds(s, f) denotes that the fluent f is true in situation s. Thus statement (1-7) can respectively be represented as:

> Holds (0, it-was-raining) (1)
>
> Holds (0, people-moving-with-umbrellas) (2)
>
> Holds (now, rain-ceased) (3)
>
> Holds (now, it-is-noon) (4)
>
> Holds (now, the-sky-is-clear) (5)
>
> Holds (now, the-sun-shining-brightly) (6)
>
> Holds (Results (now, the-sun-shining-brightly),
>
> not (anybody-keeps-umbrella-open)) (7)

The representation of the statements (1-6) in situation calculus directly follows from the definition of predicate 'Holds'. The representation of statement (7), however, requires some clarification. It means that the result (effect) of the sun shining brightly is the non-utilization of the umbrella. Further, 'not' here is not a predicate but is treated as a term (function). In other words, we cannot write not(anybody-keeps-umbrella-open) as ¬(anybody-keeps-umbrella-open).

For reasoning with the above facts, we add the following rules:

If it rains, people move with umbrellas. (8)

If the rains ceased and it is now noon then the result of sun shining brightly activates nobody to keep umbrella open. (9)

The above two rules in situation calculus are given by

$\forall$s Holds(s, it-was-raining)$\rightarrow$ Holds(s, people-moving-with-umbrellas) (8)

$\forall$s Holds(s, rain-ceased) $\land$ Holds(s, it-is-noon) $\rightarrow$ Holds (result (s, sun-shining-brightly), not (anybody-keeps-umbrella-open)) (9)

Reasoning: Let us now try to prove statement (7) from the rest of the facts and knowledge in the statements (1-9). We here call the facts axioms. So, we have:

Holds(now, rain-ceased)   Axiom (3)

Holds (now, it-is-noon)   Axiom (4)

$\forall$s Holds(s, rain-ceased) $\wedge$ Holds(s, it-is-noon)

  $\rightarrow$ Holds (results (s, sun-shining-brightly), not(anybody-keeps-

  umbrellas-open))   Rule (9)

For reasoning, we substitute s = now in (9) .

Thus we have

  $\forall$s , Holds (now, raining-ceased) $\wedge$

  Holds(now, it-is-noon) $\rightarrow$

  Holds (Results (now, the-sun-shining-brightly),

  not(anybody-keeps-umbrella-open))   (10)

Now, by modus ponens of (3) (4) and (10) ,

  Holds(Results(now, the-sun-shining-brightly),

  not(anybody-keeps-umbrella-open))

which is the desired inference.

11.5.2 The Frame Problem

The causal rules employed for reasoning in situation calculus specify the changes that occur, but they do not highlight which fluent remains unchanged from one frame (scene) to the next frame. For instance, the causal rule (9) indicates that "if the rain ceases at time s and s is noon then the result of sun shining causes nobody to keep their umbrella open." But it did not mention that the people who were moving in the earlier frame continue moving (with only closed umbrellas).

We can add one additional rule as follows to resolve this problem.

  $\forall s_1, s_2$ Holds(s_1, it-was-raining) $\wedge$

  Holds (s_1, people-moving-with-umbrellas) $\wedge$

  Holds (s2, rain-ceased) $\rightarrow$ Holds(s2, people-moving-without-

  umbrella).

11.5.3 The Qualification Problem

In many real world problems, we do not explicitly mention the conditions under which a given event will have a particular consequent. The problem of correctly stating the conditions in order to activate the consequent of an event in situation calculus is called *'qualification problem'* [3]. The problem can be solved by an appropriate use of non-monotonic reasoning. To illustrate the above problem and its possible solution, we present the following example.

Example 11.2: Suppose a computer is powered on. If we, now, press the reset key, it will start booting.

We can represent this by the following statement.

$\forall$s , Holds(s, on(power))$\rightarrow$

Holds (result (s, press-reset), booting) .

It is, however, presumed that there is no abnormality like malfunctioning of the keyboard or the system ROM or the booting software. Thus the above piece of knowledge will be more complete, if we say:

$\forall$s , Holds(s, on(power)) $\wedge\neg$abnormal (s, system)$\rightarrow$
Holds(result (s, press-reset), booting)

This obviously is a qualification over the last situation calculus expression.

Further, suppose that there is no abnormality in the system at time s but the stabilizer or the C.V.T. supplying power breaks down at time s. Thus we can write:

$\forall$s , $\neg$abnormal (s, system) $\rightarrow$

Holds (result (s, CVT-failure), not(booting)).

Suppose result (s, CVT-failure) =s'. We now want to identify the minimal abnormal interpretations such that Holds(s', not(booting)) is true. The following two interpretations in this regard are feasible. It is to be noted that s'>s and s' and s are both integers.

Interpretation I	Interpretation II
1.Holds (0,on(power))	1.Holds (0,on(power))
2.¬abnormal (0, system)	2.¬abnormal (0, system)
3.Holds (1, press-reset)	3.Holds (1, press-reset)
4.Holds (1, no(CVT-failure))	4.Holds (1, CVT-failure)
5.Holds (2, booting)	5.Holds (2, not(booting))

The facts embedded in the above two interpretations are all consistent (true); however interpretation I and II are contradictory. For instance, the inferences (4) and (5) in interpretation I and II are just opposite. A natural question then arises: which one of the interpretations is to be followed. Since abnormal predicate has the same status in both the interpretations, we can choose either of them. But in case abnormal (s, evidence) follows from one interpretation and abnormal (s, evidence) does not follow from the other interpretation, then the second one should be preferred. For more detailed treatment on this issue see Dean et al. [3].

11.6 Propositional Temporal Logic

In this section we will present an alternative form of extension of propositional logic for handling temporal variations of events. In fact, we shall use most of the formalisms of the propositional logic with two modal operators, **always (A)** and **sometimes (S)** [5]. Some authors [2], [7] denote always by $\forall$ and sometimes by ◆ . But as we already used them in non-monotonic reasoning for a different purpose, we intentionally use our own notations for these two operations.

Some elementary axioms of **propositional temporal logic** (PTL) are presented below:

1. $A (p \wedge q) \equiv A(p) \wedge A(q)$

2. $A (A(p)) \equiv A(p)$

3. $S (A(p)) \equiv A (S(p))$

4. $\neg S (p) \equiv S \neg (p)$

5. $A(p) \equiv \neg S(\neg p)$

6. $S (p) \equiv \neg A (\neg p)$

7. $A (p \rightarrow q) \rightarrow (A(p) \rightarrow A(q))$

11.6.1 State Transition Diagram
for PTL Interpretation

Consider a state transition graph where the nodes denote the temporal states and the arc denotes the transition from one state to another through passage of time. For instance, the state transition graph of fig. 11.9 describes the transition of temporal states from s_1 to s_2 and s_3, from s_2 to s_3 and s_4, from s_3 to s_2 and s_4 to s_4 itself. Further, each state s_i corresponds to a temporal value of the propositions p and q. For brevity of representation, we use the positive or negative literals like {p, q} or {¬p, q} instead of {p= true, q= true} or {p= false, q= true} respectively.

Now, suppose we want to evaluate the truth value of the formula

$$X = A(p) \vee A(q)$$

in each state.

In state s_1, X is true as its next states s_2 and s_3 both satisfy X. X is also true in s_2, as its next state s_3 satisfies A. X is also found to be true in s_3 as its next state s_2 supporting X. X is not true in s_4 as itself and its net state, which too is s_4, does not support A.

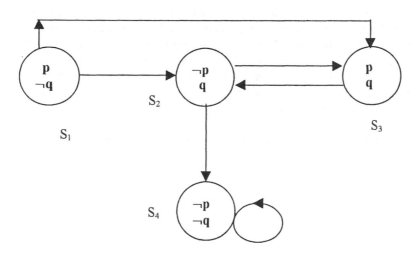

Fig. 11.9: A state transition graph representing PTL interpretation.

We now formally prove a few identities using the concept of the state transition graphs.

Example 11.3: Prove that $A(p) \equiv \neg S(\neg p)$.

Proof: Let at state s, $A(p)$ be true.
We write formally,

$$s \models A\,(p)\,. \qquad\qquad (1)$$

Let us suppose that

$$s \models S\,(\neg p). \qquad\qquad (2)$$

Then there exists a state s', the next state of s such that

$$s' \models \neg p. \qquad\qquad (3)$$

However, since $s \models A(p)$,

$$s' \models p. \qquad\qquad (4)$$

Consequently, the supposition that $s \models S(\neg p)$ is wrong, i.e. $s \not\models S(\neg p)$,

Or, $s \models \neg S(\neg p)$,

which yields

$$A(p) \equiv \neg S(\neg p)\,. \qquad\qquad \Box$$

Example 11.4: Prove that $A\ (p{\rightarrow}q) \rightarrow (\ A(p) \rightarrow A(q))$.

Proof: Given that at state, say, s, $A\ (p{\rightarrow}q)$,

$$\text{i.e., } s \models A\ (p{\rightarrow}q) \qquad\qquad (1)$$

$$\text{also given } s \models A(p) \qquad\qquad (2)$$

Let us assume that

$$s \models \neg A(q)\,, \qquad\qquad (3)$$

i.e., $A(q)$ does not follow from state s.

Further, as $A(q) \equiv \neg S \neg q$. $\qquad\qquad (4)$

$\therefore \neg A(q) = S \neg q$. $\qquad\qquad (5)$

Substituting (5) in (3) we have

$$s \models S \neg q.$$

If there exists a next state of s, say s', then

$$s' \models \neg q. \qquad (6)$$

But by the first two assumptions,

$$s' \models p \rightarrow q \qquad (7)$$

and $\quad s' \models p. \qquad (8)$

So by Modes Ponens from (7) & (8)

$$s' \models q. \qquad (9)$$

Now, (9) is a contradiction to (6).

Hence, the initial assumption $s \models \neg A(q)$ is false. Consequently $s \models A(q)$ is true. Thus

$$A(p \rightarrow q) \rightarrow (A(p) \rightarrow A(q)).$$

is a valid expression.   □

11.6.2. The 'Next-Time' Operator

Normally, time in a computer system is discrete; i.e., if the current instant is defined the third, next will be the fourth. This is due to the hardwired clocking of the system resources including processor, memory and peripheral circuit modules. Thus it makes sense to express the 'next' instant of time. The next instant of time or next-time is denoted by the 'O' symbol. Thus to represent that proposition p follows from the next instant of s_1 we write

$$s_1 \models O p.$$

From the meaning of the 'O' operator, it is thus clear that

$$\models A(p) \rightarrow O p \text{ and}$$

$$\models O p \rightarrow S p.$$

Further, linear-time temporal logic is characterized by

$$O p \models \neg O \neg p.$$

We now present a few elementary axioms using O operator.

   i) $\models O (A \rightarrow B) \rightarrow (O A \rightarrow O B)$

   ii) $\models A (A \rightarrow OA) \rightarrow (A \rightarrow A A).$

Further, for any proposition p

   iii) $\models A p \rightarrow O p$

   iv) $\models O p \rightarrow S p.$

11.6.3 Some Elementary Proofs in PTL

A few example proofs are presented below to make the readers familiar with the proof procedures in PTL.

Example 11.5: Prove that

$$\vdash O(p \wedge q) \equiv (Op \wedge Oq)$$

Proof : $\vdash p \wedge q \rightarrow p$ (1) (by propositional logic)

$\vdash O(p \wedge q) \rightarrow O p$ (2)

$\vdash (p \wedge q) \rightarrow q$ (3) (by propositional logic)

$\vdash O(p \wedge q) \rightarrow O(q)$ (4)

$\vdash O(p \wedge q) \rightarrow O(p) \wedge O(q)$ (5)

$\vdash O(p \rightarrow \neg q) \rightarrow (O p \rightarrow O \neg q)$ (6)

$\vdash \neg O p \vee \neg Oq \vee O \neg(p \rightarrow \neg q)$ (7)

$\vdash \neg Op \vee \neg Oq \vee \neg O$

$\vdash (O p \wedge O q) \rightarrow O (p \wedge q)$

$\vdash O(p \wedge q) \equiv Op \wedge Oq.$ ☐

Example 11.6: Show that

$$\vdash A(p \wedge q) \equiv A(p) \wedge A(q)$$

Proof: 1. $\vdash A(p \wedge q) \rightarrow A(p) \wedge A(q)$

2. $\vdash A(p) \wedge A(q) \rightarrow O p \wedge O q$

3. $\vdash A (p) \wedge A (q) \rightarrow O (p \wedge q)$

4. $\vdash A (p) \wedge A(q) \rightarrow A (p \wedge q)$ by induction

5. $\vdash A (p \wedge q) \equiv A(p) \wedge A(q)$.   ☐

Example 11.7: Show that

$$\vdash A(p \rightarrow q) \rightarrow (Sp \rightarrow Sq)$$

Proof :   1. $\vdash (p \rightarrow q) \rightarrow (\neg q \rightarrow \neg p)$

2. $\vdash A (p \rightarrow q) \rightarrow A(\neg q \rightarrow \neg p)$

3. $\vdash$ A(p $\rightarrow$ q) $\rightarrow$ A($\neg$q) $\rightarrow$ A($\neg$p)

4. $\vdash$ A(p $\rightarrow$ q) $\rightarrow$ ($\neg$A$\neg$p $\rightarrow$ $\neg$A$\neg$ q)

5. $\vdash$ A(p $\rightarrow$ q) $\rightarrow$ (Sp $\rightarrow$ Sq). $\square$

11.7 Interval Temporal Logic

For modeling physical processes, we often need to reason about the truth of propositions over intervals of time. The propositions that correspond to the attributes of the physical process or its world may change with time. For instance, suppose when someone rings a calling bell, people inside the house open the door, presuming some visitors have come. Once the visitors enter the house, they close the door. So, the door is kept open for a small interval of time. We may write this as follows:

$$\forall t, (0 < t \leq t_1) \wedge (t_2 \geq t_1) \wedge (t_3 > t_2) \wedge \text{closed (door, t)} \wedge \text{rings (bell, } t_1)$$

$$\rightarrow \text{open (door, } t_2) \wedge \text{closed(door, } t_3)$$

The last expression means that if the door is closed in the interval $0 < t \leq t_1$ and the bell rings at time t_1, then the door is open at $t_2 \geq t_1$ and closed at $t_3 > t_2$. In other words, the door is kept open for the interval of time t, $t_2 < t < t_3$. The above expression works well but we need formalization of the syntax. Now, we present the formal definitions of the well-formed formulae (WFF) under propositional interval logic. Let Γ be a set of time point symbols, P be a set of propositional symbols and Γv be a set of temporal variables. The WFF here are defined inductively as follows.

    1. If $t_1, t_2 \in \Gamma \cup \Gamma v$ and p$\in$ P, then $t_1 < t_2$, $t_1 \leq t_2$ and holds(t_1, t_2, t_3) are WFF.

    2. If ϕ_1 and ϕ_2 are WFF, then $\phi 1 \wedge \phi 2$ and $\neg \phi 1$ also are WFFs.

    3. If ϕ is a WFF and $t \in \Gamma v$, then $\forall t, \phi$ is a WFF.

The usual definitions of $\vee$, $\rightarrow$, $\equiv$ and $\exists$ are maintained in interval temporal logic. The following transformations [3] are often useful to derive logical proofs in interval temporal logic.

1. Holds $(t_1, t_2, \phi_1 \wedge \phi_2) \rightarrow$ Holds(t_1, t_2, ϕ_1) $\wedge$ holds(t_1, t_2, ϕ_2) .

2. Holds($t_1, t_2, \neg \phi) \rightarrow \neg$ Holds(t_1, t_2, ϕ).

The second rule is generally called **weak negation**, which is in contrast to the following rule, called **strong negation**.

$$\forall t , (t_1 \leq t) \wedge (t_2 \geq t) \wedge \text{Holds}(t_1, t_2, \neg \phi) \rightarrow \neg \text{Holds}(t, \phi) .$$

The weak and strong negation both support the following properties:

For weak negation,

$\vdash$ Holds$(t_1, t_2, \neg \neg q)$

$\vdash \neg$Holds$(t_1, t_2, \neg q)$

$\vdash \neg\neg$Holds(t_1, t_2, q)

$\vdash$ Holds (t_1, t_2, q).

For strong negation,

$\vdash \forall t , (t_1 \leq t) \wedge (t_2 \geq t) \wedge \text{Holds}(t_1, t_2, \neg \neg p)$

$\vdash \forall t , (t_1 \leq t) \wedge (t_2 \geq t) \wedge \neg \text{Holds}(t_1, t_2, \neg p)$

$\vdash \forall t , (t_1 \leq t) \wedge (t_2 \geq t) \wedge \neg \neg \text{Holds}(t_1, t_2, p)$

$\vdash \forall t , (t_1 \leq t) \wedge (t_2 \geq t) \wedge \text{Holds}(t_1, t_2, p)$.

But there exists evidence of proving Holds$(t_1, t_2, p) \vee$ Holds(t_1, t_2, q) from Holds$(t_1, t_2, p \vee q)$, which is not feasible by strong negation [3].

11.8 Reasoning with Both Space and Time

We have discussed various schemes for reasoning with space and time independently. However, there exist circumstances, when both are required concurrently. A formalization of reasoning techniques in variation of both spatial and temporal events is an open area of research till date. We here just illustrate the justification of such reasoning with an example.

Example 11.8: Suppose one summer evening, a hot burst of wind moving in the west causes a fire at the hill, which gradually started spreading in the village (Fig. 11.10). A villager reports to the fire brigade station at time t_1. A fire brigade now has to plan the route, so that it can reach the village at the earliest. There are only two roads, one the hillside road (road1) that requires longer time of traversal and the other road (road2) which requires crossing a river through a bridge to reach the village. The pilot of the fire brigade car thinks that it can reach E_1 end of the bridge within 15 minutes, but crossing the river after 15 minutes will be difficult, as by that time many villagers too will rush to cross it from other end E_2 of the bridge. But traversal through the road1 will require 20 minutes more than the time required through road2 had there been no rush. So, the main question that remains: will the villagers, who reached the end E_2 of the bridge within 15 minutes of the breaking out of fire in the village, vacate the village within 35 minutes? The decision of the pilot in selecting the right road depends solely on this answer. To resolve this, the fire brigade station-master observed the scenario in the village by moving up a high tower and found a few villagers rushing towards E_2 and instructed his men to go to the spot through road2.

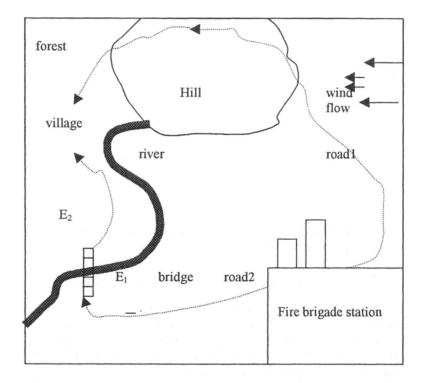

Fig. 11.10: The topological map of a hillside village and its surroundings.

The above story requires a spatial representation of the map and then reasoning about time. Here the spatial changes of the villagers and the fire brigade must be taken into account at specific time slots. Readers may try to formulate the (temporal part of the) problem by situation calculus and define the spatial part as FOL clauses. So, spatial changes can be reflected in different time frames. The non-monotonism in the problem can also be resolved by the qualification method presented under situation calculus.

11.9 Conclusions

The chapter demonstrated the scope of the extension of the predicate logic formalisms for reasoning with time and space. The principles of spatial reasoning have been covered with FOL and fuzzy logic, whereas the temporal reasoning is introduced with situation calculus and propositional temporal logic. These are active research fields and we have to wait a few more years for its complete formalization. Most important applications of spatio-temporal reasoning include co-ordination and task planning of multiple robots for handling a complex problem like machine repairing, where active participation of a number of robots is required to share time and space. It may be added here that some aspects of the co-ordination problems could have been solved with timed Petri nets. We, however, will introduce timed Petri nets and their application in co-ordination problems in chapter 24. Another interesting topic, which we could not discuss here for lack of space, is reasoning with shape [6]. This is important because, in any practical path planning problems of mobile robots, knowing the 2-D and the 3-D shapes of obstacles are useful for decision making. This also is an active area of research and will require a few more years to take final shape.

Exercises

1. Write a logic program to describe the spatial relationships among the components of a 4 legged table, assuming that at least three legs are visible in the image.

2. Justify the definitions of μ_{right} (θ) by taking $\theta = -90^0$, 0^0, 45^0 and 90^0. Can you define it in an alternative manner?

3. Graphically compute the membership of a triangle below a rectangle following the method presented in the text. Also plot the $\underline{f}$ (θ).

4. Identify the events, fluents and the situations from the following sentences: a) The teacher called John, the student, to the board. b) He then handed over the chalk to John. c) He insisted John write what he was talking to his classmates. d) Then he asked John to write what he was teaching. e) John could not write anything but started trembling. f) The teacher then advised John to leave the classroom.

5. Add the following rule to the previous facts and then show by situation calculus that 'the teacher advised John to leave the classroom' directly follows from the other facts.

Rule: If a student talks in the classroom at time t and cannot write what the teacher was teaching at time t_1 ($> t$), he advises the student to leave the classroom at time t_2 ($> t_1$).

6. Draw a state transition diagram consisting of two states that describes the facts: $p \wedge \neg q$ holds at time t_1, $q \wedge \neg p$ holds at time t_2; again: $p \wedge \neg q$ holds at time t_3, and $q \wedge \neg p$ holds at time t_4. The process thus repeats infinitely. Show that the formula $A(\neg q \vee \neg p)$ is always true following the state transition diagram.

7. Prove that $A (p \rightarrow q) \Rightarrow S (p \rightarrow q)$.

8. Represent the 'fire extinguishing problem' presented in the last example of the chapter by situation calculus and solve it to determine the right road for the fire brigade.

References

[1] Ayache, N., *Artificial Vision for Mobile Robots: Stereo Vision and Multisensory Perception*, MIT Press, Cambridge, MA, pp. 231-244, 1991.

[2] Ben-Ari, M., *Mathematical Logic for Computer Science*, Prentice-Hall, Englewood Cliffs, NJ, pp. 200-241, 1993.

[3] Dean, T., Allen, J. and Aloimonds, Y., *Artificial Intelligence: Theory and Practice*, Addison-Wesley, Reading, MA, pp. 255-288, 1995.

[4] Miyajima, K. and Ralescu, A., "Spatial organization in 2D segmented images: Representation and recognition of primitive spatial relations," *Fuzzy Sets and Systems*, vol. 65, pp. 225-236, 1994.

[5] Patterson, D. W., *Introduction to Artificial Intelligence and Expert Systems*, Prentice-Hall, Englewood Cliffs, NJ, pp. 95-97, 1990.

[6] Stefik, M., *Introduction to Knowledge Systems*, Morgan Kaufmann, San Mateo, CA, pp. 405-458, 1995.

[7] Szalas, A., " Temporal logic of programs: standard approach," in *Time and Logic: A Computational Approach*, Bolc, L. and Szalas, A., Eds., UCL Press, London, pp. 1-50, 1995.

12

Intelligent Planning

This chapter provides an exhaustive survey of the various planning strategies, employed to complex decision-making problems. It begins with a formal introduction to forward and backward chaining methods for planning. The forward or backward chaining schemes exhibit exponential growth of the search space and thus are of limited use. An alternative scheme for planning based on the principles of 'least commitment' is then introduced. The abstract classification of a planning problem by a hierarchical approach and its possible expansion to primitive (elementary) plans is covered next in the chapter. The chapter also provides a brief discussion on 'multi-agent planning'. The principles of heuristic planning have also been illustrated here with a 'flow-shop scheduling' problem.

12.1 Introduction

The word 'planning' informally refers to the generation of the sequence of actions to solve a complex problem. For instance, consider the problem of placing the furniture in your new-built house, so that i) you can fully utilize the available free space for common use and ii) the rooms look beautiful. An analysis of the problem reveals that there exist many possible alternative solutions to the problem. But finding even a single solution is not so easy. Naturally, the question arises: why? Well, to understand this, we explore the problem a little more.

Suppose, you started planning about the placement of the following furniture in your drawing room:

a) one computer table
b) one TV trolley
c) one book case
d) one corner table
e) two sofa sets and
f) one divan

We also assume that you know the dimensions of your room and the furniture. You obviously will not place the furniture haphazardly in the room as it will look unimpressive and it will not provide you with much space for utilization. But where is the real difficulty in such planning?

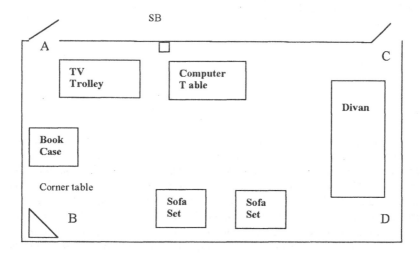

Fig.12.1: One possible plan regarding the placement of furniture in your drawing room.

To answer this, let us try to place the corner table first. Since only two corners B and D are free, you have to place it at either of the two locations. So if you do not place it first, and fill both the corners with other furniture, you will have to revise your plan. Fixing the position of your corner table at the beginning does not solve the entire problem. For example, if you fix the position of the corner table at B then the place left along AB allows you to place the bookcase or one sofa set or the TV trolley or the computer table. But as the switchboard (SB) is on the wall AC, you will prefer to keep your computer table and TV trolley in front of it. Further, you like to keep the sofa sets opposite to the TV. So they occupy the positions shown in fig. 12.1. The bookcase thus is the only choice that could be placed along the wall AB; consequently, the divan is placed along the wall CD. The following steps thus represent the schedule of our actions:

1. Place the corner table at B.

2. Place the TV trolley and computer table along the wall AC.

3. Place the two sofa sets along the wall BD.

4. Place the bookcase along the wall AB.

5. Place the divan along the wall CD.

What do we learn from the above plan? The first and foremost, with which all of us should agree, is minimizing the scope of options. This helps in reducing the possible alternatives at the subsequent steps of solving the problem. In this example, we realize it by placing the TV set and the computer close to the switchboard. Another important point to note is the 'additional constraints imposed to subsequent steps by the action in the current step'. For example, when we fix the position of the TV set, it acts as a constraint to the placement of the sofa sets. There are, however, instances, when the new constraints generated may require revising the previous schedule of actions.

The subsequent sections of the chapter will cover various issues of planning. Section 12.2 will cover the different aspects of 'linear planning' [6] by STRIPS approach [4], [9] using if-add-delete operators. In section 12.3 we shall present the principle of 'least commitment planning' [2]. The issues of 'hierarchical task network planning' will be presented in section 12.4. The principles of 'multi-agent planning' will be presented in section 12.5. The problems of scheduling are illustrated with the well-known 'flow-shop scheduling' problem in section 12.6. Conclusions are summarized in section 12.7.

12.2 Planning with If-Add-Delete Operators

We consider the problem of blocks world, where a number of blocks are to be stacked to a desired order from a given initial order. The initial and the goal state of the problem is given similar to fig. 12.2 and 12.3. To solve this type of problem, we have to define a few operators using the if-add-delete structures, to be presented shortly.

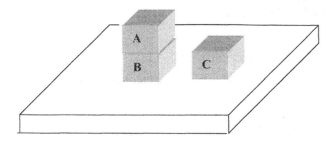

Fig.12.2: The initial state of Blocks World problem.

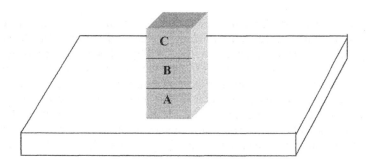

Fig: 12.3: The goal state of Blocks World problem.

The database corresponding to the initial and the goal state can be represented as follows:

The initial state:

> On (A,B)
> On (B, Table)
> On (C, Table)
> Clear (A)
> Clear (C)

The goal state:
> On (B, A)
> On (C, B)
> On (A, Table)
> Clear (C)

where On (X, Y) means the object X is on object Y and clear (X) means there is nothing on top of object X. The operators in the present context are given by the following if-add-delete rules.

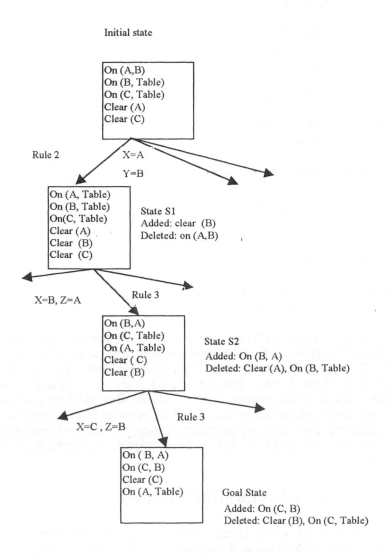

Fig. 12.4: The breadth first search of the goal state.

Rule 1: *If*   On (X, Y)
                      Clear (X)
                      Clear (Z)

        Add List: On (X, Z)
                      Clear (Y)

        Delete List: On (X,Y)
                      Clear (X)

Rule 2: *If*   On (X,Y)
                      Clear (X)

        Add List: On (X, Table)
                      Clear (Y)

        Delete List: On (X, Y)

Rule 3: *If*   On (X, Table)
                      Clear (X)
                      Clear (Z)

        Add List: On(X, Z)

        Delete List: Clear (Z)
                      On (X, Table)

     We can try to solve the above problem by the following sequencing of operators. Rule 2 is applied to the initial problem state with an instantiation of X =A and Y =B to generate state S1 (fig. 12.4). Then we apply Rule 3 with an instantiation of X =B and Z =A to generate state S2. Next Rule 3 is applied once again to state S2 with an instantiation of X =C and Z =B to yield the goal state. Generating the goal from the given initial state by application of a sequence of operators causes expansion of many intermediate states. So, forward reasoning is not appropriate for such problems. Let us try to explore the problem through backward reasoning.

12.2.1 Planning by Backward Reasoning

Much effort can be saved, if we generate a plan for the current problem by backward reasoning. While planning through backward reasoning, we should check the required preconditions to satisfy a given goal. Further, to satisfy new sub-goals generated, we should check the existence of their preconditions in the ADD-list of rules, which on firing generate the sub-goals. To illustrate this, let us consider the last problem. Here, the goal is given by On (B, A) ∧ On (C, B) ∧ On (A, Table) ∧ Clear (C).

Now to satisfy On (B, A) by Rule 3 we have three sub-goals: Clear(A), On (B, Table) and Clear(B), out of which the first two are available in the initial problem state. Further to satisfy the goal cause: On (C, B), we are required to satisfy the sub-goals: Clear (C), Clear(B) and On (B, Table), the first and third of which are available in the list. So, we are required to satisfy

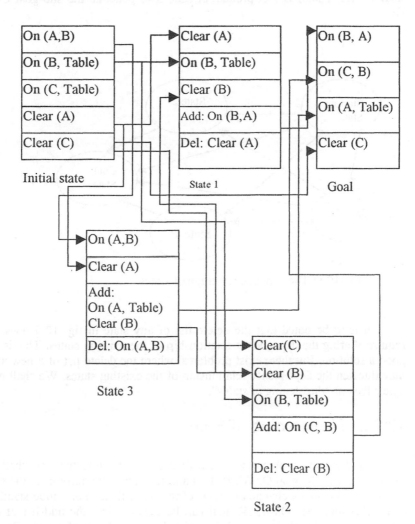

Fig 12.5: A solution by backward reasoning.

one new sub-goal: Clear(B). This can be achieved by employing rule 2. It may be noted that in the Add-list of rule 2, we have Clear (Y), where we can instantiate Y with B. The application of rule 2 in the present context gives rise to On (A, Table) also, which is required to satisfy the goal state. The dependence graph of states for the above problem is presented in fig. 12.6. Here, in state 1 and state 2, we generate the sub-goals On (B, A) and On (C, B) respectively by using rule 2. Further, for satisfying the pre-condition [Clear (B)] of rule 2, we generate state 3 by applying rule 2 on the initial state. The goal On (A, Table) is a bi-product at state 3 to generate the sub-goal Clear (B).

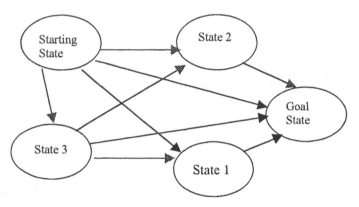

Fig 12.6: The dependence graph of states.

It is to be noted that the delete list of any state in fig. 12.5 does not require altering the dependence relationship of the existing states. This is too good a solution. But there exist problems, where the delete list of a new state may threaten the dependence relationship of the existing states. We shall now take a look at "threatening of states".

12.2.2 Threatening of States

Consider the same problem we were discussing. Assume that we shall be trying to satisfy the goal On (C, B) first and then take into account of the goal On (B, A). We do not change the name of the states to keep our understanding comprehensible. So, On(C, B) goal can be generated in the add-list of old state 2 (fig. 12.7). The pre-condition of old state 2: Clear (C) and On (B, Table) are available in the initial state. Now, to satisfy the pre-condition Clear

(B), we generate the old state 3, whose pre-conditions are all available in the initial state. Thus the goal On(C, B) is satisfied.

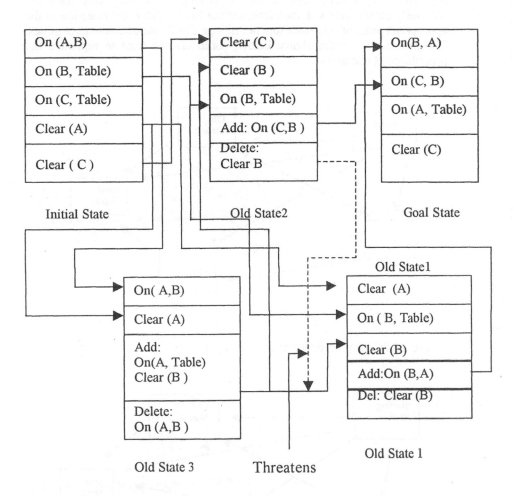

Fig.12.7: Demonstrating threatening of old state1 by old state 2.

Now, let us try to satisfy the On (B, A) in the goal. This fortunately can be made available by the add-list of old state 1. But can we satisfy the preconditions of old state 1? It is to be noted that the pre-conditions Clear(A) and On (B, Table) are available in the initial state and have not been deleted yet. But the third pre-condition Clear (B) of the state 1 has been deleted by old state 2. In other words old state 2 [or the use of rule 3 (operator 3)] **threatens** the link [Clear (B)] from old state 3 to old state 1. This has been represented by dotted lines in fig. 12.7. The dependence of states with the threatening operation is presented in fig. 12.8. The threatening of old state 2 to the link between old state 3 and old state 1 can, however, be avoided if old state 1 is generated prior to old state 2. We represent this by an additional **'before link'** [9] denoted by a dotted line in the graph (fig. 12.9). The other links are obviously before links and, therefore, are not labeled. Thus the selection of the appropriate order of the rules (operators) can avoid the threatening of states. The precedence graphs, shown below, could also be used to represent the precedence of operators.

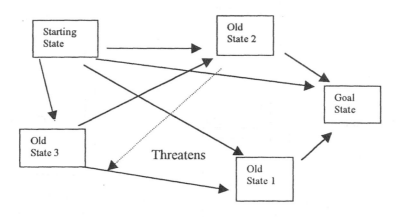

Fig. 12.8: The precedence graph for fig.12.7.

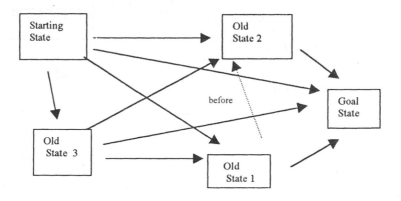

Fig. 12.9: The precedence graph of fig. 12.7 with an
                  extra before link.

12.3 Least Commitment Planning

The schemes of planning, described above, determine a list of sequence of operators, by a forward or backward reasoning in the state-space. When the number of blocks in the 'Blocks world problem' is large, determining the complete order of the sequence of operators is difficult by the proposed scheme. An alternative approach for planning is to determine 'approximate (partial) sequence of operators for each goal' separately and defer the ordering of their steps later. Such planning is referred to as **least commitment planning** [9]. In the literature of AI this is also called **non-linear planning** [6]. We now explain why it is called so. Since we delay in committing the order of operator in the partial plan of a sub-goal, it is called the least commitment planning. Further, the partial plan for each sub-goal is generated in parallel, unlike the previous state-space reasoning method for planning. It may be recollected that in the state-space approach, only after satisfying a sub-goal, the next sub-goal is considered for satisfaction. Thus in contrast to the state-space approach for linear planning, the current approach is termed non-linear planning.

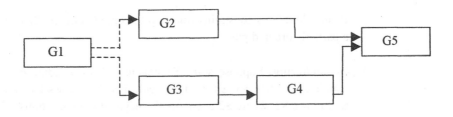

Fig. 12.10: Illustrating least commitment planning.

12.3.1 Operator Sequence in Partially Ordered Plans

Suppose realization of a goal requires 5 steps (sub-goals), denoted by operators, G1, G2, G3, G4 and G5 respectively. Let the order of the steps be represented by a graph like that in fig. 12.10. Here the firm line (——) denotes exact ordering, while dotted line (--) denotes the 'least committed' dependence relations (constraints) between two operators. Thus the above plan is an order of partially planned operators. The partially ordered plans for the problem of fig. 12.10 are listed below:

> {G1, G2, G3, G4, G5}
> {G1, G3, G2, G4, G5}   and
> {G1, G3, G4, G2, G5}

We now have to select which of the above three partially ordered plans leads to a complete plan for the goal or the sub-goal. So, in the least commitment planning we first search in the space of partially ordered plans and then select the correct complete plan among those plans.

12.3.2   Realizing Least Commitment Plans

For realizing a least commitment plan we require one or more of the following operations [6]:

a) **Step Addition:** This stands for the generation of a partially ordered plan for one sub-goal.

b) **Promotion:** This constrains one step to come before another in a partially ordered plan.

c) **Declobbering:** Suppose state S1 negated (deleted) some pre-condition of state S3. So, add S2 such that S2 follows S1 and S3 follows S2, where S2 reasserts the negated pre-conditions of S3.

d) **Simple Assignment:** Instantiate a variable to ensure pre-condition of a step.

e) **Separation:** Instantiation of variables is sometimes not done intentionally to keep the size of the plan manageable.

The following example of the well-known 'blocks world' problem, discussed earlier, will best illustrate the above definitions. Remember the problem was enlisted as follows:

Given: On (A,B) $\wedge$ Clear (C) $\wedge$ Clear(A) $\wedge$ On(C, Table) $\wedge$ On(B, Table).
Find a plan for: On (B, A) $\wedge$ On(C, B).

To start solving the problem, we first generate partial plans to achieve On (B, A) and On (C, B) separately.

The goal On (A,B) may be generated by the following rule: If X is clear and Y is clear then put X on Y. Here the pre-conditions Clear (A) and On (B, Table) are available in the in initial problem state. So, the partial plan for goal: On (B, A) can be constructed. The partial plan for this goal is presented in fig. 12.11.

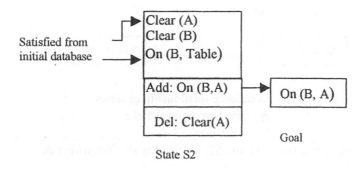

Fig. 12.11: The partial plan for the goal On (B, A).

Now, to satisfy On (C, B) we need to generate its predecessor (see fig. 12.12) as follows:

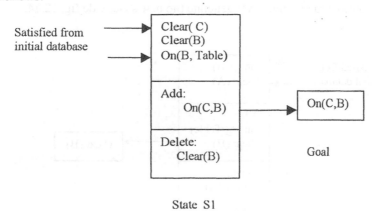

Fig. 12.12: The goal On (C, B) and its predecessor.

It may be noted that Clear (B) is a pre-condition of both the goals On(C,B) and On (B,A), but the process of generating On (C,B) deletes Clear (B). This posts an additional constraint that state S2 should follow state S1.

We denoted it by a dotted line (constraint link) in fig.12.13. Now to satisfy the pre-conditions of S1 and S2, we need to add new steps. Note that Clear (A) and On (B, Table) in both the states S1 and S2 are satisfied. So, we need to satisfy Clear (B) only in state S2 and S1.

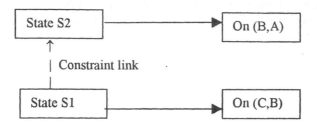

Fig. 12.13: Precedence relationship of states
by constraint (before) links.

To satisfy Clear (B) in S1 and S2, we employ the following rule:

$\quad$ *If* On (X,Y) ∧ Clear (X)
$\quad$ *Add:* On (X, Table) ∧ Clear (Y)
$\quad$ *Delete:* On(X,Y).

So, by backward reasoning, we generate the new state, vide fig. 12.14.

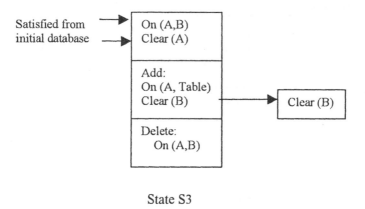

State S3

Fig.12.14: An approach to satisfy Clear (B).

We now have three partially ordered steps in our plan with one initial and one goal condition. These five partially ordered plans are presented below in a column structure.

Plan 1: *If* Clear (C) ∧ Clear (B) ∧ On (B, Table)
  Add: On (C,B)
  Delete: Clear (B)

Plan 2: *If* Clear (A) ∧ Clear (B) ∧ On (B, Table)
  Add: On (B,A)
  Delete: Clear (A)

Plan 3: *If* On (A,B) ∧ Clear (A)
  Add: On (A, Table) ∧ Clear (B)
  Delete: On (A,B)

Plan 4: *If* Nil
  Add: On (A,B) ∧ Clear (C) ∧ Clear (A)∧ On (C, Table) ∧
    On (B, Table)
  Delete: Nil

Plan 5: *If* On (B,A) ∧ On (C,B)
(goal) *Add:* Nil
  Delete: Nil

The complete order of the plans that maintain satisfiability of the pre-conditions of each partial plan is given by

plan 4 < plan 3 < plan 2 < plan 1 < plan 5

where plan j < plan k means plan j is to be executed prior to plan k.

In the above scheme for ordering a list of partially ordered plans, we demonstrated only two steps: addition of steps and promotion by adding constraints.

Let us now illustrate the principle of declobbering. Suppose, we choose the totally ordered plan as follows:

plan 4 < plan 3 < plan 1 < plan j < plan 2 < plan 5

where plan j will declobber the pre-condition (Clear (B)) of plan 2, which was clobbered by plan 1. The necessary steps in plan j are presented below:

Plan j: *If On (C,B) ∧ Clear (C)*
       Add: On (C, Table), Clear (B)
       Delete: On (C, B)

The incorporation of plan j between plan 1 and plan 2 serves the purpose of declobbering, but On (C,B) being deleted by plan j has to be executed later. Thus plan 1 has to be inserted again between plan 2 and plan 5. The new total order of the plans thus becomes:

plan 4 < plan 3 < plan 1 < plan j < plan 2 < plan 1 < plan 5

This undoubtedly is bad planning and the reader may think that declobbering has no justification. But sometimes it is useful and the only approach to refine a plan.

The operations of least commitment planning we described so far include the first three. The operation of instantiating variables to ensure preconditions of a step is also clear from our previous examples. But the last operation of intentionally deferring a non-instantiation variable is useful in planning. For example assume that there are two more blocks D and E on the table. In that case, instead of putting A on table in plan 3, we could put it on D and E as well; see our objective in plan 3 is to Clear (B). So, we employ the following rule to generate plan 3:

Rule: *If* On (X, Y) ∧ Clear (X) ∧ Clear (Z)
       Add: On (X, Z) ∧ Clear (Y)
       Delete: On (X, Y)

In the last rule Z could be a table or block D or E. We do not want to explicitly set the value of Z, because it is no longer required by other partial plans till now. Thus plan 3 could be:

Plan 3: *If* On (A,B) ∧ Clear (A) ∧ Clear (Z)
       Add: On (A,Z) ∧ Clear (B)
       Delete: On (A,B)

In this example the instantiation of Z is no longer required. However if Z is required to be instantiated later, we will then do it. It may be noted that the main benefit of deferring instantiation of variables is to keep the size of generated partial plans within limits.

12.4 Hierarchical Task Network Planning

The hierarchical task network planning, also called **hierarchical planning**, is employed in complex decision making systems. It generates a relatively abstract ordering of steps to realize the goal and then each abstract step is realized with simpler plans. A hierarchical planning scheme looks somewhat like a tree structure, where the steps at the higher level of the tree represent more abstract and complex tasks. Let us, for example, consider the plan for 'writing a book'. We, following the ABSTRIPS approach [8], first break the plan into three linear abstract plans: i) get current books and journals, ii) study them and iii) get pen and paper and write. Each abstract plan is then realized by the children under it in a sequentially ordered fashion, denoted by the dotted arrow (—-→) segment.

Fig.12.15 describes such a plan for 'writing a book'. The steps in fig. 12.15 are simple and thus need no elaboration.

The planning scheme in the present context takes care of the plan at a given level of the tree only before looking at the details in the next hierarchical level. Such a plan is often referred to as **length-first search** [6].

In the illustrative scheme of a hierarchical plan (fig. 12.15) we demonstrated only the feasible solution; but in situations we cannot guarantee the feasibility at the current level, unless we explored at lower levels, So we may generate alternative abstract plans. In fig.12.16, we describe such a plan, where the small dark rectangle denotes a primitive plan at a given level and the large rectangle ($\square$) denotes a sequential ordering of the primitive plans at a level. Let us assume that each level we select only one valid plan out of a possible number of b plans, i.e., the branching factor is b. Further, let the length of a selected plan at each layer be s. Thus, for executing such a plan, we need to consider a total of P plans [7], where

$$P = bs + bs^2 + bs^3 + \ldots + bs^{d-1}$$

$$= \sum_{j=1}^{d} b(s)^j = O(b\,s^d).$$

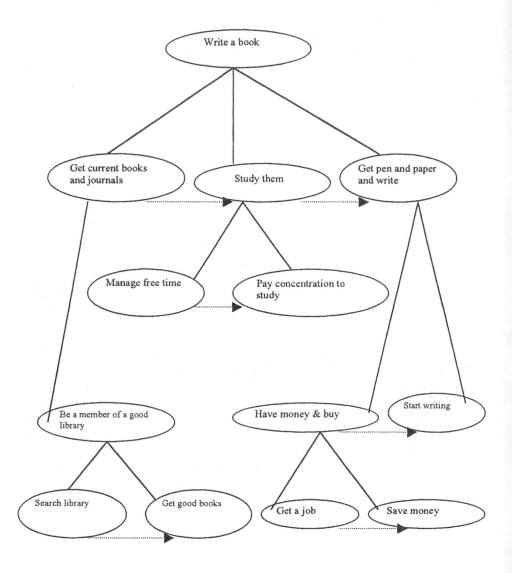

Fig. 12.15: A hierarchical plan of writing a book.

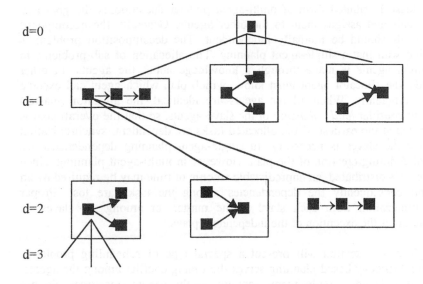

Fig. 12.16: A hierarchical plan with branching factor b=3, primitive steps s = 3 in a plan and depth (d) of the tree=3.

On the other hand, if we try to solve it by a linear planner it has to generate as many as

$$bs + (bs)^2 + (bs)^3 + \ldots + (bs)^{d-1}$$

$$= O\ (bs)^d.$$

Further for linear ordering of these plans, we require a significant amount of search among these plans. The total search complexity for linear ordering will be $O\ (bs)^{2d}$. On the other hand, in a hierarchical plan, at each level, we select 1 out of b plans. So, the time required to eliminate inconsistent plans is $O\ (b)$ and the time required to find a linear ordering at each level is $O\ (s)$. So, if there are d levels, the ordering time of plans is $O\ (s \ .d)$. Now, we can compare the ordering time of a hierarchical planner with respect to a linear planner. The factor of improvement of a hierarchical planner with respect to a linear planner can be given by $\{(b\ s)^{2d} - (s\ d)\ /\ (s\ d)\ \} = (\ b^{2\ d}\ s^{2\ d-1}\ /\ d) -1.$

12.5 Multi-agent Planning

The least distributed form of multi-agent planner decomposes the goal into sub-goals and assigns them to the other agents. Generally the decomposed sub-goals should be mutually independent. The decomposition problem is similar with that a single-agent planning. The allocation of sub-problems to different agents is made through a knowledge about the agents. In other words, the allocator agent must know: which plan which agent can execute more efficiently. When all the agents are identical, the allocator (master) should consider *load balancing* of the slave agents, so that the overall goal is executed at the earliest. If the allocated tasks are dependent, synchronization among the slaves is necessary. In single-agent planning dependencies are handled during creation of the plan. However in multi-agent planning, since the goal is distributed, an unpredictable amount of time may be required by an agent; consequently the dependencies among the tasks are lost. Proper synchronization from each slave to the master, or among the slaves, is required for the execution of the independent plans.

The next section will present a special type of scheduling problems, where a strategy-based planning serves the timing conflict among the agents. Indeed, it is not a multi-agent planning, as the agents (machines) do not participate in the planning process. It is to be noted that in a multi-agent planning the agents must be active planners. The task co-ordination problem among robots is an ideal example of such planning. For instance, two robots have to carry a large board inside a room filled with many obstacles. How will they plan interactively to transfer the object from one location to another? In this wide book, we do not have much scope to provide a solution to this problem. Interested readers, however, may attempt to solve it by assuming that both the robots have common knowledge of their world and one robot can sense the difficulty the other robot is facing in transferring the board.

12.6 The Flowshop Scheduling Problem

In flowshop scheduling there exists n different jobs, each of which has to be processed through a sequence of m different machines. The time allotment for a given job k to a machine, say, M_i is constant. To illustrate this problem, consider three machines, M_1, M_2 and M_3, and the jobs J_1 and J_2 have to pass through these machines in sequence (see fig.12.17)

The time requirement for processing the part of a job J_i in all three machines is supplied. One has to determine the job schedules ($J_1 < J_2$) or ($J_2 < J_1$), where $J_k < J_l$ for k, l $\in$ (1,2) denotes that job k is to be executed prior to job l, so that the completion time of the last job on last machine M_3 is minimized. This time is often referred to as **make-span** [3].

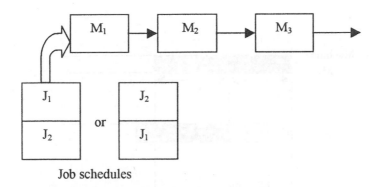

Job schedules

Fig.12.17: Illustrating the flowshop scheduling problem.

As a concrete example, let the machines M_1, M_2, M_3 be used for turning, drilling and thread cutting operations respectively. The desired jobs in this context are differently shaped wooden boxes, to be prepared from a large rectangular timber of wood. Also assume that the time involvement of the jobs on the machines is given as follows.

Table 12.1: The fixed times required for each J_k on machine M_i.

Jobs	Time Involvement in Machines		
	M_1	M_2	M_3
J_1	5	8	7
J_2	8	2	3

We now consider the two schedules of jobs $J_1<J_2$ and $J_2<J_1$, assuming that the machine sequence M_1-M_2-M_3 is fixed and identical for each job.

It is observed from the Gantt charts, fig. 12.18 (a) & (b), that the make-span for the two possible job schedules $J_1<J_2$ and $J_2<J_1$ is different. For this example the schedule $J_1<J_2$ is preferred as make-span for this case is smaller (=23) in contrast to the other schedule where it is 28.

Now, consider a problem of n jobs and m machines. Drawing Gantt chart for all possible factorial (n) schedules is difficult for large n. Such a problem is often referred to as a **non-deterministic polynomial (NP)**

complete problem [1], for which no optimal solution is not known. To solve this kind of problem we thus choose a near optimal solution.

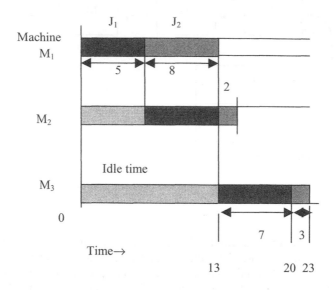

Fig. 12.18(a): Job completion time of the schedule: $J_1 < J_2$ is 23.

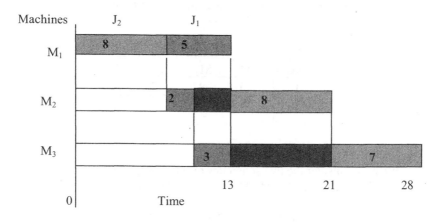

Fig. 12.18 (b): Job completion time of the schedule: $J_2 < J_1$ is 28.

One important parameter in the flowshop scheduling problem is the 'flow time'. The flow-time of job k means the absolute time of completion of job k from time 0. For example, the flowtime of job J_1 and J_2 in the schedule $J_1 < J_2$ is 20 and 23 respectively. **Total Flow Time** (TFT) is defined as the sum of the flow-time of all the jobs in the schedules, which in this case is 20+23=43 units of time. The reduction in total flow time is an active research on flowshop scheduling problem. One approach to solve the problem is through selection of a set of heuristic strategies. We call it heuristic, as it yields good results in most cases but is not guaranteed to produce good results always. Rajendran and Chaudhuri [5] developed a set of three strategies of R-C heuristics to solve this problem. The following notations will be required to understand the R-C heuristics.

Let

Σ be a partial schedule of jobs; for instance if there exists jobs J_1, J_2, J_3 one partial schedule could be $J_1 < J_3$, when the complete schedule is $J_1 < J_3 < J_2$.

$\Sigma + J_a$ denotes that job J_a is appended with the partial schedule Σ.

$q(\Sigma + J_a, k)$ denotes the completion time of a partial schedule at machine k.

Obviously, $q(\Sigma + J_a, m)$ denotes the completion time of a partial schedule ($\Sigma + J_a$) at the last machine.

12.6.1 The R-C Heuristics

A partial schedule $\Sigma + J_a$ is preferred to a partial schedule $\Sigma + J_b$ if the following heuristics are satisfied:

i) $\quad \text{Max} [q (\Sigma + J_a, j-1) - q(\Sigma, j), 0]$

$\quad <= \text{Max} [q (\Sigma + J_b) - q (\Sigma, j), 0]$.

This heuristic attempts to reduce the sum of idle time of the machines.

ii) $\quad \text{abs} [q(\Sigma + J_a, j-1) - q(\Sigma, j)]$

$\quad <= \text{abs} [q(\Sigma + J_b, j-1) - q(\Sigma, j)]$.

This criterion attempts to reduce both the idle time of machines and the waiting time of jobs.

iii) $abs[q(\Sigma +J_a, j-1) - q(\Sigma, j) + q(\Sigma + J_a, j)]$

$<= abs[q(\Sigma + J_b, j-1) - q(\Sigma, j) + q(\Sigma + J_b, j)]$.

This heuristic takes into account the criteria of the completion of the resultant partial schedule at various machines, in addition to the factors considered in heuristic 2.

The jobs like J_a are thus appended to the partial schedule, if it satisfies all the last three (or most of the) criteria. The resulting schedule of jobs has been found to be near-optimal from the point of view of TFT minimization [5].

12.7 Summary

This chapter covered an exhaustive survey of the existing methods of intelligent planning. The drawbacks of linear planning have been overcome by posting constraints in least commitment planning. The hierarchical task network planning requires minimum search and thus is useful for complex planning problems. A new type of complex task planning is illustrated with the flow-shop   scheduling problem. Here the schedule of jobs is interdependent and thus finding a complete plan for the problem is difficult. A heuristic approach to solve the problem, however, saves much computational time but cannot guarantee the optimal solution for the problem.

Exercises

1. Given the following initial and the goal state for the Blocks world problem. Construct a set of operators (Rules) and hence generate a plan to reach the goal state from the initial state.

    Initial State: On (C, A),
            Clear (C),
            On (B, Table) ,
            Clear (B).

    Goal State: On (B, A),
            On (C, B).

2. Realize the above plan by the least commitment planning.

3. Design a hierarchical plan for the construction of a house building. Clearly mark at least two sub-plans, which cannot be realized at the next level of the tree.

4. Given three machines M_1, M_2, M_3 and four jobs J_1, J_2, J_3, J_4. The technological order of the machines is M_1, M_2, M_3 in order. Assign a fixed time to each job on a given machine and represent it in matrix form. Use the RC heuristics to generate the schedule of jobs.

References

[1]   Bender, E. A., *Mathematical Methods in Artificial Intelligence*, IEEE Computer Society Press, Los Alamitos, p.601, 1996.

[2]   Dean, T., Allen, J. and Aloimonds, Y., *Artificial Intelligence: Theory and Practice*, Addison-Wesley, Reading, MA, pp. 297-351, 1995.

[3]   Laha, D. and Konar, A., "A simple heuristic method for the n-job, m-machine flow-shop scheduling problem," *Proc. of 3rd Int. Conf. On Computer Integrated Manufacturing*, Singapore, 1995.

[4]   Nilson, N. J., *Principles of Artificial Intelligence*, Springer-Verlag, Berlin, pp. 275-357, 1996.

[5]   Rajendra, C. and Choudhury, D., "An efficient heuristic approach to the scheduling of jobs in a flow-shop," *European Journal of Operational Research*, vol. 61, pp. 318-325, 1991.

[6]   Rich, E. and Knight, K., *Artificial Intelligence*, McGraw-Hill, New York, pp.329-357, 1996.

[7]   Russel, S. and Norvig, P., *Artificial Intelligence: A Modern Approach*, Prentice-Hall, Englewood Cliffs, NJ, pp. 337-390, 1995.

[8]   Sacerdoti, E. D., "Planning in a hierarchy of abstraction spaces," *Artificial Intelligence*, vol. 5, No. 2, pp. 115-135, 1974.

[9]   Winston, P. H., *Artificial Intelligence*, Addison-Wesley, Reading, MA, pp. 323-346, 1993.

13

Machine Learning Techniques

The chapter outlines four different categories of machine learning techniques, such as i) supervised learning, ii) unsupervised learning, iii) reinforcement learning and iv) learning by inductive logic programming. Among the supervised class of machine learning much stress is given to 'decision tree learning' and 'versions space-based learning'. The unsupervised class of learning is introduced briefly with an example classification problem. The reinforcement learning covered in the chapter includes Q-learning and temporal difference learning. The principle of inductive logic programming is introduced from the first principle and illustrated with an example, involving common family relations. The chapter ends with a discussion on the computational theory of learning.

13.1 Introduction

Learning is an inherent characteristic of the human beings. By virtue of this, people, while executing similar tasks, acquire the ability to improve their performance. This chapter provides an overview of the principle of learning that can be adhered to machines to improve their performance. Such learning is usually referred to as 'machine learning'. Machine learning can be broadly classified into three categories: i) Supervised learning, ii) Unsupervised learning and iii) Reinforcement learning. Supervised learning requires a trainer, who supplies the input-output training instances. The learning system adapts its parameters by some algorithms to generate the desired output patterns from a given input pattern. In absence of trainers, the desired output for a given input instance is not known, and consequently the learner has to adapt its parameters autonomously. Such type of learning is termed 'unsupervised learning'. The third type called the reinforcement learning bridges a gap between supervised and unsupervised categories. In reinforcement learning, the learner does not explicitly know the input-output instances, but it receives some form of feedback from its environment. The feedback signals help the learner to decide whether its action on the environment is rewarding or punishable. The learner thus adapts its parameters based on the states (rewarding / punishable) of its actions. Among the supervised learning techniques, the most common are inductive and analogical learning. The inductive learning technique, presented in the chapter, includes decision tree and version space based learning. Analogical learning is briefly introduced through illustrative examples. The principle of unsupervised learning is illustrated here with a clustering problem. The section on reinforcement learning includes Q-learning and temporal difference learning. A fourth category of learning, which has emerged recently from the disciplines of knowledge engineering, is called 'inductive logic programming'. The principles of inductive logic programming have also been briefly introduced in this chapter. The chapter ends with a brief discussion on the 'computational theory of learning'. With the background of this theory, one can measure the performance of the learning behavior of a machine from the training instances and their count.

13.2 Supervised Learning

As already mentioned, in supervised learning a trainer submits the input-output exemplary patterns and the learner has to adjust the parameters of the system autonomously, so that it can yield the correct output pattern when excited with one of the given input patterns. We shall cover two important types of supervised learning in this section. These are i) inductive learning and ii) analogical learning. A number of other supervised learning techniques using neural nets will be covered in the next chapter.

13.2.1 Inductive Learning

In supervised learning we have a set of $\{x_i, f(x_i)\}$ for $1 \leq i \leq n$, and our aim is to determine 'f' by some adaptive algorithm. The inductive learning [7-12] is a special class of the supervised learning techniques, where given a set of $\{x_i, f(x_i)\}$ pairs, we determine a hypothesis $h(x_i)$ such that $h(x_i) \approx f(x_i)$, $\forall i$. A natural question that may be raised is how to compare the hypothesis h that approximates f. For instance, there could be more than one $h(x_i)$ where all of which are approximately close to $f(x_i)$. Let there be two hypothesis h_1 and h_2, where $h_1(x_i) \approx f(x_i)$ and $h_2(x_i) = f(x_i)$. We may select one of the two hypotheses by a preference criterion, called **bias** [10].

When $\{x_i, f(x_i)\}$, $1 \leq \forall i \leq n$ are numerical quantities we may employ the neural learning techniques presented in the next chapter. Readers may wonder: could we find 'f' by curve fitting as well. Should we then call curve fitting a learning technique? The answer to this, of course, is in the negative. The learning algorithm for such numerical sets $\{x_i, f(x_i)\}$ must be able to adapt the parameters of the learner. The more will be the training instance, the larger will be the number of adaptations. But what happens when x_i and $f(x_i)$ are non-numerical? For instance, suppose given the truth table of the following training instances.

Truth Table 13.1: Training Instances

Input Instances	Output Instances
a_1 , $a_1 \rightarrow b_1$	b_1
a_2, $a_2 \rightarrow b_2$	b_2
.	
.	
a_n, $a_n \rightarrow b_n$	b_n

Here we may denote $b_i = f(a_i, a_i \rightarrow b_i)$ for all $i=1$ to n. From these training instances we infer a generalized hypothesis h as follows.

$$h \equiv \forall i \ (a_i, a_i \rightarrow b_i) \Rightarrow b_i.$$

We shall now discuss two important types of inductive learning: i) learning by version space and ii) learning by decision tree.

13.2.1.1   Learning by Version Space

This is one of the oldest forms of inductive learning, which was popularized by Mitchell [8] in the early 80's. An application package LEX was also built that works on the principles of version space. LEX was designed to perform symbolic integration. We start this section with a few definitions and then cover the well-known '**candidate elimination algorithm**' [9] and finally demonstrate the application of the algorithm in LEX.

Definition 13.1: An **object** is an entity that can be described by a set of attributes. For instance, a box is an object with attribute length, breadth, height, color and the material of which it is made.

Definition 13.2: A **class** is a subset of a universe of objects, which have common attributes. For example, 'easy chair' could be a class in the universe of object 'chair'.

Definition 13.3: A **concept** is a set of rules that partitions the universe of objects into two sets; one set must satisfy the rules and the others should not.Thus the concept of easy chairs should support the attributes of easy chair, and should not match with the attributes of other chairs.

Definition 13.4: A **hypothesis** is a candidate or tentative concept that is asserted about some objects in the universe. For example, a hypothesis of 'easy chair' should be a candidate concept that must be supported by most of the attributes of the objects in its class.

Definition 13.5 : The **target concept** is a concept that correctly classifies all the objects in the universe [11].

Definition 13.6: The exemplary objects and their attributes that support the target concept are called **positive instances**.

Definition 13.7: The exemplary objects and their attributes that do not support (or contradict) the target concept are called **negative instances**.

Definition 13.8: A rule that is true for all positive instances and false for all negative instances is called a **consistent classification rule** (concept).

Definition 13.9: 'Induction' refers to the process of class formation. In other words it includes the steps by which one can construct the target concept.

Definition 13.10: Selective induction is a form of induction by which class descriptions are formed by employing the attributes and their relations that appear only in the positive instances [11].

Definition 13.11: Constructive induction is a form of induction by which new descriptors not found in any of the instances are constructed.

Definition 13.12: A **specialization rule** (or concept) is a rule that can classify one sub-class of an object from its class. For instance, the specialization rule for easy chair can easily isolate them from the class "chair". A specialization rule may employ one or more of the following operators:

i)   Replacing variables by constants; for instance, color (red) could be specialized to color (ball, red).

ii)   Adding condition to a conjunctive expression; for example, in the concept of chair, we may add one or more conjunctive terms like has-a-slant-surface-of (chair).

iii)   Eliminating a disjunctive literal from an expression; for example, one can eliminate Color (x, red) from Color (X, red) $\vee$ Size (X, large), to specialize a rule.

iv)   Replacing a property with its child in its class hierarchy; for example, if we know primary color is a super-class of (green) then we can replace color (x, primary color) by color (x, green), for specializing a rule.

Definition 13.13: A **generalization** of a set of rules is a rule that can group a set of classes (sub-classes) into a super-class (class) by employing inverse operators, corresponding to those defined in specialization rule. Formally, the operators are:

     i)   Replacing constants by variables,
     ii) Deleting a condition from conjunctive expressions,
     iii) Adding a disjunctive literal to an expression, and
     iv) Replacing a property with its parent in a class hierarchy.

Definition 13.14: The **constructive generalization** is a generalization of the rules with the additional relations, not available in the existing instances. For example, if block A is on the table, block B is on A and block C is on B, we can write

$$On (A , table) \wedge On (B, A) \wedge On (C, B),$$

from which we can generalize, topmost block (C) $\wedge$ bottommost block (A), which are not present in the previous descriptions.

Definition 13.15: Inductive bias is defined as a set of factors that influence the selection of hypothesis, excluding those factors that are directly related to the training instances. There exist two types of bias in general: i) restricting hypothesis from the hypotheses space and ii) the use of a preferential ordering among the hypotheses [11].

The restriction of hypotheses can be implemented by adding a conjunct to the previous hypothesis. The preferential ordering among the hypotheses can be made by a heuristic evaluation function that excludes some objects from the target concept. The hypothesis that is not supported by the excluded class is preferred to the hypothesis that is supported by it .

The Candidate Elimination Algorithm

The candidate elimination algorithm is employed to reduce the concept (version) space from both general to specific and from specific to general form. It, thus, is a bi-directional search. Positive instances are used for generalization and negative instances are utilized to prevent the algorithm from over-generalization. The learned concept will, therefore, be general enough to include all positive instances and exclude all negative instances. We now present the procedure candidate-elimination [5], [9].

Procedure Candidate-Elimination
Begin
    Initialize G to be the most general concept in the space;
    Initialize S to be the first positive training instance;

    For each new positive instance p do
    Begin
        Eliminate the members of G that do not match with p;
        for all $s \in S$, if s does not match with p, replace s with its most specific generalization that match with p;
        Eliminate from S any hypothesis that is more general than some other in S;
        Eliminate from S any hypothesis, which is no more specific than some hypothesis in G;
    End For;

    For each negative instance n do
    Begin
        Eliminate all members of S that match with n;
        for each $g \in G$ that matches with n, replace g by its most general specialization that does not match with n;
        Eliminate from G any hypothesis, which is more specific in some other hypothesis in G;
        Eliminate from G any hypothesis, which is more specific than some other hypothesis in S;
    End For
    If $G=\varnothing$ and $S=\varnothing$ **Then** report "no concept supports all positive and refutes all negative instances";

If G=S and both contain one element, **Then** report "a single concept, that supports all positive and refutes all negative instances, has been found".

End.

Example 13.1: This example illustrates the candidate elimination procedure with the following positive and negative training instances.

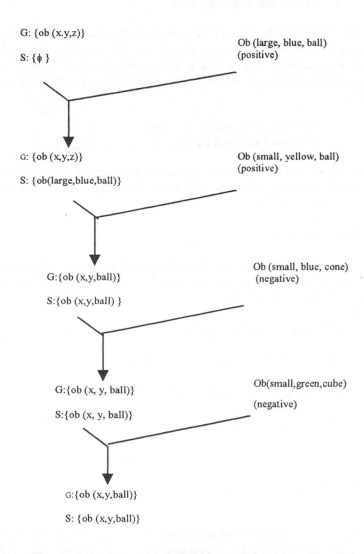

G: {ob (x.y,z)}

S: {φ }

Ob (large, blue, ball)
(positive)

G: {ob (x,y,z)}

S: {ob(large,blue,ball)}

Ob (small, yellow, ball)
(positive)

G:{ob (x,y,ball)}

S:{ob (x,y,ball) }

Ob (small, blue, cone)
(negative)

G:{ob (x, y, ball)}

S:{ob (x, y, ball)}

Ob(small,green,cube)
(negative)

G:{ob (x,y,ball)}

S: {ob (x,y,ball)}

Fig. 13.1: Illustrating the candidate elimination algorithm.

Positive instances:

1.   Object (large, blue, ball)
2.   Object (small, yellow, ball)

Negative instances:

1.   Object (small, blue, cone)
2.   Object (small, green, cube)

Fig. 13.1 illustrates the trace of the candidate elimination algorithm. In the first step G is assigned to ob (x, y, z) and S to {∅} set. Then because of the first positive instance, S is updated to {ob (large, blue, ball)}, but G remains same. In the next state, the second positive instance is supplied and G and S become

$$G = \{ \ ob\ (\ x, y, ball\ )\ \} \qquad \text{and}$$
$$S = \{ \ ob\ (\ x, y, ball\ )\ \}$$

It is to be noted that G=S, but the algorithm is continued further to take into account the negative instances. The next two steps, however, caused no changes in either G or S.

The LEX System

LEX learns heuristic rules for the solution of symbolic integration. The system possesses about 40 rules (operators) of integration. A few rules of integration are presented below.

OP1: 1*f(x) → f(x), ∀ real f(x)

OP2: ∫c*f(x)dx → c∫f(x)dx, ∀ constant c

OP3: ∫{f₁ (x) + f₂ (x)}dx → ∫f₁(x)dx + ∫f₂(x)dx

OP4: ∫cos(x)dx → sin(x)

OP5: ∫sin(x)dx → -cos(x)

OP6: ∫ u dv → uv - ∫ v du

When any of the left hand side occurs in an expression, LEX replaces it by the right hand side of the '→' operator. This is called a heuristic. The most significant feature of LEX is that it can control the bounds of the version (concept) space by generalization and specialization rules. Lex generalizes expressions by replacing a symbol by its ancestors following a generalization tree grammar. A portion of the generalization tree grammar is given in fig. 13.2.

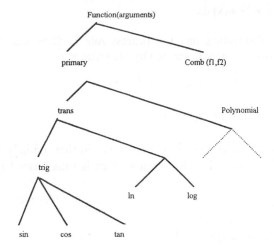

Fig. 13.2: A segment of the LEX generalization tree.

The architecture of LEX comprises four basic modules, as shown in fig.13.3. For the sake of training, the problem generator generates sample training problems and the problem solver attempts to solve it by employing available heuristics and operators. A solution is obtained when the operator yields an expression free from integration. The critic analyses the solution trace and produces positive and negative training instances from the solution trace. The generalizer performs candidate elimination to learn new heuristics.

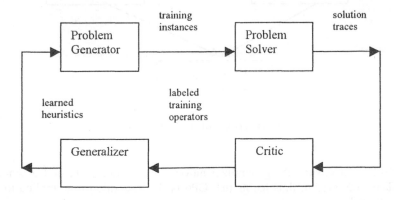

Fig. 13.3: The architecture of LEX.

To illustrate how LEX performs integration, let us assume that the problem generator submits the integral I, where

$$I = \int e^{2x} \sin(x) \, dx.$$

The problem solver determines that the nearest rule is OP6 and thus employs that rule. Thus two instances generated by the critic are

     i) $f_1(x) = e^{2x}, \; f_2(x) = \sin(x)$

     ii) $f_1(x) = \sin(x), \; f_2(x) = e^{2x}.$

The generalizer first initializes G= $\int f_1(x) * f_2(x) \, dx$. Further, initially the specialized function S= $\int e^{2x}\sin(x) \, dx$. The version space is thus formed like fig. 13.4.

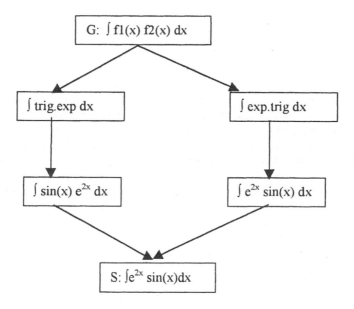

Fig. 13.4: Version space of $\int e^{2x} \sin(x) dx$.

G=S occurs when the generalizer moves up to $\int \sin(x)e^{2x}dx$ or $\int e^{2x}\sin(x) \, dx$. The LEX system thus learns that OP6 is the best rule to be applied to some tree problem.

13.2.1.2 Learning by Decision Tree

A decision tree receives a set of attributes (or properties) of the objects as inputs and yields a binary decision of true or false values as output. Decision trees, thus, generally represent Boolean functions. Besides a range of {0,1} other non-binary ranges of outputs are also allowed. However, for the sake of simplicity, we presume the restriction to Boolean outputs. Each node in a decision tree represents 'a test of some attribute of the instance, and each branch descending from that node corresponds to one of the possible values for this attribute' [8], [10].

To illustrate the contribution of a decision tree, we consider a set of instances, some of which result in a true value for the decision. Those instances are called **positive instances**. On the other hand, when the resulting decision is false, we call the instance 'a negative instance' [12]. We now consider the learning problem of a bird's flying. Suppose a child sees different instances of birds as tabulated below.

Table 13.2: Training Instances

Instances	No. of wings	Broken wings if any	Living status	Wing area/ weight of bird	Fly
1.	2	0	alive	2.5	True
2.	2	1	alive	2.5	False
3.	2	2	alive	2.6	False
4.	2	0	alive	3.0	True
5.	2	0	dead	3.2	False
6.	0	0	alive	0	False
7.	1	0	alive	0	False
8.	2	0	alive	3.4	True
9.	2	0	alive	2.0	False

It is seen from the above table that Fly = true if (no. of wings=2)∧(broken wings=0)∧(living status-alive)∧((wing area / weight) of the bird ≥2.5) is true.

Thus we can write:

Fly = (no. of wings = 2) ∧ (broken wings = 0) ∧ (living status = alive) ∧ (wing area / weight (A/W) ≥ 2.5)

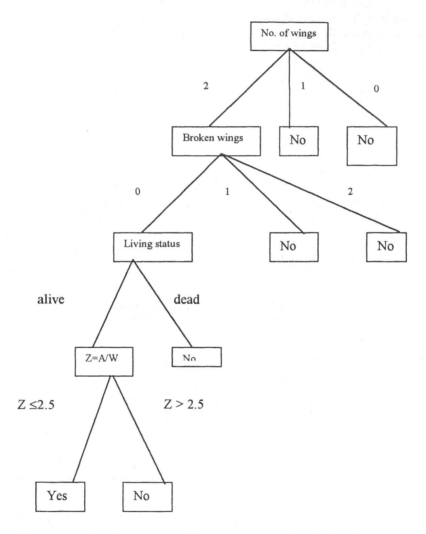

Fig. 13.5: A decision tree describing the instance for a flying bird;
            each leaf describes the binary value of 'Fly'.

The decision tree corresponding to the last expression is given in
fig. 13.5. In this figure if no. of wings is less than 2, Fly = false appears at the
next level of the root. If no. of wings =2, then we test: how many wings are
broken. If it is 0, we check whether the living status = alive, else we declare

Fly =false. Further if living status =alive, we check whether $Z=A/W \geq 2.5$. If yes, Fly =true, else Fly is declared false.

A question then naturally arises: in what order should we check the attribute value of the instances in a decision tree? In answering this question we need to know: which attribute is more vital? To measure the vitality of the attributes we require a statistical property called **information gain**, which will be defined shortly. The information gain depends on a parameter, called **entropy**, which is defined as follows:

Given a collection S of positive and negative instances for a target concept (decision). The entropy S with respect to this Boolean classification is:

Entropy $(S) \equiv$ -pos $\log_2$ (pos) – neg $\log_2$ (neg)   (13.1)

where 'pos' and 'neg' denote the proportion of positive and negative instances in S. While calculating entropy, we define $0\log_2 (0) = 0$.

For illustration, let us consider S that has 3 positive and 6 negative instances. We, following Mitchell [10], adopt the notation [3+, 6-] to summarize this sample of data. The entropy of S with respect to the Boolean classifier Fly is given by

Entropy [3+, 6-]

$= $ -(3 / 9) $\log_2$ (3 / 9) - (6 / 9) $\log_2$ (6 / 9)

$= $ -(1/3) $\log_2$ (1/3) - (2/3) $\log_2$ (2/3)

$= $ -(1/3) $\log_2$ (1/3) - (1/3)$\log_2$ (4/9)

$= $ - (1/3) [$\log_2$ (1/3) + $\log_2$ (4/9)]

$= $ - (1/3) [$\log_2$ (4 /27)]

$= $ 0.9179.

It is to be noted that when all the instances are either positive or negative, entropy is zero, as neg or pos =1. Further, when neg = pos = 0.5, entropy is 1; when neg $\neq$ pos, entropy must lie in the interval [0,1].

When the classifier has an output range that takes c different values, then entropy of S with respect to this c-wise classification will be

$$\text{Entropy } (S) = \sum_{i=1}^{c} -P_i \log_2 (P_i)$$ (13.2)

where P_i denotes the proportion of S belonging to class i.

It has already been pointed out that the order of checking the attributes requires a measure of information gain for those attributes. Information gain of an attribute A in a set of training instances S is given by

Gain (S, A)

$\equiv$ Entropy (S) - Σ ($|S_v|$ / $|S|$) Entropy (S_v)   (13.3)
          $v \in$ values of A

where $| S_v|$ denotes the subset of S, for which attribute A has value v. The $|X|$ denotes cardinality of x for $x \in \{ S_v , S \}$.

For the sake of illustration, let us compute the gain (S, living status). Here,

S= [3+, 6-] , S_{alive} = [3+, 5-] and S_{dead} = [0+,1-].

 Therefore,

Gain (S, living status)

=Entropy (S) - Σ ($|S_v|$ / $|S|$) Entropy (S_v)
           $v \in \{$alive, dead$\}$

= Entropy (S) – ($|S_{alive}|$ / $|S|$) Entropy (S_{alive}) – ($|S_{dead}|$ / $|S|$) Entropy (S_{dead})
                                                                (13.4)
where

   $| S_{alive} | = 8,$ $| S_{dead} | = 1,$ $|S| = 9.$

Entropy (S_{alive})

= Entropy [3+, 5-]

= - (3/8)$\log_2$ (3/8) - (5/8) $\log_2$ (5/8)

= 0.5835

and Entropy (S_{dead})

  = Entropy [0+,1-]

  = -(0/1)$\log_2$ (0/1) – (1/4)$\log_2$ (1/1) = 0-0 = 0.

Substituting the above values in (13.4) we found:

Gain (S, living status)

= 0.9179 – (8/ 9) x 0.5835

=0.3992.

ID3 is one of such systems that employs decision trees to learn object classifications from labeled training instances. For obtaining a decision tree of smaller size (lesser no. of nodes), ID3 measured information gain of all attributes and expanded the nodes of the tree based on the order of the attributes, following the descending sequence of their information gain.

13.2.2   Analogical Learning

In inductive learning we observed that there exist many positive and negative instances of a problem and the learner has to form a concept that supports most of the positive and no negative instances. This demonstrates that a number of training instances are required to form a concept in inductive learning. Unlike this, analogical learning can be accomplished from a single example. For instance, given the following training instance, one has to determine the plural form of bacilus.

Problem Instance

Input (singular)	Output (plural)
fungus	fungi

Obviously, one can answer that the plural form of bacillus is bacilli. But how do we do so? From common sense reasoning, it follows that the result is because of the similarity of bacillus with fungus. The analogical learning system thus learns that to get the plural form of words ending with 'us' is to replace it with 'i'. We demonstrate this in fig.13.6.

The main steps in analogical learning are now formalized below.

1.   **Identifying Analogy:** Identify the similarity between an experienced problem instance and a new problem.

2.   **Determining the Mapping Function:** Relevant parts of the experienced problem are selected and the mapping is determined.

3.   **Apply Mapping Function:** Apply the mapping function to transform the new problem from the given domain to the target domain.

4.   **Validation:** The newly constructed solution is validated for its applicability through its trial processes like theorem or simulation [11].

5.   **Learning:** If the validation is found to work well, the new knowledge is encoded and saved for future usage.

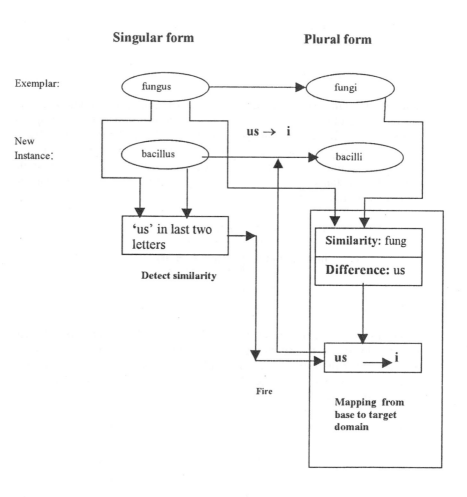

Fig. 13.6: Learning from an example.

Analogical reasoning has been successfully realized in many systems. Winston's analogical system [14] was designed to demonstrate that the relationship between acts and actors in one story can explain the same in another story. Carbonell's [1] transformational analogy system employs a new approach to problem solving. It solves new problems by modifying existing solutions to problems until they may be applied to new problem instances.

13.3 Unsupervised Learning

In supervised learning the input and the output problem instances are supplied and the learner has to construct a mapping function that generates the correct output for a given input pattern. Unsupervised learning, however, employs no trainer. So, the learner has to construct concepts by experimenting on the environment. The environment responds but does not identify which ones are rewarding and which ones are punishable activities. This is because of the fact that the goals or the outputs of the training instances are unknown; so the environment cannot measure the status of the activities of the learner with respect to the goals. One of the simplest ways to construct concept by unsupervised learning is through experiments. For example, suppose a child throws a ball to the wall; the ball bounces and returns to the child. After performing this experiment a number of times, the child learns the 'principle of bouncing'. This, of course, is an example of unsupervised learning. Most of the laws of science were developed through unsupervised learning. For instance, suppose we want to learn Ohm's law. What should we do? We construct a simple circuit with one cell, one potentiometer, one voltmeter and one ammeter. Each time we set the potentiometer to a position and measure the reading of the voltmeter and ammeter. After taking around 100 readings, suppose we plot 'the current' against the 'voltage drop across the potentiometer'. Then we find that the voltage across the potentiometer is proportional to the current passing through it, what all of us know to be the standard Ohm's law. It is to be noted that we do not perform experiments to learn a specific law. Rather the experimental results reveal a new concept/law. Let us take another example to illustrate the principles of concept formation by unsupervised learning. Suppose, we want to classify animals based on their speed and height/weight ratio. We, for example, take sample animals and measure the above features and then plot these on a two dimensional frame.

What result should we derive? It may be found from fig.13.7 that cows, dogs, tigers and foxes form different classes. Further, there is an overlap in the classes of foxes and dogs.

Now, if we are given a measured value of the speed and height/ weight ratio of an unknown animal, we can easily classify it, unless it does not coincide with the overlapping classes. An overlapped region cannot correctly determine the animals because it does not include the sufficient features to describe the animal. For illustration, both foxes and dogs have comparable speed and height/weight ratio. So, other features like shape of face, etc. are required to differentiate them.

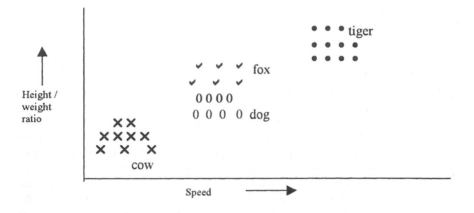

Fig. 13.7: Learning animals through classification.

In fact, the identification of the features itself is a complex problem in many situations, especially when the problem characteristics are not clearly known. We, for instance, consider an example from the biological classification problems. 'Phospholipid vesicles' are the major components of biological cell membranes and the classification of these vesicles is a significant problem for the biologists. My colleague Lahiri [3-4] took up this problem as part of his Ph. D. thesis and finally reached a conclusion that a 2-dimension classification of the phospholipids can be carried out by two features: i) porosity of the vesicles and ii) their mass fractal dimension. The formal definitions [3-4] of these features are beyond the scope of the book. Informally, porosity can be expressed as the normalized area of the porous regions on the 2-D images of the vesicles. On the other hand, fractals are special type mathematical functions $f(x)$ that are applied recursively on a randomly selected point x in a given space. After a large finite number of recursions, the graphical plot of the resulting points: $\{x, f(f(f(..x)))\}$, for real x and f, sometimes form interesting geometric shapes. Informally, the fractal dimension has relevance with the smallest dimension of the repetitive structures in the recursive functional plot. A formal definition of fractal dimension can be given following Haussdorff-Besicovitch dimension [2], [6] which, however, is not presented here to maintain continuity of the present context.

The microscopic view of the three common vesicular aggregates, studied by Lahiri [3], is presented below. It demonstrates that the 2-D spatial periodicity

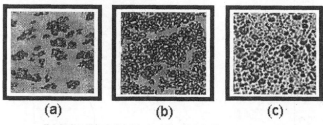

(a)   (b)   (c)

DMPC = Dimyristoyl phosphatidylcholine

Fig. 13.8: Digitized images for the vesicular clusters made by (a) DMPC, (b) DMPC in presence of spectrin and (c) a mixed lipid system made by DMPC and cholesterol, obtained from a Phase Contrast Microscope.

(structural repetitiveness) of the texture of these vesicular clusters has considerable differences in microscopic observation, and as a consequence their classification from 2-D features is feasible. A non-linear 2-D classification of these three vesicular aggregates has been presented in fig. 13.9. It is observed from the figure that the boundaries of the classes (b) and (c) have a small separation, while that of (a) & (b), and (a) & (c) have a large spatial gap. The results also intuitively follow from the microscopic view.

Not only in Biology, but in almost every branch of Science, pattern classification is of immense importance. In Psychology, for example, pattern classification is used to classify people of different mental diseases for treatment by a common therapy. In criminology the fingerprint of a suspect is classified into typical classes prior to matching it with known fingerprint databases of that class. Whatever be the problem, the main task in pattern classification is to extract the features, for which no automated methods is known to date.

13.4 Reinforcement Learning

In reinforcement learning, the learner adapts its parameters by determining the status (reward / punishment) of the feedback signal from its environment. The simplest form of reinforcement learning is adopted in learning automata. Currently Q-learning and temporal difference learning have been devised based on the reward/ punishment status of the feedback signal.

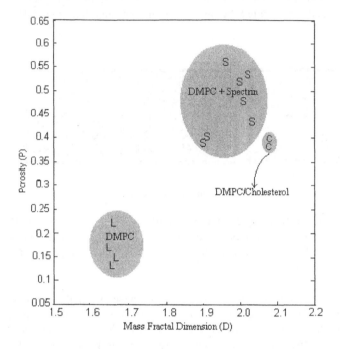

Fig.13.9: A classification scheme is made by constructing a two dimensional feature space by fractal dimension and porosity to classify the phospholipid vesicles from the images of their aggregates.

13.4.1   Learning Automata

Among the well-known reinforcement learning schemes, the most common is the *learning automata.* The learning mechanism of such a system includes two modules: *the learning automation* and *the environment*. The learning cycle starts with the generation of a stimulus from the environment. The automation on receiving a stimulus generates a response to the environment. The environment receives and evaluates the response and offers a new stimulus to the automation. The learner then automatically adjusts its parameters based on the last response and the current input (stimulus) of the automation.   A scheme for learning automation is presented in fig. 13.10.

Here the delay unit offers unit delay, to ensure that the last response and the current stimulus enter the learner concurrently.

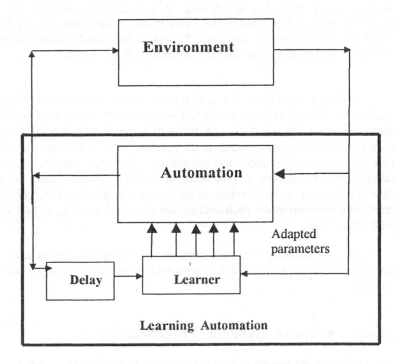

Fig. 13.10: The learning scheme in a learning automation.

The principles of learning automata can be extended for applications to many real world problems. One illustrative example, realized with the above principle, is the NIM game. In a NIM game there are three sets of tokens placed on a board, as shown in fig. 13.11. The game requires two players. Each player, in his/ her turn, has to remove at least one token but cannot access tokens from more than one row. The player who has to remove the last token is the loser, and obviously the opponent is the winner.

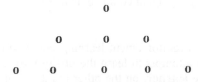

Fig. 13.11: The NIM game.

Let us now assume that the game is being played between a computer and a person and the computer keeps a record of the moves it has chosen in all its turns in a game. This is recorded in a matrix, where the (i, j)-th element of the matrix stands for the probability of success, if the computer in its turn makes a change from j-th to i-th state. It is to be noted that the sum of all elements under each column of the above matrix is one. This directly follows intuitively, since the next state could be any of the possible states under one column. The structure of the matrix is presented in fig. 13.12.

It should be added that the system learns from a reward-penalty mechanism. This is realized in the following manner. After completion of a game, the computer adjusts the elements of the matrix. If it wins the game, then the elements corresponding to all its moves are increased by δ and the rest of the elements under each column are decreased equally, so that the column-sum remains one. On the other hand, if the computer loses the game, then the elements corresponding to its moves are decreased by δ and the remaining elements under each column are increased equally, so that column sum remains one.

After a large number of such trials, the matrix becomes invariant and the computer in its turn selects the state with highest probability under a given column.

13.4.2   Adaptive Dynamic Programming

The reinforcement learning presumes that the agent receives a response from the environment but can determine its status (rewarding/punishable) only at the end of its activity, called the terminal state. We also assume that initially the agent is at a state S_0 and after performing an action on the environment, it moves to a new state S_1. If the action is denoted by a_0, we say

$$S_0 \xrightarrow{a_0} S_1,$$

i.e., because of action a_0, the agent changes its state from S_0 to S_1. Further, the reward of an agent can be represented by a **utility function**. For example, the points of a ping-pong agent could be its utility.

The agent in reinforcement learning could be either passive or active. A **passive learner** attempts to learn the utility through its presence in different states. An **active learner**, on the other hand, can infer the utility at unknown states from its knowledge, gained through learning.

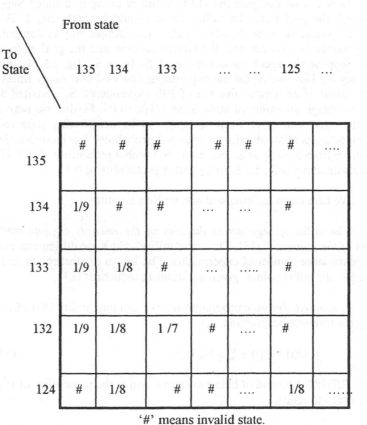

'#' means invalid state.

Fig. 13.12: A part of the state transition probability matrix in NIM game.

		Goal
S_3	S_4	S_7
S_2	S_5	S_8
Start S_1	S_6	S_9

S_i denotes the i-th state.

Fig. 13.13: A simple stochastic environment.

How can we compute the utility value of being in a state? Suppose, if we reach the goal state, the utility value should be high, say 1. But what would be the utility value in other states? One simple way to compute static utility values in a system with the known starting and the goal state is given here. Suppose the agent reaches the goal S_7 from S_1 (fig. 13.13) through a state say S_2. Now we repeat the experiment and find how many times S_2 has been visited. If we assume that out of 100 experiments, S_2 is visited 5 times, then we assign the utility of state S_2 as 5/100=0.05. Further we may assume that the agent can move from one state to its neighboring state (diagonal movements not allowed) with an unbiased probability. For example, the agent can move from S_1 to S_2 or S_6 (but not to S_5) with a probability of 0.5. If it is in S_5, it could move to S_2, S_4, S_8 or S_6 with a probability of 0.25.

We here make an important assumption on utility.

"The utility of sequence is the sum of the rewards accumulated in the states of the sequence" [13]. The static utility values are difficult to extract as it requires large number of experiments. The key to reinforcement learning is to update the utility values, given the training sequences [13].

In adaptive dynamic programming, we compute utility U(i) of state i by using the following expression.

$$U(i) = R(i) + \sum_{\forall j} M_{ij} U(j) \tag{13.5}$$

where R(i) is the reward of being in state i, Mij is the probability of transition from state i to state j.

In adaptive dynamic programming, we presume the agent to be passive. So, we do not want to maximize the $\sum M_{ij} U(j)$ term.

For a small stochastic system, we can evaluate the U(i), $\forall i$ by solving the set of all utility equations like (13.5) for all states. But when the state space is large, it becomes somewhat intractable.

13.4.3   Temporal Difference Learning

To avoid solving the constraint equations like (13.5), we make an alternative formulation to compute U(i) by the following expression.

$$U(i) \leftarrow U(i) + \alpha[R(i) + (U(j) - U(i))] \tag{13.6}$$

where α is the learning rate, normally set in [0,1].

In the last expression, we updated U(i) by considering the fact that we should allow transition to state j from state i, when U(j) >> U(i). Since we consider temporal difference of utilities, we call this kind of learning temporal difference (TD) learning.

It seems that when a rare transition occurs from state j to state i, U(j)-U(i) will be too large, causing U(i) large by (13.6). However, it should be kept in mind that the average value of U(i) will not change much, though its instantaneous value seems to be large occasionally.

13.4.4 Active Learning

For passive learner, we considered M to be a constant matrix. But for an active learner, it must be a variable matrix. So, we redefine the utility equation of (13.5) as follows.

$$U(i) = R(i) + Max_a \Sigma_{\forall j} M_{ij}^a U(j) \tag{13.7}$$

where M_{ij}^a denotes the probability of reaching state j through an action 'a' performed at state i. The agent will now choose the action a for which M_{ij}^a is maximum. Consequently, U(i) will be maximum.

13.4.5 Q-Learning

In Q-learning, instead of utility values, we use q-values. We employ Q(a, i) to denote the Q-value of doing an action 'a' at state i. The utility values and Q-values are related by the following expression:

$$U(i) = max_a Q(a, i). \tag{13.8}$$

Like utilities, we can construct a constraint equation that holds at equilibrium, when the Q-values are correct [13].

$$Q(a, i) = R(i) + \Sigma M_{ij}^a . max_{a'} Q(a',i). \tag{13.9}$$

The corresponding temporal-difference updating equation is given by

$$Q(a, i) \leftarrow Q(a, i) + \alpha[R(i) + \max_{a'} Q(a', j) - Q(a, i)] \tag{13.9(a)}$$

which is to be evaluated after every transition from state i to state j.

The Q-learning continues following expression 13.9 (a) until the Q-values at each state i in the space reaches a steady value.

13.5   Learning by Inductive Logic Programming

Inductive logic programming (ILP) employs an inductive method on the first order clauses through a process of **inverse resolution**. The ILP can generate new predicates and is therefore called **constructive induction**.

Let us first start with the resolution theorem of predicate logic. Suppose we have two clauses C_1 and C_2 given by

$$C_1 = \text{Male}(X) \leftarrow \text{Boy}(X)$$
and $$C_2 = \text{Boy}(\text{ram}).$$

So, $C_1 = \neg \, \text{Boy}(X) \vee \text{Male}(X).$ $\hspace{2cm}$ (13.10)

Let literal $L_1 = \neg \, \text{Boy}\,(X)$ $\hspace{3cm}$ (13.11)

and $L_2 = \text{Boy}\,(\text{ram})$ $\hspace{3.5cm}$ (13.12)

Let the unifying substitution $\theta = \{\text{ram} \,/\, X\}$. We can now write

$L_1\theta = \neg \, L_2\theta = \neg \, \text{Boy}\,(\text{ram})$. Now, after resolving C_1 and C_2, suppose we obtain the resolvent C, where

$C = \text{Male}(\text{ram})$. Thus C is the union of $(C_1 - \{L_1\})\theta = \text{Male}(\text{Ram})$

and $(C_2 - \{L_2\})\theta = \phi$.

Formally, we can always write $C = (C_1 - \{L_1\})\theta \cup (C_2 - \{L_2\})\theta$ $\hspace{1cm}$ (13.13)

The unifying substitution θ can be factored to θ_1 and θ_2 [10], i.e.

$$\theta = \theta_1\theta_2 \hspace{5cm} (13.14)$$

where θ_1 contains all the substitutions involving variables from clause C_1 and θ_2 contains all the substitutions involving variables from C_2. Thus

$$C = (C_1 - \{L_1\})\theta_1 \cup (C_2 - \{L_2\})\theta_2$$

where '–' denotes set difference operator. We can re-express the above as

$$C - (C_1 - \{L_1\})\theta_1 = (C_2 - \{L_2\})\theta_2 \hspace{2cm} (13.15)$$

which finally yields

$$C_2 = (C-(C_1-\{L_1\})\theta_1)\ \theta_2^{-1} \cup L_2. \tag{13.16}$$

Further, $L_2 = \neg\ L_1\theta_1\theta_2^{-1}$. $\qquad$ (13.17)

Substituting (13.17) in (13.16) we find

$$C_2 = (C-(C_1-\{L_1\})\theta_1)\theta_2^{-1} \cup (\neg L_1\theta_1\theta_2^{-1}) \tag{13.18}$$

The last expression describes an inverse resolution process. It is to be noted that θ_2^{-1} stands for inverse substitution, i.e. replacing constants by terms.

As an example, consider the two step inverse resolution shown in fig. 13.14, which ultimately yields a new piece of knowledge:

$$\text{Grandfather } (X, Z) \leftarrow \text{Mother } (Y, Z) \wedge \text{Father } (X, Y).$$

Let us now verify the steps of the computation .

First C_1 = Father (janak, sita) $=L_1$ and

C = Grandfather (janak, lob).

Here, $\theta_1=\{\ \}$ and $\theta_2^{-1}=\{\ X\ /\ janak\ \}$.

So, $[C- (C_1-\{L_1\})\ \theta_1]\ \theta_2^{-1}$

$= (C\theta_1)\ \theta_2^{-1}$

$= \text{Grandfather } (X, lob)$

Further, $\neg L_1\theta_1\ \theta_2^{-1} = \neg\text{Father } (X, sita).$

Thus the derived clause by (13.18) is Grandfather $(X, lob) \vee \neg\text{Father}(X, sita)$.

The second step of inverse resolution can be verified easily in a similar manner.

13.6 Computational Learning Theory

The main question on machine learning is: how does one know that his learning algorithm has generated a concept, appropriate for predicting the future correctly? For instance, in inductive learning how can we assert that our hypothesis h is sufficiently close to the target function f, when we do not know f? These questions can be answered with the help of computational learning theory.

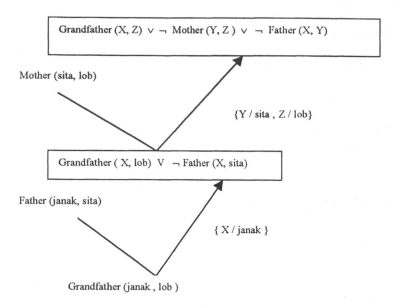

Fig. 13.14: A two step inverse resolution to derive: Mother (Y, Z) ∧ Father (X, Y) →Grandfather (X,Z).

The principle of computational learning theory states that any hypothesis, which is sufficiently incorrect, will be detected with a high probability after experimenting with a small number of training instances. Consequently, a hypothesis that is supported by a large number of problem instances will be unlikely to be wrong and hence should be **Probably Approximately Correct (PAC).**

The PAC learning was defined in the last paragraph w.r.t. training instances. But what about the validity of PAC learning on the test set (not the training set) of data. The assumption made here is that the training and the test data are selected randomly from the population space with the same probability distribution. This in PAC learning theory is referred to as **stationary assumption.**

In order to formalize the PAC learning theory, we need a few notations [13].

Let X = exhaustive set of examples,

D = distribution by which the sample examples are drawn,
m = cardinality of examples in the training set,
H = the set of possible hypothesis,
and f = function that is approximated by the hypothesis h.

We now define an error of hypothesis h by

$$\text{Error}(h) = P(h(x) \neq f(x) \mid x \varepsilon D)$$

A hypothesis h is said to be **approximately correct** when

$$\text{error}(h) \leq \varepsilon,$$

where ε is a small positive quantity.

When an approximate hypothesis h is true, it must lie within the ε-ball around f. When h lies outside the ε-ball we call it a bad hypothesis. [13]

Now, suppose a hypothesis $h_b \in H_{bad}$ is supported by first m examples. The probability that a bad hypothesis is consistent with an example $\leq(1-\varepsilon)$. If we consider m examples to be independent, the probability that all m samples will be consistent with hypothesis h_b is $\leq(1-\varepsilon)^m$. Now, if H_{bad} has to contain a consistent hypothesis, at least one of the hypothesis of H_{bad} should be consistent. The probability of this happening is bounded by the sum of individual probabilities.

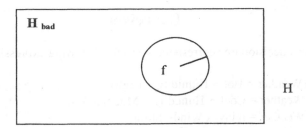

Fig. 13.15: The ball around f.

Thus $P(\text{a consistent } h_b \in H_{bad}) \leq \mid H_{bad} \mid (1 - \varepsilon)^m$

$$\leq |H| (1 - \varepsilon)^m$$

where $\mid H_{bad} \mid$ and $|H|$ denotes the cardinality of them respectively. If we put a small positive upper bound δ to the above quantity: we find

$$|H| (1-\varepsilon)^m \leq \delta$$

$$\Rightarrow m \geq (1/\varepsilon) [\ln (1/\delta) + \ln (H)].$$

Consequently, if a learning algorithm asserts an hypothesis that is supported by m number of examples, then it must be correct with a probability $\geq(1-\delta)$, when the error is $\leq \varepsilon$. So, we can call it probably approximately correct.

13.7 Summary

The principles of learning have been introduced in the chapter through version space, decision trees, inductive logic programming and learning automata. The reinforcement learning is one of the most modern aspects of learning, which has been covered thoroughly by Q-learning and temporal difference learning. The concepts of computational learning theory have also been outlined briefly to emphasize its significance to the readers. Computational learning theory and inductive logic programming are two open areas of research on machine learning. For instance, in inductive logic programming, selection of one clause, say C_1, and the resolvent C for proving C_2 from a given set of clauses are not unique, and most selections will result in old findings (existing in the knowledge base). Further, many possible C_2s can be generated from given clauses C and C_1. We hope that in the future the theory could be extended to discover knowledge in practical systems.

Exercises

1. Draw a decision tree corresponding to the following expression:

    If   (Weather = Hot $\wedge$ Humidity = High) $\vee$
         (Weather = Cool $\wedge$ Humidity = Moderate) $\vee$
         (Weather = rainy $\wedge$ Wind= Strong)
    Then start reading a storybook.

2. Make a list of the entries: Weather, Humidity, Wind and Read storybook of problem 1 in a tabular form and hence find the information gain for weather.

3. Show the trace of the candidate elimination algorithm on the following data sets.

    Positive instances: a) object (red, round, apple),
                          b) Object (green, round, mango);

Negative instances: a) Object (red, large, banana),
b) Object (green, round, guava).

4. Using the principles of inductive logic programming, determine the clause C_2, when C = Son (lob, ram) and C_1 = $\neg$ Daughter (lob, ram).

5. Given the following clauses:

a) Father (dasaratha, ram)
b) Grandchild (lob, dasaratha)
c) Father (ram, lob).

Derive the following rule by the inverse resolution principle.

Grandchild (Y, X) $\vee$ $\neg$ Father (X, Z) $\vee$ $\neg$ Father (Z, Y)

References

[1] Carbonell, J. G., "Learning by analogy: Formulating and generalizing plans from past experience," In *Machine Learning: An Artificial Intelligence Approach*, Michalski, R. S., Carbonell, J. G. and Mitchell, T. M., Eds., Tioga, Palo Alto, 1983.

[2] Feder, J., *Fractals*, Plenum Press, New York, 1988.

[3] Lahiri, T., Chakrabarti, A. and Dasgupta, A. K., "Multilamellar vesicular clusters of phosphatidylcholine and their sensitivity to spectrin: A study by fractal analysis," *Journal of Structural Biology*, vol. 123, pp. 179-186, 1998.

[4] Lahiri, T., Chakrabarti, A. and Dasgupta, A. K., "Onset of percolation and fractal classification scheme for multilamellar lipid vesicles," *Journal of Colloid and Interface Science*, vol. 211, pp. 89-95, 1999.

[5] Luger, G. F. and Stubblefield, W. A., *Artificial Intelligence: Structures and Strategies for Complex Problem Solving*, The Benjamin/ Cummins Publishing Company, Menlo Park, CA, pp. 462-532, 1993.

[6] Mandelbrot B. B., *The Fractal Geometry of Nature*, W. H. Freeman, New York, 1982.

[7]   Michalski, R. S., "A theory and methodology of inductive learning," *Artificial Intelligence*, vol. 20, no. 2, pp. 111-161, 1983.

[8]   Mitchell, T. M., "Generalization and search," *Artificial Intelligence*, vol. 18, no. 2, pp. 203-226, 1982.

[9]   Mitchell, T. M., "Version spaces: A candidate elimination approach to rule learning," *Proc. of the Fifth IJCAI*, pp. 305-310, 1977.

[10]   Mitchell, M . M., *Machine Learning*, McGraw-Hill, New York, pp. 52-385, 1997.

[11]   Patterson, D. W., *Introduction to Artificial Intelligence and Expert Systems*, Prentice-Hall, Englewood Cliffs, NJ, pp. 357-431, 1990.

[12]   Quinlan, J. R., "Induction of decision trees," *Machine Learning*, vol. 1, no. 1, pp. 81-106.

[13]   Russel, S. and Norvig, P., *Artificial Intelligence: A Modern Approach*, Prentice-Hall, Englewood Cliffs, NJ, pp. 598-644, 1995.

[14]   Winston, P., *Learning Structural Descriptions from Examples*, Ph.D. dissertation, MIT Technical Report AI-TR-231, 1970.

14

Machine Learning Using Neural Nets

The chapter presents various types of supervised, unsupervised and reinforcement learning models built with artificial neural nets. Among the supervised models special emphasis has been given to Widrow-Hoff's multi-layered ADALINEs and the back-propagation algorithm. The principles of unsupervised learning have been demonstrated through Hopfield nets, and the adaptive resonance theory (ART) network models. The reinforcement learning is illustrated with Kohonen's self-organizing feature map. The concepts of fuzzy neural nets will also be introduced in this chapter to demonstrate its application in pattern recognition problems.

14.1 Biological Neural Nets

The human nervous system consists of small cellular units, called neurons. These neurons when connected in tandem form nerve fiber. A biological neural net is a distributed collection of these nerve fibers.

A neuron receives electrical signals from its neighboring neurons, processes those signals and generates signals for other neighboring neurons attached to it. The operation of a biological neuron, which decides the nature of output signal as a function of its input signals is not clearly known to date.

However, most biologists are of the opinion that a neuron, after receiving signals, estimates the weighted average of the input signals and limits the resulting amplitude of the processed signal by a non-linear inhibiting function [6]. The reason for the non-linearity as evident from current literature [26] is due to the concentration gradient of the Potassium ions within a neuronal cell with respect to the Sodium-ion concentration outside the cell membrane. This ionic concentration gradient causes an electrical potential difference between the inner and outer portion of the neuronal cell membrane, which ultimately results in a flow of current from outside to inside the cell. A neuron, thus, can receive signals from its neighboring neurons. The variation of weights of the input signals of a neuron is due to the differences in potential gradient between a neuron and its surrounding cells. After the received signals are processed in a nerve cell, an invasion in diffusion current occurs due to the synaptic inhibiting behavior of the neuron. Thus, the processed signal can propagate down to other neighboring neurons.

A neuron has four main structural components [1]-[2]: the dendrites, the cell body, the axon and the synapse. The dendrites act as receptors, thereby receiving signals from several neighborhood neurons and passing these on to a little thick fiber, called dendron. In other words, dendrites are the free terminals of dendrons. The received signals collected at different dendrons are processed within the cell body and the resulting signal is transferred through a long fiber named axon. At the other end of the axon, there exists an inhibiting unit called synapse. This unit controls the flow of neuronal current from the originating neuron to receiving dendrites of neighborhood neurons. A schematic diagram, depicting the above concept, is presented in fig.14.1.

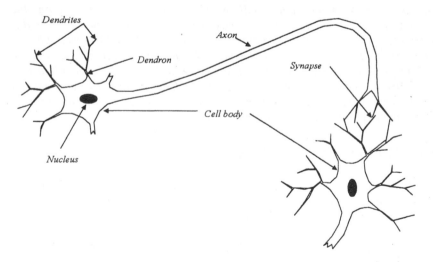

Fig. 14.1: A biological neuron.

14.2 Artificial Neural Nets

An artificial neural net is an electrical analogue of a biological neural net [23]. The cell body in an artificial neural net is modeled by a linear activation function. The activation function, in general, attempts to enhance the signal contribution received through different dendrons. The action is assumed to be signal conduction through resistive devices. The synapse in the artificial neural net is modeled by a non-linear inhibiting function, for limiting the amplitude of the signal processed at cell body.

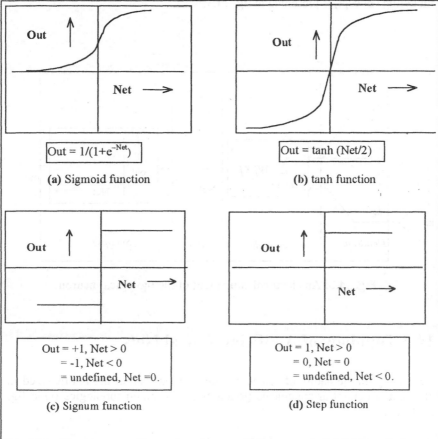

$$Out = 1/(1+e^{-Net})$$

(a) Sigmoid function

$$Out = \tanh (Net/2)$$

(b) tanh function

$$Out = +1, Net > 0$$
$$= -1, Net < 0$$
$$= undefined, Net = 0.$$

(c) Signum function

$$Out = 1, Net > 0$$
$$= 0, Net = 0$$
$$= undefined, Net < 0.$$

(d) Step function

Fig.14.2: Common non-linear functions used for synaptic inhibition. Soft non-linearity: (a) Sigmoid and (b) tanh; Hard non-linearity: (c) Signum and (d) Step.

The most common non-linear functions used for synaptic inhibition are:

- sigmoid function
- tanh function
- signum function
- step function

Sigmoid and tan hyperbolic (tanh) functions are grouped under soft non-linearity, whereas signum and step functions are under hard type non-linearity. These functions are presented graphically in fig. 14.2 for convenience.

The schematic diagram of an artificial neuron, based on the above modeling concept, is presented in fig. 14.3.

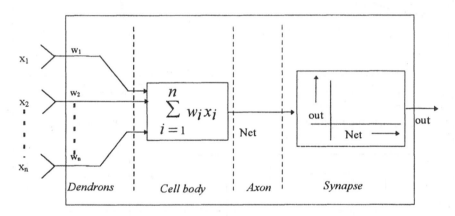

Fig. 14.3: An electrical equivalent of the biological neuron.

14.3 Topology of Artificial Neural Nets

Depending on the nature of problems, the artificial neural net is organized in different structural arrangements (topologies). Common topologies (vide fig. 14.4) are:

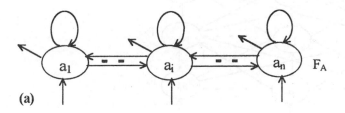

(a)

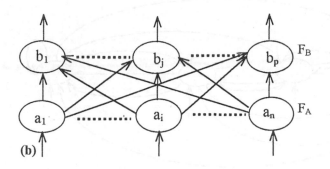

(b)

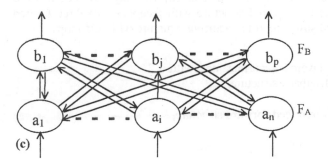

(c)

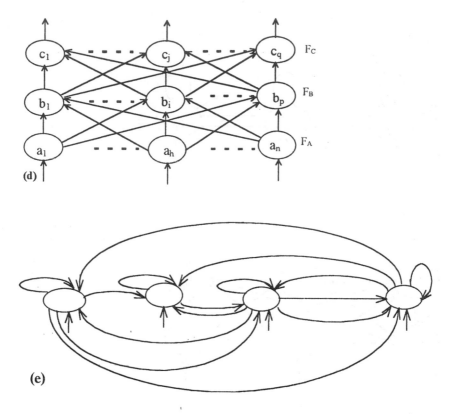

Fig.14.4: Common topologies of artificial neural net: (a) single layered recurrent net with lateral feedback, (b) two layered feed-forward structure, (c) two layered structure with feedback, (d) three layered feed-forward structure, (e) a recurrent structure with self-loops.

- Single layered recurrent net with lateral feedback structure,

- Two layered feed-forward structure,

- Two layered feedback structure,

- Three layered feed-forward structure and

- Single layered recurrent structure.

The single layered recurrent net with lateral feedback topology was proposed by Grossberg [3], which has successfully been applied for classifying analog patterns. The feed-forward structured neurals are the most common structures for the well-known back-propagation algorithm [20]. Two layered feedback structure, on the other hand, has been used by Carpenter and Grossberg [3] for realization of adaptive resonance theory [4] and Kosko [13]-[14] for realization of bi-directional associative memory. The last class of topology shown in fig. 14.4(e) represents a recurrent net with feedback. Many cognitive nets [12] employ such topologies. Another interesting class of network topologies, where each node is connected to all other nodes bi-directionally and there is no direct self-loop from a node to itself, has been used by Hopfield in his studies. We do not show the figure for this topology, as the readers by this time can draw it themselves for their satisfaction.

14.4 Learning Using Neural Nets

Artificial neural nets have been successfully used for recognizing objects from their feature patterns. For classification of patterns, the neural networks should be trained prior to the phase of recognition process. The process of training a neural net can be broadly classified into three typical categories, namely,

- Supervised learning
- Unsupervised learning
- Reinforcement learning.

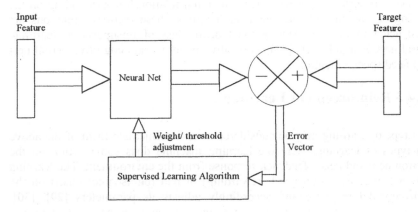

Fig. 14.5: The supervised learning process.

14.4.1 Supervised Learning

The supervised learning process (vide fig. 14.5) requires a trainer that submits both the input and the target patterns for the objects to get recognized. For example, to classify objects into "ball", "skull", and "apple", one has to submit the features like average curvature, the ratio of the largest solid diameter to its transverse diameter, etc. as the input feature patterns. On the other hand, to identify one of the three objects, one may use a 3-bit binary pattern, where each bit corresponds to one object. Given such input and output patterns for a number of objects, the task of supervised learning calls for adjustment of network parameters (such as weights and non-linearities), which consistently [19] can satisfy the input-output requirement for the entire object class (spherical objects in this context). Among the supervised learning algorithms, most common are the back-propagation training [20] and Widrow-Hoff's MADALINEs [28].

14.4.2 Unsupervised Learning

The process of unsupervised learning is required in many recognition problems, where the target pattern is unknown. The unsupervised learning process attempts to generate a unique set of weights for one particular class of patterns. For example, consider a neural net of recurrent topology (vide fig. 14.4 e) having n nodes. Assume that the feature vector for spherical objects is represented by a set of n descriptors, each assigned to one node of the structure. The objective of unsupervised learning process is to adjust the weights autonomously, until an equilibrium condition is reached when the weights do not change further. The process of unsupervised learning, thus, maps a class of objects to a class of weights. Generally, the weight adaptation process is described by a recursive functional relationship. Depending on the topology of neural nets and their applications, these recursive relations are constructed intuitively. Among the typical class of unsupervised learning, neural nets, Hopfield nets [24], associative memory, and cognitive neural nets [15] need special mention.

14.4.3 Reinforcement Learning

This type of learning may be considered as an intermediate form of the above two types of learning. Here the learning machine does some action on the environment and gets a feedback response from the environment. The learning system grades its action good (rewarding) or bad (punishable) based on the environmental response and accordingly adjusts its parameters [29], [30]. Generally, parameter adjustment is continued until an equilibrium state occurs, following which there will be no more changes in its parameters. The self-organizing neural learning may be categorized under this type of learning.

14.5 The Back-propagation Training Algorithm

The back-propagation training requires a neural net of feed-forward topology [vide fig. 14.4 (d)]. Since it is a supervised training algorithm, both the input and the target patterns are given. For a given input pattern, the output vector is estimated through a forward pass [21] on the network. After the forward pass is over, the error vector at the output layer is estimated by taking the component-wise difference of the target pattern and the generated output vector. A function of errors of the output layered nodes is then propagated back through the network to each layer for adjustment of weights in that layer. The weight adaptation policy in back-propagation algorithm is derived following the principle of **steepest descent** approach [16] of finding minima of a multi-valued function [5]. A derivation of the algorithm is given in appendix B.

The most significant issue of a back-propagation algorithm is the propagation of error through non-linear inhibiting function in backward direction. Before this was invented, training with a multi-layered feed-forward neural network was just beyond imagination. In this section, the process of propagation of error from one layer to its previous layer will be discussed shortly. Further, how these propagated errors are used for weight adaptation will also be presented schematically.

Typical neurons employed in back-propagation learning contain two modules (vide fig. 14.6(a)). The circle containing $\sum w_i x_i$ denotes a weighted sum of the inputs x_i for i= 1 to n. The rectangular box in fig. 14.6(a) represents the sigmoid type non-linearity. It may be added here that the sigmoid has been chosen here because of the continuity of the function over a wide range. The continuity of the nonlinear function is required in back-propagation, as we have to differentiate the function to realize the steepest descent criteria of learning. Fig. 14.6(b) is a symbolic representation of the neurons used in fig. 14.6(c).

In fig. 14.6(c), two layers of neurons have been shown. The left side layer is the penultimate (k −1)-th layer, whereas the single neuron in the next k-th layer represents one of the output layered neurons. We denote the top two neurons at the (k-1)-th and k-th layer by neuron p and q respectively. The connecting weight between them is denoted by $w_{p,q,k}$. For computing $W_{p,q,k}(n+1)$, from its value at iteration n, we use the formula presented in expression (14.2-14.4).

We already mentioned that a function of error is propagated from the nodes in the output layer to other layers for the adjustment of weight. This functional form of the back-propagated error is presented in expression (14.4) and illustrated in fig. 14.7. It is seen from expression (14.4) that the contribution of the errors of each node at the output layer is taken into account in an exhaustively connected neural net.

For training a network by this algorithm, one has to execute the following 4 steps in order for all patterns one by one.

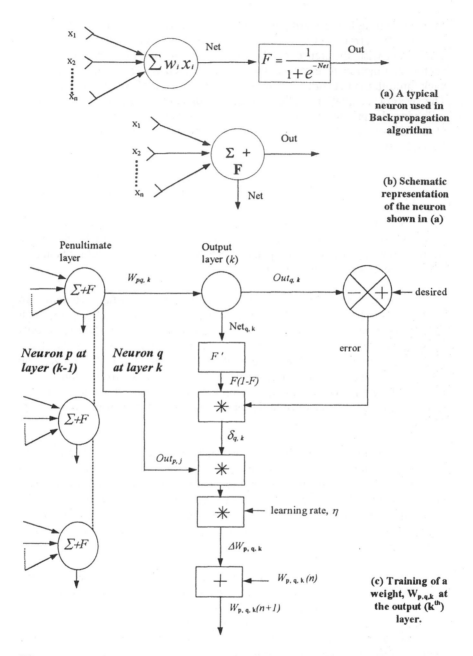

(a) A typical neuron used in Backpropagation algorithm

(b) Schematic representation of the neuron shown in (a)

(c) Training of a weight, $W_{p,q,k}$ at the output (k^{th}) layer.

Fig. 14.6: Attributes of neurons and weight adjustments by the back-propagation learning algorithm.

For each input-output pattern **do begin**

1. Compute the output at the last layer through forward calculation.
2. Compute δs at the last layer and propagate it to the previous layer by using expression (14.4).
3. Adjust weights of each neuron by using expression (14.2) and (14.3) in order.
4. Repeat from step 1 until the error at the last layer is within a desired margin.

End For;

The adaptation of the weights for all training instances, following the above steps, is called a learning epoch. A number of learning epochs are required for the training of the network. Generally a performance criterion is used to terminate the algorithm. For instance, suppose we compute the square norm of the output error vector for each pattern and want to minimize the sum. So, the algorithm will be continued until the sum is below a given margin.

The error of a given output node, which is used for propagation to the previous layer, is designated by δ, which is given by the following expression:

$$\delta = F' * (target - Out) = Out\ (1 - Out)\ (target - Out) \tag{14.1}$$

Weight adaptation in back-propagation

The weight adaptation policy[28] is described by the following expressions:

$$\Delta w_{p,\,q,\,k} = \eta\, \delta_{q,\,k}\ Out_{p,\,j} \tag{14.2}$$

$$w_{p,\,q,\,k}\,(n+1) = w_{p,\,q,\,k}\,(n) + \Delta w_{p,\,q,\,k} \tag{14.3}$$

where $w_{p,\,q,\,k}\,(n)$ = *the weight from neuron p to neuron q, at n^{th} step, where q lies in the layer k and neuron p in $(k-1)^{th}$ layer counted from the input layer;*

$\delta_{P,\,k}$ = *the error generated at neuron q, lying in layer k;*

$Out_{p,\,j}$ = *output of neuron p, positioned at layer j.*

For generating error at neuron p, lying in layer j, we use the following expression:

$$\delta_{p,\,j} = Out_{p,\,j}\ (1-Out_{p,\,j})(\sum_{q} \delta_{q,\,k}\, w_{p,\,q,\,k}) \tag{14.4}$$

where

$q \in \{q_1,\ q_2,\ q_3\}$ in fig. 14.7.

Drawbacks of back-propagation algorithm

The back-propagation algorithm suffers from two major drawbacks, namely network paralysis and trapping at local minima. These issues are briefly outlined below.

Network paralysis: As the network receives training, the weights are adjusted to large values. This can force all or most of the neurons to operate at large Outs, i.e., in a region where $F'(Net) \to 0$. Since the error sent back for training is proportional to $F'(Net)$, the training process comes to a virtual standstill. One way to solve the problem is to reduce η, which, however, increases the training time.

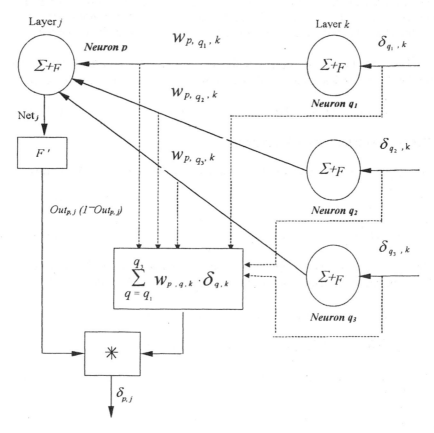

Fig. 14.7: The computation of δ_p at layer j.

Trapping at local minima: Back-propagation adjusts the weights to reach the minima (vide fig. 14.8) of the error function (of weights). However, the network can be trapped in local minima. This problem, however, can be solved by adding momentum to the training rule or by statistical training methods applied over the back-propagation algorithm [26].

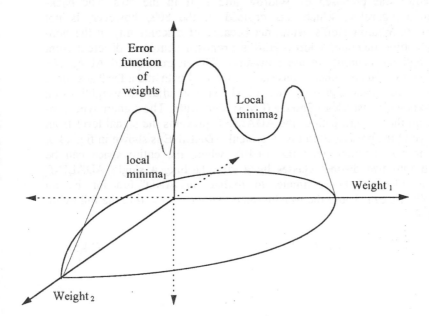

Fig. 14.8: Valleys in error function cause back-propagation algorithm trapped at local minima.

Adding momentum to the training rule: To eliminate the problem of trapping at local minima, recently a momentum term was added to the right hand side of the adaptation rule [17]. Formally,

$$W_{p,q,k}(n+1) = W_{p,q,k}(n) + \eta\, \delta_{q,k}\, Out_{p,j} + \alpha\, \Delta\, w_{p,q,k}\, (n-1).$$

The last term in the right hand side corresponds to momentum. The addition of the momentum term forces the system to continue moving in the same direction on the error surface, without trapping at local minima. A question may naturally arise: why call this momentum term? The answer came from the analogy of a rolling ball with high momentum passing over a narrow hole.

14.6 Widrow-Hoff's Multi-layered ADALINE Models

Another classical method for training a neural net with a sharp (hard limiting) non-linearity was proposed by Widrow and Hoff in the 60's. The back-propagation algorithm, which was devised in the 80's, however, is not applicable to Widrow-Hoff's neural net because of discontinuity in the non-linear inhibiting function. Widrow-Hoff's proposed neuron is different from the conventional neurons we use nowadays. The neuron, called ADALINE (ADAptive LINEar combiner), consists of a forward path and a feedback loop. Given a desired scalar signal d_k, the ADALINE can adjust its weights using the well-known **Least Mean Square** (LMS) **algorithm**. The signum type non-linearity on the forward path of the ADALINE prevents the signal level from going beyond the prescribed limits. A typical ADALINE is shown in fig. 14.9. For linear classification (vide fig. 14.10), where the entire space can be classified into two distinct classes by a straight line, a single ADALINE neuron is adequate. For example, to realize an 'AND' function by an ADALINE, we can choose the weights such that $w_1 = w_2 > 0$ in the following expression:

$$w_1 x_1 + w_2 x_2 > 0; \quad x_1, x_2 \in \{-1, +1\} \tag{14.5}$$

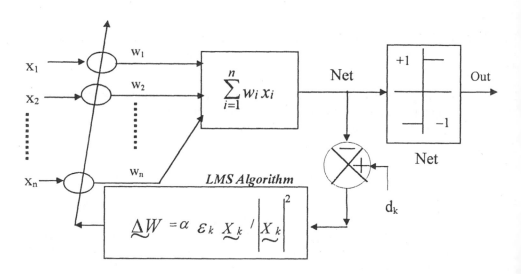

Fig. 14.9: A typical ADALINE neuron.

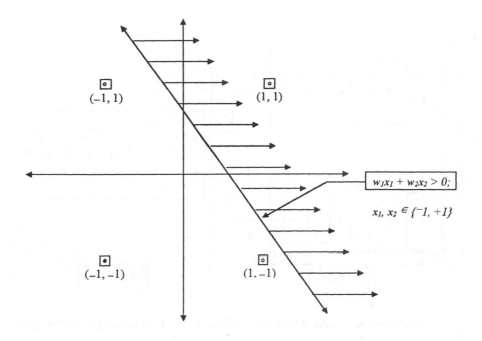

Fig. 14.10: Linear classification by ADALINE.

However, a single ADALINE cannot be used for non-linearly separable classes. For example, consider the problem of realizing an 'XOR' function, where the two distinct classes are {(-1, -1), (1,1)} and {(-1,1), (1,-1)}. For realizing such functions, we require three ADALINE neurons, where two neurons should be organized to form the first layer, while the third neuron forms the second (the last) layer, vide fig. 14.11 below.

The first layered neurons can do partial classification of the region (by two line segments **ac** and **bc**) (fig. 14.12) and the third neuron satisfies the joint occurrence of the convex region formed by these lines. In other words the third region is an AND logic. The design of the ADALINEs [18] is left as an exercise for the students.

ADALINEs are not only used for classifying patterns, but they have application also in recognizing unknown patterns. In fact, for pattern recognition by ADALINEs, one has to consider the topology of a feed-forward neural net of Fig.14.4 (d), with the replacement of the neurons referred to therein by ADALINEs.

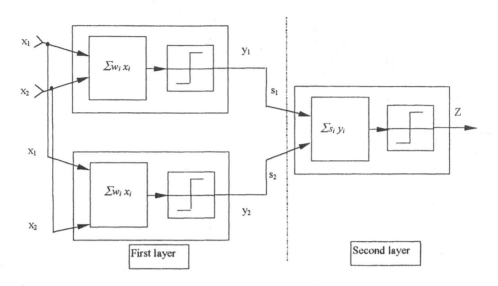

Fig. 14.11: A network of three ADALINES in two layers for realizing an 'XOR' gate.

The training process of an ADALINE neuron is realized by the Least Means Square (LMS) algorithm presented below.

Procedure LMS

Input d_i *and* $\underset{\sim}{X}_k$

Output weight vector $\underset{\sim}{W}_k$

Begin

Randomize initial weight vector W_0 ;

Repeat

$$\in_k = d_k - \underset{\sim}{X}_k^T \underset{\sim}{w}_k ;$$

$$\underset{\sim}{w}_{k+1} = \underset{\sim}{w}_k + \alpha \in_k \underset{\sim}{X}_k / |\underset{\sim}{X}_k|^2 ;$$

Until mod $|(\underset{\sim}{w}_{k+1} - \underset{\sim}{w}_k)| <$ pre – defined limit;

Pr int $\underset{\sim}{w}_k$;
End.

It can be easily proved that the LMS algorithm converges to stable points when $0 \leq \alpha \leq 2$. For solution, see exercise of this chapter.

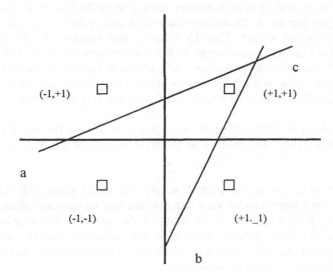

Fig. 14.12: A convex region represented by three ADALINEs of fig. 14.11.

For training the neural net with ADALINE neurons (fig. 14.4(d)), one has to make a trial adaptation starting from the 1st layer based on a principle, called **Minimum Disturbance Principle,** which is presented below.

- Identify a neuron at a given (i^{th}) layer whose Net value is closest to zero, compared to all other neurons at the same layer.
- Reverse its Out value and check whether the norm of error vector at the output layer (target vector minus computed output vector) decreases.

$$\in_k = d_k - \underset{\sim}{X}_k^T \underset{\sim}{w}_k ;$$

- If so, adjust the weights of that ADALINE using the LMS algorithm described above.

- If not, select another ADALINE, whose value is next closest to zero. Repeat the process for all ADALINEs in the same layer.

The above principle is applied to each neuron starting from the first to the last (output) layer. Then one has to choose two neurons together from each layer, whose outputs are closest to zero. Then by flipping their outputs, one has to check whether there is any improvement in the square norm of error at the output layer. If so, train them by procedure LMS; else select two other neurons from the same layer, whose outputs are next closest to zero. Such a process has to be repeated from the first to the last layer, until error norm is minimized.

The principle is so called because it attempts to disturb the weights minimally so as to get the near-zero error on the error surface (of weights).

Once the training is over, the trained weights are saved along with the input vector for each input pattern in a file. During the recognition phase, when an unknown pattern appears at the input of the network, the weights corresponding to the input pattern, closest to the unknown pattern, are assigned to the appropriate arcs in the network and the output pattern vector is generated through a forward pass in the network.

Another interesting issue to be discussed is the capability of Widrow-Hoff's neural net in recognizing translation, rotation and size-variant patterns. For recognition of translation, rotation and size invariant patterns, Widrow designed a specialized Majority Voter (MAJ) circuit, which receives signals from a number of ADALINEs and generates a single bit output. The MAJ output is 1, when majority of the ADALINEs connected to the input of the MAJ generates 1 as output. The ADALINEs with a MAJ are placed on a plane and there exist a number of such planes, located around the object from which retinal signals are received by these planes. A two layered trained neural net is used to decode the one-bit output of such planes of ADALINEs. For maintaining translational invariance, the weights of the ADALINES in one plane, called the reference plane, are assumed arbitrarily, and the weights of other planes of ADALINEs are selected based on the weight of the reference plane [27]. In fact, the possible shift of the object in the retinal plane is taken care of by judiciously selecting the weight matrices of the planes. The weight matrices of all planes are constructed by translating horizontally or vertically, whichever is appropriate, the weight matrix of the reference plane. Analogously, for rotational invariance, the weight matrices of each four planes are grouped and the weight matrices of them are set by rotating the weight matrices of one of them (around a pivot) by 0^0, 90^0, 180^0 and 270^0. Thus each group of planes can take care of 0^0, 90^0, 180^0, and 270^0 rotations. It may be added that since four planes are grouped, and output of such four planes is connected to the input of a MAJ circuit, a 4:1 reduction takes place in the number of signal carrying output lines of the invariance network. The output of the invariance network, thus, can represent the scrambled version of the rotated input patterns and, the scrambled pattern can be de-scrambled by a two layer trained neural network. For maintaining size invariance, Widrow devised an interesting means to adjust the dimension of the weight matrices and the value of the weights in each plane of ADALINEs. A complete implementation of this issue, however, remains a far cry to this date.

14.7 Hopfield Neural Net

Among the unsupervised learning algorithms, Hopfield neural nets [7] need special mention. In fact, in the nineties, nearly one third of the research papers on neural network include works on Hopfield nets. Hopfield nets can be of two common types [24], namely,

- **Binary Hopfield net**
- **Continuous Hopfield net**

In a binary Hopfield net, the input and output of the neurons are binary, whereas in a continuous Hopfield net, they could assume any continuous values between 0 and 1. Hopfield neural nets find application in many problems, especially in realizing ADCs as highlighted by John Hopfield in his early papers. The principle of weight adjustment in Hopfield net is done by optimizing a function of weights and signal values at nodes, popularly known as Liapunov functions. Liapunov functions are energy functions, used to identify system states, where the function yields the minimum value of energy.

Binary Hopfield Net

In a binary Hopfield model, each neuron can have two output states: *'0'* and *'1'*, (sometimes denoted as n_i^0 and n_i^1). Let us consider a neuron *'i'*; then the total input to neuron *'i'*, denoted by H_i, is given below:

$$H_i = \sum_{j \neq i} w_{ij} n_j + l_i \tag{14.6}$$

where, l_i = *external input to neuron 'i'*

n_j = *inputs from neuron 'j'*

w_{ij} = *synaptic interconnection strength from neuron 'j' to* neuron *'i'*

n_i^0 = *'0' output (non-firing)*

$$n_i^l = \text{'}1\text{' } output \ (firing).$$

Each neuron has a fixed threshold, say, th_i for neuron 'i'. Each neuron readjusts its states randomly by using the following equations.
For neuron i :

Output n_i $= 1$; $if \ \sum_{j \neq i} w_{ij} \, n_j > th_i$

              $= 0$;   *otherwise*   (14.7)

The information storage for such a system is normally described by

$$w_{ij} = \sum_{s} (2 \, n_i^s - 1)(2 \, n_j^s - 1)$$

where n_i^s represents set of states for s being an integer $1,2,\ldots , n$.

To analyze the stability of such a neural network, Hopfield proposed a special kind of Liapunov energy function given in expression (14.8)

$$E = - (1/2) \ \sum_{i \neq j} w_{ij} \, n_i \, n_j - \sum_{i} I_i \, n_i + \sum_{i} th_i \, n_I . \qquad (14.8)$$

The change in energy ΔE due to change in the state of neuron 'i' by an amount Δn_i is given by

$$\Delta E = - \sum_{i \neq j} (w_{ij} \, n_j + I_I - th_i \,) \, \Delta \, n_i \qquad (14.9)$$

As Δn_i is positive only when the bracketed term is positive, thus any change in E under the expression (14.9) is negative. Since E is bounded, so the iteration of the Hopfield neural algorithm, given by the expression (14.8), must lead to stable states.

Continuous Hopfield net

The continuous Hopfield net can be best described by the following differential equation,

$$C_i \frac{du_i}{dt} = \sum_{j} w_{ij} \, n_j - u_i / R_i + I_i \qquad (14.10)$$

where,

> $w_{i\ j}$ = *synaptic interconnection strength (weight) from neuron 'j' to neuron 'i',*
> n_j = *output variable for neuron 'j',*
> u_i = *instantaneous input to neuron 'i',*
>
> u_i = *instantaneous input to neuron 'i',*
> l_i = *external input to neuron 'i',*
> $g_i = n_i / u_i$: *output-input sigmoidal relationship,*
> C_i and R_i *represent dimensional constants.*

Such symbols are used in the above nomenclature to keep these comparable with parameters of electric networks. The continuous Hopfield model uses an energy like function, the time derivative of which is given by:

$$\frac{dE}{dt} = \sum_{i=1}^{n} \frac{dn_i}{dt} \left(\sum_j w_{ij} n_j - u_i / R_i + l_i \right)$$

$$= \sum_{j=1}^{n} C_i \frac{dn_i}{dt} \frac{du_j}{dt}$$

$$= -\sum_{j=1}^{n} C_i g_i^{-1} (n_i) \left(\frac{dn_i}{dt} \right)^2$$

(14.11)

where, g_i^{-1} (n_i) is a monotonically increasing term and C_i is a positive constant. It can be shown easily that when $w_{ij} = w_{ji}$, the Hopfield net can evolve towards an equilibrium state.

14.8 Associative Memory

The models of associative memory pioneered by Kosko [14] is a by-product of his research in fuzzy systems. Associative memory to some extent has resemblance with the human brain, where from one set of given input output pattern, the brain can determine analogously the output (input) pattern when the input (output) pattern is known. In Kosko's model, he assumed the fuzzy membership distribution of the output patterns for a given input pattern and designed the weights *(W)* of a single-layered feed-forward neural net by taking the product of input feeder *(I)* and the transpose of the output feeder *(O)* :

$$\underset{\sim}{W} = \underset{\sim}{I} \, o \, \underset{\sim}{O}^{T}$$, where "*o*" denotes the max-min composition operator.

Once the weight matrix between the output and the input layer is formed, one can estimate an output vector O' when the input vector $\underset{\sim}{I'}$ is given:

$$\underset{\sim}{O'} = W o \underset{\sim}{I'} \, .$$

Analogously, the input vector I' can also be estimated, when O' is given

by $$\underset{\sim}{I'} = W^{T} o \underset{\sim}{O'} \, .$$

Recently, Saha and Konar [22] designed a new methodology for estimation of inversion of fuzzy matrix, *W*. By using their definition, one can compute *I'* by

$$\underset{\sim}{I'} = W^{-1} o \, O' \, ,$$ where W^{-1} is defined as $W^{-1} o W \rightarrow I$, the '$\rightarrow$' symbol

represents 'tends to' and I is the identity matrix. There exist n^{n} matrices for a (n × n) *W* matrix. The best inverse, which is defined as W_{best}^{-1}, is where

$$\sum_{i} \sum_{j} \delta_{ij}^{2}$$ is minimum for $$\left[\delta_{ij} \right] = \left[\left(W_{best}^{-1} o \, W \right) - I \right]$$ and

is generally used for estimation of *I'* using $$\underset{\sim}{I'} = W_{best}^{-1} o \, \underset{\sim}{O'} \, .$$

It may be added that in very special cases $W_{best}^{-1} = W^{T}$, and thus the inverse matrix support Kosko's model [14].

14.9 Fuzzy Neural Nets

Conventional sets include elements with membership values zero or one. However, for many real world problems, the membership value of an element in a set need not be always binary but may assume any value between zero and one. In fuzzy logic Max, Min, composition, t- norm, s- norm, and other operators are used along with fuzzy sets to handle real world reasoning problems. Fuzzy logic has been successfully used for modeling artificial neural nets, [19] where the input and output parameters of the neural net are the membership distributions.

Among the classical fuzzy neural nets, the OR-AND neuron model of Pedrycz [19] needs special mention. In an OR-AND neuron, we have two OR nodes and one AND node [vide fig. 14.13].

Here, $Z_1 = \wedge \ (W_{ij} \vee x_j)$

$$1 \leq j \leq n$$

$$Z_2 = \vee \ (V_{ij} \wedge x_j)$$

$$1 \leq j \leq n$$

and $y = (S_{11} \wedge Z_1) \vee (S_{12} \wedge Z_2)$

where $'\wedge'$ and $'\vee'$ denote fuzzy $'t'$ and $'s'$ norm operators. The above two operations are executed by using expressions (14.12) and (14.13).

$$x_1 \wedge x_2 = x_1.x_2 \tag{14.12}$$

$$x_1 \vee x_2 = (x_1 + x_2 - x_1.x_2) \tag{14.13}$$

where '.' and '+' are typical algebraic multiplication and summation operations.

Pedrycz devised a new algorithm for training an AND-OR neuron, when input membership distributions x_i for $i = 1$ to n and target scalar y are given. The training algorithm attempts to adjust the weights, w_{ij}, so that a pre-defined performance index, is optimized.

Pedrycz also designed a pseudo-median filter for using AND-OR neurons, which has many applications in median filtering under image processing. The typical organization of a pseudo-median filter for five delta points is given in fig. 14.14.

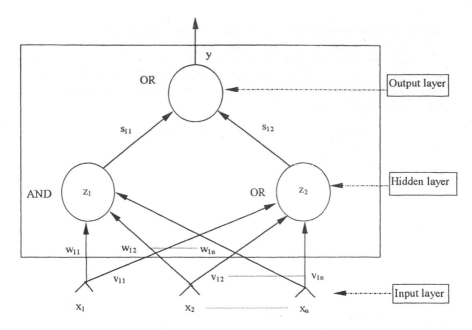

Fig. 14.13. Architecture of an AND-OR neuron

In the figure, the pseudo-median is defined as:

pseudomedian (χ)

$= (1/2)$ Max [Min (x_1, x_2, x_3), Min (x_2, x_3, x_4), Min (x_3, x_4, x_5)] $+$

$(1/2)$ Min [Max (x_1, x_2, x_3), Max (x_2, x_3, x_4), Max (x_3, x_4, x_5)],

which loses little accuracy w.r.t. typical median filter.

A set of 10 input-output patterns have been used to train pseudo-median filter. The weights w_{ij}, after training the net, are saved for a subsequent recognition phase. In the recognition phase, the neurons can compute the pseudo-median, when the input data points are given.

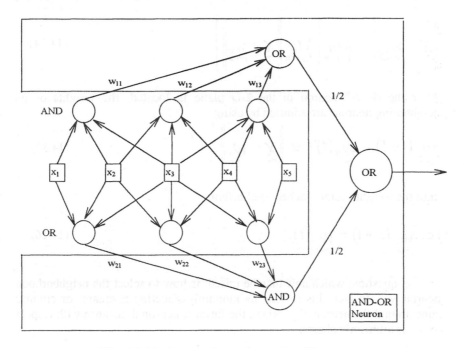

Fig. 14.14: A typical pseudo-median filter.

14.10 Self-Organizing Neural Net

Kohonen [9] proposed a new technique for mapping a given input pattern onto a 2-dimensional spatial organization of neurons. In fact, Kohonen considered a set of weights, connected between the positional elements of the input pattern and a given neuron, located at position *(i, j)* in a two dimensional plane. Let us call the weights' vector for neuron N_{ij} to be $\underset{\sim}{W}_{ij}$. Thus for a set of *(n × n)* points on the 2-D plane, we would have n^2 such weight vectors, denoted by $\underset{\sim}{W}_{ij}$, $1 \le i, j \le n$. In Kohonen's model, the neuron with minimum distance between its weight vector $\underset{\sim}{W}_{ij}$ and the input vector $\underset{\sim}{X}$ is first identified by using the following criterion [10].

Find the *(k, l)*th neuron, where

$$\left\| \underset{\sim}{X} - \underset{\sim}{w_{k,l}} \right\| = \underset{1 \le j < n}{Min} \left[\underset{1 \le j < n}{Min} \left\| \underset{\sim}{X} - \underset{\sim}{w_{ij}} \right\| \right]. \tag{14.14}$$

After the $(k, l)^{th}$ neuron in the 2-D plane is located, the weights of its neighboring neurons are adjusted by using

$$\underset{\sim}{w_{ij}}(t+1) = \underset{\sim}{w_{ij}}(t) + \alpha \left\| \underset{\sim}{X} - \underset{\sim}{w_{ij}} \right\| \tag{14.15}$$

until the weight vector reaches equilibrium,

i.e., $\underset{\sim}{w_{ij}}(t+1) = \underset{\sim}{w_{ij}}(t).$ (14.16)

A question, which now may be raised, is how to select the neighborhood neuron $N_{i,j}$. In fact, this is done by randomly selecting a square or circular zone around the neuron $N_{i,j}$, where the furthest neuronal distance with respect to $N_{i,j}$ is arbitrarily chosen.

Once the training of the neighborhood neuron around $N_{i,j}$ is over, the process of selection of the next neuron by criterion (A) is carried out. After repetition of the selection process of a neuron, followed by weight adaptation of the neighborhood neurons for a long time, the system reaches an equilibrium with all weight vectors $\underset{\sim}{w_{ij}}$ for $1 \le i, j \le n$, being constant for a given dimension of the neighborhood of neurons. The neighborhood of the neurons is now decreased and the process of selection of neuron and weight adaptation of its neighboring neurons is continued until the equilibrium condition is reached. By gradually reducing the neighborhood of the neuron, selected for weight adaptation, ultimately, a steady-state neighborhood of neurons in the 2-D plane is obtained. The plane of neurons thus obtained represents a spatial mapping of the neurons, corresponding to the input pattern.

The algorithm for a self-organizing neural adaptation for a input vector $\underset{\sim}{X}$ and the corresponding weight vector $\underset{\sim}{w_{ij}}$ is presented below.

Procedure Self-organization ($\underset{\sim}{X}, \underset{\sim}{w_{ij}}$)

Begin
      Repeat

For i:= 1 to n **do**
Begin

    For j:= 1 to n **do**
    Begin

$$\text{If } \left\| \underset{\sim}{X} - \underset{\sim}{w}_{k,l} \right\| = \underset{1 \le j < n}{Min} \left[\underset{1 \le j < n}{Min} \left\| \underset{\sim}{X} - \underset{\sim}{w}_{ij} \right\| \right]$$

Then Adjust weights of neighboring neurons of $N_{k,l}$
by the following rule:
For i':= (k − δ) to (k + δ) **do**
Begin

    For j':= (l − δ) to (l + δ) **do**
    Begin

$$\underset{\sim}{w}_{i',j'}(t+1) = \underset{\sim}{w}_{i',j'}(t) + \alpha \left\| \underset{\sim}{X} - \underset{\sim}{w}_{i',j'}(t) \right\|$$

    End For;
End For;
δ := δ − ε; // **Space-organization** //
    End For;
End For;
Until δ ≤ pre-assigned quantity;
End.

    The above algorithm should be repeated for all input patterns. Consequently, all the input patterns will be mapped on the 2-D plane as a n-dimensional point, each having a $\underset{\sim}{w}_{ij}$.

14.11 Adaptive Resonance Theory (ART)

Another unsupervised class of learning, which takes into account the stability [3] of the previously learnt pattern classes and plasticity [4] of storing a new pattern class, is the ART network model, proposed by Carpenter and Grossberg [3]. It is a two-layered network, where the input layer transforms the received new pattern ($\underset{\sim}{X}$) into state vector ($\underset{\sim}{V}$) with the help of stored patterns ($\underset{\sim}{Q}$), and the output layer identifies a weight vector ($\overline{\underset{\sim}{W}}$) among a stored set of competitive weight vectors ($\overline{\underset{\sim}{W}}$). The new states V_j, which are components of $\underset{\sim}{V}$, are given by

$$V_j = Sgn\left(x_j + \sum_{\forall i} w_{ji}O_i + A - 1.5\right)$$

$$= \begin{cases} x_j, \text{when all } O_i = 0 \\ x_j \wedge \sum_{\forall i} w_{ji} O_i, \text{otherwise} \end{cases}$$

$$\text{where } A = Sgn\left(\sum_j x_j - N\sum_i O_i - 0.5\right)$$

$$\text{and } Sgn(y) = +1, \ y > 0$$
$$= 0, \ y < 0.$$

The weight vector $\overline{\underset{\sim}{W}}^*$ is identified by the output layer using the following **Competition Rule:**

Select $\overline{\underset{\sim}{W}}^*$ if

$$\overline{\underset{\sim}{W}}^* \cdot \underset{\sim}{X} \geq \overline{\underset{\sim}{W}} \cdot \underset{\sim}{X}, \text{ for all } \overline{\underset{\sim}{W}}.$$

The schematic architecture of the ART network model is given in fig.14.15.

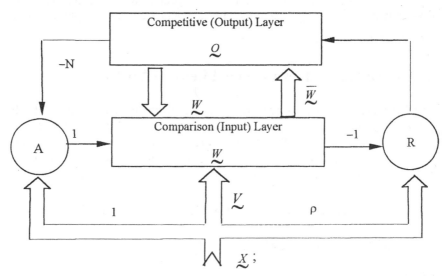

Fig. 14.15: An ART network model.

The algorithm for weight adaptation (learning) in an ART network is given below.

Procedure ART-learning;

Begin

        Enable all $\underset{\sim}{Q}$ vectors;

        Receive a new pattern vector $\underset{\sim}{X}$;

        Repeat

                For $\overline{\underset{\sim}{W}} \in$ set of stored weight vectors

$$\text{If } \overset{*}{\overline{\underset{\sim}{W}}} \cdot \underset{\sim}{X} \geq \overline{\underset{\sim}{W}} \cdot \underset{\sim}{X}$$

                        Then select $\overset{*}{\overline{\underset{\sim}{W}}}$;

        End For;

$$\text{If } r = \frac{\overset{*}{\overline{\underset{\sim}{W}}} \cdot \underset{\sim}{X}}{\sum_j X_j} \geq \text{predefined } \rho$$

        Then (resonance occurs)

$$\underset{\sim}{\Delta W} = \eta \, (\underset{\sim}{V} - \overset{*}{\underset{\sim}{W}}) ;$$

                Else disable the $\underset{\sim}{Q}$ vector which yields $r < \rho$;

        Until all $\underset{\sim}{Q}$ vectors are disabled;

End.

14.12 Applications of Artificial Neural Nets

Artificial neural nets have applications in a wide range of problems like pattern classification, machine learning, function realization, prediction, knowledge acquisition and content addressable memory. In this book we have demonstrated various application of the tools and techniques of artificial neural nets in the fields mentioned above. For instance, in chapter 23 we used the back-propagation learning for speaker identification from their speech phoneme. Chapter 24 demonstrates the use of self-organizing neural nets and the back-propagation learning algorithm in path planning of mobile robots. Further, chapter 20 illustrates the use of the Hebbian type unsupervised learning for acquisition of knowledge in an expert system. Self-organizing neural nets have also been applied in chapter 17 for recognition of human faces from their facial images. Some recent developments in cognitive learning have been presented in chapter 16 for applications in psychological modeling, which in the coming future will find applications in intelligent expert systems. The complete list of applications of artificial neural nets would be as large as this book. We thus do not want to make it complete any way.

Currently, fuzzy logic and genetic algorithms [25] are being used together for building complex systems. The aim of their joint use is to design systems that can reason and learn from inexact data and knowledge [11] and can efficiently search data or knowledge from their storehouses. Such autonomous systems will be part and parcel of the next generation of intelligent machines.

Exercises

1. Draw a neural network from the specification of their connection strengths (weights):

    $W_{xy} = W_{yx} = -1, \ W_{xz} = W_{zx} = +1, \ W_{zv} = W_{vz} = +2,$

    $W_{vw} = W_{wv} = +1, \ W_{wm} = W_{mw} = -1, \ W_{mu} = W_{um} = +3,$

    $W_{uy} = W_{yu} = +3, \ W_{xu} = W_{ux} = -1, \ W_{zu} = W_{uz} = -1,$

    $W_{zw} = W_{wz} = +1, \ W_{uw} = W_{wu} = -2$

    where the suffixes denote the two nodes between which a weight is connected. Use the concept of binary Hopfield net to compute the next states from the current state. The pattern P of the state variables and its corresponding 4 instances P_1, P_2, P_3 and P_4 are given below.

    $P = [x \ y \ z \ v \ w \ m \ u]$

    $P_1 = [0 \ 0 \ 0 \ 0 \ 0 \ 0 \ 0]$

    $P_2 = [1 \ 0 \ 1 \ 1 \ 0 \ 0 \ 0]$

    $P_3 = [0 \ 1 \ 0 \ 0 \ 0 \ 1 \ 1]$

$P_4 = [0 \ 1 \ 1 \ 1 \ 0 \ 0 \ 1]$

Compute the next instances of each pattern by using the rules of binary Hopfield nets, set threshold = 0.

Repeat the above process by replacing 0 in the patterns by -1 ; do you get any changes in the results?

Can you call these 4 patterns stable states (states that do not change)? If so, in which of the two cases?

Take any other arbitrary patterns, comprising of 1 and 0s. Are they also stable states?

2. Suppose the 4 patterns are only given. Can you compute the weights in the network by using the formula given in the text?

3. Show that the learning rule in Widrow-Hoff's model, given by the following expression, is stable, when $0 < \alpha < 2$, unstable for $\alpha > 2$ and oscillatory when $\alpha = 2$.

Learning rule: $\Delta w_k = \alpha \in_k \underset{\sim}{X}_k / | \underset{\sim}{X}_k |^2$,

where ΔW_k = change in weight vector, X_k = the input vector, whose components are $x_1, x_2, \ldots x_n$.

[**Hints:** $\in_k = d_k - \underset{\sim}{X}_k^T \underset{\sim}{w}_k$;

So, $\Delta \in_k = - \underset{\sim}{X}_k^T \cdot \Delta \underset{\sim}{w}_k$

$= -\alpha \in_k \underset{\sim}{X}_k / | \underset{\sim}{X}_k |^2$

$= -\alpha \in_k$

Now, $(E - 1 + \alpha) \in_k = 0$,

or, $\in_k = (1 - \alpha)^k$.

Now, put $0 < \alpha < 2$, $\alpha > 2$ and $\alpha = 2$ to prove the results.]

4. Derive the gradient descent learning rule for the following error function:

$E = (1 / 2) \sum (t_k - Out_k)^2 + \sum w_{ji}^2$

where the first summation is over all components of the output patterns, t_k for the scaler target and Out_k for the scaler computed output at node k. W_{ij} is the weight from neuron i to neuron j.

5. Realize an Ex-or function by three ADALINEs.

References

[1]   Anderson, B. D. O, and Moore, J. B., *Optimal Filtering*, Prentice-Hall, Englewood Cliffs, NJ,1979.

[2]   Anderson, J. A., " A simple neural network generating an associative memory," *Mathematical Biosciences*, vol. 14, pp. 197-220, 1972.

[3]   Carpenter, G. A. and Grossberg, S., "A massively parallel architecture for   a self-organizing neural pattern recognition machine," *Computer Vision,   Graphics and Image Processing*, vol. 37, pp. 54-115, 1987.

[4]   Carpenter, G. A. and Grossberg, S., "ART2: Self-organization of stable category recognition codes for analog input patterns," *Applied Optics*, vol. 23, pp. 4919-4930, Dec. 1987.

[5]   Fu, Li M, *Neural Networks in Computer Intelligence*, McGraw-Hill, NewYork, 1994.

[6]   Hertz, J., Krogn, A. and Palmer, G. R., *Introduction to the Theory of Neural Computation*, Addison-Wesley, Reading, MA, 1990.

[7]   Hopfield, J. "Neural nets and physical systems with emergent collective computational abilities," *Proc. of the National Academy of Sciences*, vol. 79, pp. 2554-2558, 1982.

[8]   Hopfield, J. J., "Neural networks with graded response have collective computational properties like those of two state neurons," *Proc. of the National Academy of Sciences*, vol. 81, pp. 3088-3092, May 1984.

[9]   Kohonen, T., *Self-organization and Associative Memory*, Springer-Verlag, Berlin, 1989.

[10]   Kohonen, T., Barna, G. and Chrisley, R., "Statistical pattern recognition using neural networks: Benchmarking studies," *IEEE Conf. on Neural Networks*, San Diego, vol. 1, pp. 61-68.

[11] Konar, A., *Uncertainty Management in Expert Systems Using Fuzzy Petri Nets*, Ph. D. thesis, Jadavpur University, 1994.

[12] Konar, A. and Pal, S., Modeling cognition with fuzzy neural nets, in *Neural Network Systems: Techniques and Applications*, Leondes, C. T., Ed., Academic Press, New York, 1999.

[13] Kosko, B., "Adaptive bi-directional associative memories," *Applied Optics*, vol. 26, pp. 4947-4960, 1987.

[14] Kosko, B., "Bi-directional associative memories," *IEEE Trans. on Systems, Man and Cybernetics*, vol. SMC-18, pp. 49-60, Jan 1988.

[15] Kosko, B., *Neural Networks and Fuzzy Systems: A Dynamical Systems Approach to Machine Intelligence*, Prentice-Hall, Englewood Cliffs, NJ, 1991.

[16] Luo, Fa- Long and Unbehauen, R., *Applied Neural Networks for Signal Processing*, Cambridge University Press, London, pp. 1-31, 1997.

[17] Mitchell, M. M., *Machine Learning*, McGraw-Hill, New York, pp. 81-127, 1997.

[18] Paul, B., Konar, A. and Mandal A. K., "Fuzzy ADALINEs for gray image recognition," *Neurocomputing*, vol. 24, pp. 207-223, 1999.

[19] Pedrycz, W., *Fuzzy Sets Engineering*, CRC Press, Boca Raton, FL, pp. 73-106,1996.

[20] Rumelhart, D. E. and McClelland, J. L., *Parallel Distributed Processing: Exploring in the Microstructure of Cognition*, MIT Press, Cambridge, MA, 1986.

[21] Rumelhart, D. E., Hinton, G. E. and Williams R. J., "Learning representations by back-propagation errors, " *Nature*, vol. 323, pp. 533-536, 1986.

[22] Saha, P. and Konar, A., "A heuristic approach to computing inverse fuzzy relation," communicated to *Information Processing Letters*, 1999.

[23] Schalkoff, R. J., *Artificial Neural Networks*, McGraw-Hill, New York, pp. 146-188, 1997.

[24] Tank, D. W. and Hopfield, J. J., "Simple neural optimization networks: An A/D converter, signal decision circuit and a linear programming

circuit," *IEEE Trans. on Circuits and Systems*, vol. 33, pp. 533-541, 1986.

[25]   Teodorescu, H. N., Kandel, A. and Jain, L. C., Eds., *Fuzzy and Neuro-Fuzzy systems in Medicine*, CRC Press, London, 1999.

[26]   Wasserman, P. D., *Neural Computing: Theory and Practice*, Van Nostrand Reinhold, New York, pp. 49-85, 1989.

[27]   Widrow, B., "Generalization and information storage in networks of ADALINE neurons," in *Self-Organizing Systems*, Yovits, M. C., Jacobi, G. T. and Goldstein, G. D., Eds., pp. 435-461, 1962.

[28]   Widrow, B. and Hoff, M, E., Adaptive Switching Circuits, in *1960 IRE WESCON Convention Record*, Part 4, pp. 96-104, NY, 1960.

[29]   Williams, R. J., "On the use of back-propagation in associative reinforcement learning," in *IEEE Int. Conf. on Neural Networks*, NY, vol. 1, pp. 263-270, 1988.

[30]   Williams, R. J. and Peng, J., "Reinforcement learning algorithm as function optimization," Proc. of *Int. Joint Conf. on Neural Networks*, NY, vol. II, pp. 89-95, 1989.

15

Genetic Algorithms

The chapter presents a new kind of classical algorithm that emulates the biological evolutionary process in intelligent search, machine learning and optimization problems. After a brief introduction to this algorithm, the chapter provides an analysis of the algorithm by the well-known Schema theorem and Markov Chains. It also demonstrates various applications of GA in learning, search and optimization problems. The chapter ends with a discussion on Genetic Programming.

15.1 Introduction

Professor John Holland in 1975 proposed an attractive class of computational models, called Genetic Algorithms (GA) [1]-[19], that mimic the biological evolution process for solving problems in a wide domain. The mechanisms under GA have been analyzed and explained later by Goldberg [10], De Jong [7], Davis [6], Muehlenbein [16], Chakraborti [3]-[5], Fogel [8], Vose [19] and many others. GA has three major applications, namely, intelligent search, optimization and machine learning. Currently, GA is used along with neural nets and fuzzy logic for solving more complex problems. Because of

449

their joint usage in many problems, these together are often referred to by a generic name: "soft-computing".

A GA operates through a simple cycle of stages [9]:

i)   Creation of a "population" of strings,
ii)   Evaluation of each string,
iii) Selection of best strings and
iv)   Genetic manipulation to create new population of strings.

The cycle of a GA is presented below in fig. 15.1.

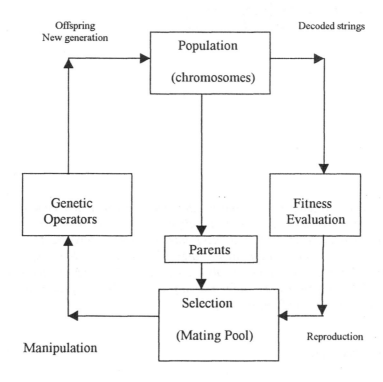

Fig. 15.1: The cycle of genetic algorithms.

Each cycle in GA produces a new generation of possible solutions for a given problem. In the first phase, an initial population, describing representatives of the potential solution, is created to initiate the search process. The elements of the population are encoded into bit-strings, called chromosomes. The performance of the strings, often called fitness, is then evaluated with the help of some functions, representing the constraints of the problem. Depending on the fitness of the chromosomes, they are selected for a subsequent genetic manipulation process. It should be noted that the **selection** process is mainly responsible for assuring survival of the best-fit individuals. After selection of the population strings is over, the genetic manipulation process consisting of two steps is carried out. In the first step, the **crossover** operation that recombines the bits (genes) of each two selected strings (chromosomes) is executed. Various types of crossover operators are found in the literature. The single point and two points crossover operations are illustrated in fig. 15.2 and 15.3 respectively. The **crossover points** of any two chromosomes are selected randomly. The second step in the genetic manipulation process is termed **mutation**, where the bits at one or more randomly selected positions of the chromosomes are altered. The mutation process helps to overcome trapping at local maxima. The offsprings produced by the genetic manipulation process are the next population to be evaluated.

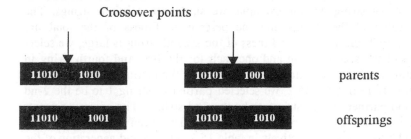

Fig 15.2: A single point crossover after the 3-rd bit position from the L.S.B.

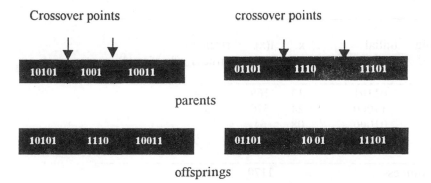

Fig. 15.3: Two point crossover: one after the 4th and the other after the 8th bit positions from the L.S.B.

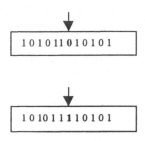

Fig. 15.4: Mutation of a chromosome at the 5th bit position.

Example 15.1: The GA cycle is illustrated in this example for maximizing a function $f(x) = x^2$ in the interval $0 \leq x \leq 31$. In this example the fitness function is f (x) itself. The larger is the functional value, the better is the fitness of the string. In this example, we start with 4 initial strings. The fitness value of the strings and the percentage fitness of the total are estimated in Table 15.1. Since fitness of the second string is large, we select 2 copies of the second string and one each for the first and fourth string in the mating pool. The selection of the partners in the mating pool is also done randomly. Here in table 15.2, we selected partner of string 1 to be the 2-nd string and partner of 4-th string to be the 2nd string. The crossover points for the first-second and second-fourth strings have been selected after o-th and 2-nd bit positions respectively in table 15.2. The second generation of the population without mutation in the first generation is presented in table 15.3.

Table 15.1: Initial population and their fitness values

string no.	initial population	x	f(x)	strength fitness (% of total)
1	01101	13	169	14.4
2	11000	24	576	49.2
3	01000	08	64	5.5
4	10011	19	361	30.9
Sum-fitness			1170	100.00

Table 15.2: Mating pool strings and crossover

String No.	Mating Pool	Mates string	Swapping	New population
1	01101	2	0110[1]	01100
2	11000	1	1100[0]	11001
2	11000	4	11[000]	11011
4	10011	2	10[011]	10000

Table 15.3: Fitness value in second generation

Initial population	x	f(x) (fitness)	strength (% of total)
01100	12	144	8.2
11001	25	625	35.6
11011	27	729	41.5
10000	16	256	14.7
Sum-fitness =		1754	100.00

A **Schema** (or schemata in plural form) / **hyperplane** or **similarity template** [5] is a genetic pattern with fixed values of 1 or 0 at some designated bit positions. For example, S = 01?1??1 is a 7-bit schema with fixed values at 4-bits and don't care values, represented by ?, at the remaining 3 positions. Since 4 positions matter for this schema, we say that the schema contains 4 **genes**.

A basic observation made by Prof. Holland is that *"A schema with an above average fitness tends to increase at an exponential rate until it becomes a significant portion of the population."*

15.2 Deterministic Explanation of Holland's Observation

To explain Holland's observation in a deterministic manner let us presume the following assumptions [2].

i)   There are no recombination or alternations to genes.

ii)   Initially, a fraction f of the population possesses the   schema S and those
   individuals reproduce at a fixed rate r.

iii)   All other individuals lacking schema S reproduce at a rate   s < r.

Thus with an initial population size of N, after t generations, we find $Nf\,r^{\,t}$ individuals possessing schema S and the population of the rest of the individuals is $N(1 - f)\,s^{t}$. Therefore, the fraction of the individuals with schema S is given by

$$(N f\, r^{t})\,/\,[\,N(1\text{-}f)\,s^{t} + N f r^{t}\,]$$

$$= f\,(r/s)^{t}\,/\,[\,1 + f\,\{(r/s)^{t} - 1\}\,]. \tag{15.1}$$

For small t and f, the above fraction reduces to $f\,(r/s)^{\,t}$, which means the population having the schema S increases exponentially at a rate (r / s). A stochastic proof of the above property will be presented shortly, vide a well-known theorem, called the *fundamental   theorem of Genetic algorithm.*

15.3 Stochastic Explanation of GA

For presentation of the fundamental theorem of GA, the following terminologies are defined in order.

Definition 15.1: The **order of a schema H**, denoted by **O(H)**, is the number of fixed positions in the schema. For example, the order of schema H = ?001?1? is 4, since it contains 4 fixed positions.

Definition 15.2: The **defining length of a schema**, denoted by d(H), is the difference between the leftmost and rightmost specific (i.e., non-don't care) string positions.

For example, the schema ?1?001 has a defining length d(H) = 4 - 0 = 4, while the d(H) of ???1?? is zero.

Definition 15.3: The schemas defined over L-bit strings may be geometrically interpreted **as hyperplanes in an L- dimensional hyperspace** (a binary vector space) with each L-bit string representing one corner point in an n-dimensional cube.

The Fundamental Theorem of Genetic Algorithms
(Schema theorem)

Let the population size be N, which contains $m_H(t)$ samples of schema H at generation t. Among the selection strategies, the most common is the *proportional selection*. In proportional selection, the number of copies of chromosomes selected for mating is proportional to their respective fitness values. Thus, following the principles of proportional selection [17], a string i is selected with probability

$$f_i / \sum_{i=1}^{n} f_i \qquad (15.2)$$

where f_i is the fitness of string i. Now, the probability that in a single selection a sample of schema H is chosen is described by

$$\sum_{i=1}^{m_H(t)} f_i / \sum_{i=1}^{n} f_i$$

$$= m_H(t) \, f_H / \sum_{i=1}^{n} f_i \qquad (15.3)$$

where f_H is the average fitness of the $m_H(t)$ samples of H.

Thus, in a total of N selections with replacement, the expected number of samples of schema H in the next generation is given by

$$m_H(t+1) = N \, m_H(t) \, f_H / \sum_{i=1}^{n} f_i$$

$$= m_H(t) \, f_H / f_{av} \qquad\qquad (15.4)$$

$$\text{where } f_{av} = \sum_{i=1}^{n} f_i / N \qquad\qquad (15.5)$$

where f_{av} is the population average fitness in generation t. The last expression describes that the Genetic algorithm allocates over an increasing number of trials to an above average schema. The effect of crossover and mutation can be incorporated into the analysis by computing the schema survival probabilities of the above average schema. In crossover operation, a schema H survives if the cross-site falls outside the defining length d(H). If p_c is the probability of crossover and L is the word-length of the chromosomes, then the disruption probability of schema H due to crossover is

$$p_c \, d(H) / (L-1). \qquad\qquad (15.6)$$

A schema H, on the other hand, survives mutation, when none of its fixed positions is mutated. If p_m is the probability of mutation and O(H) is the order of schema H, then the probability that the schema survives is given by

$$(1 - p_m)^{O(H)}$$

$$= 1 - p_m \, O(H). \qquad\qquad (15.7)$$

Therefore, under selection, crossover and mutation, the sample size of schema H in generation (t + 1) is given by

$$m_H(t+1) \geq (m_H(t) \, f_H / f_{av}) \, [1 - p_c \, d(H) / (L-1) - p_m \, O(H)]. \quad (15.8)$$

The above theorem is called the **fundamental theorem of GA or the Schema theorem.** It is evident from the Schema theorem that for a given set of values of d(H), O(H), L, p_c and p_m, the population of schema H at the subsequent generations increases exponentially when $f_H > f_{av}$. This, in fact, directly follows from the difference equation:

$$m_H(t+1) - m_H(t) \geq (f_H / f_{av} - 1 / K) \ K \ m_H(t) \tag{15.9}$$

where $K = 1 - p_c \ d(H) / (L-1) - p_m \ O(H)$. $\tag{15.10}$

$$\Rightarrow \Delta m_H(t) \geq K (f_H / f_{av} - 1/K) m_H(t). \tag{15.11}$$

Replacing Δ by $(E-1)$, where E is the extended difference operator, we find

$$(E - 1 - K_1) m_H(t) \geq 0 \tag{15.12}$$

where $K_1 = K (f_H / f_{av} - 1/K)$. $\tag{15.13}$

Since $m_H(t)$ in equation (15.12) is positive, $E \geq (1 + K_1)$. Thus, the solution of (15.12) is given by

$$m_H(t) \geq A (1 + K_1)^t \tag{15.14}$$

where A is a constant. Setting the boundary condition at $t = 0$, and substituting the value of K_1 by (14.13) therein, we finally have:

$$m_H(t) \geq m_H(0) (K f_H / f_{av})^t \tag{15.15}$$

Since K is a positive number, and $f_H / f_{av} > 1$, $m_H(t)$ grows exponentially with iterations. The process of exponential increase of $m_H(t)$ continues until some iteration r, when f_H approaches f_{av}. This is all about the proof of the schema theorem. $\square$

15.4 The Markov Model for Convergence Analysis

To study the convergence of the GA, let us consider an exhaustive set of population states, where 'state' means possible members (chromosomes) that evolve at any GA cycle. As an illustration, let us consider 2-bit chromosomes and population size = 2, which means at any GA cycle we select only two chromosomes. Under this circumstance, the possible states that can evolve at any iteration are the members of the set S, where

S= { (00, 00), (00,01), (00,10), (00,11), (01, 00), (01, 01),

    (01, 10), (01, 11), (10, 00), (10, 01), (10, 10), (10, 11),

    (11, 00), (11, 01), (11, 10), (11, 11)}

    For the sake of understanding, let us now consider the population size = 3 and the chromosomes are 2-bit patterns, as presumed earlier. The set S now takes the following form.

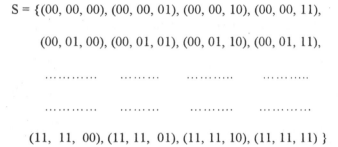

S = {(00, 00, 00), (00, 00, 01), (00, 00, 10), (00, 00, 11),

    (00, 01, 00), (00, 01, 01), (00, 01, 10), (00, 01, 11),

    . . . .

    . . . .

    (11, 11, 00), (11, 11, 01), (11, 11, 10), (11, 11, 11) }

    It may be noted that the number of elements of the last set S is 64. In general, if the chromosomes have the word length of m bits and the number of chromosomes selected in each GA cycle is n, then the cardinality of the set S is 2^{mn}.

    The Markov transition probability matrix P for 2-bit strings of population size 2, thus, will have a dimension of (16 x 16), where the element p_{ij} of the matrix denotes the probability of transition from i-th to j-th state. A clear idea about the states and their transitions can be formed from fig. 15.5.

    It needs mention that since from a given i-th state, there could be a transition to any 16 j-th states, therefore the row sum of P matrix must be 1. Formally,

$$\sum_{\forall j} P_{ij} = 1, \tag{15.16}$$

for a given i.

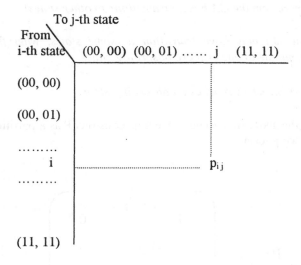

Fig. 15.5: The Markov state-transition matrix P.

Now, let us assume a row vector π_t, whose k-th element denotes the probability of occurrence of the k-th state at a given genetic iteration (cycle) t; then π_{t+1} can be evaluated by

$$\pi_{t+1} = \pi_t . P \tag{15.17}$$

Thus starting with a given initial row vector π_0, one can evaluate the state probability vector after n-th iteration π_n by

$$\pi_n = \pi_0 . P^n \tag{15.18}$$

where P^n is evaluated by multiplying P matrix with itself (n-1) times.

Identification of a P matrix for a GA that allows selection, crossover and mutation, undoubtedly, is a complex problem. Goldberg [10], Davis [6], Fogel [8] and Chakraborty [3] have done independent work in this regard. For simplicity of our analysis, let us now consider the GA without mutation.

The behavior of GA without mutation can be of the following three types.

i)   *The GA may converge to one or more absorbing states (i.e., states wherefrom the GA has no transitions to other states).*

ii)   *The GA may have transition to some states, wherefrom it may terminate to one or more absorbing states.*

iii)   *The GA never reaches an absorbing state.*

Taking all the above into account, we thus construct P as a partitioned matrix of the following form:

$$P = \begin{pmatrix} I & 0 \\ R & Q \end{pmatrix}$$

where I is an identity matrix of dimension (a x a) that corresponds to the absorbing states; R is a (t x a) transition sub-matrix describing transition to an absorbing state; Q is a (t x t) transition sub-matrix describing transition to transient states and not to an absorbing state and 0 is a null matrix of dimension (t x t).

It can be easily shown that P^n for the above matrix P can be found to be as follows.

$$P^n = \begin{pmatrix} I & 0 \\ N_n R & Q^n \end{pmatrix}$$

where the n-step transition matrix N_n is given by

$$N_n = I + Q + Q^2 + Q^3 + \ldots + Q^{n-1} \tag{15.19}$$

As n approaches infinity,

$$\underset{n \to \infty}{Lt} \ N_n = (I - Q)^{-1}. \tag{15.20}$$

Consequently, as n approaches infinity,

$$\underset{n \to \infty}{Lt} \ P^n = \begin{pmatrix} I & 0 \\ \\ (I-Q)^{-1}R & 0 \end{pmatrix}$$

Goodman has shown [8] that the matrix $(I - Q)^{-1}$ is guaranteed to exist. Thus given an initial probability vector π_0, the chain will have a transition to an absorbing state with probability 1. Further, there exists a non-zero probability that absorbing state will be the globally optimal state [8].

We now explain: why the chain will finally terminate to an absorbing state. Since the first 'a' columns for the matrix P^n, for $n \to \infty$, are non-zero and the remaining columns are zero, therefore, the chain must have transition to one of the absorbing states. Further, note that the first 'a' columns of the row vector π_n for $n \to \infty$ denote the probability of absorption at different states, and the rest of the columns denote that the probability of transition to non-absorbing states is zero. Thus probability of transition to absorbing states is one. Formally,

$$\sum_{i=1}^{a} \ \underset{n \to \infty}{Lt} \ (\pi_n)_i$$

$$= \sum_{i=1}^{a} \, Lt_{n \to \infty} \, (\pi_0 P^n)_i$$

$$= \sum_{i=1}^{a} \, Lt_{n \to \infty} \, \left(\pi_0 \cdot \begin{pmatrix} I \\ (I-Q)^{-1} R \end{pmatrix} \right)_i$$

$$= 1.$$

15.5 Application of GA in Optimization Problems

Genetic algorithms have been successfully used in many optimization problems. For instance, the classical travelling salesperson problem, the flowshop and the jobshop scheduling problems and many of the constraint satisfaction problems can be handled with GA. In this section, we illustrate the use of GA in antenna design. There exist various types of antennas. We here illustrate the design of a specialized antenna, called a *monopole loaded with a folded dipole*. The whole antenna has 6 components (fig. 15.6), namely Z_1, Z_2, Z_3, Z_4, X_1 and X_2. We want the electric field vector E_θ to be optimized in one half of the hemisphere. A non-specialist reader can easily understand the meaning of E_θ from fig. 15.7.

Altshuler and Linden [1] used an NEC$_2$ package that computes E_θ for given arm-lengths of the antenna (fig. 15.8). The GA program, used by them, on the other hand evolves new chromosomes with 6 fields vide fig. 15.9. Each field is represented by 5 bits, thus having a possible length in the range of 0 and 31 units. The selection criterion in the present context [1] is given by

$$\text{Minimize} \quad Z = \sum_{-180 \le \forall \varphi \le 0} \left(E_\theta - E_{desired} \right)^2.$$

For realizing this problem, Altshuler et al. considered an initial population of 150 and selected 75 in each iteration. The normal single bit crossover and mutation is used for mating in their scheme. For the best results, the mutation probability is varied between 0 to 0.9.

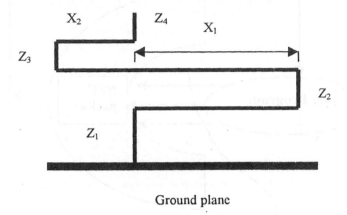

Fig. 15.6: A monopole loaded with a folded dipole.

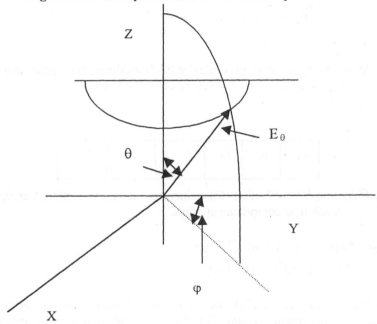

Fig. 15.7: The electric field vector E_θ for a given elevation angle θ is kept closer to its desired value for the variation of ϕ from 0 to −180 degrees.

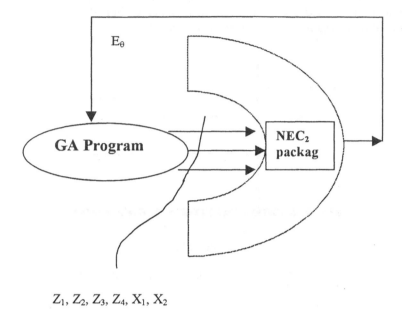

$Z_1, Z_2, Z_3, Z_4, X_1, X_2$

Fig. 15.8: The NEC_2 package evaluates E_θ from the evolved parameter set: $Z_1, Z_2, Z_3, Z_4, X_1, X_2$ generated by the GA program.

Z_1	Z_2	Z_3	Z_4	X_1	X_2

Fig. 15.9: The field definition of the chromosome, used in antenna design, each field comprising of 5 bits.

15.6 Application of GA in Machine Learning

Machine learning is one of the key application fields of Genetic algorithms. A survey of the recent literature [14] on GA reveals that a major area of its applications is concerned with artificial neural nets. It can work with neural nets in three basic ways [15]. First it can adjust the parameters, such as weights and non-linearity of a neural net, when the training instances are supplied. Here it serves the same purpose of the neural learning algorithm. For instance, we can replace the well- known back-propagation algorithm by a GA based scheme. Secondly, GA can be employed to determine the structure of a neural net. Thus when the number of neurons in one or more hidden layer cannot be guessed properly, we may employ GA to solve this problem. Thirdly, GA may be employed to automatically adjust the parameters of a prototype learning equation. This has many useful applications in adaptive control, where the adaptation of the control law is realized with GA.

15.6.1 GA as an Alternative to Back-propagation Learning

The back-propagation learning adjusts the weights of a feed-forward neural net by employing the principles of steepest descent learning. One main drawback of this classical algorithm is trapping at local minima. Due to mutation in a GA, it has the characteristics of hill climbing, and thus can overcome the difficulty of trapping at local minima. The principle of using GA for neural learning is presented below.

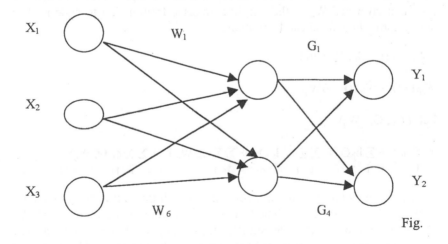

Fig.

Fig. 15. 10: Illustrating the use of GA in neural learning.

Here, we considered a three layered neural net with weights (fig. 15.10). Let the input pattern and the output patterns be [X_1 X_2 X_3]^T and [Y_1 Y_2]^T respectively. Let the weights of the first layer and the second layer be $[W_1 \ W_2 \ W_3 \ W_4 \ W_5 \ W_6]^T$ and $[\ G_1 \quad G_2 \ ]^T$ respectively. Let the non-linearity of each neuron be F. Also, the desired output vector $[d_1 \quad d_2 \ ]^T$ is given, where d_1 corresponds to the top output neuron and d_2 corresponds to the bottom output neuron. The selection function in the present context is to minimize Z where

$$Z = [\ (d_1 - Y_1)^2 + (d_2 - Y_2)^2\ ]^{1/2}.$$

The chromosome in the present context comprises of 10 fields, such as W_1, W_2, W_3, W_4, W_5, W_6, G_1, G_2, G_3, and G_4. Each field may be represented by a signed real number, expressed in 2's complemented arithmetic. The typical crossover and mutation operators may be used here to evolve the weights. The algorithm terminates when the improvement in Z ceases. Thus for a given input-output training instance, we find a set of neural weights. The idea can be easily extended for training multiple input-output patterns.

15.6.2 Adaptation of the Learning Rule / Control Law by GA

A supervised learning system has to generate a desired output problem instance from a given input problem instance. Let us assume that a multi-layered feed forward neural net is used as the learning agent. Let O_k be the output node at node o_k in the output layer, I_t be the input at the t-th node in the input layer and W_{ij} be the weight connected from node i to node j. The learning rule in general can be written as

$$\Delta W_{ij} = f\ (I_t,\ O_j,\ W_{ij}) \text{ and}$$

$$W_{ij}(t+1) = W_{ij}\ (t) + \Delta W_{ij}.$$

Let $f\ (I_t,\ O_j,\ W_{ij})$

$$= \sum_t a_t\ I_t + \sum_j b_j\ O_j + \sum_t \sum_i c_{ti}\ I_t\ O_i + \sum_i \sum_t d_{it}\ W_{it}\ I_t + \sum_i \sum_j e_{ij}\ O_j\ W_{ij}.$$

Here, the chromosomes are constructed with the following parameters: a_t, b_i, c_{ti}, d_{it}, e_{ij} as the fields. The fitness of the chromosomes is measured by the square norm of the error signals (target-output) at the output layered nodes. The smaller the norm, the better are the chromosomes. The crossover and mutation operators are comparable with their standard use. After a number of genetic evolutions, GA determines the near optimal values of the parameters. Since the parameters: a_t, b_i, c_{ti}, d_{it}, e_{ij} govern the learning rule, their adaptation constructs new learning rules. Fig. 15.11 describes the adaptation process of the learning rules.

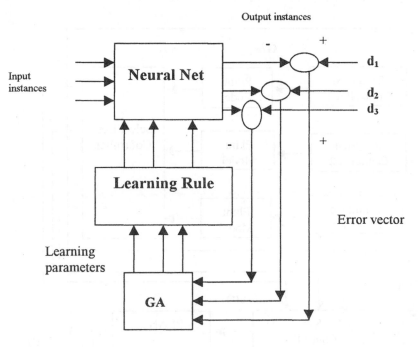

Fig. 15.11: Adaptation of the learning rule by using GA.

GA can also be used for the adaptation of the control laws in self-tuning adaptive control systems. For instance, consider a self-tuning P-I-D controller. Here, the control law can be expressed as

$$u(t) = K_P\, e(t) + K_I \int e(t)\, dt + K_D\, (de/dt)$$

where e (t) denotes the error (desired output — computed output), u (t) the control signal and K_P, K_I and K_D are the proportional, integral and derivative co-efficients. The optimization of K_P, K_I and K_D is required to satisfy some criteria, like minimization of the integral square error:

$$ISE = \int e^2(t)\, dt.$$

GA here may be employed to emulate various control laws by randomly selecting different vectors $[K_P \quad K_I \quad K_D]^T$. In other words these vectors represent the population. The ISE, here, has been used as the fitness function. So, GA in each evolution cycle (iteration) selects better chromosomes from the population by minimizing the ISE. After each finite interval of time (typically of the order of minutes) the GA submits the control law, described by the optimal vector $[K_P^{\bullet} \quad K_I^{\bullet} \quad K_D^{\bullet}]$. The schematic architecture of the overall system is presented in fig. 15.12.

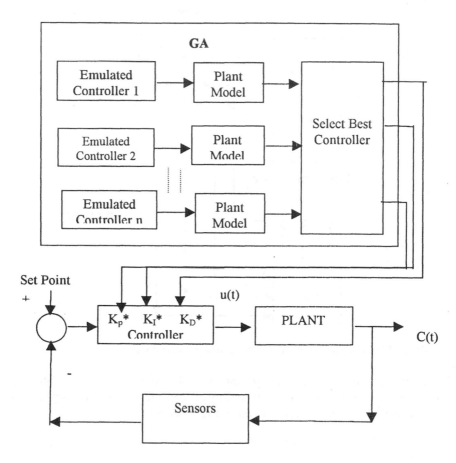

Fig. 15.12: The working of GA-based P-I-D tuner.

15.7 Applications of GA in Intelligent Search

Most of the classical AI problems such as n-Queen, the water-jug or the games are search problems, where GA has proved its significance. Other applied search problems include routing in VLSI circuits and navigational planning for robots [17] in a constrained environment. We just introduce the scheme for navigational planning of robots here and will explain the details in chapter 24.

15.7.1 Navigational Planning for Robots

The navigational planning for robots is a search problem, where the robot has to plan a path from a given starting position to a goal position. The robot must move without hitting an obstacle in its environment (fig. 15.13). So, the obstacles in robot's work-space act as constraints to the navigational planning problem. The problem can be solved by GA by choosing an appropriate fitness function that takes into account the distance of the planned path- segments from the obstacles, length of the planned path and the linearity of the paths as practicable.

Michalewicz [17] has formulated the navigational planning problem of robots by GA and simulated it by a new type of crossover and mutation operators. An outline of his scheme is presented in chapter 24.

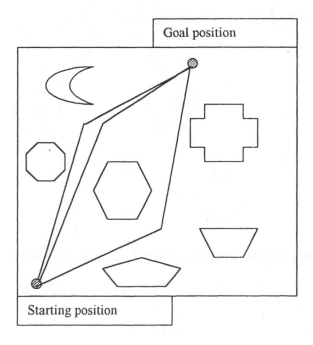

Fig. 15.13: Path planning by a robot amidst obstacles.

15.8 Genetic Programming

Koza [13] applied GA to evolve programs, called Genetic Programming. He represented the program by structure like a parse tree. For instance a function

$$f(x) = Sin\ (x) + [x * x + y]^{1/2}$$

can be represented by a tree, presented in fig. 15.14.

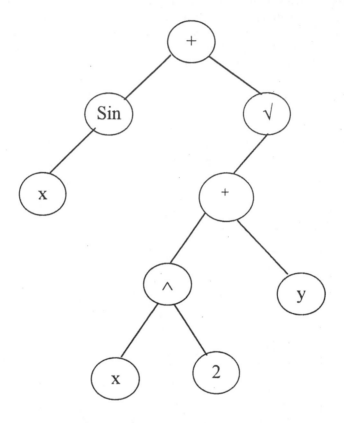

Fig. 15.14: A program tree representing the function $f(x)= Sin(x)+\sqrt{(x^2+y)}$.

The crossover operator here is applied on two trees as illustrated in fig. 15.15. From the parent program for $x^2 + y$ and $xy+ y^2$ the offsprings produced by crossover are $y^2 +y$ and $xy+x^2$.

Koza applied genetic programming to solve the blocks world problem. He considered 9 blocks marked with U,N,I,V,E,R,S,A and L respectively and placed a few of these blocks on a table and the remaining blocks at a block stack in a random order. He employed genetic programming to place the blocks on the stack in an ordered manner, so that the word 'UNIVERSAL' was generated.

Koza defined some operators for block placements and initialized the population with programs for block movements. He considered 166 training samples and then by his GP he discovered a new program that could solve all the 166 training problems. His resulting program is given below.

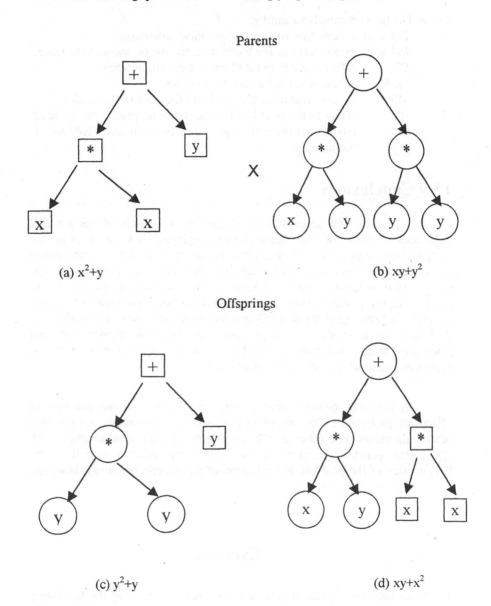

Parents

X

(a) x^2+y

(b) $xy+y^2$

Offsprings

(c) y^2+y

(d) $xy+x^2$

Fig. 15.15: Crossover between 2 genetic programs (a) and (b) yields new programs (c) and (d).

```
(EQ (DU(MT CS)  (NOT CS))
    (DU(MS NN) (NOT NN)))
```

where DU (x, y) means Do x until y;
        EQ(x, y) returns true when x=y and false otherwise;
        MT x   means transfer the block from the top of stack to the table;
        CS   refers to the topmost block in the current stack;
        NOT x   is true if x is false and vice versa;
        MS x   means 'move the block x from the table to the stack top';
        NN   refers to the next block required to be placed on the stack top so that the ordering of the letters in UNIVERSAL is maintained.

15.9 Conclusions

GA has proved itself successful in almost every branch of science and engineering. Most of the applications employed GA as a tool for optimization. In fact, when there exist no guidelines to optimize a function of several variables, GA works as a random search and finds an optimal (or at least a near optimal) solution. There exists a massive scope of GA in machine learning, but no significant progress has been observed in this area to date. The next generation intelligent machines are likely to employ GA with neuro-fuzzy models to reason and learn from incomplete data and knowledge bases. Such systems will find immense applications in robotics, knowledge acquisition and image understanding systems.

A common question that is often raised is: can we use GA in 'discovery problems'? The answer to this is in the affirmative. As GA is a search algorithm, like other search algorithms, it may in association with specialized genetic operators explore some undiscovered search space. Formulation of the problem and selection of the genetic operators, however, play a vital role in this regard.

Exercises

1. Show the first 3 cycles of genetic evolution for optimizing the function y = x^3 –27 in the interval $0 \le x \le 12$. Use set P = { 1100, 1010, 1011, 0011} as the initial population.

2. How can you realize crossover probability = 0.7 (say) in the GA program?

   [**Hints:** Generate a random number in the interval [0, 1]. If the number generated > 0.7, then allow crossover between the two selected chromosomes.]

3. List the possible states for population size =2 and length L of chromosomes = 3. How many states do you obtain? What is the dimension of the Markov state transition matrix P in this context?

4. Suppose in expression (15.8), d(H) = L-1, p_m O (H) $\rightarrow$ 0. Under this setting,

   $$m_H(t+1) \geq m_H(t) (f_H / f_{av}) (1 - p_c).$$

   It is evident from the above inequality that as $p_c \rightarrow 0$, $m_H(t)$ grows at a faster rate. Can you give a logical justification to this observation? If you feel comfortable to answer this, then try to explain the situation when $p_c \rightarrow 1$.

5. Unlike the standard GA, suppose you devise a new type of algorithm, where in each cycle you randomly pick up M chromosomes and select good chromosomes from these by a selection function, and keep copies of the good schemata from the last cycle, the selected population size taking into consideration of the two types = N. Further, assume there is no crossover or mutation in your algorithm. Can you analyze the performance of your algorithm? Can you compare the performance with that of a standard GA?

References

[1] Altshuler, E. E. and Linden, D. S., " Wire–Antenna designs using genetic algorithms," *IEEE Antennas and Propagation Magazines*, vol. 39, no. 2, April 1997.

[2] Bender, E. A., *Mathematical Methods in Artificial Intelligence*, IEEE Computer Society Press, Los Alamitos, pp. 589-593, 1996.

[3] Chakraborty, U. K., Deb, K. and Chakraborty, M., "Analysis of selection algorithms: A Markov chain approach," *Evolutionary Computation*, vol. 4, no. 2, pp. 133-167, 1996.

[4]   Chakraborty, U. K. and Muehlenbein, H., "Linkage equilibrium and genetic algorithms," *Proc. 4th IEEE Int. Conf. On Evolutionary Computation*, Indianapolis, pp. 25-29, 1997.

[5]   Chakraborty, U. K. and Dastidar, D. G., "Using reliability analysis to estimate the number of generations to convergence in genetic algorithm," *Information Processing Letters*, vol. 46, pp. 199-209, 1993.

[6]   Davis, T. E. and Principa, J. C., " A Markov chain framework for the simple genetic algorithm," *Evolutionary Computation*, vol. 1, no. 3, pp. 269-288, 1993.

[7]   De Jong, K. A., *An analysis of behavior of a class of genetic adaptive systems*, Doctoral dissertation, University of Michigan, 1975.

[8]   Fogel, D. B., *Evolutionary Computation*, IEEE Press, Piscataway, NJ, 1995.

[9]   Filho, J. L. R. and Treleven, P. C., *Genetic Algorithm Programming Environment*, IEEE Computer Society Press, pp. 28-43, June 1994.

[10] Goldberg, D. E., *Genetic Algorithms in Search, Optimization and Machine Learning*, Addison-Wesley, Reading, MA, 1989.

[11] Gupta, B., "Bandwidth enhancement of Microstrip antenna through optimal feed using GA," *Seminar on Seekers and Aerospace Sensors*, Hyderabad, India, 1999.

[12] Holland, J. H., *Adaptation in Natural and Artificial Systems*, University of Michigan Press, Ann Arbor, 1975.

[13] Koza, J. R., *Genetic Programming: On the Programming of Computers by Means of Natural Selection*, MIT Press, Cambridge, MA, 1992.

[14] Mc Donell, J. R., "Control" in   *Handbook of Evolutionary Computation*, Back, T., Fogel, D. B. and Michalewicz, Z., Eds., IOP and Oxford University Press, New York, 1998.

[15] Mitchell, M., *An Introduction to Genetic Algorithms*, MIT Press, Cambridge, MA, 1996.

[16] Muehlenbein, H. and Chakraborty, U. K., "Gene pool recombination genetic algorithm and the onemax function," *Journal of Computing and Information Technology*, vol. 5, no. 3, pp. 167-182, 1997.

[17] Michalewicz, Z, *Genetic Algorithms + Data Structures = Evolution Programs*, Springer-Verlag, Berlin, 1992.

[18] Srinivas, M. and Patnaik, L. M., "Genetic search: Analysis using fitness moments," *IEEE Trans. on Knowledge and Data Engg.*, vol. 8, no. 1, pp. 120-133, 1996.

[19] Vose, M. D. and Liepins, G. E., "Punctuated equilibrium in genetic search," *Complex Systems*, vol. 5, pp. 31-44, 1991.

16

Realizing Cognition Using Fuzzy Neural Nets

The chapter presents various behavioral models of cognition, built with fuzzy neural nets for applications in man-machine interface, automated coordination and learning control systems. It starts with a simple model of a fuzzy neural net that mimics the activities of "long term memory" of the biological cognitive system. The model has the potential of both reasoning and automated learning on a common structure. The model is then extended with Petri nets to represent and reason with more complex knowledge of the real world. A second model presented in this chapter is designed for application in a learning control system. It emulates the task of the motor controller for the limb movements in the biological cognitive system. An example of the automated eye-hand coordination problem for robots has also been presented here with timed Petri net models. The chapter ends with a discussion of the possible application of the proposed models in a composite robotic system.

16.1   Cognitive Maps

The hippocampus region [2] of the human brain contains a specialized structure, responsible for reasoning, recognition and control of the cognitive activities in the biological brain. This structure is usually called the **cognitive map.** The biological cognitive map stores spatial, temporal, incidental and factual relationship among events. The cognitive maps are generally represented by graphs, where the nodes denote the events and the directed arcs denote the causal dependence of the events. The graph may include cycles as well. Generally, a token representing the belief of the fact is assigned at the corresponding node, describing the fact. The directed arcs in a cognitive map are weighted. The weight associated with an arc in a cognitive map represents the degree by which the "effect node" is influenced for a unit change in the "causal node". Sometimes, a plus (+) or a minus (-) sign [15] is attached with a directed arc to describe whether the cause has a growing or decreasing effect on the directed node.

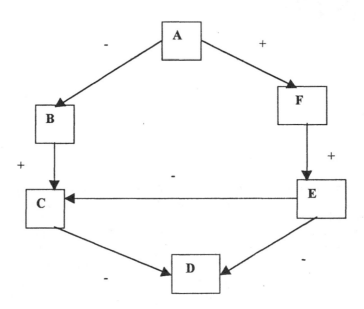

A= Islamic Fundamentalism, B = Soviet Imperialism, C= Syrian Control on Lebanon, D= Strength of Lebanese Government, E= PLO Terrorism, F =Arab Radicalism

Fig. 16.1: A cognitive map representing political relationship, describing the Middle East Peace.

Cognitive maps are generally used for describing soft knowledge [6]. For example, political and sociological problems, where a very clear formulation of the mathematical models is difficult, can be realized with cognitive maps. The cognitive map describing the political relationship to hold the middle east peace is presented in fig. 16.1 [13]. It may be noted from this figure that the arcs here are labeled with + or − signs. A positive arc from the node A to F denotes an increase in A will cause a further increase in F. Similarly, a negative arc from A to B represents an increase in A will cause a decrease in B. In the cognitive map of fig. 16.1, we do not have weights with the arcs. Now, suppose the arcs in this figure have weights between −1 to +1. For instance, assume that the arc from A to B has a weight −0.7. It means a unit increase in A will cause a 0.7 unit decrease in B. There exist two types of model of the cognitive map. One type includes both positive and negative weighted arcs and the events are all positive. The other type considers only positive arcs, but the events at the nodes can be negated. In this chapter, we will use the second type representation in our models.

The principles of cognitive mapping [12] for describing relationships among facts were pioneered by Axelord [1] and later extended by Kosko [5]. Kosko introduced the notion of fuzzy logic for approximate reasoning with cognitive maps. According to him, a cognitive map first undergoes a training phase for adaptation of the weights. Once the training phase is completed, the same network may be used for generating inferences from the known beliefs of the starting nodes. Kosko's model is applicable in systems, where there exists a single cause for a single effect. Further, the process of encoding of weights in Kosko's model is stable for acyclic networks and exhibits limit cycle behavior for cyclic nets. The first difficulty of Kosko's model has been overcome by Pal and Konar [8], who used fuzzy Petri nets to represent the cognitive map. The second difficulty, which refers to limit cycles in a cyclic cognitive map, has also been avoided in [3] by controlling one parameter of the model. In the forgoing sections, we will present these models with their analysis for stability. Such analysis is useful for applications of the models in practical systems.

16.2 Learning by a Cognitive Map

A cognitive map can encode its weights by unsupervised learning [7] like the Long Term Memory (LTM) in the biological brain. In this chapter, we employ the **Hebbian learning** [5] following Kosko [6], which may be stated as follows.

The weight W_{ij} connected between neuron N_i and N_j increases when the signal strength of the neurons also increase over time.

In discrete timed system, we may write

$$W_{ij}(t+1) = W_{ij}(t) + \Delta W_{ij} \qquad (16.1)$$

where $\Delta W_{ij} = f(n_i) f(n_j)$ (by Hebbian learning)   (16.2)

and f is the nonlinear sigmoid function, given by

$$f(n_i) = 1/(1+\exp(-n_i)). \qquad (16.3)$$

The model of *forgetfulness of facts* can be realized in a cognitive map by a gradual decay in weights, which may be formally written as

$$\Delta W_{ij} = -\alpha W_{ij}. \qquad (16.4)$$

Taking into account the superposition of (16.2) and (16.4), we find

$$\Delta W_{ij} = -\alpha W_{ij} + f(n_i) f(n_j). \qquad (16.5)$$

The weight adaptation equation in a cognitive map may thus follow from expression (16.1) and (16.5) and is given by

$$W_{ij}(t+1) = (1-\alpha) W_{ij} + f(n_i) f(n_j). \qquad (16.6)$$

16.3   The Recall in a Cognitive Map

After the encoding is over, the recall process in a cognitive map is started. The recall model is designed based on the concept that the next value of belief n_i of node N_i can never decrease below its current value. Further, the belief of node N_i is influenced by the nodes N_j, when there exists signal flow from nodes N_j to N_i. Thus formally,

$$n_i(t+1) = n_i(t) \quad \vee [\ \vee (W_{ij} \wedge n_j(t))\ ]. \qquad (16.7)$$
$$\forall j$$

It may be added here that the above model is due to Pal and Konar [8]. In Kosko's recall model, the term $n_i(t)$ is absent from the right hand side of expression 16.7. This means that the recall model has no memory.

16.4   Stability Analysis

For the analysis of stability of the encoding and the recall model, we use the following properties.

Theorem 16.1: *With initial values of $n_i(o) \leq 1$ and $W_{ij} \leq 1$, the n_i remains bounded in the interval (0,1).*

Proof: Proof of the theorem directly follows from the recursive definition of expression (16.7). □

Theorem 16.2: *The recall model described by expression (16.7) is unconditionally stable.*

Proof: For oscillation, the value of a function f should increase as well as decrease with time. Since $n_i(t+1)$ can only increase or remain constant, but cannot decrease (vide expression (16.7)), therefore $n_i(t+1)$ cannot exihibit oscillations. Further, since $n_i(t+1)$ is bounded and is not oscillatory, thus it must be stable. □

It may be added here that the recall model of Kosko, which excludes $n_i(t)$ from the right hand side of expression (16.7), is oscillatory for a cyclic cognitive net.

Theorem 16.3: *The encoding model represented by expression (16.6) is*
 stable, when $0 \leq \alpha \leq 2$
 unstable, when $\alpha \geq 2$
 oscillatory, when $\alpha = 2$.

Proof: Replacing Δ by E - 1, we have

$$(E - 1 + \alpha) W_{ij} = f(n_i) f(n_j) \tag{16.8}$$

Since at steady-state f (n_i), f (n_j) become constant, then let f(n_i) f(nj) at steady-state be denoted by f $(n_i)^*$ and f$(n_j)^*$ respectively.

Thus, $(E - 1 + \alpha) W_{ij}(t) = f(ni)^* f(n_j)^*$. $\tag{16.9}$

The complementary function for the above equation is

$(E - 1 + \alpha) W_{ij} = 0$

which yields $W_{ij}(t) = C(1-\alpha)^t$. $\tag{16.10}$

The particular integral for equation (16.9) is given by

$$W_{ij}(t) = (1/\alpha) f(n_i)^* f(n_j)^*. \tag{16.11}$$

Combining expression (16.10) and (16.11), we find:

$$W_{ij}(t) = C (1 - \alpha)^t + (1/\alpha) f(n_i)^* f(n_j)^*. \qquad (16.12)$$

Satisfying the initial condition we find from the last equation that

$$W_{ij}(t) = [W_{ij}(0) - (1/\alpha) f(n_i)^* f(n_j)^*] (1- \alpha)^t + (1/\alpha) f(n_i)^* f(n_j)^*. \qquad (16.13)$$

The condition for stability, limit cycles and instability directly follows from the last expression. □

It may be noted that from the last expression that for a stable cognitive map $(0<\alpha<2)$, $W_{ij}(t)$ is nonzero when $W_{ij}(0)$ is nonzero. So, *structural stability* of a cognitive map is ensured.

16.5   Cognitive Learning with FPN

The model of cognition, we present now, can represent one-to-one causal dependence.   There are, however, instances when many-to-one causal dependence exists in a problem. The FPN, which we covered in chapter 10, can represent such many-to-one cause-effect relationships. In this section, we thus realize the cognitive map with FPN. The model of FPN we shall use here is somewhat different from the model we came across in chapter 10. The modifications are as follows:

i)   *The places here have single valued tokens, unlike in chapter 10, where the tokens were fuzzy distribution.*

ii)   *Instead of relational matrices, each transition possesses a scaler weight. Thus W_{ij} is a weight associated with the arc connected between transition tr_i and place p_j.*

The modified belief updating equation, hereafter called the recall equation, in vector-matrix form becomes

$$N(t+1) = N(t) V [W^T o T(t+1)] \qquad (16.14)$$

where $T(t+1) = (Q o N^C(t))^C \wedge U [(Q o N^C(t))^C - Th]. \qquad (16.15)$

Here $\mathbf{W}^T$ is a transposed weight matrix, whose (i, j)th element denotes the weight from transition tr_i to place p_j. 'Th' is the threshold vector, the i-th component of which is a scaler denoting the threshold of transition tr_i. 'U' is a unit step vector, which applies step function to all its arguments. 'N' is the belief vector, whose i-th component is a scaler, representing the fuzzy belief of proposition d_i. 'T' is the fuzzy truth token (FTT) vector, whose i-th component is a scaler denoting the FTT of transition tr_i. 'Q' is a binary place to transition connectivity matrix, whose (i, j)-th element denotes the existence (or no existence) of connectivity from the place p_j to transition tr_i.

A new weight adaptation rule, presented below, is used for the encoding of the weights:

$$\frac{dW_{ij}}{dt} = -\alpha\, W_{ij} + t_i(t) \cdot n_j(t) \tag{16.16}$$

where $t_i(t)$ and $n_j(t)$ denote the FTT and the belief of the transition tr_i and the place p_j respectively. The above equation can be discretized into the following form

$$W_{ij}(t+1) = (1-\alpha)\, W_{ij} + t_i(t) \cdot n_j(t). \tag{16.17}$$

The vector-matrix form of the above equation is given by

$$\mathbf{W}(t+1) = (1-\alpha)\,\mathbf{W}(t) + [\,(\mathbf{N}(t) \cdot \mathbf{T}^T(t))\wedge\mathbf{P}\,] \tag{16.18}$$

where $\mathbf{P}$ is a binary connectivity matrix from transition to places in a FPN.

The recall model, represented by expression 16.14 and 16.15, is stable, when

$$N(t+1) = \mathbf{N}(t) = \mathbf{N}^* \quad \text{at } t = t^*,\text{ the equilibrium time.}$$

Now, the above relation holds good for $(\mathbf{Q} \circ \mathbf{N}^{*C})^C > \mathbf{Th}$ if

$$\mathbf{N}^* \geq [\,\mathbf{W}^T \circ (\mathbf{Q} \circ \mathbf{N}^{*C})^C\,]. \tag{16.19}$$

The steady-state values of the components of $\mathbf{N}^*$ thus can be easily evaluated, with the known values of the $\mathbf{Q}$ and $\mathbf{W}$ matrix.

The conditional convergence: $0 < \alpha < 2$ of the encoding model can be easily proved by solving (16.18) using E-operator. We omit the proof, as it is similar with the proof of stability we derived from expression (16.8).

16.6   Application in Autopilots

Fig. 16.2 describes a cognitive map for an automated pilotless car driving system. A vision system attached to the car receives visual information about the road traffic and the pedestrians crossing the road. The belief of the received signals is then mapped to the appropriate nodes (places) in the cognitive map. The weights trained through unsupervised learning cycles are also mapped to the appropriate arcs before the reasoning process is initiated. The reasoning process continues updating the belief strength of the nodes based on the signal strength of the received information and the weights. The action is taken based on the concluding node with the highest belief.

Let the initial weights be as follows.

$w_{41}(0) = 0.95$, $w_{43}(0) = 0.85$, $w_{46}(0) = 0.75$, $w_{84}(0) = 0.8$, $w_{85}(0) = 0.4$, $w_{11,7}(0) = 0.9$, $w_{12,8}(0) = 0.85$.

Let us also assume that the thresholds associated with all transitions = 0.1 and the mortality rate $\alpha = 1.8$. The P and Q matrices are now constructed and the equations 16.15, 16.14 and 16.18 are recursively executed in order until the steady-state in weights occurs. The steady-state values of weights, presented below, are saved for use in the recognition phase.

Steady-state weights: $w_{41}{}^* = 0.25$, $w_{43}{}^* = 0.25$, $w_{46}{}^* = 0.25$, $w_{56}{}^* = 0.17$, $w_{84}{}^* = 0.31$, $w_{85}{}^* = 0.33$, $w_{11,7}{}^* = 0.2$, $w_{12,8}{}^* = 0.2$ and all other $w_{ij} = 0$.

In recall phase suppose we submit the belief vector $N(0)$ and the cognitive network generates an N^*, where

$N(0) = [0.2\ \ 0.3\ \ 0.4\ \ 0.0\ \ 0.0\ \ 0.3\ \ 0.35\ \ 0.0\ \ 0.4\ \ 0.3\ \ 0.0\ \ 0.0]^T$ and

$N^* = [0.2\ \ 0.3\ \ 0.4\ \ 0.25\ \ 0.17\ \ 0.3\ \ 0.35\ \ 0.33\ \ 0.4\ \ 0.3\ \ 0.17\ \ 0.17]^T$.

It is observed from N^* that among the concluding places $\{p_4, p_5, p_8, p_{11}, p_{12}\}$ $p_8 = $ 'rear car speed decreases' has the highest steady-state belief. So, this has to be executed.

16.7   Generation of Control Commands
by a Cognitive Map

With gradual learning and self-adaptation [5], the cognitive memory in human brain builds up a control model for muscle movements and determines the valuation space of the control signals for the execution of a task. For example, for moving one's arm, the control model for arm movement is encoded through learning (experience) by the person in his early life and the same model is used for the estimation of neuronal control signals required for proper positioning of the arm.

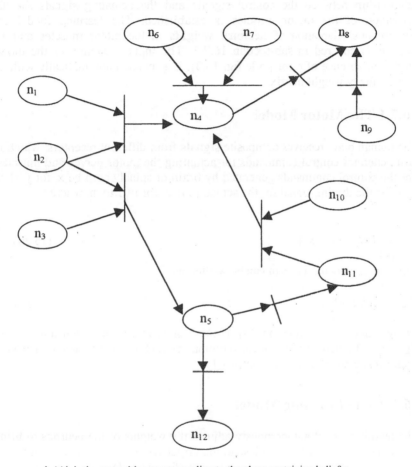

d_j, $\forall$ i, is the proposition corresponding to the place containing belief n_i.

n_1 = Car behind the side car narrow in breadth, n_2 = Side car of the front car too close, n_3=Car behind side car is wide, n_4= Front car speed decreases, n_5 = Front car speed increases, n_6 = Passer-by changes her direction, n_7= Passer-by crosses the road, n_8 =Rear car speed decreases, n_9 = Front car changes direction, n_{10} = Rear car changes direction, n_{11} =Rear car speed increases, n_{12} =Rear car keeps safe distance with respect to the front car. n_i = proposition associated with place p_i.

Fig. 16.2: A fuzzy cognitive map representing an autopilot.

In this section, the encoding model, to be preserved by the cognitive system, is presumed and a fuzzy relational algebra based approach [11] is used for finding the valuation space of the control signals with the help of a specialized *"inverse function"* model [11] for discrete fuzzy relations.

The section is organized as follows. In sub-section 16.7.1, a fuzzy relationship between the control signals and the actuating signals for the activation of the motor muscles is established. The learning model for autonomous adaptation of neuronal weights in the motor muscles and its stability is covered in sub-section 16.7.2. The input excitation of the motor model, in absence of error [vide fig. 16.3], is generated automatically with an inverse fuzzy weight matrix.

16.7.1 The Motor Model

The human brain receives composite signals from different receptors, which in turn generates control commands for actuating the motor nerves and muscles. Let the control commands generated by brain or spinal chord be x_j for j =1 to n, while the driving signal for the actuators is y_i for i= 1 to m, where

$$y_i = \overset{n}{\underset{j=1}{V}} (w_{ij} \wedge x_j) \qquad (16.20)$$

which in vector-matrix form can be written as

$$Y_K = W_k \, o \, X_k, \qquad (16.21)$$

where k denotes the iteration, W_k is the system matrix of dimension (n x n), X_k and Y_k denote the control (command) vector and the actuating signal vector respectively of dimension (n x 1).

16.7.2 The Learning Model

The learning model autonomously adjusts the weights of the neurons to bring the actuator signal vector Y_k close to the target vector D, which corresponds to the desired signals for the motor muscles (for example, to lift an arm). The intuitive learning equation for the model is given by

$$\Delta W_k = \alpha \, E_k \, o \, X_{k+1}^T / ((X_{k+1}) \, o \, (X_{k+1}^T)) \qquad (16.22)$$

where

Error vector $E_k = D - Y_k$ $\qquad (16.23)$

and the weight adaptation equation is given by

$$\mathbf{W_{k+1}} = \mathbf{W_k} + \Delta\mathbf{W_k}.$$ (16.24)

The feedback loop in fig. 16.3 first generates $\mathbf{Y_0}$ arbitrarily and later makes correction through a change in $\mathbf{W_k}$, which subsequently helps to determine controlled $\mathbf{X_{k+1}}$ and hence $\mathbf{Y_{k+1}}$. $\mathbf{W_k^{-1}}$ denotes the fuzzy compositional inverse (strictly speaking, *pre-inverse*) for matrix W_k. The stability of the learning model is guaranteed, vide theorem 16.3.

Theorem 16.3: *The error vector E_k in the learning model converges to a stable point for $0 < \alpha < 2$ and the steady-state value of error is inversely proportional to α.*

Proof: We have

$\mathbf{Y_k} = \mathbf{W_{k+1}} \text{ o } \mathbf{X_{k+1}}$

$\quad = (\mathbf{W_k} + \Delta\mathbf{W_k}) \text{ o } \mathbf{X_{k+1}}$

$\quad = \{\mathbf{W_k} + \alpha\, \mathbf{E_k} \text{ o } \mathbf{X_{k+1}^T} / ((\mathbf{X_{k+1}^T}) \text{ o } (\mathbf{X_{k+1}}))\} \text{ o } \mathbf{X_{k+1}}$

$\quad = \{\mathbf{W_k} + \alpha\, \mathbf{E_k} \text{ o } \mathbf{X_{k+1}^T} / ((\mathbf{X_{k+1}^T}) \text{ o } (\mathbf{X_{k+1}}))\} \text{ o } (\mathbf{W_k^{-1}} \text{ o } \mathbf{Y_k})$

$\quad \leq \mathbf{Y_k} + \{\alpha\, \mathbf{E_k} \text{ o } \mathbf{X_{k+1}^T} / ((\mathbf{X_{k+1}^T}) \text{ o } (\mathbf{X_{k+1}})\} \text{ o } (\mathbf{W_k^{-1}} \text{ o } \mathbf{Y_k})$

$\quad = \mathbf{Y_k} + \{\alpha\, \mathbf{E_k} \text{ o } \mathbf{X_{k+1}^T} / ((\mathbf{X_{k+1}^T}) \text{ o } (\mathbf{X_{k+1}}))\} \text{ o } \mathbf{X_{k+1}}$

$\quad \approx \mathbf{Y_k} + \alpha\, \mathbf{E_k}.$

Thus, $\mathbf{Y_{k+1}} \leq \mathbf{Y_k} + \alpha\, \mathbf{E_k}$ (16.25)

$\quad\quad = \mathbf{Y_k} + \alpha\,(\mathbf{D} - \mathbf{Y_k})$

$\Rightarrow \mathbf{Y_{k+1}} \leq \alpha\mathbf{D} + (\mathbf{I} - \alpha\mathbf{I})\,\mathbf{Y_k}$ (16.26)

$\Rightarrow [\mathbf{E\,I} - (\mathbf{I} - \alpha\,\mathbf{I})]\,\mathbf{Y_k} = (\alpha - \alpha')\mathbf{D}, \quad \text{for} \quad 0 < \alpha' \leq \alpha$ (16.27)

where E is the extended difference operator and $\mathbf{I}$ is the identity matrix.

The complementary function for the equation (16.27) is given by

$$\mathbf{Y}_k = \mathbf{A} \, (\mathbf{I} - \alpha \mathbf{I})^{\,k}, \tag{16.28}$$

where A is a constant matrix, to be determined from the boundary condition.

The particular integral for equation (16.27) is given by

$$\mathbf{Y}_k = [\, \mathbf{EI} - (\mathbf{I} - \alpha \mathbf{I})]^{-1} \, (\alpha - \alpha')\mathbf{D}. \tag{16.29}$$

Since $(\alpha - \alpha')\mathbf{D}$ is a constant vector, we substitute $E = I$ in (16.29), and thus find the particular integral as follows:

$$\mathbf{Y}_k = [\alpha \mathbf{I}]^{-1} \, (\alpha - \alpha')\mathbf{D}$$

$$= \mathbf{D} - (\alpha' / \alpha) \, \mathbf{D}. \tag{16.30}$$

The complete solution for $\mathbf{Y}_k$ is thus given by

$$\mathbf{Y}_k = \mathbf{A} \, (\mathbf{I} - \alpha \mathbf{I})^{\,k} + \mathbf{D} - (\alpha'/\alpha) \, \mathbf{D} \tag{16.31}$$

$$\mathbf{E}_k = \mathbf{D} - \mathbf{Y}_k$$

$$= (\alpha'/\alpha) \, \mathbf{D} - \mathbf{A} \, (\mathbf{I} - \alpha \mathbf{I})^{\,k}. \tag{16.32}$$

For $0 < \alpha < 2$, as evident from expression (16.32), $\mathbf{E}_k$ converges to a stable point with a steady-state value of $(\alpha'/\alpha) \, \mathbf{D}$. The steady-state value of $\mathbf{E}_k$ is thus inversely proportional to α.

16. 7.3 Evaluation of Input Excitation by Fuzzy Inverse

The input neuronal excitation / control signal vector $\mathbf{X}_{k+1}$ for the motor actuation signal vector $\mathbf{Y}_k$ is evaluated autonomously with the help of the following relation:

$$\mathbf{X}_{k+1} = \mathbf{W}_k^{-1} \text{ o } \mathbf{Y}_k.$$

The estimation of $\mathbf{W}_k^{-1}$ from $\mathbf{W}_k$ can be carried out by a new formulation of AND-OR compositional inverse, as outlined in chapter 10.

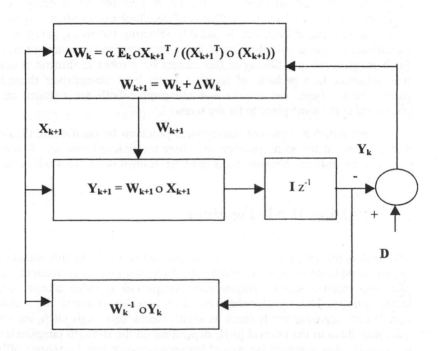

Fig. 16.3: An autonomous fuzzy learning ontroller.

16.8 Task Planning and Co-ordination

Planning a sequence of tasks and their co-ordination, as evident from the model of cognition cycle in chapter 2, is an intermediate mental state of perception and action. Timed Petri net [3] models are generally employed to realize this state of cognition. In a timed Petri net, tokens are time-tagged and the transitions are assigned with firing delays. A transition in a timed Petri net fires, when all its input places possess tokens at least for a time T, where T denotes its firing delay. On firing of the transition, the tokens of its input places are transferred to its output places with new time labels, obtained by adding the time last attached with the token plus the firing delay of the transition [3]. The well known problems of *deadlock* and *conflict* [3] can be overcome in a timed Petri net by suitably selecting the firing delays of the transitions. A typical co-ordination problem between the movement of an arm (with gripper) and positioning of the camera by a robot is illustrated below with reference to a problem of fixing a screw by a screwdriver through a narrow hole. Here, the camera and the gripper both are required to be positioned at the same place to fix the screw.

Human beings, however, handle such problems by positioning their eye and the hand at the same position in a time-multiplexed manner. A similar model for co-ordination among tasks has been framed in fig. 16.4 with a timed Petri net.

16.9 Putting It All Together

The models presented so far have been used compositely in this section for eye-ear-hand co-ordination of a robot (fig. 16.5) working in an assembly-line. The robot receives sensory information through video, audio/ ultrasonic and tactile sensors. The signal conditioner and fuzzifier converts the received signals into linguistic levels (such as small, round, soft, high pitch, etc.) and quantifies them in the interval [0,1], depending on the strength (amplitude) of the signals. Any standard supervised learning network may be trained offline with the fuzzified signals as the input and the recognized objects as the output pattern. For instance, we can employ an FPN based supervised learning algorithm [9], [3] to classify and recognize the objects. After the object is classified, an FPN based model may be used for reasoning and a timed Petri net model connected in cascade with the reasoning model may be used for generating control commands (like move arm up, turn head left, etc.) for motor activation. It may be noted that a single block in fig. 16.5, for brevity, has represented the reasoning and the co-ordination models. Once the control commands are issued, the control model generates actuation signals for the motors through an unsupervised learning. The knowledge refinement model in the fig., covered in chapter 20, adjusts the parameters of the FPN based reasoning model through Hebbian type unsupervised learning.

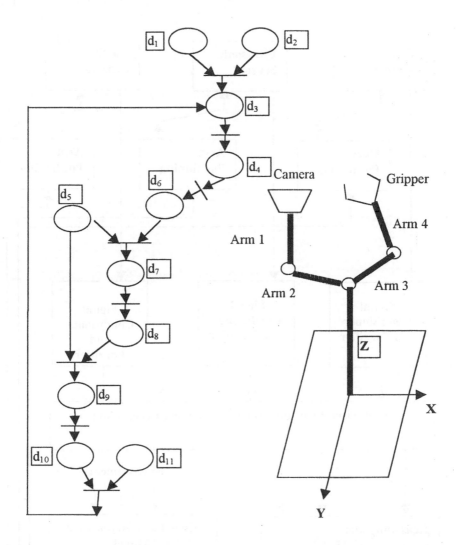

d_1=Front view of object ≠ desired object, d_2 =Total azimuth shift < 180 degrees, d_3 = Rotate camera in azimuth clockwise by angle φ, d_4 = Move motor of arm$_2$ by angle φ, d_5= Gripper placement needed, d_6 = Obstacled gripper, d_7 = Rotate camera in azimuth anticlockwise by φ, d_8 = Gripper position free ?, d_9= Position gripper at desired location and grip and move, d_{10}= camera positioning needed ?, d_{11}= Gripper function over ?

Fig. 16.4: Co-ordination of sensor (camera) and gripper by a robot.

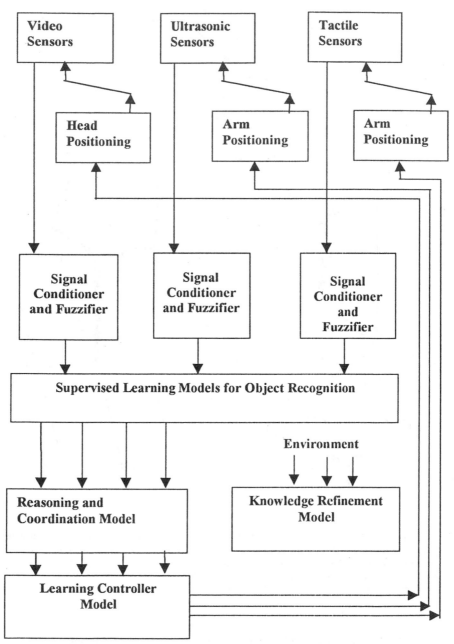

Fig.16.5: The schematic architecture for eye-ear-hand co-ordination of a robot.

16.10 Conclusions and Future Directions

The chapter presented a simple mental model of cognition and demonstrated its scope of application in intelligent systems capable of reasoning, learning, co-ordination and control. The integrated view of designing complex systems like co-ordination of eye-hand-ear of a robot for assembly-line applications has been illustrated in the chapter in detail.

Much emphasis has been given in the chapter for the development of the behavioral states of cognition and the analysis of their dynamism with fuzzy network models. The condition for stability of the system states for the models has been derived and the estimated range of system parameters for attaining stability has been verified through computer simulation for illustrative problems.

The fuzzy networks used in the model of cognition have a distributed architecture and thus possess a massive power of concurrent computing and fault tolerance [4]. It also supports pipelining [4] of rules, when used for reasoning in knowledge-based systems. Further, its functional capability of modeling neural behavior made it appropriate for applications in both reasoning and learning systems. The methodology for building an intelligent system of the above two types by FPN, to the best of the author's knowledge, is the first work of its kind in Artificial Intelligence. The added advantage of timed Petri nets in modeling co-ordination problems further extends the scope of the proposed network in designing complex systems.

It may be noted that besides FPN, many other formalisms could also be used to model cognition. For instance, Zhang et al. used negative-positive-neutral (NPN) logic for reasoning [13-16] and co-ordination among distributed co-operative agents [16] in a cognitive map. However, their model cannot be utilized for handling as many states of cognition as an FPN can. Kosko's model [6] for cognitive map, however, may be considered as a special case of the FPN-based model presented here.

Realization of the proposed systems in practical form is in progress. For example, Konar and Mandal [4] designed an Expert System for criminal investigation with the proposed models. Their system contains approximately 200 rules and an inference engine realized with FPN. It was implemented in Pascal and tested with simulated criminology problems. Much work, however, remains for field testing of such systems. The learning, co-ordination and control models of cognition have yet to be implemented for practical systems. The use of the proposed models in an integrated system like coordination of the eye, hand and ear of a robot is an open problem for future research and development.

Exercises

1. Starting with the point-wise equations for belief and FTT updating, derive the vector-matrix representation of the belief updating equations 16.14 and 16.15.

2. Verify the co-ordination model with appropriate timed tokens at the places of fig. 16.4.

3. Design a reasoning model of fuzzy timed Petri nets that can handle fuzziness of timed tokens. [*open ended for researchers*]

4. Take an example from your daily life that includes concurrency and temporal dependence of activities. Order them according to their time schedule (absolute / relative) and dependence. Map them onto a timed Petri net and test whether the Petri net model can handle this problem.

5. Design a supervised learning algorithm for a fuzzy Petri net. [*open ended for researchers*]

    [**Hints:** Represent the AND/ OR operator by smooth function, so that their derivative with respect to the arguments of the function exists. Then use a back-propagation style derivation (see Appendix B).]

References

[1]   Axelrod, R., *Structure of Design*, Princeton University Press, Princeton, NJ, 1976.

[2]   Downs, R. M. and Stea, D., *Cognitive maps and Spatial Behavior: Process and Products*, Aldine Publishing Co., 1973.

[3]   Konar, A. and Pal, S., "Modeling cognition with fuzzy neural nets," in *Fuzzy Theory Systems: Techniques and Applications*, Leondes, C. T., Ed. Academic Press, New York, 1999.

[4]   Konar, A. and Mandal, A. K., "Uncertainty management in expert sytems using fuzzy Petri nets," *IEEE Trans. on Knowledge and Data Engg.*, vol. 8, no. 1, pp. 96-105, 1996.

[5]   Kosko, B., "Virtual world in fuzzy cognitive map," In *Fuzzy Engineering*, Prentice-Hall, Englewood Cliffs, NJ, pp. 499-527, 1997.

[6]   Kosko, B., "Fuzzy cognitive maps," *Journal of Man-Machine Studies*, vol. 24, pp. 65-75, 1986.

[7]   Kosko, B., "Bidirectional associative memories," *IEEE Trans. on Systems, Man and Cybernetics*, Vol. SMC-18, pp. 42-60, 1988.

[8]   Pal, S. and Konar, A., "Cognitive reasoning using fuzzy neural nets," *IEEE Trans. on Systems, Man and Cybernetics*, Part B, vol. 26, no. 4, 1996.

[9]   Pedrycz, W. and Gomide, F., "A generalized fuzzy Petri net model," *IEEE Trans. on Fuzzy Systems*, vol. 2, no. 4, pp. 295-301, 1994.

[10]   Pedryz, W., *Fuzzy Sets Engineering*, CRC Press, Boca Raton, FL, pp. 253-306, 1995.

[11]   Saha, P. and Konar, A., "Backward reasoning with inverse fuzzy relational matrices," *Inter. Conf. on Automation, Robotics and Computer Vision*, Singapore, 1996.

[12]   Taber, W. R. and Siegl, M., "Estimation of expert weights with fuzzy cognitive maps, *Proc. of the First IEEE Int. Conf. On Neural Networks (ICNN'87)*, Vol. II, pp. 319-325, 1987.

[13]   Zhang, W. R, Chen, S. S. and Bezdek, J.C., "Pool2: A generic system for cognitive map development and decision analysis," *IEEE Trans. on Systems, Man and Cybernetics*, vol. 19, no. 1, Jan-Feb 1989.

[14]   Zhang, W. R., " Pool- A semantic model for approximate reasoning and its application in decision support," *Journal of Management Information System, Special issue: Decision support and knowledge-based system*, vol. 3, no. 4, pp. 65-78, Spring 1987.

[15]   Zhang, W. R., "NPN fuzzy sets and NPN qualitative algebra: A computational network for bipolar cognitive modeling and multi-agent decision analysis," *IEEE Trans. on Systems, Man and Cybernetics*, vol. 26, no. 4, August 1996.

[16]   Zhang, W. R., Chen, S. S, Wang, W. and King, R., "A cognitive map based approach to the co-ordination of distributed co-operative agents," *IEEE Trans. on Systems, Man and Cybernetics*, vol. 22, no. 1, pp. 103-114, 1992.

17

Visual Perception

Human beings can form perceptions about their 3-D world through many years of their experience. Building perception for machines to recognize their 3-D world by visual information is difficult, as the images obtained by cameras can only represent 2-D information. This chapter demonstrates one approach for understanding the 3-D world from 2-D images of a scene by Kalman filtering. Before employing Kalman filters, the images, however, have to be pre-processed and segmented into objects of interest. The chapter thus begins with low and medium level image processing that deals with pre-processing and segmentation of images. Next the principles of perspective projection geometry are covered to explain many interesting phenomena of the 3-D world. The recognition of images with self-organizing feature maps and the principal component analysis have then been taken up as case studies. The chapter ends with reconstructing the 3-D world from multiple 2-D images by using Kalman filtering.

17.1 Introduction

The phrase '*Visual perception*' generally refers to construction of knowledge to understand and interpret 3-dimensional objects from the scenes that humans and machines perceive through their visual sensors. The human eye can receive a wide spectrum of light waves, and discriminate various levels of intensity of light reflected from an object. Many lower class animals, however, have a narrow spectrum of vision in the visible (400-700 nM) or infrared (> 700 nM) wavelength. Modern robots are equipped with sonar (ultrasonic) or laser sensors and video cameras to capture the information around its world for constructing the model of the neighborhood. The knowledge about its neighboring world, derived from the sensory information, also called the '*perception of the robot*', is useful for generating navigational plans and action (like arm movement) sequences by the robot.

In order to construct the models of visual perception, we, in this chapter, will presume the camera to be the sensor. Though there exists some similarity between the human eye and a camera with respect to image (reflected light from an object grabbed by a camera) formation, the resolution of a camera cannot be compared with the human eye. In fact, the resolution of the human eye is more than 100 times the resolution of a good camera. Another basic difference between the human eye and a camera lies in the dimensionality of the devices. While a human eye can feel the third dimension (the depth), a camera can extract only the 2-dimensional view of an object. Getting the third dimension requires integration of the multiple camera images taken from different directions. This is generally referred to as the 3-D re-construction problem, which is a frontier area of research in imaging. There are many approaches to re-construct 3-D objects from their 2-D partial images. The 3-D reconstruction problem and its solution will be covered in this chapter by Kalman filtering, which dates back to the early 60's and remains a useful tool for signal recovery to date.

Another interesting and useful issue of image understanding is the 'recognition problem'. Kalman filtering can be used for recognizing and interpreting objects with planer surface boundaries. But it cannot identify objects having complex 3-D surfaces. Alternatively, the features of an object are extracted from its images and then a 'feature matching algorithm' may be employed to search the extracted features in the model feature-space of the objects. There exist many model-based feature matching algorithms in the existing literature of Artificial intelligence, a detailed account of which is beyond the scope of this book. Artificial neural nets, on the other hand, can also be used for recognizing objects from their features. We, in this chapter, will employ the self-organizing neural nets to recognize human faces from their images.

From our background of chapter 1, we remember that the entire class of vision problems can be subdivided into three typical classes: low, medium and high level vision. Though there exist no clear boundaries of these three levels, still the low and the medium levels include steps like pre-processing, enhancement [5] and segmentation [6] of images, while the high level corresponds to recognition and interpretation of scenes from their 2-D images.

Among these three levels, the **high** level vision utilizes many of the tools and techniques of Artificial Intelligence. This chapter thus briefly outlines the principles of the first two levels and demonstrates the application of AI and soft computing tools in high level vision systems.

17.1.1 Digital Images

The reflected light from the surface of an object is received by a camera and mapped onto a grid of cells in a Cartesian plane, called the **image plane**. The light received by the image plane is spatially sampled, quantized and encoded in binary codes. The encoded information thus obtained is represented by a two dimensional array of sampled points, called **pixels**, and the encoded intensity level of each pixel is called its **gray level**. For instance, if 5 bits are used to represent the coding, we can have gray levels ranging from 0 to 31. For the sake of illustration, let us consider a (4 x 4) image, comprising of four gray levels. Let us define the darkest gray level to be 0 and the brightest to be 3 and the rest are in between.

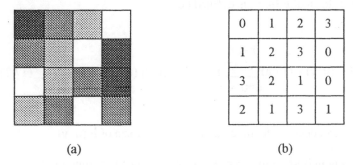

0	1	2	3
1	2	3	0
3	2	1	0
2	1	3	1

(a)   (b)

Fig. 17.1: (a) An illustrative gray image and (b) its encoding
of gray levels in the range [0,3].

A digital image is a two dimensional representation of pixels like that of fig. 17.1. The sampling and quantization process in a digital camera adds noise to the image. Elimination of noise from an image is a prerequisite step for understanding the image. In the next section, we discuss some elementary methods to eliminate noise from a digital image.

17.2 Low Level Vision

Low level vision pre-processes the image for subsequent operations. The most common type of pre-processing includes filtering the image from the noise contaminated with it. However, for filtering an image, it is to be first represented in frequency domain. Since an image is a two dimensional array of pixels, we require two dimensional **Fourier transforms** to get the frequency domain representation of the image. Let x and y denote the spatial distance of a pixel (x, y) along x- and y-direction of the image. Let u and v be the frequency of gray levels along x- and y-direction respectively. The Fourier transform of an image now can be stated as

$$F (u, v) = (1/ n) \sum_{x=0}^{n-1} \sum_{y=0}^{n-1} f (x, y) \exp [-2j (u x + v y) / n] \qquad (17.1)$$

where n denotes the number of pixels in x- and y-directions, and j is the well-known complex number $\sqrt{(-1)}$.

Let G (u, v) be the transfer function of a two dimensional filter of known characteristics. For the reconstruction of the image after filtering it by G (u, v), we need to compute the inverse transform of the product. The inverse Fourier transform of F (u, v) is defined below.

$$f (x, y) = F^{-1}(u, v) = (1 / n) \sum_{u=0}^{n-1} \sum_{v=0}^{n-1} f (u, v) \exp [2j (u x + v y) / n] \qquad (17.2)$$

where F^{-1} (u, v) denotes the inverse Fourier transform of F (u, v).

For retrieving noise-free image h (x, y), we thus need to compute

$$h(x, y) = F^{-1} [F (u, v) . G (u, v)] . \qquad (17.3)$$

Sometimes impulse response of G(u, v) are known. Under this circumstance, one may use the **convolution theorem** to compute h(x, y).

$$h (x, y) = \sum_{u=-\infty}^{\infty} \sum_{v=-\infty}^{\infty} f (u, v) g (x - u, y - v). \qquad (17.4)$$

The convolution of f(u, v) with g(u, v) is denoted by '*', using the standard nomenclature. Thus h (x, y) is written as h(x, y) = f(u, v) * g (u, v). (17.5)

Thus one can obtain the noise-free image, when the impulse response of the filter transfer function is known. A gaussian filter is one of the most commonly used devices, represented by

$$g(x, y) \;=\; (\,1\,/\,2\,\pi\,\sigma^2\,)\exp[\,-(x^2 + y^2\,)\,/\,2\,\sigma^2\,] \qquad (17.6)$$

where the standard deviation σ is a parameter of the user's choice.

The digital filtering by gaussian function attempts to smooth the frequency response F (u, v) of the image and is thus referred to as smoothing. We describe different smoothing techniques in the next sub-section.

17.2.1 Smoothing

Smoothing, as evident from its name, is required to reduce the variations in the amplitude of F(u, v) over a wide range of frequencies u and v. Gaussian filter undoubtedly is a good scheme for smoothing. Besides gaussian filter, there exist masking techniques for smoothing. Two such masks, referred to as the 4-point and the 8-point masks, are presented below.

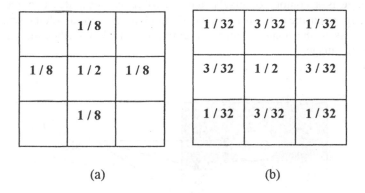

(a) (b)

Fig. 17.2: (a) 4-point and (b) 8-point masks used for smoothing.

It may be noted from fig. 17.2 that the sum of the weights assigned to the pixels of a mask must add to unity. A question that naturally arises is how to employ these masks for smoothing an image. The following steps may be executed to solve the problem.

1. *Move the mask over the image, so that the central (x, y) pixel of the mask is merged with a pixel of the image. The boundary pixels of the image, however, cannot not be merged with the central pixel of the mask.*

2. *Evaluate the changed gray level at pixel (x, y) by taking the sum of the product of the mask weights and the gray level at the corresponding pixels and save it.*

3. *If all feasible pixels on the image have been visited, replace the old gray levels by the corresponding new (saved) values and stop, else continue from step 1.*

Among the other smoothing algorithms, most common are mean and median filters. The **mean filter** takes the average of the gray levels of all 8-neighboring pixels to compute the changed gray level at pixel (x, y). The median filter, on the other hand, replaces the current gray level of pixel (x, y) by the median of its 8 neighborhood pixels. Experimental evidences show that mean filter **blurs** the image, but the median filter **enhances** the **sharpness** of the image.

17.2.2   Finding Edges in an Image

An **edge** is a contour of pixels that   (artificially) separates two regions of different intensities. It also can be defined as a contour along which the brightness in the image changes abruptly. Fig. 17.3 describes the edges in a synthetic image.

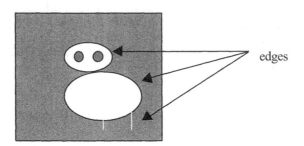

Fig. 17.3: Edges in a synthetic image.

There exists a vast literature concerned with edge findings in an image. The simplest method to find edges is to evaluate the directional derivatives of g(x, y) in x- and y-directions. Let us call them g_1 and g_2 respectively. Thus,

$$g_1 = \partial g(x, y) / \partial x \qquad (17.7)$$

$$\text{and} \quad g_2 = \partial g(x, y) / \partial y. \qquad (17.8)$$

The resulting gradient can be evaluated by the vector addition of g_1 and g_2 and is given by

$$\text{gradient } g = [g_1{}^2 + g_2{}^2]^{1/2} \qquad (17.9)$$

$$\text{and phase } \phi = \tan^{-1} (g_2 / g_1). \qquad (17.10)$$

A pixel is said to lie on an edge if the gradient g is over a given threshold. An extension of this principle has been realized by **Sobel mask**, presented below.

-1	0	1
-2	0	2
-1	0	1

1	2	1
0	0	0
-1	-2	-1

mask for g_1 mask for g_2

Fig. 17. 4: The Sobel masks.

One can thus evaluate g_1 and g_2 by moving the Sobel masks over the image and evaluating them by taking the sum of the products of the weights and the gray level of the corresponding pixels. The gradient g at each pixel then can be evaluated and a threshold may be used to check whether g exceeds the threshold. If yes, that pixel is supposed to lie on an edge.

An alternative way to determine edges is to use gaussian function. For example, the Laplacian operator

$$\nabla^2 = \partial^2 / \partial x^2 + \partial^2 / \partial y^2$$

may be applied on the convolution of g* f to evaluate

$$\nabla^2 (g * f)$$

$$= (\nabla^2 g) * f.$$

With $g(x, y) = (1 / 2 \pi \sigma^2) \exp [- (x^2 + y^2) / 2 \sigma^2]$, vide expression (17.6), we can now evaluate $\nabla^2 (g * f)$, which will be non-zero at the pixels lying on edges in an image and will be zero as we go off the edges. The present method of edge determination, thus, is a two step process that involves smoothing and then application of Laplacian over the resulting image.

We now say a few words on **texture** of an image before closing this section.

17.2.3 Texture of an Image

A **texture** [6] is a repeated pattern of elementary shapes occurring on an object's surface. It may be regular and periodic, random, or partially periodic. For instance, consider the image of a heap of pebbles. It is partially periodic as the pebbles are not identical in shape. But a heap of sand must have a regular and periodic texture. Textures are useful information to determine the objects from their images. Currently, fractals are being employed to model textures and then the nature of the texture is evaluated from the function describing the fractals. Various pattern recognition techniques are also used to classify objects from the texture of their surfaces.

17.3   Medium Level Vision

After the edges in an image are identified, the next major task is to segregate the image into modules of interest. The process of partitioning the image into modules of interest is informally called **segmentation**. The modules in the image are generally segmented based on the homogeneous features of the pixel regions and/ or the boundaries created by the connected edges in low level processing. The intermediate level requires combining the pieces of edges of contiguous regions to determine the object boundaries and then attaching a **label** to each of these boundaries. We now discuss some of the well-known techniques of image segmentation and **labeling**.

17.3.1 Segmentation of Images

The most common type of segmentation is done by a specialized plot of frequency versus intensity levels of an image, called a **histogram**. Fig. 17.5 describes a typical histogram. The peaks in the histogram correspond to regions of the same gray levels. The regions of same intensity levels, thus, can be easily isolated by judiciously selecting a threshold such that the gray level of one module in the image exceeds the threshold. This is referred to as **thresholding**. Selection of threshold, however, is a difficult issue as the range of gray levels that exceed the thresholds may be present in more than one disjoint modules.

In fig. 17.5, we find three peaks, each likely to correspond to one region. So, we select threshold at gray levels th_1, th_2 and th_3 following the peaks, so that we can expect to identify one or more regions. Since a gray level may occupy more than one region, segmentation by histogram cannot always lead to identification of the regions.

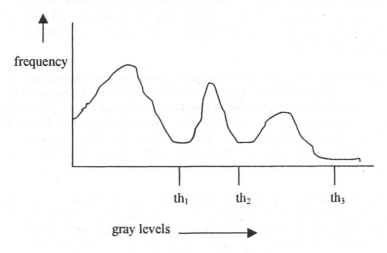

Fig. 17.5: A typical histogram of an image.

Regional segmentation is accomplished by growing or splitting regions. In the first instance, neighboring regions having some form of homogeneity such as uniformity in gray level or color or texture are grouped together to form a larger region. The process of building a larger region is continued recursively, until a specific type of known structure is identified. **Region growing** is generally called a bottom up approach, as the regions, representing leaves of a tree, are grouped to get the intermediate node (describing complex regions), which after grouping at several levels form the root, describing the complete image.

Region splitting starts with the entire image and splits it into large regions of more or less uniform features including gray levels, texture or color. The regions thus obtained are sub-divided recursively until regions describing some known 2-D shapes are identified. Region splitting is called a top down approach for regional segmentation.

It may be noted that segmenting an image into regions without any knowledge of the scene or its modules is practically infeasible. Generally, the partial knowledge about the image such as an outdoor scene or a classroom or a football tournament helps the segmentation process.

17.3.2 Labeling an Image

After segmenting an image into disjoint regions their shapes, spatial relationships and other characteristics can be described and labeled for subsequent interpretation. Typically a region description includes size of the area, location of the center of mass, minimum bounding rectangles, boundary contrast, shape classification number, chain code [6], position and type of vertices and the like. We here explain the method for describing the boundary of a region by a chain code.

A **chain code** is a sequence of integers, corresponding to direction of traversal of the segments around a region, starting from a fixed position. The direction of traversals of a segment is compared with the set of primitive directions and the nearest one is selected. The traversed segment is encoded following the standard integer code of that direction. The traversal is continued until the chain repeats. The minimum index in the closed chain is identified to match it with known strings, constructed in the same manner.

One important point that needs mention is that the consecutive chain indices may be identical, when we move along a large linear segment. Under this circumstance the portion of the chain describing the same index is replaced by the index, taken only once. Constructing a chain code for the boundary of a region is illustrated below (vide fig. 17.6 and 17.7).

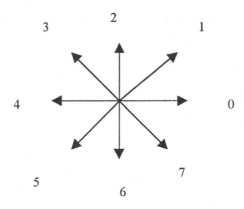

Fig. 17.6: The direction indices for chain coding: the angle between two consecutive directions is 45 degrees.

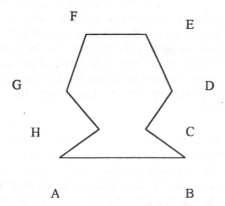

Fig. 17.7: A region boundary to illustrate the construction
of the chain code.

Suppose we arbitrarily start coding from point A along AB. Suppose
we have a small fixed length stick, by which we measure line AB and write
the code for horizontal right (0) (vide fig. 17.6) as many times as the number
of the measuring stick. Last time the stick length may have been a little
larger/smaller than the actual length of the remnant line. Let us assume that
AB is 4 times the length of our measuring stick; so we write 4 zeroes to
describe AB. Proceeding in this manner, we find the chain code of the closed
boundary as follows.

Chain code= { 00003311333445557755}

Rewriting repeated consecutive symbols once only, we find the
modified code = {03134575}. Since it is a closed chain, we can start
anywhere. But the code must start with the minimum index. Here, it is
already in the required form. For a specific example let the chain code be
{2304568}; we would express it in the form {0456823}, which starts with the
minimum index. Such representation helps matching the code of an unknown
boundary with known ones. It may be noted that for two distinct shapes of
boundaries the chain code must be different. If it is not the case, then direction
indices need to be increased.

17.4   High Level Vision

The major task of high level vision is to recognize objects from their feature
space, obtained at earlier levels, and interpret the scene comprising of one or
more objects. For instance, suppose a football tournament is going on and the

machine has to narrate the game online. This undoubtedly is a very complex problem and present day systems are incapable to do so. But where are the hurdles? Suppose low and medium level image processing are done very fast in real time. So, the difficulty may not be at these steps. In fact for narrating the game, the machine has to see the position of the ball, the players and recognize them. In many occasions, the players may be visible partially and the machine has to identify them from their partial side images. Translating the scene into language for interpretation also is not an easy task. This section will be devoted to presenting recent techniques for recognition and interpretation of images.

The feature-space of objects sometimes comprises of geometric information like parallelism or perpendicularity of lines. Detecting these is also a complex task. For instance parallel rail lines at a far distance do not look parallel. The highway at a far end seems to be unparallel. But really they are not so. This happens due to a inherent characteristic of the lens in our eye or camera. This characteristic is called **perspective projection**. We will discuss these issues briefly in this section.

17.4.1 Object Recognition

Recognition problems are concerned with two major issues: feature extraction and classification. Features are essential attributes of an object that distinguishes it from the rest of the objects. Extracting features of an object being a determining factor in object recognition is of prime consideration. The well-known supervised, unsupervised and reinforcement algorithms for pattern classification are equally useful for recognition of objects. The choice of a particular technique depends on the area of application and the users' experience in that problem domain. In this section, we present an unsupervised (more specifically, reinforcement) learning by employing a self-organizing neural network. Experimental evidences show that the pixel gray levels are not the only features of an object. The following case study illustrates the **principal component analysis** to determine the first few **eigen values** and hence eigen vectors as image attributes. These vectors are subsequently used as the features of a self-organizing neural net for classification of the object.

17.4.1.1 Face Recognition by Neurocomputing
Approach

This section describes two approaches for face recognition. The first is accomplished by principal component analysis, and the latter by self-organizing neural nets. There feature extraction parts for both the methods are common, as evident from fig. 17.8.

Principal component analysis: The detail of the analysis [4] is beyond the scope of the book. We just illustrate the concept from a practical standpoint. We, in this study, considered 9 facial images of (32 x 32) size for 40 individuals. Thus we have 360 images each of (32 x 32) size. The average gray value of each image matrix is then computed and subtracted from the corresponding image matrix elements. One such image after subtraction of the average value is presented in fig. 17.9. This is some form of a normalization that keeps the image free from illumination bias of the light source. Now, we construct a matrix $\mathbf{X}$ of $((32 \times 32) \times 360) = (1024 \times 360)$ dimension with the above data points. We also construct a covariance matrix by taking the product $\mathbf{X}\,\mathbf{X}^T$ and evaluate the eigen vectors of $\mathbf{X}\,\mathbf{X}^T$. Since $\mathbf{X}$ is of dimension (1024×360), $\mathbf{X}\,\mathbf{X}^T$ will have a dimension of (1024×1024). $\mathbf{X}\,\mathbf{X}^T$ thus will have 1024 eigen values, out of which we select the first 12, on experimental basis. The eigen vectors corresponding to these 12 eigen values are called the first 12 **principal components**. Since the dimension of each principal component is (1024×1), grouping these 12 principal components in column-wise fashion, we construct a matrix of dimension (1024×12). We denote this matrix by $\mathbf{EV}$ (for eigen vector).

To represent an image in the eigen space, we first represent that image in (1×1024) format and then project it onto the face space by taking the dot product of the image matrix $\mathbf{IM}$, represented in (1×1024) format and the $\mathbf{EV}$ and call it a point $(\mathbf{PT})$ in the face space. Thus

$$(\mathbf{PT})_{1 \times 12} = (\mathbf{IM})_{1 \times 1024} \cdot (\mathbf{EV})_{1024 \times 12.} \qquad (17.11)$$

The projection of images onto face space as 12-dimensional points is presented in fig. 17.10. Representing all 360 images by the above expression, we thus get 360 points in 12-dimensional image space.

Now, suppose given a test image and we want to classify it to one of the 40 persons. One simple way to solve this problem is to determine the corresponding image point $(\mathbf{PT})$ in 12-dimensional space and then determine the image point (out of 360 points) that has the least Euclidean distance w.r.t the test image point. The test image thus can be classified to one of 360 images.

The principal component analysis (PCA) thus reduces the dimension of matching from (32×32) to (1×12) but requires computing the distance of a test point with all image points. An alternative scheme that reduces less computing time is by a self-organizing neural net. The self-organizing scheme inputs the (1×12) points for all of the 360 images, constructs a network and searches a test point by the best first search paradigm in the search space. A schematic diagram, briefly outlining the two approaches, is presented in fig. 17.8 below.

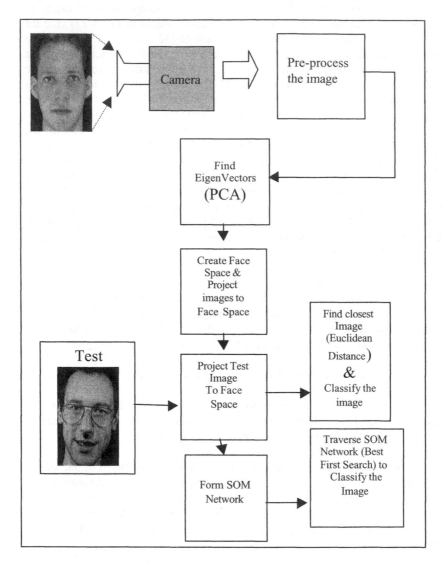

Fig. 17.8: A schematic architecture of the overall system.

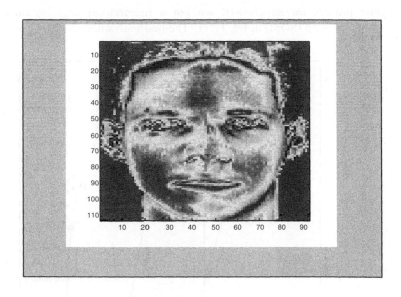

Fig. 17.9: A facial image after subtracting the average value from the pixels of the image.

Self-organizing neural nets for face recognition: Let us assume that 12-dimensional vectors, one each for the 360 images, are already available by principal component analysis. We now consider each 12-dimensional point as input and adapt the weights of the neurons on a two-dimensional plane. Remember that for mapping a feature vector (here the 12-dimensional vector), we need to find the neuron on the 2-D plane, where the error $\| \mathbf{IP} - \mathbf{W} \|$ is minimum for some weight vector $\mathbf{W}$ of a neuron [7]. We then have a selected small square region centering that neuron and adjust the weights of all neurons within that encirclement. Now, we find a next point, where again the error $\| \mathbf{IP} - \mathbf{W} \|$ is minimum. We now consider a smaller square around the point on the 2-D space, where this error too is minimum. The process of adaptation of weights of neurons is thus continued, until we reach a situation, when the weights of a node do not change further. Thus the 12-dimensional point is mapped to one point on the 2-D neuronal plane. Sometimes instead of a single point, a small cluster of points is found to have almost no change in weights. In that case the 12-dimensional point is mapped onto the cluster. In the same manner, we map all the 360 points onto a 2-dimensional surface.

It may so happen that more than one projected image point corresponds to the same cluster on the 2-D surface. In that case the weight vectors of those neurons are again mapped hierarchically to another 2-D surface. Repeating the same process for all clusters, we got a hierarchical organization of self-organizing map (SOM) in fig. 17.11. The main advantage of this structure is

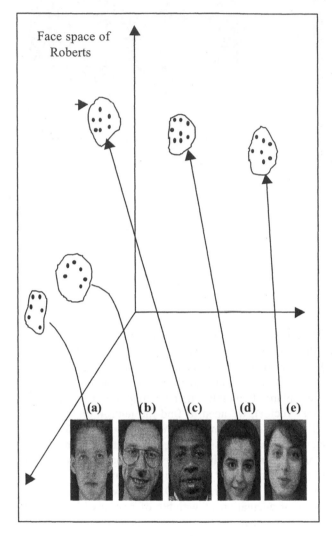

(a) Jim, (b) Jack, (C) Robert, (d) Lucy and (e) Margaret

Fig. 17.10: Projection of different facial poses of persons mapped on 12-dimensional space as points; we considered 9 poses per person that together form a cluster; all 9 points in a cluster, however, have not been shown for clarity.

that it can always map a new projected image point on to the leave surfaces of the structure. Here is the significance of SOM over the back-propagation learning. For a new input-output pattern the back-propagation net is to be re-trained, whereas for a new input pattern the incremental classification of SOM just maps the new point onto one of the existing leaves of the 2-D surfaces.

After the structural organization of the SOM is over, we can classify a test image by the structure. Let **IP** be the projected point corresponding to the test image. Let **W** be the weight vector of a neuron. We first compute the $\|$**IP** $-$**W** $\|$ for all neurons mapped on the surface at level 0 and identify the neuron where the measure is minimum. We then compute $\|$**IP** $-$**W** $\|$ for all neurons of a surface in level 1, which is pointed to by the winning neuron at level 0. The process is thus continued until one of the leaves of the structure is reached. The search procedure in a SOM thus is analogous to best first search.

17.4.1.2 Non-Neural Approaches for Image Recognition

There exist quite a large number of non-neural approaches for pattern recognition. For instance Bayes' classifier, rule based classifier, and fuzzy classifier are some of the well-known techniques for pattern classification, which have also applications in recognition of human faces. The simplest among them is the rule based classification. The range of parameters of the feature space is mentioned in the premise part of the rules. When all the preconditions are jointly satisfied, the rule fires classifying the object. One illustrative rule for classification of sky is presented below.

Rule: If (the location-of-the-region = upper) and
(Intensity-of-the-region-is-in {0.4,0.7}) and
(Color-of-region = (blue or grey)) and
(Texture-of-the-region-is-in {0.8 , 1})
Then (region = sky).

One drawback of the rule based system is that in many occasions the precondition of no rules are supported by the available data space. Fuzzification of the data space is required under this circumstance to identify the rules having partially matched preconditions. The application of such fuzzy rule based systems for facial image matching will be presented in a subsequent chapter of this book. Bayes' classification rule and fuzzy c-means classification schemes are beyond the scope of this book.

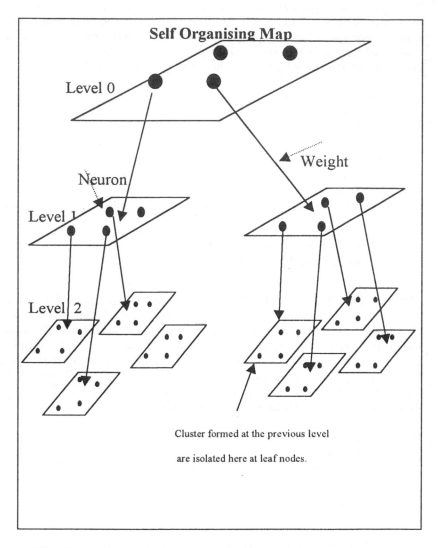

Fig. 17.11: Hierarchical structure of self-organizing neural systems.

17.4.2 Interpretation of Scenes

Interpretation of scenes includes a wide class of problems. The simplest form of the problem is to interpret an object from the labels of its recognized components. For instance, if we can identify the back plane, the seat and the legs of a chair, we should interpret it to be a chair. Even the simplest problem of this type is difficult to realize. Naturally the question arises: why? In fact, we cannot always grab all components of a chair in its image. So, for determining it to be a chair, we need to grab several images of it and then design schemes to fuse these images or their attributes for interpretation of the object. Further, the parallel lines of the objects may not be parallel in its image, especially when the lines are long and away from the camera. For understanding these, we need to learn the geometric aspects of image formation. The most fundamental issue in geometric aspects of image formation is the **perspective projection**.

17.4.2.1 Perspective Projection

The concept of perspective projection stems from the principle of image formation in a pinhole camera. Light passes through a pinhole of a camera and forms an inverted image in the back plane, also called the image plane of the camera. Generally, the image plane is defined as a plane perpendicular to the direction of light and located at a distance of f, the focal length of the camera lens from the pinhole. A schematic diagram of the geometry of image formation in a camera is presented in fig. 17.12. It shows two sets of axes: one representing the global axes denoted by X, Y and Z, and the other corresponding to the camera axes denoted by X_c, Y_c, and Z_c. In fig. 17.12, however, all the axes of the global coordinate system and the camera coordinate system are aligned.

A light ray passing through the optical center of the camera forms an image point (u, v) on the image plane of the camera corresponding to point P having a global coordinate (x, y, z). From the properties of similarity of triangles, the following relations are evident:

$$x/z = -u/f$$

$$\Rightarrow u = -x \cdot f/z \tag{17.12}$$

$$y/z = -v/f$$

$$\Rightarrow v = -y \cdot f/z \tag{17.13}$$

The expressions (17.12) and (17.13) are jointly referred to as the **perspective projection relations** that describe the mapping of point (x, y, z) to (u, v) on the image plane. The negative signs in the right hand side of these expressions indicate that the image at (u, v) will be inverted, both left-right and up-down.

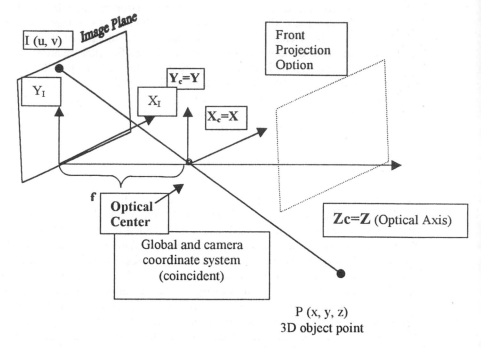

Fig. 17.12: Camera model of 3-D perspective projection.

The reason: why parallel rail lines at the furthest end seem to be unparallal (converging) can now be ascertained. Let us consider a point P_m on the line passing through (x, y, z) in the direction (p, q, r). The coordinate of the point then will be $(x + m p, y + m q, z + m r)$. The perspective projection of P_m on the image plane is given by

$$(u, v) = (\,-f\,(x + m\,p)\,/\,(z + m\,r),\ -f\,(y + m\,q)\,/\,(z + m\,r)). \qquad (17.14)$$

When m approaches infinity, (u, v) approaches $(-f\,p\,/\,r,\ -f\,q\,/\,r)$, then we call P_m at $m \rightarrow \infty$ to be the vanishing point. It may be noted that as $m \rightarrow \infty$, the projected image point does not depend on the point (x, y, z). Thus parallel lines are mapped at a given point $(-f\,p\,/\,r,\ -f\,q\,/\,r)$ and consequently the parallel lines seem to be converging at the furthest end.

17.4.2.2 Stereo Vision

The objective of stereo vision is to recover the geometric structure of objects in a scene from multiple images. Generally the images of a scene are taken from different viewpoints and the correspondence between the points in two images is accomplished. Sometimes all the points in one image are not available in the other images because of different viewpoints. In this section we illustrate the concept of triangulation that helps in identifying the corresponding points in images. Suppose we see an object by a camera. The location of the point must lie on the line passing through the center of projection and the image point. So, if we see a point through two cameras, vide fig. 17.12, the 3-D point must lie on the intersection of the lines passing through the center of projections and the image points.

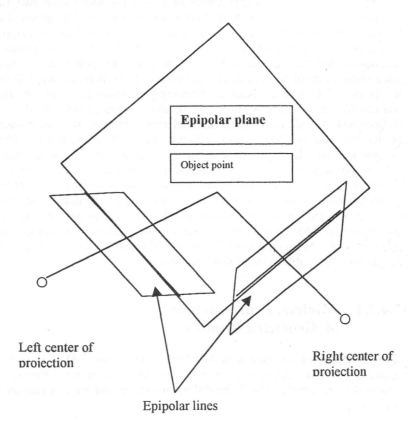

Fig. 17.13: The epipolar geometry showing that the epipolar lines are formed by the intersection of the epipolar plane with the image planes.

It must be added here that when two cameras are used to see an object from different viewpoints, a hypothetical plane passes through the center of projections and the object point. This plane is called **epipolar plane**. The epipolar planes meet the image planes at **epipolar lines**. From the epipolar geometry of fig. 17.13, it is clear that for a given image point, there must be a corresponding point on the second image. Thus if the image point in the first image is shifted, the corresponding point will also be shifted on the epipolar line of the second image.

Finding the corresponding points in images is generally referred to as the **correspondence problem** [3]. There are many approaches to handle the correspondence problem. In this book we will first determine the 3-D points of an object from its 2-D image points and then for more than two images determine the correspondence by measuring the shortest Euclidean distance between the point sets of the two or more images. If the 2-D to 3-D mapping of the points are satisfactory and the images consist of common points, then determining the correspondence between the points of two or more images is not a complex problem. For determining the 3-D points from their 2-D images we, in this book, will use **Kalman filtering** [1]. However, before introducing Kalman filtering, we briefly outline the minimal representation of 2-D lines, 3-D lines and 3-D planes. With a minimal representation, we can extract 3-D points from multiple 2-D points in different images by using Kalman filtering. We use Kalman filtering because it has an advantage of recursively operating on incoming data stream like 2-D points from n number of images. The more is the value of n, the better will be the accuracy of the results. The other least square estimators, unlike Kalman filtering, demand all the data set together; so the user has no choice to control the level of accuracy at the cost of computational time. In Kalman filtering, one can observe the improvement in accuracy in the estimation of the parameter of lines or planes and accordingly decide about the submission of new data points.

17.4.2.3   Minimal Representation
of Geometric Primitives

For estimation of parameters of 2-D lines, 3-D points, 3-D lines and 3-D planes, we first represent them with minimal parameters. Further the selected representation should be differentiable, so that we can employ the principles of Linear Kalman filtering.

Representation of Affine lines in R^2: A 2-D line can be represented by at least two independent parameters. The simplest form of representation of a 2-D line is given by the following expressions.

Case 1: When the lines are not parallel to the Y-axis, they are represented by

$$a\, x + y + p = 0 \tag{17.15a}$$

Case 2: When the lines are not parallel to the X-axis, they are represented by:

$$x + a y + p = 0 \qquad\qquad\qquad (17.15b)$$

In brief, the representation of a 2-D line is given by a vector (a, p), where the line passing through $(0, 0)$ and $(a, 1)$ is normal to the line under consideration, which also passes through the point $(0, -p)$. This is illustrated in fig. 17.14.

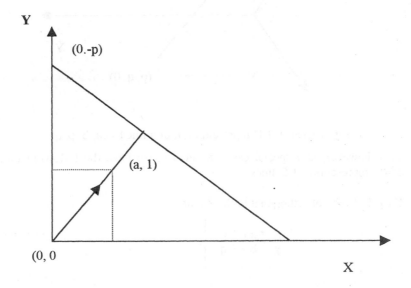

Fig. 17.14: A 2-D line represented by (a, p).

Representation of Affine lines in R^3: The 3-D affine line can be represented minimally by four parameters given by (a, b, p, q). Other minimal representations are possible but there exists scope of ambiguity in other representations [1]. For example, when the line is parallel to the direction vector $(a, b, 1)^T$ and touches the x-y plane at $(p, q, 0)^T$, it can be represented by the following two expressions (vide fig. 17.15).

$$\left. \begin{array}{l} x = a z + p \\ y = b z + q \end{array} \right\}$$

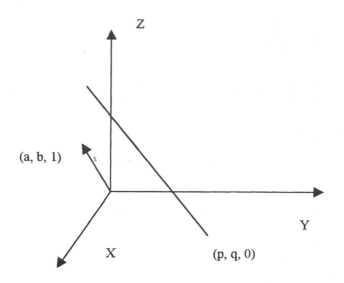

Fig. 17.15: A 3-D representation of a line by (a, b, p, q).

This, however, is a special case. In general, we have the following three cases representing 3-D lines

Case I: Line not orthogonal to the Z axis:

$$\left.\begin{array}{l} x = a\,z + p \\ y = b\,z + q \end{array}\right\} \qquad (17.16a)$$

Case II: Line not orthogonal to the X axis:

$$\left.\begin{array}{l} y = a\,x + p \\ z = b\,x + q \end{array}\right\} \qquad (17.16b)$$

Case III: Line not orthogonal to the Y axis:

$$\left.\begin{array}{l} z = a\,y + p \\ x = b\,y + q \end{array}\right\} \qquad (17.16c)$$

The representation is preferred by the following counts:

i)   It imposes no constraints on the parameters (a, b, p, q).

ii)   Parametric representations of the lines remain linear, which are advantageous to Kalman filtering optimization.

Representation of Affine planes in R^3: One way of representing 3-D planes is by a 3-D vector (a, b, p) such that points (x, y, z) of the plane are defined by the following equation.

   $$a x + b y + z + p = 0$$

Here the vector $(a, b, 1)^T$ is the normal to the plane and the point $(0, 0, -p)^T$ is the point of intersection of the plane with the Z-axis.

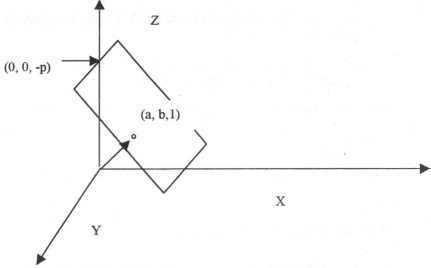

Fig. 17.16: A 3-D plane representation by (a, b, p).

The limitation of this notation is that planes parallel to the Z axis can not be represented. More formally, we have three cases:

Case I: Planes not parallel to the Z axis
   $$a x + b y + z + p = 0 \qquad (17.17a)$$

Case II: Planes not parallel to the X axis
   $$x + a y + bz + p = 0 \qquad (17.17b)$$

case III: Planes not parallel to the Y axis

$$b\,x + y + a\,z + p = 0 \qquad\qquad (17.17c)$$

17.4.2.4 Kalman Filtering

A Kalman filter is a digital filter that attempts to minimize the measurement noise from estimating the unknown parameters, linearly related with a set of measurement variables. The most important significance of this filter is that it allows recursive formulation and thus improves accuracy of estimation up to users' desired level at the cost of new measurement inputs.
Let

$f_i(x_i, a) = 0$ be a set of equations describing relationships among a parameter vector a and measurement variable vector x_i,

$x_i{}^* = x_i + l_i$, such that $E[l_i] = 0$, $E[l_i\,l_i{}^T] =$ positive symmetric matrix Λ_i, and $E[l_i\,l_j{}^T] = 0$,

$a_{i-1}{}^* = a + s_{i-1}$, such that $E[s_{i-1}] = 0$, $E[S_{i-1}\,S_{j-1}{}^T] =$ positive symmetric matrix S_{i-1}, $E[S_{i-1}\,S_{i-1}{}^T] = 0$.

Expanding $f_i(x_i, a)$ by Taylor's series around $(x_i{}^*, a_{i-1})$, we find

$f_i(x_i, a)$

$= f_i(x_i{}^*, a_{i-1}{}^*) + (\partial f_i / \partial x)(x_i - x_i{}^*) + (\partial f_i / \partial a)(a - a_{i-1}{}^*)$

$= 0.$

After some elementary algebra, we find

$y_i = M_i\,a + w_i$

where $y_i = \quad - f_i(x_i{}^*, a_{i-1}) + (\partial f_i / \partial a)(a - a_{i-1}{}^*)$

is a new measurement vector of dimension $(p_i \times 1)$.

$M_i = (\partial f_i / \partial a)$ and

$w_i = (\partial f_i / \partial x)(x_i - x_i{}^*)$ is a measurement noise vector of

dimension $(p_i \times 1)$.

We also want that E [w_i] =0 and define

$$W_i = E [w_i \, w_i^T] = (\partial f_i / \partial x) \, \Lambda_i \, (\partial f_i / \partial x)^T.$$

Let $S_i = E [(a_i - a_{i*}) (a_i - a_i^*)^T]$

An attempt to minimize S_i yields the filter equations, given by:

$$a_i^* = a_{i-1}^* + K_i (y_i - M_i \, a_{i-1}^*)$$

$$K_i = S_{i-1} \, M_i^{\,T} (W_i + M_i \, S_{i-1} M_i^{\,T})^{-1} \qquad\qquad (17.18)$$

$$S_i = (I - K_i \, M_i) \, S_{i-1}.$$

Given S_0 and a_0, the Kalman filter recursively updates a_i, K_i, S_i until the error covariance matrix S_i becomes insignificantly small, or all the number of data points have been submitted. The a_i obtained after termination of the algorithm is the estimated value of the parameters.

The Kalman filter has been successfully used for determining

i)   affine 2-D lines from a set of noisy 2-D points,
ii)   3-D points from a set of noisy 2-D points,
iii) affine 3-D lines from noisy 2-D points and
iv) 3-D planes from 3-D lines.

In the next section, we illustrate the formation of 2-D lines and 3-D lines from a set of noisy points by Kalman filtering.

17.4.2.5 Construction of 2-D Lines from Noisy 2-D Points

We shall directly apply the filter equations for the construction of affine 2-D lines from noisy 2-D points. Here, given the set of points $x_i^* = (u_i^*, v_i^*)$, we have to estimate the parameters $a = (a, p)^T$. The $f_i (x_i, a)$ in the present context is given by

$$f_i (x_i, a) = a \, u_i + v_i + p = 0.$$

$$y_i = M_i \, a + w_i,$$

where

$$y_i = -f_i \, (x_i,^* \, a_{i-1}^*) \; + (\partial \, f_i / \partial \, a) \, (a - a_{i-1}^*)$$

$$= - \, v_i$$

$$M_i = \; (\partial \, f_i / \partial \, a)$$
$$= (u_i, \; 1)$$

The measurement noise w_i is given by

$$w_i = \; (\partial \, f_i / \partial \, x_i) \, (x_i - x_i \, ^*),$$

where $\; (\partial \, f_i / \partial \, x_i) = [a_{i-1}^* \, , \; 1]$

The covariance matrix W_i is given by

$$W_i = (\partial \, f_i / \partial \, x_i) \, \Lambda_i \; (\partial \, f_i / \partial \, x_i)^T$$

where $\Lambda_i = 1$.

Computer simulation: We now present a computer simulation to construct 2-D lines from noisy 2-D points [7].

Procedure Construct 2-D-Lines (2D-image-points: u, v)
Input: *the coordinates of the image points from a input file 2d.dat.*
Output: *the state estimate A (2×1), along with the covariance error S (2×2), associated with the estimate.*

Begin
  Initialize the Initial Covariance matrix S and the state estimate:
      $S_0 \leftarrow$ Very large initial value;
      $a_0 \leftarrow$ Arbitrary value preferably $[0 \; 0 \; 0]^T$;

      For (no-of -points = 1 to n)
          Compute the perspective matrix from the
          input camera parameters;
          Compute the measurement vector y (1×1),
          the linear transformation M (2×1) obtained after linearizing
          measurement equation from input parameters at each iteration;
          Initialize all the matrices involved in matrix multiplication;
          Compute gain K using previous S and M values;

Compute covariance matrix S recursively using its value at previous iteration;
Compute the state estimate 'a' recursively using its value at previous iteration;
End For,
End.

Traces of the Procedure: Construct-2-D-Lines

This program is taking the input data from the 2d.dat, which is given below. Each row of this file contains (x, y) coordinate of a point.

File: 2d.dat

9.500000	0.500000
9.400000	1.000000
9.500000	1.500000
9.000000	2.000000
8.700000	2.300000
8.800000	2.800000
8.400000	3.000000
8.500000	3.500000
7.800000	4.000000
8.000000	4.500000
7.500000	5.000000
7.500000	5.500000
7.300000	6.000000
6.800000	6.200000
6.800000	6.600000
6.500000	7.000000
6.500000	7.500000
6.200000	7.800000
6.200000	8.200000
5.800000	8.300000
5.900000	8.900000
5.500000	9.000000
5.500000	9.500000
4.900000	10.000000
5.000000	10.500000

The sample output of this program in terms of the parameters of the extracted line is given for only 6 iterations and the graphical output is also presented in fig. 17.17.

Sample Execution:

This is a program for constructing 2-D line from 2-D points
2-D points are stored in an input file Enter a:2d.dat to get
data from the input file

Enter the input file name:a:\2d.dat
value of u entered for iteration 1 is 9.500000
value of v entered for iteration 1 is 0.500000
value of 'a' for iteration 1 is -0.052579
value of 'p' for iteration 1 is -0.000498
Press enter to get the next data from the input file

value of u entered for iteration 2 is 9.400000
value of v entered for iteration 2 is 1.000000
value of 'a' for iteration 2 is 0.166446
value of 'p' for iteration 2 is -2.318713
Press enter to get the next data from the input file

value of u entered for iteration 3 is 9.500000
value of v entered for iteration 3 is 1.500000
value of 'a' for iteration 3 is -0.090538
value of 'p' for iteration 3 is -0.134702
Press enter to get the next data from the input file

value of u entered for iteration 4 is 9.000000
value of v entered for iteration 4 is 2.000000
value of 'a' for iteration 4 is 1.271064
value of 'p' for iteration 4 is -13.124207
Press enter to get the next data from the input file

value of u entered for iteration 5 is 8.700000
value of v entered for iteration 5 is 2.300000
value of 'a' for iteration 5 is 1.399287
value of 'p' for iteration 5 is -14.335988
Press enter to get the next data from the input file

value of u entered for iteration 6 is 8.800000
value of v entered for iteration 6 is 2.800000
value of 'a' for iteration 6 is 1.644827
value of 'p' for iteration 6 is -16.664473
Press enter to get the next data from the input file

.

.

continued.

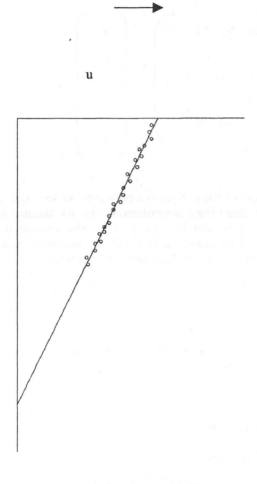

Fig. 17.17: The 2-D line from noisy 2-D points, obtained through simulation.

17.4.2.6 Construction of 3-D Points
Using 2-D Image Points

The 3-D object points are mapped onto an image plane by using the principle of perspective projection. Let the 3-D object point be P having co-ordinates $(x,y,z)^T$, which is mapped onto the image plane at point $(U, V, S)^T$. Let T be the perspective projection matrix. Then

$$\begin{pmatrix} U \\ \\ V \\ \\ S \end{pmatrix} = \begin{pmatrix} t_{11} & t_{12} & t_{13} & t_{14} \\ \\ t_{21} & t_{22} & t_{23} & t_{24} \\ \\ t_{31} & t_{32} & t_{33} & t_{34} \end{pmatrix} \begin{pmatrix} x \\ y \\ z \\ 1 \end{pmatrix}$$

where t_{ij} is the (i, j) th element of the perspective projection matrix. Let $u = U/S$ and $v = V/S$. Now, after elementary simplification, let us assume for brevity that $t_i = (t_{i1}\ t_{i2}\ t_{i3}\ t_{i4})^T$ and P is $(x, y, z)^T$. Also assume that $a = (t_1 t_{14} t_2 t_{24} t_3)^T$. For a match of an image point I with an associated scene point P, we now have the following relationships between P, u and v.

$$P^T t_1 + t_{14} \quad -u\,(P^T t_3 + 1) = 0 \quad \text{and}$$

$$P^T t_2 + t_{24} \quad -v\,(P^T t_3 + 1) = 0.$$

Now suppose we have $x_i = (u_i, v_i)^T$ and we have to evaluate $a = (x, y, z)^T$. The measurement equation is given by

$f_i(x_i, a) = 0$ yields

$$(t_1^i - u_i t_3^i)\,a + t_{14}^i - u_i t_{34}^i = 0$$

and $\quad (t_2^i - v_i t_3^i)\,a + t_{24}^i - v_i t_{34}^i = 0$

where t_j^i comes from perspective matrix T_i from camera i.

Further, $y_i = M_i\,a + w_i$
where

$$y_i = \begin{pmatrix} t_{34}^i\,u_i * - t_{34}^i \\ \\ t_{34}^i\,v_i * - t_{24}^i \end{pmatrix}$$

and $\quad \mathbf{M_i} = \begin{pmatrix} -\left(u_i\, t_3{}^i - t_1{}^i\right)^{\mathrm{T}} \\[2em] -\left(v_i\, t_3{}^i - t_2{}^i\right)^{\mathrm{T}} \end{pmatrix}$

$\partial f / \partial\, \mathbf{x_i} = \begin{pmatrix} -t_3{}^i\ a_{i-1} *\ -t_{34}{}^i & 0 \\[2em] 0 & -t_3{}^i\ a_{i-1} *\ -t_{34}{}^i \end{pmatrix}$

The following algorithm may be used for the construction of 3-D points from noisy 2-D points.

Procedure 3-D-Point-Construction (2-D image points: u,v; camera parameters: $x_o, y_o, z_o,$ A,B,C)

Input: *coordinates of the image points along with the six camera parameters determined by its position $(x_o, y_o, z_o\,)$ and orientation (A,B,C) w.r.t global coordinate system.*
Output: *the state estimate a (3×1), along with the covariance error S (3×3), associated with the estimate.*
Begin
For (no. of points: = 1 to n) do
        Initialize the Initial Covariance matrix **S** and the state estimate
                $S_0\leftarrow$ Very large initial value;
                $a_0\leftarrow$ Arbitrary Value preferably $[0\ 0\ 0]^{\mathrm{T}}$;

      For (j: =1 to no. of iterations) do
          Compute the perspective matrix from the input camera
          Parameters;
          Compute the measurement vector **y** (2×1), the linear
          transformation **M** (2×3) and **W** (2×2) the weight matrix

obtained after linearizing measurement equation from the input
parameters at each iteration ;
Initialize all the matrices involved in matrix multiplication;
Compute gain **K** using previous **S** and **M** values;
Compute covariance matrix **S** recursively using its value at
previous iteration;
Compute the state estimate 'a' recursively using its value at
previous iteration; j:= j + 1;
     End For;
End For;
End.

Traces of Procedure 3-D-Point-Construction

Input file for the above program: In this input file first two columns
contains the (u, v) co-ordinates of the image points. Next three columns
correspond to the (x_0, y_0, z_0) co-ordinates of the camera position and last three
columns represent orientation of the camera co-ordinate system (A, B, C)
w.r.t. some user selected reference co-ordinate systems. Here, A   represents
the pan angle of the camera, B represents the tilt angle and C represents the
skew angle of the camera w.r.t. the global co-ordinate system. The whole data
set comprises of 6 blocks, each of 6 rows. The first row corresponds to the
point1 from image 1, the second row for point1 from image 2 and so on for
six different images. Similarly the second block is for point 2 and so on. The
output datafile, on the other hand (row-wise), presents the estimated (x, y, z)
points. All the input and output data files are given below.

Input file: Set1.dat

-1.8	1.7	9.0	14.0	28.5	-0.262	-1.974	0.0
-1.8	2.0	18.0	15.5	28.5	0.0	-1.974	0.0
-4.4	1.7	24.0	13.5	28.5	0.0	-1.974	0.0
-1.7	1.0	26.5	13.7	28.5	0.227	-1.974	0.0
-3.2	0.8	32.5	14.8	28.5	0.262	-1.974	0.0
-5.0	-0.5	28.0	3.0	28.5	0.0	-1.974	0.0
5.5	0.7	9.0	14.0	28.5	-0.262	-1.974	0.0
6.5	1.7	18.0	15.5	28.5	0.0	-1.974	0.0
3.8	1.4	24.0	13.5	28.5	0.0	-1.974	0.0
6.3	1.7	26.5	13.7	28.5	0.222	-1.974	0.0
4.5	1.8	32.5	14.5	28.5	0.262	-1.974	0.0
1.5	-0.7	28.0	3.0	28.5	0.0	-1.974	0.0

-1.5	4.9	9.0	14.0	28.5	-0.262	-1.974	0.0
-1.5	5.2	18.0	15.5	28.5	0.0	-1.974	0.0
-3.8	4.7	24.0	13.5	28.5	0.0	-1.974	0.0
-1.5	4.2	26.5	13.7	28.5	0.227	-1.974	0.0
-2.95	3.8	32.5	14.8	28.5	0.262	-1.974	0.0
-4.6	2.3	28.0	3.0	28.5	0.0	-1.974	0.0
5.2	3.6	9.0	14.0	28.5	-0.262	-1.974	0.0
6.0	4.5	18.0	15.5	28.5	0.0	-1.974	0.0
3.7	4.4	24.0	13.5	28.5	0.0	-1.974	0.0
6.0	4.5	26.5	13.7	28.5	0.227	-1.974	0.0
4.40	4.6	32.5	14.8	28.5	0.262	-1.974	0.0
1.7	2.0	28.0	3.0	28.5	0.0	-1.974	0.0
3.8	-1.0	9.0	14.0	28.5	-0.262	-1.974	0.0
5.4	-0.5	18.0	15.5	28.5	0.0	-1.974	0.0
3.1	0.6	24.0	13.5	28.5	0.0	-1.974	0.0
5.84	-0.44	26.54	13.74	28.54	0.227	-1.974	0.0
4.7	-0.5	32.5	14.8	28.5	0.262	-1.974	0.0
1.4	-2.15	28.0	3.0	28.5	0.0	-1.974	0.0
-2.5	0.0	9.0	14.0	28.5	-0.262	-1.974	0.0
-1.5	0.0	18.0	15.5	28.5	0.0	-1.974	0.0
-4.0	-0.2	24.0	13.5	28.5	0.0	-1.974	0.0
-0.8	-0.6	26.5	13.7	28.5	0.227	-1.974	0.0
-1.8	-0.8	32.5	14.8	28.5	0.262	-1.974	0.0
-4.0	-1.8	28.0	3.0	28.5	0.0	-1.974	0.0

□

Corresponding Output file: Out1.dat

```
13.558480    48.728981    8.969839
33.009026    46.612656    10.612383
13.892322    48.551067    -1.117031
32.706348    46.120842    3.406096
33.843052    57.854473    11.443270
13.166441    60.997398    8.871586
```
□

17.4.2.7 Fusing Multi-sensory Data

Generally images of a scene are taken from different orientation w.r.t the camera co-ordinate systems. Since many cameras are used from different angles, the images need to be represented in a global co-ordinate system. The 2-D points on the image, therefore, are now transformed to the global co-ordinate systems. The correspondence between each two points on different images is then determined. One simple way to handle the correspondence problem is to first represent all 2-D points into 3-D points and then identify which point in one image has a close resemblance in another image. There are many other approaches for multi-sensory data fusion. Dempster-Shafer (D-S) theory, for instance, is one significant tool to handle this problem. The mapping of real image data onto basic probability assignments in D-S theory, however, is a problem of practical interest. Neural tools can also be used for eliminating uncertainty in multi-sensory data, but here too mapping from image domain to neural topology is a practical problem.

17.5   Conclusions

Vision systems can be functionally classified into three main levels, namely, low, medium and high level vision. AI is mostly used in the high level vision. The high level vision mainly deals with recognition and interpretation of 3-D objects from their 2-D images. There exist many approaches to interpret a scene from more than one image. Kalman filtering is one of such techniques. Its main advantage is that it employs a recursive algorithm and thus can update an estimator from input data in an incremental fashion. The vertices of points in a 2-D image can be first mapped to their 3-D locations by supplying the same 2-D points from multiple images. Now, we can construct the equation of 3-D lines from 3-D points by a second stage of Kalman filtering. Lastly, we can determine the equation of the planes containing more than one line. The spatial relationships among the planes are then analyzed to determine the 3-D planer object.

Exercises

1. Draw on a graph paper a 2 level (binary) image and verify the edge detection algorithms manually. Note that here one pixel is equal to one smallest cell on the graph paper.

2. Write a program to find edges from a given (64 x 64) gray image using Sobel masks.

3. Wrte a program to verify the gaussian filter.

4. Can you derive the gaussian filter mask from the supplied filter equations?

5. For a small image of (16 x 16) determine the eigen values and eigen vectors. Do you know any other numerical methods to compute the eigen values of large matrices? If yes, can you compute it for large images of dimension (1024 x 1024)?

6. Verify the image recognition scheme by principal component analysis.

7. Using MATLAB can you verify the self-organizing scheme for face recognition?

8. List the parameters required to compute the estimator 'a' (slope and y-intercept) for construction of a 2-D line from supplied 2-D points by Kalman filtering. Also write a program to realize this.

9. Do the same thing with the least square method. Can you identify any benefits of Kalman filter over the least square estimation?

References

[1] Ayache, N., *Artificial Vision for Mobile Robots: Stereo Vision and Multisensory Perception*, MIT Press, Cambridge, MA, pp. 7-298, 1991.

[2] Das, S. S., *3D Reconstruction by Extended Kalman Filtering*, M.E. Thesis, Jadavpur University, 1999.

[3] Dean, T., Allen, J., and Aloimonds, Y., *Artificial Intelligence: Theory and Practice*, Addison-Wesley, Reading, MA, pp. 409-469, 1995.

[4] Luo, Fa-Long and Unbehauen, R., *Applied Neural Networks for Signal Processing*, Cambridge University Press, London, pp. 188-236, 1997.

[5] Pratt, W. K., *Digital Image Processing*, John Wiley and Sons, New York, pp. 201-279, 1978.

[6] Patterson, D. W., *Introduction to Artificial Intelligence and Expert Systems*, Prentice-Hall, Englewood Cliffs, NJ, pp. 285-324, 1990.

[7] Schalkoff, R. J., *Artificial Neural Networks*, McGraw-Hill, New York, pp. 308-334, 1997.

18
Linguistic
Perception

Building perception about the syntactic, semantic and pragmatic usage of natural language is a complex and tedious process. We require several years to learn and use them. Realizing 'linguistic perception' on machines too is a hard problem. This chapter presents various levels of analysis of natural language with special reference to English and formalizes a few concepts through illustrations to make the readers understand their importance in building natural language understanding programs.

18.1 Introduction

The process of building perception by human beings involves construction of an internal representation of knowledge (and concepts) from the sensory information acquired by them. The phrase 'linguistic perception' refers to building perception to enhance our capability to understand and reason with linguistic information. Linguistic information could be of two basic forms: i) spoken and ii) written. Written information generally is more structured and free from much noise than spoken words. The spoken information includes many semantically incomplete clauses than the listener can follow because of his subjective knowledge on the topics being discussed. Written semantics, on the other hand, when meant for non-specialized readers, require less semantic skills of the readers. While talking of the linguistics, we would often use the words 'syntax' and 'semantics', where the former implies the grammar involved and the latter corresponds to the association of the words in a sentence for its understanding. The act of understanding a sentence in a natural language thus requires background knowledge of the syntax of the language and the concept on the subject. For instance, a sentence: 'A cow ate a tiger' is syntactically correct but semantically wrong. A child having no perception about a cow and a tiger cannot determine the semantic weakness of this sentence. Forming perception of the world comprising of animals like cows and tigers, therefore, should precede the act of understanding the sentence.

This chapter covers in detail the principles of representing syntax and semantics of natural languages by specialized data structures and computational tools. For the purpose of illustrating the concepts, we take a fragment of the English language as the domain of our case study. Besides syntax and semantics, a natural language requires the following types of analysis for its comprehension. These are i) prosody, ii) phonology, iii) morphology, iv) pragmatics and v) world knowledge. An overview to these types of analysis is outlined below.

Prosody: It deals with the rhythm and intonation of a language. Rhythm is often used in the babbling of infants and children's wordplay. In religious ceremony and poetic competition, the importance of rhythm is felt. Unfortunately this type of analysis is difficult and is thus ignored in most natural language understanding programs.

Phonology: It examines the sounds of the words that are combined to form sentences in a language.

Morphology: A 'morphene' is the smallest component of words. Morphology deals with the rules that allow adding prefix and suffix to already known words. Morphological analysis is useful for identifying the role of a word in a sentence including tenses, number and part of speech.

Pragmatics: It describes the study of the ways by which the language is expressed. It also considers the effect of words on the listener. As an example, if someone asks "What is your date of birth?", one should answer a date and this is expected.

World Knowledge: The world knowledge stands for the domain knowledge of the environment, without which the semantic understanding of the sentences is difficult.

The types of analysis mentioned above do not have clear-cut boundaries. In fact, this partition is artificial with a motivation to represent the psychological aspects of understanding the natural languages. The first step to understand natural languages is parsing, which analyses the syntactic structure of sentences. It serves two basic purposes. First it determines the linguistic relations such as subject-verb, etc. and finally checks the syntactical correctness of the sentences. Generally, the parser learns the sentence as a structured object (tree/graph) for subsequent semantic analysis.

18.2 Syntactic Analysis

In this section we shall discuss three distinct methods for syntactic analysis of sentences. These are: i) by using context free grammar, ii) by employing transition network and iii) by using context sensitive grammar.

18.2.1 Parsing Using Context
Free Grammar

A context free grammar G is generally defined by a 4 tuple, given by G = $\{V_n, V_t, S, P\}$, where V_n is a set of non-terminal symbols, V_t denotes a set of terminal symbols, S denotes the starting symbol and P represents a set of production rules that cause a change. It may further be noted that $V_n \cup V_t$ denotes the entire set of symbols. The above definition of context free grammar is prevalent in automata theory and compilers. The tuples in G are illustrated below with an example.

Example 18.1: Let us consider the following sentence and illustrate the parameters involved in the definition of G in connection with the problem of natural language understanding.

Sentence: *The man killed a deer.*

Here, V_t = {the, man, killed, a, deer}

   V_n = {noun-phrase, verb-phrase, article, noun, ---}.

   S = initial symbol

   P: 1) S → noun-phrase verb-phrase,

      2) noun-phrase →noun,

      3) noun-phrase → article noun,

      4) verb-phrase → verb,

      5) verb-phrase → verb noun-phrase,

6) article → a,

7) article → the,

8) noun → man,

9) noun → deer,

10) verb → killed,

11) verb → likes.

It is to be noted from the above set of production (re-write) rules that the terminals do not appear in the left-hand side of the rules.

The sentence: '*The man killed a deer*' now can be analyzed by the following sequence of re-write rules. We here onwards abbreviate noun-phrase as NP, verb-phrase as VP, noun as N, verb as V and article as ART. Now, starting with re-write rule 1, we have:

S → NP VP	(by 1)
→ ART N VP	(by 3)
→ the man VP	(by 7 & 8)
→ the man V NP	(by 5)
→ the man killed NP	(by 10)
→ the man killed ART N	(by 3)
→ the man killed a deer	(by 6 & 10)

The above example describes a **top-down** derivation. It starts with the sentence symbol S and continues replacing the left-hand side of the selected re-write rules by their right-hand side, so that ultimately the complete sentence appears at the end of the derivation. An alternative form called the **bottom-up** derivation starts with the string of the sentential representation and continues replacing the right-hand side of the re-write rules by their left-hand side until the starting symbol S is reached.

A tree, called the **parse tree**, can also represent the derivations presented above. The parse tree for the sentence under consideration is presented in fig. 18.1.

Fig. 18.1 has similarity with the derivation of the sentence, presented in example 18.1. It is however observed from fig. 18.1 that VP and NP occur more than once at different levels of the tree. This happened so as we did not expand VP or NP at its first occurrence. Had it been so, the resulting tree would look like fig. 18.2.

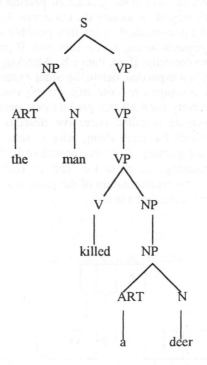

Fig. 18.1: The parse tree for the sentence '*the man killed a deer*'.

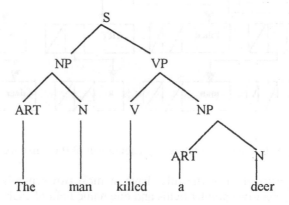

Fig. 18.2: The modified parse tree.

One important issue in the process of parsing is the selection of the re-write rules at any step of expanding the parse tree. One simple way to explore this is to expand a non-terminal by the first possible means and check whether the expansion supports the part of the sentence. If yes, the selection of the re-write rule is done correctly. If no, then a backtracking has to be made upto the node in the tree, just expanded, delete the older expansions from that node and select a second alternative re-write rule for expansion. The process is to be continued recursively until all the productions terminate to terminal nodes. Such type of parsing is called **recursive descent parsing**. An alternative approach is to check the input string prior to selection of the re-write rule, called **look-ahead parsing**. It is also prevalent in the literature [6]. For the purpose of computing, the parse tree can be realized by a structure with pointers and list. The representation of the parse tree of fig.18.2 by pointers to structures and list is presented below.

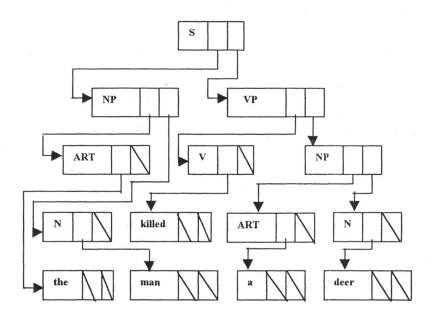

Fig. 18.3: Pointer to structure representation of the parse tree of fig. 18.2.

Since the re-write rules have at most two symbols at the right hand side, we kept two pointer fields and one name field of each structure. The null pointers have been denoted by crossed lines within the field.

The list representation of the parse tree, shown in fig.18.2 is presented now.

```
( S ( NP ( ( ART  the )
        (N  man ) )
   ( VP ( V killed )
        ( NP ( ART a )
            ( N deer ) ) ) ) )
```

The re-write rules we presented earlier can handle only a narrow class of English sentences. A further extension of the grammar is possible by the inclusion of the following determiners (DET), adjectives (ADJ), auxiliary verbs (AUX), adverbs (ADV), prepositions (PREP) and prepositional phrases (PP).

$$PP \rightarrow PREP\ NP$$
$$VP \rightarrow V\ NP\ PP$$
$$VP \rightarrow V\ PP$$
$$VP \rightarrow V\ ADV$$
$$VP \rightarrow AUX\ V\ NP$$
$$DET \rightarrow ART\ DET \rightarrow ART\ ADJ$$

The augmentation of these re-write rules in P helps constructing the parse tree for the following type of sentences. '*The man killed the deer with a gun*' or '*The beautiful girl danced Kathak to receive appreciation from the audience*'. One point should be added before concluding this section. Parsing can be carried out for semantically wrong sentences as well. For instance, the sentence like: '*The deer killed the man with a stick*' could be parsed correctly, though it does not have logically correct semantics.

18.2.2 Transition Network Parsers

Transition networks are used in 'automata theory' [4] to represent grammar. The process of parsing can be done analogously by using transition nets. A transition net, used for parsing, comprises of nodes and labeled arcs where the nodes denote states and arcs correspond to a symbol, based on which a transition is made from a given state (node) to the next state. The re-write rules and the corresponding transition networks are presented below.

Sentence:

Fig. 18.4(a): Transition network for 'sentence' constructed using
S → NP VP.

NP:

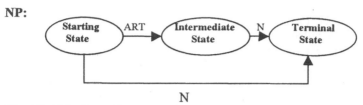

Fig. 18.4(b): Transition network for noun phrase, constructed
from re-write rules: NP→ART N and NP→N.

VP:

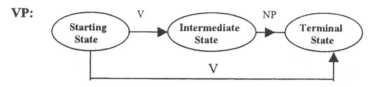

Fig. 18.4(c): Transition network for verb phrase using the re-write
rules: VP→V. NP, VP→V.

For analyzing sentences with transition networks, one has to first start
with the transition networks for sentence S. In case the sentential
representation matches with the part (noun-phrase) of the transition network,
the control jumps to the transition network for NP. Suppose, there exists an
article and noun in the beginning clause of the sentence. So, the transition
network for NP succeeds, transferring the control to the transition network for
sentence at the intermediate state and then proceeds to check the verb-phrase.
In the transition network for VP, suppose the entire VP matches. The control
then jumps to the terminal state of the sentence and stops. The transition
diagram for the sentence: '*The man laughed*' thus looks like as in fig.18.5.

The transition network, shown in fig.18.5, includes 10 outlined line
segments, depicting the flow of control within the states of the network. First,
when the control finds NP at S_0, it moves to NP transition network via dotted
arc 1. Now, again finding ART at S_0 of NP transition network, it moves to
ART network via dotted arc 2. It selects 'the' from the ART transition
network and returns to S_1 of NP network via dotted arc 3. Now, at S_1 of NP, it
finds N and accordingly moves to the starting state S_0 of transition network N
via arc 4. The process continues in this manner, until S_2 of 'sentence transition
network' is reached via arc 10. The traversal on the transition networks is not
always smooth, as shown in fig.18.5. There exist circumstances, when the
control moves to wrong states and backtracks to the previous state, when it
fails to match symbols. To illustrate backtracking in the transition network,
we attempt to represent the sentence '*dogs bark*' by this nomenclature.

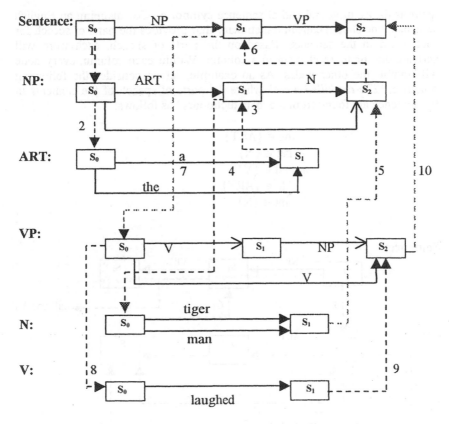

Number against the dotted arc denotes order of transversal.

Fig. 18.5: Trace of '*the man laughed*' by transition network parsing.

The backtracking is needed in fig. 18.6 to return the control to S_0 of NP from S_0 of ART, as no articles are found in the sentence: 'dogs bark'.

18.2.3 Realizing Transition Networks with Artificial Neural Nets

This section presents an alternative representation of transition networks by artificial neural networks (ANN). For the simplicity of design, we consider exhaustive connectivity between neurons of each of two successive layers. Fig. 18.7 describes a multi-layered neural net with such exhaustive connectivity. The i-th layer of neurons in this figure corresponds to the determiner d_i, adjective a_i, noun n_i, verb v_i and start /end of sentence symbol e_i. The zeroth layer, for instance, comprises of d_0, a_0, n_0, v_0 and the start of sentence symbol e_0. The output layer (here the 6-th layer), here, comprises of neurons d_6, a_6, n_6, v_6 and end of sentence symbol e_6. For mapping an English sentence, one should consult a dictionary that provides the parts of speech for each word in the sentence. Based on the parts of speech, each word will activate one or more element in a column. Within each column, every node will inhibit the other nodes. As an example, let us consider the following sentence: '*The cat killed a rat*', where the parts of speech of the elements in the sentence are found from the 'data dictionary' as follows:

$$the \in \{ART\}$$
$$cat \in \{N\}$$
$$killed \in \{V\}$$
$$a \in \{ART\}$$
$$rat \in \{N\}$$

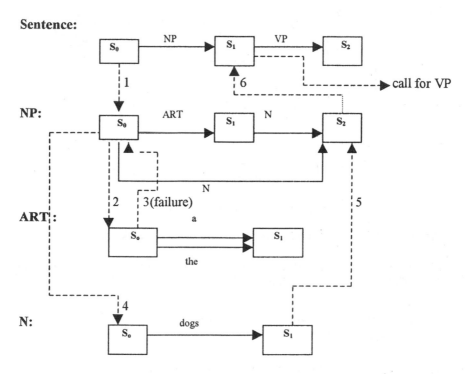

Fig. 18.6: Backtracking from S_0 of ART transition network to S_0 of NP due to failure in symbol matching for articles in '*dogs bark*'.

Here, e_0 in the zeroth layer, a_1 in the first layer, n_1 in the second layer, v_3 in the third layer, a_4 in the fourth layer, n_5 in the fifth layer and e_6 in the sixth layer are activated by the input parts of speech of the words in the sentence.

18.2.3.1 Learning

For adaptation of weights in the neural net, shown in fig.18.7, one simple method is to first map the patterns from different sentences onto the nodes of the neurons and then update the weights by employing the principles of bi-directional associative memory (BAM) [5]. For instance, suppose we have two sentences:

1. *A cat killed a rat.*
2. *A man threw a stone.*

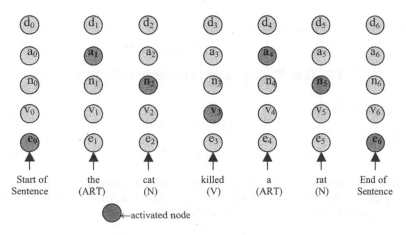

Fig. 18.7: A neural net with exhaustive connectivity (not shown for clarity) between nodes of each two successive layers.

The activated nodes in the neural net of fig.18.7 for these patterns may be evaluated. The activated node i will have a node value $a_i=1$ and the rest of the non-activated nodes in the net will have node value=-1. The weight w_{ij} from neuron n_i to n_j is now given by

$$w_{ij} = \sum_{\forall \text{ pattern k}} (a_i. a_j)_k \qquad (18.1)$$

In this example we have only two patterns; so k has a maximum value 2. Thus

$$w_{ij} = (a_i.a_j)_1 + (a_i.a_j)_2 \qquad (18.2)$$

18.2.3.2 Recognition

For recognizing the parts of speech of a missing word in a sentence, the following recognition rules may be employed.

$$net_j(t) = \sum_{\forall i} w_{ij}.a_i(t) \qquad\qquad (18.3)$$

where $net_j(t)$ denotes the net input to node j at time t, $a_i(t)$ is the activation level of node i at time t and w_{ij} is the weight connected between neuron i and j [8].

The non-linearity associated with the neurons is presented in fig.18.8.

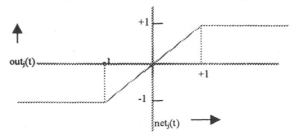

Fig. 18.8: The nonlinearity of neurons in fig.18.7.

When $net_j(t) > 0$,
$$a_j(t+1) = a_j(t) + out_j(t)(1 - a_j(t)) \qquad\qquad (18.4)$$
When $net_j(t) < 0$,
$$a_j(t+1) = a_j(t) + out_j(t) a_j(t) \qquad\qquad (18.5)$$

We now present the recognition process with an example. Suppose we have already trained the neural net for the two sentences, mentioned earlier. Now, we present a sentence with two missing words:

*A **man** kicked a ball.*

where the bold words are missing from the above sentence. In the sentence with those two missing words, when mapped onto the proposed neural network, the parts of speech of the dropped words can easily be recognized [8]. The recognition process will result in a large positive value (close to one) for the nodes n_2 and a_4 in the neural net of fig.18.7.

18.2.4 Context Sensitive Grammar

It should be pointed out once again that the context free grammar parses sentences based on the syntax of the rewrite rules supplied. Thus syntactically correct but semantically ill sentences are also parsed by the context free grammar. A context sensitive grammar, on the other hand, can check the singularity/plurality of a noun and its corresponding verbs or the persons ($1^{st}/2^{nd}/3^{rd}$) used in a noun and the corresponding verbs in a sentence. The context sensitive grammar is thus more useful than its context free counterpart.

One important point needs mention about the context sensitive grammar. Unlike the context free grammar, the context sensitive counterpart can have more than one literal in the left-hand side of the re-write rules. One typical re-write rule that illustrates this is: x y z→x w z, which means if y is between x and z, replace it by w.

For the sake of simplicity, we present below a fragment of the context sensitive grammar that enforces number agreement between article and noun and between subject and verb, as presented below:

1. S → NP VP
2. NP → ART NUM N   //NUM=Number//
3. NP → NUM N
4. NUM → singular
5. NUM → plural
6. ART singular → a singular
7. ART singular → the singular
8. ART plural → the plural
9. ART plural → some plural
10. Singular N → bird singular
11. Plural N → birds plural
12. singular VP → singular V
13. plural VP → plural V
14. singular V → flies
15. singular V → loves
16. plural V → fly
17. plural V → love

Now, with this context sensitive grammar that checks singularity/plurality, we want to parse the sentence: *'Birds fly'*. The derivation includes the following steps.

1.	S	
2.	NP VP	(by rule(1))
3.	NUM N VP	(by rule(3))
4.	Plural N VP	(by rule(5))
5.	birds plural VP	(by rule(11))
6.	birds plural V	(by rule(13))
7.	birds fly	

The context sensitive grammar presented in this section is meant for a limited vocabulary and includes checking of persons, numbers, etc. Then matching time of the re-write rules' left part with sentential representation (in top-down parsing) will be significantly large. An alternative approach to parsing is to employ augmented transition networks that can define context sensitive languages, but is more powerful than the context sensitive grammar.

18.3 Augmented Transition Network Parsers

Augmented transition networks (ATN) [6] extend transition networks by attaching procedures to the arcs of the networks. When an arc is traversed, the ATN parser executes the procedure, attached to that arc. These procedures i) assign values to grammatical features, and ii) check whether the number or person (1st, 2nd, 3rd) conditions are satisfied and accordingly allow or fail a transition. For instance, the ATN that may be employed for parsing part of the sentence 'the boy' is presented below. Suppose the grammatical characteristics or features of the words 'NP', 'the' and 'boy' are available in a data dictionary, as shown in fig.18.9. Also assume that the procedures are attached to the arcs of the article and noun, as described below.

NP:

DET	Noun	Number

(a)

the:

Part of Speech	Root	Number
ART	the	Singular/ Plural

(b)

boy:

Part of Speech	Root	Number
N	boy	Singular

(c)

Fig. 18.9: The grammatical characteristics of the word (a) NP, (b) the and (c) boy, represented by structures.

Now, for parsing the string sequence: 'the boy', suppose we use the ATN for NP. So, at state S_0 of NP, it first assigns ART:= the and checks whether the part-of-speech = ART. If yes, it assigns NP.DET:=ART; otherwise it fails and the control returns to wherefrom it was called.

Generally, we start the sentence S that contains NP and VP. So, the control from NP ATN should return to starting state of the ATN for S.

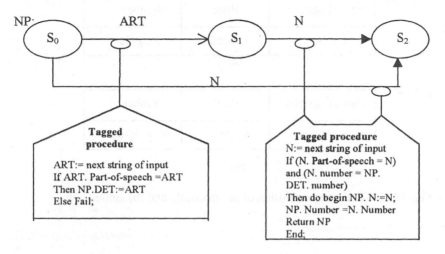

NP:

S_0 — ART → S_1 — N → S_2

N

Tagged procedure

ART:= next string of input
If ART. Part-of-speech =ART
Then NP.DET:=ART
Else Fail;

Tagged procedure
N:= next string of input
If (N. **Part-of-speech** = N)
and (N. number = NP.
DET. number)
Then do begin NP. N:=N;
NP. Number =N. Number
Return NP
End;

Fig.18.10: The ATN grammar for NP used to check number agreement to build parse trees.

Thus, if we have the structural definitions of the ATNs for sentence, NP,VP, data dictionary definitions of the words 'the', 'crocodile' and 'smiled' and known procedures at the arcs of the transition nets for NP and VP, we shall be able to parse the tree employing them. An example is given below to parse the tree for the sentence: *the crocodile smiled.*

Some of the required definitions have already been presented; the rest are presented in fig.18.11, 18.12 and 18.13 below.

Sentence:
(S)

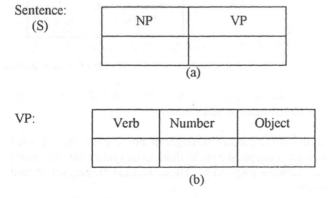

NP	VP

(a)

VP:

Verb	Number	Object

(b)

Fig. 18.11: Structure definitions for (a) sentence and (b) VP.

Crocodile:

Part of speech	Root	Number
N	Crocodile	Singular

(a)

Smiled:

Part of speech	Root	Number
V	smile	Sing./Plural

(b)

Fig. 18.12: Dictionary definition of (a) crocodile and (b) smiled.

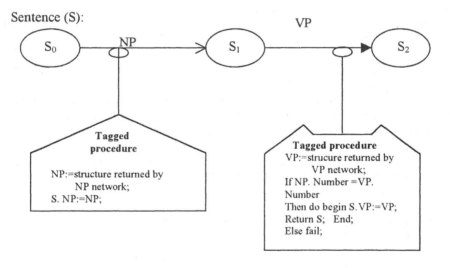

Sentence (S):

Fig. 18.13 (a): The ATN grammar of a sentence that checks number agreement.

The parse tree for the sentence 'the crocodile smiled' can now be drawn (fig. 18.14) following the above definitions.

In constructing the ATN parsers throughout our discussion, we used conceptual graphs. Besides conceptual graphs there exist quite a large number of other techniques for constructing ATN. These include frame, script and level logic representations.

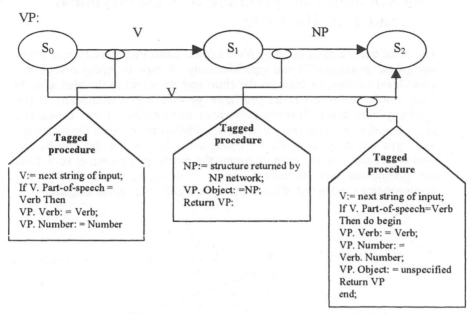

Fig.18.13(b): The ATN grammar for VP that checks number agreement.

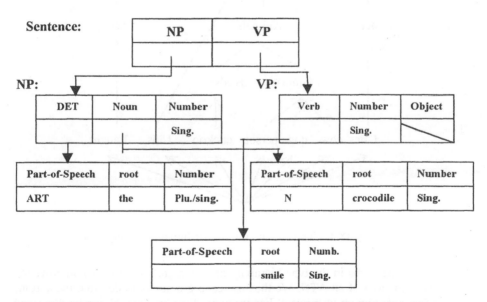

Fig. 18.14: The parse tree for the sentence 'the crocodile smiled', built with ATN grammar.

18.4 Semantic Interpretation by Case Grammar and Type Hierarchy

The parse tree we constructed by ATN can be semantically interpreted [3] by using case grammars [2] and type hierarchy. A type hierarchy describes a subsethood relationship between the child and its parent in the hierarchy. In other words, the parent nodes are more generic than the children. It has similarity with lattice diagrams [1], where there exists a lower bound and upper bound to such hierarchy and a strict relation to construct the hierarchy. The 'type hierarchy' is useful for constructing knowledge about verbs and their object space, if any, or to determine the meaning of a noun or verb, from the generalized and specialized events / items in the hierarchy. For demonstrating the concept of hierarchy, we consider the animal kingdom.

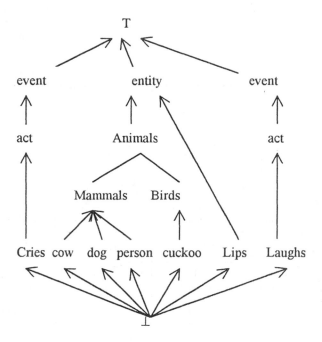

Fig. 18.15: The type hierarchy for the animal kingdom.

The verbs in the type hierarchy have a significant role for interpreting the semantics of an ATN. We thus use case grammars to describe the actions of an agent by an instrument. For instance, the verb 'laugh' can be described as an action of an agent by his lips to give himself and others enjoyment. Thus we can represent laugh as a case frame as presented below in fig. 18.16.

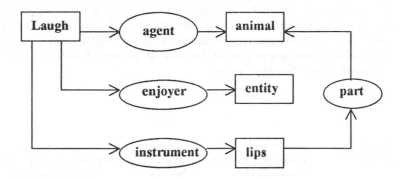

Fig. 18.16: Case frame for the verb 'laugh'.

In fig. 18.16, we presented the description of the world by rectangular boxes, while the relationship between each of two boxes is done by a relation represented by an 'ellipse'.

The following points may be noted for describing semantics from an ATN.

i)   While starting with the sentence S in the ATN, determine the noun phrase and verb phrase to get a representation of the noun and the verb.   Bind the noun concept with the subject (agent) of the corresponding case frame.

ii) While processing the noun phrase, determine the noun; the singularity/plurality of the article and bind marker to noun concept.

iii) While processing the verb phrase, determine the verb. If the verb is transitive, then find its corresponding noun phrase and declare this as object of the verb.

iv) While processing the verb, retrieve its case frame.

v)   While processing the noun retrieve the concept of noun.

The following example illustrates the principle of generating semantic inferences from a dynamically expanding ATN.

Example 18.2: The sentence calls the noun phrase and the noun phrase in turn calls the noun. The noun returns a concept for the noun crocodile. As the article is definite, the NP instantiates an individual to the concept. Next the VP is called, which in turn calls verbs, the case frame of which is known. Since the verb is intransitive, we need not bind an object to it. The semantics derived from the parse tree of fig.18.4 is presented here [fig.18.17], by replacing 'smile' there by 'laugh'.

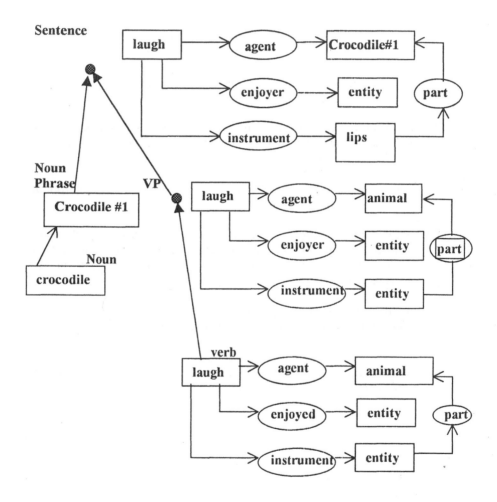

Fig.18.17: Semantic derivation from the parse tree of fig.18.14, with the replacement of smile by laugh.

18.5 Discourse and Pragmatic Analysis

Syntactic and semantic analysis are essential but not sufficient to understand natural languages. The discourse and pragmatic concept [7] in which a sentence is uttered are equally useful for understanding the sentence. For instance, consider the following dialogs.

Dialog 1: *Did you read the AI book by Konar? The last chapter on robotics is interesting.*

Dialog 2: *Mr. Sen's house was robbed last week. They have taken all the ornaments Mrs. Sen possessed.*

Dialog 3: *John has a red car. Jim wanted it for a picnic.*

In the last three examples, the first sentence asserts a fact or a query and the second sentence refers to it directly or indirectly. In example 1, 'the last chapter' refers to the last chapter of Konar's book. The subject 'they' in the second example corresponds to the robbers. The word 'it' in example 3 refers to John's car. It is thus evident that to represent part or whole of entities and actions, we often refer to it by a word or phrase.

Programs that can accomplish multiple sentence understanding rely on large knowledge bases, which is difficult to construct for individual problems. Alternatively, a set of strong constraints on the domain of discourse may be incorporated in the program, so that a more limited knowledge base is sufficient to solve the problems. To realize the pragmatic context in the programs, the following issues may be considered.

a) **Focussing the relevant part in the dialog:** While understanding natural language, the program should be able to focus on the relevant parts of the knowledge base available to it. These knowledge bases may then be employed to resolve ambiguity among the different parts of the uttered message. For example, in the first noun phrase in the sentence: 'the last chapter on robotics is interesting', the knowledge base should identify the phrase 'the last chapter' and determine its significance in connection with a book (see dialog 1). The 'part-whole' relationship thus should be stressed and the related rules are to be checked for firing.

b) **Modeling individual beliefs:** In order to participate in a dialog the program should be efficient to model the beliefs of the individuals. The **modal logic** should be used to represent such beliefs. We illustrate the use of two modal operators in this regard. The first one is Believe (A,P), which is true, when A believes that the proposition P is true. The other operator is Know (A,P), which means Know (A,P) is true when A knows P is true. A formal relationship between the two modal operators is: Believe (A,P) $\wedge$ P $\rightarrow$ Know(A,P), which means that if A believes P to be true and P is true then 'A knows P' is true. It is to be noted that Believe(A,P) may be true, even when P is false.

An alternative way to represent individuals' belief is to partition the knowledge base in a manner so that it would use those rules for firing that instantiate the shared belief and ignore the contradictory beliefs of the participants. One part of the knowledge base that support shared beliefs thus will be active.

c)   **Identifying goals and plans for understanding:** The third pragmatic context to understand natural language is to identify the other participants' goals and the plans to realize the goals. The understanding system should keep track of the other's goal throughout the dialog. For example, suppose the participant says 'I am planning to visit Delhi for an interview. Tickets, however, have not been booked yet.' From these sentences opponents should realize that the speaker's goal is to visit some place and he is talking of air tickets. But if the goal is missed the second sentence will be difficult to understand, as it can have many alternative meanings. In this example the word 'plan' was explicitly mentioned. However, there are situations, when one has to determine the goal and the plan from implicit description of the problem. For instance, consider the following story *"John's mother was severely ill. She was admitted to a nursing home. John will go to see his mother this evening."*

Here the goal is to visit a nursing home, which is not explicitly mentioned in the paragraph. The program used to understand the above spoken sentences, should thus take care to identify the goal and then continue reasoning in the direction of the goal.

d)   **Speech acts:** This refers to the communicative acts that the speaker wants to realize. For instance, suppose the speaker wants to get his pen from the drawer to sign his banker's cheque. He requests his spouse to bring the pen for him. This speech act presumes that his spouse knows where the pen is kept. Let A be the person, S be the spouse and P be the pen. Then by modal logic [7] we represent it

Request (A, S, P)
Precondition: Believe (A, knows-whereabout (S, P))
                Believe (A, willing-to-do (S, P)

Thus the elements of communicative plans can be formalized and used for executing a task.

18.6 Applications of Natural Language Understanding

Natural language understanding programs have wide usage starting from commercial software to high performance VLSI CAD tools design. In the long run it will be part and parcel of most of the software. It is mainly required as a front-end tool for communication with the users. One of the first successful applications of the natural language understanding program was in a 'question answering' system. In this system, a paragraph is first supplied to the machine, which constructs an internal representation of the paragraph and then matches the question part with the internal representation to answer it. Winograd's SHRDLU [9] system in this regard needs special mention. In database management systems, the natural language understanding program is equally useful for building front-end tools. The user may submit his query in plain English and the machine will answer him in English as well. In expert systems, natural language processing is very much useful for submitting many facts to the database /working memory. For augmenting knowledge in the expert systems, this will also play a vital role. In VLSI CAD tool design, the users are expert electronic or electrical engineers. They do not want to remember rigid format of the syntax of the end-users' command. So, if a natural language interface is present as a front tool, they will be able to save their valuable time and consequently will improve their productivity. In banking, library management, post office and telephone diagnostic software the natural language understanding program will be employed as a front-end tool in the short run.

Exercises

1. Draw the parse trees for the following sentences:

    a) The boy smoked a cigarette.
    b) The cat ran after a rat.
    c) She used a fountain pen to write her biography.

2. Parse the following sentence by a transition network parser: The man reacted sharply.

3. Design a formalism to represent each word of a given vocabulary by a numerical code, so that any two distinct words should not have the same code. Now, with this coding arrangement, can you use the Back-propagation algorithm to understand sentences?

4. Using the ATN grammar, show how to check the number agreements between the noun phrase and the verb phrase in the following sentence: He swims across the river.

5. Represent the following sentence in modal logic: I know that you know that he does not believe it.

References

[1]   Dougherty, E. R. and Giardina, C. R., *Mathematical Methods for Artificial Intelligence and Autonomous Systems*, Prentice-Hall, Englewood Cliffs, NJ, chapter 2,1988.

[2]   Filimore, C., The case for case, In *Universals in Linguistic Theory*, Bach, E. and Harms, R. T., Eds., Holt, New York, 1968.

[3]   Hendrix, G. G., Sacerdoti, D., Sagalowicz, D. and Slocum, J., "Developing a natural language interface to complex data, *ACM Trans. on Database Systems*, vol. 3, pp. 105-147, 1978.

[4]   Kohavi, Z., *Switching Circuits and Finite Automata Theory*, McGraw-Hill, New York, 1983.

[5]   Kosko, B., "Adaptive bidirectional associative memories," *IEEE Trans. on Systems, Man and Cybernetics*, vol. SMC-18, pp. 42-60, 1988.

[6]   Luger, G. F. and Stubblefield, W. A., *Artificial Intelligence: Structures and strategies for complex problem solving*, The Benjamin / Cummings Publishing Company, Menlo Park, CA, pp. 396-425, 1993.

[7]   Rich, E. and Knight, K., *Artificial Intelligence*, McGraw-Hill, New York, pp. 377-426, 1996.

[8]   Tveter, D. R., *The Pattern Recognition Basis of Artificial Intelligence*, IEEE Computer Society Press, Los Alamitos, pp. 261-319, 1998.

[9]   Winograd, T., *Language as a Cognitive Process: Syntax*, Addison-Wesley, Reading, MA, 1983.

19

Problem Solving by Constraint Satisfaction

The chapter presents three different types of constraint satisfaction problems, dealing with algebraic, logical and geometric constraints. The propagation of constraints through constraint nets to solve a given algebraic problem is illustrated with examples. An algorithm for testing constraint satisfiability is introduced. The principles of constraint logic programs have been outlined with examples. The junction labeling problem of trihedral objects has also been covered in sufficient details. Examples have been taken from electrical circuits, crypto-arithmetic and map coloring problems.

19.1 Introduction

Many of the real world problems demand the solution to satisfy a set of algebraic or logical conditions, called constraints. The class of problems, whose search space is constrained by a set of conditions, is generally referred to as 'Constraint Satisfaction Problems (CSP)' [1]-[8]. We have come across quite a large number of CSP problems since our childhood. Linear optimization problems where we want to maximize $Z= \sum c_i x_i$ subject to a set of linear constraints of the form: $\sum a_i x_i \le b$, for instance, is a CSP.

Here a_i, b, c_i, and $x_i \in \Re$ and one has to determine x_i s for known constants a_i and b, so that $Z=\sum c_i x_i$ is maximized. The search space of the solution x_i s, here, is constrained by the set of linear inequalities. Making a change for a 10$ note is also a CSP problem, where the constraints for the solution x_i s are given by $\sum c_i x_i = 10\$$ and $\exists x_i \geq 0$, where x_i denotes the valid notes 1$, 2$ and 5$ and c_i denotes the count of x_i. For example, if we consider x_1,x_2 and x_5 to correspond to 1$, 2$ and 5$ notes and c_1, c_2, c_5 be their respective counts, then a few possible solutions are $\{x_5 =2\}$ or $\{x_5= 1 ,x_2 =2,x_1= 1\}$ or $\{x_2=5\}$ or $\{x_1 =10\}$ or $\{x_2= 4,x_1= 2\}$. The last two examples involve the algebraic constraints. Many formulations of the problems, however, require satisfying logical or geometric constraints. For example, finding the grand-fatherhood relationship between X_3 (=d) and X_1 (=l) from the definition of grandfather and father, discussed below, is a CSP. To be more specific, suppose, given the following set of constraints:

        1.Grandfather (X_3,X_1) :-
            Father (X_2,X_1),
            Father (X_3,X_2).
        2. Father (d, r).
        3. Father (r, l).

We want to determine: Grandfather (X_3, X_1) ?

    The above problem can be solved by constructing an SLD-resolution tree, which finally yields the solution: Grand-father(d, l). Problems that deal with such logical constraints also belong to the category of CSP. Before switching to the next section, we present an example of geometric constraint satisfaction problem. Suppose, we want a set of three dimensional objects like a i) Box , ii) Pyramid, iii) Cone (see fig.19.1) and we recognize the box among them by a set of geometric constraints, presented below.

    1.   (Object = Box) :-
        Has-no-of- vertices = 8, No-of-planes-meeting-at-each-vertex =3,
        Angle-between-any- two-planes-meeting- long-an-edge =90°.

    2.   (Object = Pyramid) :-
        Has-no-of- vertices =4,
        No-of-planes- meeting- at-each-vertex $\geq$ 3,
        Angle-between-any-two-planes-meeting-along-an-edge $<90^\circ$.

    3.   (Object = Cone) :-
        Has-no-of -vertices =1,
        Has-no-of-circular-plane-surface =1,
        Has-curved-surface = 1.

The recognition problem is the present content needs to check the geometric feature of the object under test and determine the object by satisfying the premises of one of the above three constraints.

While dealing with CSP we should consider the following three issues:

i) whether the constraints are satisfiable, i.e., does the problem has solutions that satisfy all the constraints?

ii) simplification of the constraints by equivalent simpler ones,

iii) optimization of the solution based on some other criteria, if more than one solution exists.

19.2 Formal Definitions

We now present a few definitions, which will be referred to throughout the chapter.

Definition 19.1: Constraints are mathematical/ logical relationships among the attributes of one or more objects.

To understand the above definition, let us consider the 'make-changes of 10$' problem, where the 1$, 2$, 5$ notes are objects and their counts are the attributes in the problems. Thus the mathematical relationship among them is the expression $\sum_{\exists x_i} c_i x_i = 10\$ $, which is the constraint.

Definition 19.2: The legitimate set of operators like '+' , '* ', '∧' or '∨' and the type of the variables and their domains, the type of functions used and their range and the arguments of the operators are specified by a domain, called **constraint domain**.

Formally, the constraint domain D is a 5 tuple, given by

$D=\{v, f, op, d, r\}$

where   v denotes the set of variables,
        f denotes the set of functions,
        op stands for the set of legal operators to be used on variables or functions,
        d is the domain of variable, and
        r is the range of the functions employed in the constraints.

Definition 19.3: A **primitive constraint** comprises of one or more constraint relation symbols together with their arguments, but cannot have a conjunction ($\wedge$) in it.

For example, $x \leq 2$, $(x + y) \leq 2$ are primitive constraints.

Definition 19.4: A **non-primitive** or **generic constraint** is a conjunction ($\wedge$) of primitive constraints.

For example, $(x + y \leq 2) \wedge (y \leq 3)$ is a non-primitive constraint.

Thus formally a constraint of the form $C_1 \wedge C_2 \wedge \ldots\ldots\ldots\ldots \wedge C_n$, $n > 1$ is a non-primitive constraint, when C_1, C_2, $\ldots\ldots\ldots C_n$ all are primitive constraints [4].

Definition 19.5: A **valuation** θ for the set of variables v is an assignment of the values from the constraint domain to the variables v [4]. An expression E having variables v is given a value $\theta(E)$, computed by replacing the variable by its assigned value in the expression.

For example, let $v = \{v_1, v_2\}$; the domain of v_1, v_2 being $1 \leq v_1$, $v_2 \leq 2$; let $E = v_1^2 + v_2^2$; now $\theta(E) = [v_1^2 + v_2^2]_{v1 = x, v2 = y}$ where $1 \leq x, y \leq 2$.

Definition 19.6: A constraint is called **satisfiable**, if there exists at least one solution satisfying the constraint.

For instance, $\{(x + y \leq 2) \wedge (0 \leq x, y \leq 2)\}$ are satisfiable constraints for integer x, y as there exists a solution $x = y = 1$ that satisfies the constraints. On the other hand, $\{(x + y \leq 2) \wedge (x > 2) \wedge (y > 2)\}$ does not have a solution for integer x, y. Consequently, constraint $(x + y \leq 2) \wedge (x > 2) \wedge (y > 2)$ is called **unsatisfiable.**

Definition 19.7: Two constraints C_1 and C_2 having same solution set are called **equivalent** and the equivalent relation between C_1 and C_2 is denoted by

$C_1 \leftrightarrow C_2$.

For example, $C_1 = \{(x + y \leq 2) \wedge (x = 1) \wedge (y < 2) \wedge (y > 0)\}$

and $C_2 = \{(x + y \leq 2) \wedge (0 < x) \wedge (x \leq 2) \wedge (0 < y) \wedge (y \leq 2)\}$

are equivalent when x, y are elements of positive integer sets.

CSP deals with two classical problems: i) *solution problem* and ii) *satisfaction problem*. The solution problem is concerned with finding the solutions satisfying all the constraints. The satisfaction problem requires clarification about the existence of a solution. An algorithm that is capable of determining the satisfaction of a constraint is called a **'Constraint Solver'**. Since, a satisfaction problem also constructs a solution as a bi-product, we emphasize the satisfaction problem over the solution problem.

Definition 19.8: A **linear Constraint** is a conjunction of linear equation / inequalities.

For example, $\{(x + y \leq 2) \wedge (x \leq 1) \wedge (y \leq 1)\}$ is a linear constraint as it is a conjunction of linear inequalities.

Definition 19.9: A Constraint is called a **Boolean constraint**, if it comprises of Boolean variables, having only two values: true and false.

Boolean constraints include operators like AND (&), OR ($\vee$), implication ($\rightarrow$), bi-directional implication ($\leftrightarrow$) and exclusive–OR ($\oplus$). It is to be noted that we deliberately used & for AND instead of '$\wedge$'as it is used for conjunction of constraints in an expression.

Definition 19.10: A variable x is **determined** by a set of constraints C, if every solution of C is a solution of x= e, where e is a variable-free expression. We illustrate it in the next section.

19.3 Constraint Propagation in Networks

Determination of the value of variables and their substitution in other constraints simplify them, which subsequently lead to the solution of the CSP. For example, let us consider a simple circuit comprising of two resistances and a D.C. cell (fig.19.1).

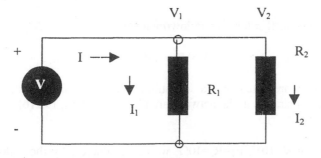

Fig.19.1: A simple circuit.

By applying elementary network theorems, we derive the following constraints:

$$V_1 = I_1 * R_1$$
$$V_2 = I_2 * R_2$$
$$V = V_1$$
$$V = V_2$$
$$I = I_1 + I_2$$

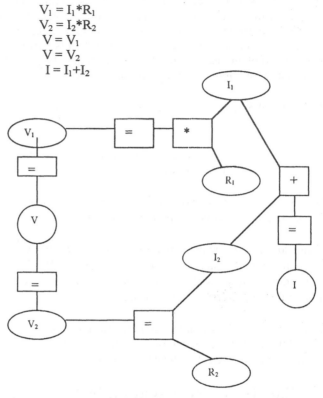

Fig. 19.2 (a): The constraint network corresponding to fig. 19.1.

These constraints together take the following form:

$$C = (V_1 = I_1 * R_1) \wedge (V_2 = I_2 * R_2) \wedge (V = V_1) \wedge (V = V_2) \wedge (I = I_1 + I_2),$$

which can be represented by a constraint network (fig.19.2(a)). The propagation of constraints in the network are illustrated in fig.19.2(b) and (c) for computation of I.

Now, given V=10V, $R_1 = R_2 = 10\Omega$, we can evaluate I by the following steps using the constraint network.

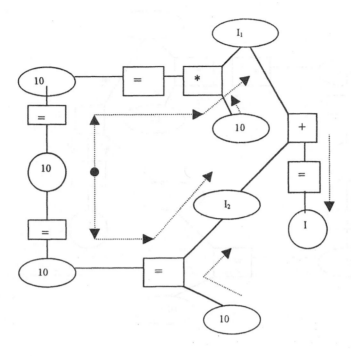

Fig. 19.2 (b): Propagation of constraint: First V=10 is propagated
to V_1 and V_2 and next $V_1=10$, $R_1 = 10$ is propagated
to I_1; and $V_2 =10$, $R_2 = 10$ is propagated to I_2.

In this example, V_1 and V_2 are 'determined' (by definition 19.10) as their
variable free values were found by the constraints $V_1=V$ and $V_2=V$. The
variables I_1 and I_2 were also 'determined' as they were not replaced by other
variables. In fact we found I_1 from 10 $I_1=10$ and I_2 from 10 $I_2=10$. So, I_1 and I_2
yield constant values, which were propagated to I for its evaluation. There,
however, exist situations when the constraints containing variables need to be
propagated. We call it simplification of constraints. The following example
illustrates simplification of constraints by using constraint networks.

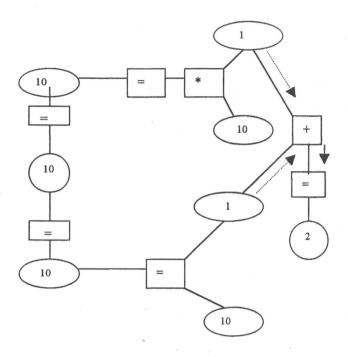

Fig. 19.2 (c): Propagation of constraints from I_1 and I_2 to I.

Consider the circuitry, shown in fig.19.3. Here, the list of constraints are

$$V_1 = I_1 * R_1$$

$$V_2 = I_2 * R_2$$

$$V_3 = I_3 * R_3$$

$$I_1 = I_2 + I_3$$

$$V_2 = V_3$$

$$V = V_1 + V_2$$

$$V = V_1 + V_3,$$

which can, on conjunction, yield C, given by

$$C = (V_1 = I_1 * R_1) \wedge (V_2 = I_2 * R_2) \wedge (V_3 = I_3 * R_3) \wedge (I_1 = I_2 + I_3) \wedge (V_2 = V_3) \\ \wedge (V = V_1 + V_2) \wedge (V = V_1 + V_3)$$

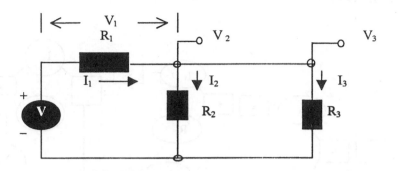

Fig. 19.3: A more complex circuit.

Since, $V_2=V_3$, we find from the above relations

$$I_2*R_2 = I_3*R_3$$

$$\Rightarrow \quad I_3 \quad = I_2.\,R_2/R_3$$

Substituting this value of I_3 in the constraint $I_1 = I_2+I_3$, we find

$I_2 = I_1.R_3/(R_2+R_3)$.

We may thus use only the following constraints to find the value of I_1.

$$V = V_1+V_2$$

$$V_1 = I_1*R_1$$

$$V_2 = I_2*R_2$$

$$I_2 = I_1.R_3/(R_2+R_3).$$

However, if we want to find a constraint C', equivalent to C, we may take C' as follows,

$$C' = (V=V_1+V_2) \wedge (V_1=I_1*R_1) \wedge (V_2=I_2*R_2) \wedge (V_3=I_3*R_3) \wedge$$

$$(I_1=I_2+I_3) \wedge (I_2=I_1.R_3/(R_2+R_3)).$$

C' need not be unique. In fact, we could replace $I_1=I_2+I_3$ from the last expression by $I_3= I_1R_2/(R_2+R_3)$.

For computing I_1, we may simplify further as given in the constraint nets (vide **fig**. 19.4).

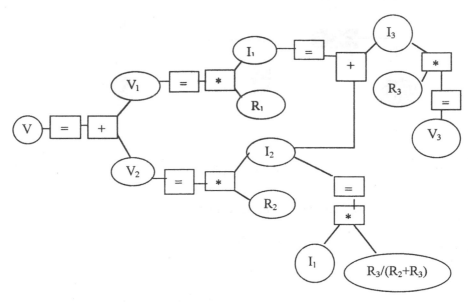

Fig. 19.4 (a): The representation of C'.

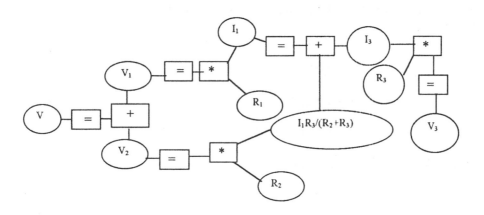

Fig. 19.4 (b): A simplified 19. 4 (a).

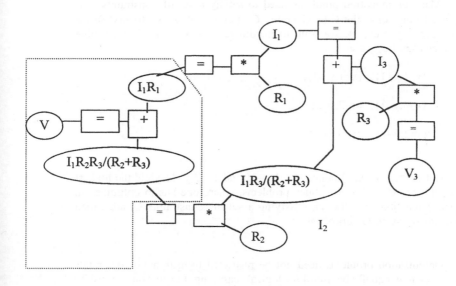

Fig. 19.4 (c): A further simplified constrained network of fig. 19.4.

Now assuming the values of V, R_1, R_2 and R_3, we find the dotted box of fig. 19.4 (c) that yields

$$10 = 5I_1 + (100\, I_1\, /20)$$

$$\Rightarrow I_1 = 1\ A.$$

It may be added that for finding the solutions, the constraint network should be simplified with the substitution of the given variables as early as possible. Also note that some part of the constraint network may not be useful for finding the solution of the problem.

Here $V_3 = I_3 \cdot R_3$ part of the network has not been used in fig. 19.4 (c) for evaluating I_1. However, if V_3 was to be computed then we had to use this constraint at the last step. Lastly, it is needless to say that the simplified constraint networks are equivalent.

Many optimization problems need to satisfy a set of constraints. For instance let the constraint C be given by $C= (x+ y \leq 4) \wedge (x >0) \wedge (y>0)$ for all integer x, y. Suppose we want to maximize $z = x^2 + y^2$. The possible assignments of (x ,y) are

{x=1,y=1}, {x=1,y =3},

{x=3,y=1}, {x=2,y=2}

{x=2,y=1}, {x=1,y=2}.

Out of these 6 (x, y) points {x=1,y=3} and {x=3,y=1} yield the highest value of $z = x^2 + y^2 = 10$. Such problems, where one has to optimize an objective function z = f(x, y), satisfying a set of constraints C, are called *constraint optimization problems.*

Optimization problems need not be restricted to arithmetic constraints only. They may equally be useful for logical reasoning. For instance, consider the blocks world problem we considered in the chapter of intelligent planning. Suppose there are 4 objects on a table: 2 boxes, one cone, and one sphere.

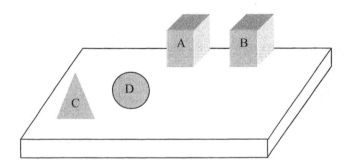

Fig. 19.5: The initial state of a blocks world problem.

Suppose, we want to keep the cone at the highest position from the surface of the table. What should be the plan for object positioning that would lead us to this requirement?

The initial state for the problem is given in fig. 19.5. We can represent this state by

On (A, Ta) ∧ On (B, Ta) ∧On (C, Ta) ∧ On (D, Ta) ∧ Box (A)
∧Box (B) ∧Cone (C) ∧ Sphere (D).

Suppose we have the following rules to define the height of an object from the table.

1. If On (X, Y) ∧ On (Y, Z) ∧On (Z, Ta) Then Height-of (X, 3).
2. If On (X, Y) ∧On (Y, Ta) Then Height-of (X, 2).
3. If On (X, Ta) Then Height-of (X, 1).

We could define height by a generic approach. But to illustrate the concept of the optimization, the present definition is adequate.

The optimization problem in the present context is: Maximize h, such that On(c,_) ∧ Height-of (c, h) is true, where '_' denotes something. The solution of the problem can be evaluated easily by considering all the possible stacking, such that C is on top. Note that D being spherical is excluded from the stack.

The possible stackings are:
i) On (C, Ta) ∧ Height-of (C,1).
ii) On (C,A) ∧On (A, Ta) ∧ Height-of (C, 2)
iii) On (C, B) ∧On(B, Ta) ∧ Height-of (C,2)
iv) On(C, B) ∧ On(B, A) ∧On(A, Ta) ∧ Height-of (C,3).
v) On(C, A) ∧On(A, B) ∧On (B, Ta) ∧ Height-of (C,3).

Out of these 5 possible stackings, the last two yield Height (=3) for C. So, we may select any of them as the solution (see fig 19.6 (a) &(b)).

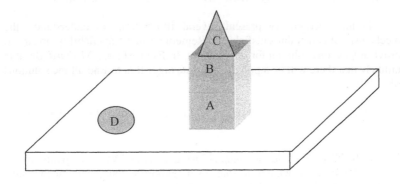

(a) On (C,B) ∧ On (B,A) ∧ On (A, Ta) ∧ Height- of (C, 3).

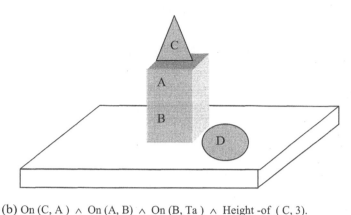

(b) On (C, A) $\wedge$ On (A, B) $\wedge$ On (B, Ta) $\wedge$ Height -of (C, 3).

Fig. 19.6: Stacking in blocks word problem to find the greatest height of C.

19.4 Determining Satisfiability of CSP

A CSP now can be defined as a problem, consisting of a constraint C over a set of variables x_1, x_2 , ...x_n and a domain D that describes the finite valuation space of each variable x_i. This finite valuation space of x_i is denoted, hereafter, by $D(x_i)$. Thus a CSP represents

$$C \wedge x_1 \wedge x_2 \wedge x_3 .... \wedge x_n \wedge x_1 \in D(x_1) \wedge x_2 \in D(x_2)......... x_n \in D(x_n).$$

In this section, we present several illustrations to understand the mechanism of constraint satisfiability. Remember that 'satisfiability' means to prove at least one solution for the problem. In fact CSPs are *NP-hard*. So, it is unlikely that there will be a polynomial time algorithm to find all the solutions of a CSP.

The following examples illustrate the way of testing satisfiability of a CSP.

Example 19.1: Consider the '*map coloring problem*' where a given map is to be colored in a manner so that no neighboring states contain the same color. Fig. 19.7 describes a hypothetical map, consisting of 6 states.

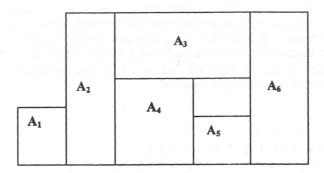

Fig. 19.7: Map of a hypothetical country having 6 states.

We now represent the neighboring relationship by a graph (fig. 19.8), where the nodes denotes the states and the arcs between nodes correspond to the neighborhood relationship between the nodes. Further, we include the constraint at the arcs.

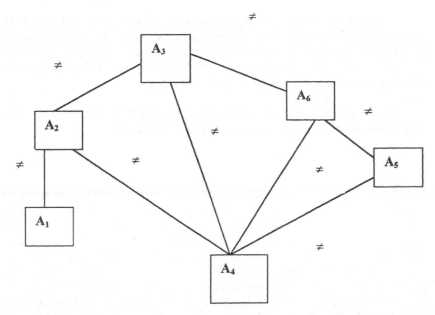

Fig. 19.8: The connectivity graph among states A_i and their adjacency by Arcs. '≠' against the arcs denote that color of the adjacent nodes should be different.

Suppose there exist 3 colors {red, green, yellow} by which we have to color the map satisfying the constraint that no two adjacent states have the same color. We start with an arbitrary state and arbitrary color assignment for it. Let us choose r, g, and y for red, green and yellow respectively in our subsequent discussions. The constraints we get after assigning r to A_1 are also recorded in the list. Thus we find step 1. In our nomenclature $A_i \neq r$ denotes color of A_i is different from red.

Step 1: $\{A_1 \leftarrow r, A_2 \neq r\}$

Step 2: $\{A_1 \leftarrow r, (A_2 \leftarrow g, A_3 \neq g, A_4 \neq g)\}$

Step 3: $\{A_1 \leftarrow r, A_2 \leftarrow g, (A_3 \leftarrow y, A_4 \neq y, A_6 \neq y), A_4 \neq g\}$

Step 4: $\{A_1 \leftarrow r, A_2 \leftarrow g, A_3 \leftarrow y, (A_4 \leftarrow r, A_5 \neq r, A_6 \neq r), A_6 \neq y\}$

Step 5: $\{A_1 \leftarrow r, A_2 \leftarrow g, A_3 \leftarrow y, A_4 \leftarrow r, (A_5 \leftarrow g, A_6 \neq g), A_6 \neq r\}$

Step 6: $\{A_1 \leftarrow r, A_2 \leftarrow g, A_3 \leftarrow y, A_4 \leftarrow r, A_5 \leftarrow g, A_6 \leftarrow y\}$

In step 2, we replaced $A_2 \neq r$ of step 1 by $A_2 \leftarrow g$ and added the constraints with it in the parenthesis. From fig. 19.8, the constraints imposed on A_3 and A_4 are $A_3 \neq g$, $A_4 \neq g$.

In step 3, we replaced $A_3 \neq g$ of step 2 by $A_3 \leftarrow y$ and added the constraints $A_4 \neq y$ and $A_6 \neq y$ and lastly copied the remaining entries from the last step. The process continues until an inconsistency occurs at a step, when one has to backtrack to the last step for alternate solutions. Fortunately, in our proposed solutions we did not observe inconsistency.

If at least one solution is obtained we say that CSP is satisfiable. Here step 6 yields one of the possible solutions for the problem.

Example 19. 2: In this example we consider a problem with a constraint set which exhibits a partial satisfaction w.r.t primitive constraints but total satisfiability is not feasible. The problem is: given the constraint $C=(x<y) \wedge (y<z)$ where x, y, z $\in$ {1, 2}, test the satisfiability of the constraint. Here $(x<y)$ and $(y<z)$ are two primitive constraints and their individual (partial) satisfiability holds. For instance if x=1,y=2, then x<y holds. But (total) satisfiability that requires $(x<y) \wedge (y<z)$ does not hold for x, y, z $\in$ {1,2}. So, backtracking will take place, when the total satisfiability fails. Unfortunately, the whole search space will be exhausted for this problem with no feasible solution. The steps of satisfiability testing are presented below.

Step 1:
$$(x<y) \wedge (y<z)$$
$$x = 1 \searrow$$
$$(1<y) \wedge (y<z)$$

(a)

Step 2:
$$(x<y) \wedge (y<z)$$
$$x = 1 \searrow$$
$$(1<y) \wedge (y<z)$$
$$y = 1 \swarrow \quad \nearrow \text{backtracking}$$
$$(1<1) \wedge (1<z)$$
false

(b)

Step 3:
$$(x<y) \wedge (y<z)$$
$$x = 1 \searrow$$
$$(1<y) \wedge (y<z)$$
$$y = 1 \swarrow \qquad \searrow y = 2$$
$$(1<1) \wedge (1<z) \qquad (1<2) \wedge (2<z)$$
false

(c)

Step 4:
$$(x < y) \wedge (y < z)$$
$$x = 1 \searrow$$
$$(1<y) \wedge (y<z)$$
$$y = 1 \swarrow \qquad \searrow y = 2$$
$$(1<1) \wedge (1<z) \qquad (1<2) \wedge (2<z)$$
false
$$z = 1 \swarrow \quad \nearrow \text{backtracks}$$
$$(1<2) \wedge (2<1)$$
false

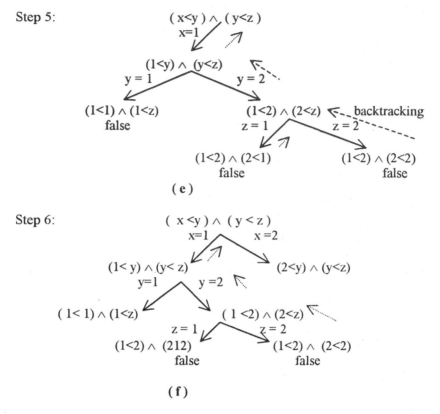

Fig. 19.9: The first 6 steps: (a)-(f) of constraint satisfiability testing.

The first 6 steps for testing constraint satisfiability are given in fig.19.9(a)-(f). Two more steps will be required to complete the search process for a solution. The next two steps are for y=1 and y=2 respectively. This is left as an exercise for the students to test that they too yield false condition.

Let us now try to construct an algorithm that will check the satisfiability of the constraints. How should we proceed? First, we should list the constraint C and assign one of the variables a value from its domain and then check whether that assignment causes an inconsistency. Note that the first variable assignment cannot cause an inconsistency. After a variable x_i is assigned a value, see what other variables are linked through some constraints with the variable x_i . If there is no such variable, do nothing, But if there exists such variable x_j, x_k, etc., attach the constraints with them. If we could achieve this one pass will be over.

In the next pass, repeat the above by assigning a value to x_j that does not violate the constants over x_j we defined in the last pass. If we could find a suitable value foe x_j then we will check the variable (say x_j', x_k',etc.), which are linked with x_j through some constraints and mark the constraints for them. Suppose the constraints together are consistent. Then we will move to the next step to assign values to x_j'. But if the constraints are inconsistent themselves or inconsistent with the current assignment to x_j, then we should backtrack to the previous step for generating an alternative solution.

The process will continue until either of the following occurs:
(i) no consistent variable bindings are possible to move further,
(ii) a solution has been arrived at with consistent bindings of all the variables.

To formalize the above concept, let us define S to be the set of variables in the constraint C. The decision is a binary variable, stating the status of satisfiability of C.

Procedure Satisfiability-test (C, S, decision)
Begin
   For each variable $x_i \in$ S **do**
     Begin
       Repeat
            $x_i \leftarrow$ Next (Domain (x_i)) do
            Substitute x_i in C ;
            If C is not partially satisfiable
               Flag $\leftarrow$ false;
            End;
         Until Next (Domain(x_i)) = Nil
       If flag = false report "unsatisfiable", return;
    End For;
   report 'decision = satisfiable';
End;

In the above algorithm, Next (domain(x_i)) gives the first element from Domain(x_i). After one call of function Domain(x_i), its first element is lost and the next element comes to the position of the first element.

One major modification in the last algorithm w.r.t our previous discussion is that we selected variables randomly in the algorithm to keep it simplified. In our previous discussion, we however had a strategy to select the variables for assignment from their domain.

19.5 Constraint Logic Programming

We already introduced Constraint logic programming (CLP) in chapter 6. Here, we will show how efficiently we can code a problem in CLP. We start with the well-known Crypto-arithmetic problem that needs to find the integer values of the variables {S, E, N, D, M, O, R, Y} in the range [0,9], so as to add to produce

SEND

+

MORE

———————

MONEY

We could solve the problem by the method covered in the last section. But a simple and efficient approach is to code in CLP language.

Crypto-arith (S, E, N, D, M,O, R,Y):-

[S,E,N,D,M,O,R,Y] : : [0...9],

Constraint ([S,E,N,D,M,O,R,Y]),

 labeling ([S,E,N,D,M,O,R,Y]).

Constraint ([S,E,N,D,M,O,R,Y]):-
        $S \neq 0$, $M \neq 0$,
All-unequal ([S,E,N,D,M,O,R,Y]),

 $(1000*S + 100 * E + 10 * N + D) +$
 $(1000* M + 100 * O + 10 * R + E)$
 $= 10000 * M + 1000 * O + 100 * N + 10 * E + Y .$

Goal: Crypto-arith (S,E,N,D,M,O,R,Y) →

Given the above goal, the program correcting determines

{ S=9, E=5, N=6, D=7, M=1, O=0, R=8, Y=2}.

The predicate labeling in the above program is a library function that ensures the search in [0, 9] for each variable. The rest of the algorithm is simple and clear from the statements themselves.

When no satisfiable solution for the constraint is available, the constraint is broken into priority-wise components. Let C be the constraint, which is broken into C_1 ,C_2 and C_3 say. Let the priority of C_1> priority of C_2 >priority of C_3. If the constraints are listed in the program in a hierarchical manner according to their priority, then a solution that satisfies the minimal constraint from the priority list may be accepted. This is the basis of *hierarchical constraint logic programming*.

19.6 Geometric Constraint Satisfaction

The CSP is not only useful for algebraic or logical problems but it may also be employed to recognize objects from their geometric features [9-10]. The features in the present context are junction labeling. In this book, we will restrict our discussion to trihedral objects, i.e., objects having three intersecting planes at each junction (vertex). However the concept may easily be extended for polyhedral objects as well. For labeling a junction of a trihedral object (vide fig.19.10), we require two types of formalism to mark the edges meeting at the junctions.

i) Mark the boundary edges of the objects by arrowheads

ii) For each non boundary edge of the object, check whether it is convex or concave. If convex, mark that edge by '+' else by '-'.

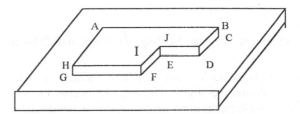

Fig 19.10: A typical trihedral object.

One question then naturally arises: how to test the convexity or concavity of an edge. The following principle, in this regard, is worthwhile mentioning.

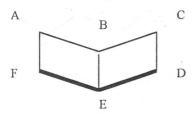

Fig. 19.11: Testing convexity of edge BE.

Look at the planes that form an edge, whose convexity/concavity you want to determine. Find the outer angle between the planes that form the boundary edges. If the angle $\geq$ 90 degrees, call the edge **convex,** else it is **concave.** In fig.19.11, BE is a convex edge as the outer angle between the planes ABEF and BCDE is greater than 90 degrees around the edge BE.

Let us now try to mark the edges of fig.19.10. We show the marking in fig.19.12.

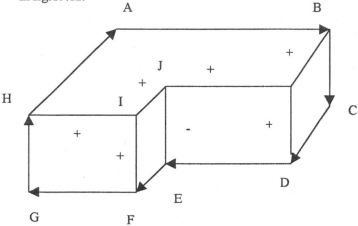

Fig. 19.12: The edge labeling of the (expanded) block of fig.19.10.

Here the boundary is ABCDEFGHA, which is denoted by a clockwise traversal around the object. The edge JE is concave, and the inner edges are convex. This is represented according to our nomenclature ('+' for convex, '-' for concave) mentioned above. We can now label each junction by one of the 18 possible types (vide fig.19.13)

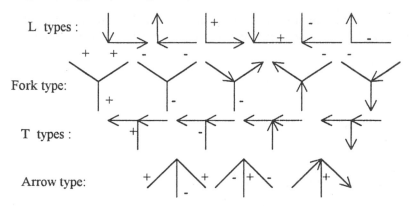

Fig. 19.13: Typical 18 junction types.

The possible labels of the junctions of fig.19.12 are presented below (Table 19.1) based on their nearest matching with those in fig.19.13.

Table 19.1: Possible labels of the junctions of fig. 19.12.

Index	Junction of fig.19.12	Possible type
1.	A	
2.	B	
3.	C	
4.	D	
5.	E	
6.	F	
7.	G	
8.	H	
9.	I	
10.	J	

In the above table, we considered rotation of the elementary types. For instance, the exact form of the junction A is not present in the prescribed types. However, by rotating the standard L types, we got two possible matches (a) and (b). Similarly, we labeled other junctions. Now, in many junctions we have more than one possible label. We can overcome it by considering the *'constraints of neighboring junctions'* [8], so that the type of each junction is unique for a realistic object. Careful observation yields that since A and B are neighboring junctions, trying to connect (a) with (c) and (b) with (c) shows that (a)-(c) connection is feasible. The constraints of neighborhood junctions are applied to each label to reduce it to a single type per junction. The result of the type reduction is presented below.

Table 19.2: Junction type reduction.

Junctions	possible name of label
A	(a)
C	(d)
E	(g)
G	(k)

The algorithm for junction labeling, also called the waltz algorithm [9], is presented below.

Procedure junction-labeling (edges, labels)
Begin

1. Label boundary by clockwise encirclement;
2. Label interior edges by '+' or '-' as convenient;
3. Label junction types from the possible match with fig. 19.13. You may allow rotation of types and the types need not be uniqu;
4. Reduce the types of each junction to a single one by considering the types of the neighboring junctions;
5. **Repeat** step 4 **until** each junction of the object has a single type.
End.

19.7 Conclusions

The chapter introduced three basic types of CSP dealing with algebraic, logical and geometric constraints. The concept of constraint simplification and equivalence is discussed and constraint propagation in networks has been illustrated. The principles of constraint logic programming are also presented with examples. An algorithm for constraint satisfiability testing is discussed. The concepts of recognizing trihedral objects from their junction labeling have also been covered briefly.

The main part of CSP is the formulation of the problem, as the solution is done automatically by searching in the entire domain. Improving the search efficiency by employing heuristics may give a new flavor to the CSP.

Exercises

1. Draw a map of your country and shade the states carefully with a minimum number of colors, so that no two neighbors have the same color. Draw an equivalent constraint graph for the problem and hence determine theoretically the minimum number of colors you require to color the map.

2. Draw a passive electrical circuit with at least two loops, and then construct the constraint net for your circuit. Instantiate the input variables and show how the constraints propagate through the network to compute the output variables.

3. Write three ineqalities, each of 3 variables, and solve them by the approach presented in the chapter.

4. Draw a right angled pyramid with 4 planes and label its junctions: convex or concave. Is there any uncertainty in your labeling? If yes, then how will you resolve it?

References

[1] Cohen, J., "Constraint logic programming languages," *Communications of the ACM*, vol. 33, no. 7, pp. 52-68, 1990.

[2] Freuder, E. and Mackworth, A., *Constraint Based Reasoning*, MIT Press, Cambridge, MA, 1994.

[3] Jaffer, J. and Maher, M., Constraint logic programming: A survey, *Journal of Logic Programming*, vol. 19/20, pp. 503-582, 1994.

[4] Marriot, K. and Stuckey, P. J., *Programming with constraints: An introduction*, MIT Press, Cambridge, MA, 1998.

[5] Stefik, M., Planning with constraints (MOLGEN: Part 1), *Artificial Intelligence*, vol. 16, pp. 111-139, 1981.

[6]   Tsang, E., *Foundations of Constraint Satisfaction*, Academic Press, New York, 1993.

[7]   Hentenryck, P. Van, *Constraint Satisfaction in Logic Programming*, MIT Press, Cambridge, MA, 1989.

[8]   Rich, E. and Knight, K., *Artificial Intelligence*, McGraw-Hill, New York, pp. 367- 375, 1991.

[9]   Waltz, D. L., "Understanding the drawing of scenes with shadows," In *The Psychology of Computer Vision*, Winston, P., Ed., McGraw-Hill, New York, 1975.

[10]   Winston, P. H., *Artificial Intelligence*, Addison-Wesley, Reading, MA, pp. 231-280, 1992.

20

Acquisition of Knowledge

Acquisition of knowledge is equally hard for machines as it is for the human beings. The chapter provides various tools and techniques for manual and automated acquisition of knowledge. Special emphasis is given to knowledge acquisition from multiple experts. A structured approach to knowledge refinement using fuzzy Petri nets has also been presented in the chapter. The proposed method analyzes the known case histories to refine the parameters of the knowledge base and combines them for their usage in a new problem. The chapter concludes with the justification of the reinforcement learning and the inductive logic programming in automated acquisition of knowledge.

20.1 Introduction

The phrase acquisition (elicitation) [2]-[7], [9], [12]-[17] of knowledge, in general, refers to collection of knowledge from knowledge-rich sources and its orderly placement into the knowledge base. It also allows refinement of knowledge in the existing knowledge base. The process of acquisition of knowledge could be carried out manually or automatically. In manual mode, a knowledge engineer receives knowledge from one or more domain experts, whereas in automatic mode, a machine learning system is used for

585

autonomous learning and refining knowledge from the external world. One main difficulty with manual acquisition of knowledge is that the experts often fail to correctly encode the knowledge, though they can easily solve a complex problem of their domain. Further, the ordering of the pieces of knowledge carried out by the experts, being sometimes improper, causes a significant degradation in search efficiency of the inference procedure. Lastly, the certainty factor of the pieces of knowledge, set by the experts, too, is not free from human bias and thus may lead to inaccurate inferences. The need for automated knowledge acquisition is, therefore, strongly felt by the scientific community of AI.

20.2 Manual Approach for Knowledge Acquisition

Knowledge acquisition is a pertinent issue in the process of development of expert systems. A good expert system should contain a well-organized, complete and consistent knowledge base. An incomplete or inconsistent knowledge base may cause instability in reasoning, while a less organized system requires quite a significant time for search and matching of data. The malfunctioning of the above forms originates in an expert system generally due to the imperfections in i) the input resources of knowledge and ii) their encoding in programs. The imperfection in the input resources of knowledge can be overcome by consulting proved knowledge-rich sources, such as textbooks and experts of respective domains. The encoding of knowledge could be erroneous due to either incorrect understanding of the pieces of knowledge or their semantic misinterpretation in programs. A knowledge engineer, generally, is responsible for acquiring knowledge and its encoding. Understanding knowledge from experts or textbooks, therefore, is part of his duties. A clear understanding of the knowledge base, however, requires identification of specific knowledge from a long narration of the experts. The knowledge engineer, who generally puts objective questions to the expert, therefore, should allow the expert to answer them in sufficient detail, explaining the points [6]. The semantic knowledge earned from the experts could be noted point-wise for   subsequent   encoding   in   programs. Occasionally, the experts too are not free from bias. One way to make the knowledge base bias-free is to consult a number of experts of the same problem domain and take the view of the majority of the members as the acquired knowledge.

20.3 Knowledge Fusion from Multiple Experts

Fusion of knowledge from multiple number of experts can be implemented in either of the following two ways. First the knowledge engineer may invite the experts to attend a meeting and record the resulting outcome of the meeting. Alternatively, she may visit the office of the experts and record his view about the concerned problem and finally combine their views together by some principle. The former scheme suffers from the following limitations [6], [13]:

a) Less participation of an expert because of dominance of his supervisor or senior experts.

b) Compromising solutions generated by a group with conflicting opinions.

c) Wastage of time in group meetings.

d) Difficulties in scheduling the experts.

All the above limitations of the former scheme, however, can be overcome by the latter scheme. But how can we integrate the views of the multiple number of experts in the latter scheme? An answer to this question is presented here following Rush and Wallace [17]. In a recent publication, Rush and Wallace devised a scheme to combine 'the influence diagrams' (ID) of several experts for constructing a 'multiple expert influence diagram' (MEID). The MEIDs represent the causal dependence relationship of facts, supported by a majority of the experts, and thus may be used as a collective source of knowledge, free from human biases and inconsistencies.

20.3.1 Constructing MEIDs from IDs

Formally, "*an ID is a directed graph which displays the decision points, relevant events and potential outcomes of a decision situation* [17]." The rectangular nodes in an ID represent facts or decisions, while the directed arcs denote causal dependencies. For understanding the definition, let us consider the "oil wildcatter decision problem [17]". Here the decision is whether to drill at a given geographical location for possible exploration of natural oil. An ID for the above problem, provided by an expert, is presented in fig. 20.1.

The IDs can be represented by *incidence matrix* with (i, j)-th element denoted by w_{ij}, where

$w_{ij} = 1$, if there exists an arc from rectangular component i to j,

= 0, otherwise.

For instance, the ID presented in fig. 20.1 has the following incidence matrix W.

	To	A	B	C	D	E	F	G
From								
	A	0	1	0	0	0	0	0
	B	0	0	1	0	0	0	0
	C	0	0	0	1	0	0	0
W =	D	0	0	0	0	0	0	0
	E	0	1	0	0	0	0	0
	F	0	0	0	1	1	0	0
	G	0	0	0	1	0	0	0

For comparing the IDs from two sources ID_1 and ID_2, let us call their incidence matrices W_1 and W_2 respectively. Now, the symmetric difference metric $d(W_1, W_2)$ is given by

$$d(W_1, W_2) = Tr[(W_1 - W_2)^T (W_1 - W_2)] \tag{20.1}$$

The function $d(W_1, W_2)$ denotes the count of the number of discrepant edges between two influence diagrams.

The following properties directly follow from the definition of $d(W_1, W_2)$, vide expression (20.1).

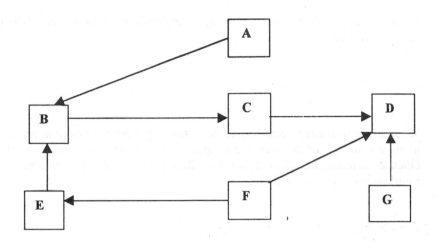

A= test ?, B = test results, C = drill ?, D = profit ?,
E= seismic structure, F= amount of oil ? , G = cost of drilling.

Fig. 20.1: An influence diagram for testing significance of oil exploration.

i) $d(W_1, W_2) = 0$, iff $W_1 = W_2$ (reflexive)   (20.2)

ii) $d(W_1, W_2) = d(W_2, W_1)$ (symmetric)   (20.3)

iii) $d(W_1, W_2) \leq d(W_1, W_3) + d(W_3, W_2)$ (triangle inequality)

     (20.4)

Bank and Carley [1] defined the following probability measure:

$P(w) = c(s) \exp(-s \cdot d(w, W))$, for each matrix w in W,   (20.5)

where s is a dispersion parameter and c(s) is a normalizing constant.

Given the above expression, we can determine W^* and s^*, the estimate of W and s respectively, by means of classical likelihood techniques. For a sample of expert influence diagram of size n, the log likelihood of the sample [17] is given by

$$L [W, s] = n \log c(s) - s \sum_{\forall w \in W} d (w, W).\qquad(20.6)$$

It can be easily shown that L [W, s] is maximized in W by W*, which maximizes the quantity

$$\sum d (w, W), \quad \forall w \in W.$$

W* is first constructed according to the following rule: *if an edge is present in more than 50% of the influence diagrams, then it should be present in W*.* Once W* is found, one can obtain s* by differentiating L[W, s] with respect to s and setting it to zero, which yields

$$s^* = - \ln \{ [(r\ n)^{-1}\ Z] / [1 - (r\ n)^{-1}\ Z] \}\qquad(20.7)$$

where $Z = \sum d (w, W), \forall\ w\ \varepsilon\ W$

and $r = m (m -1)$,

given that m is the number of nodes in each influence diagram. It may be noted that s* denotes a measure of disagreement among the experts.

20.4 Machine Learning Approach to Knowledge Acquisition

Manual acquisition of knowledge is difficult for two main reasons. First the knowledge engineer has to remain in constant touch with the experts for a significant amount of time, which sometimes may be of the order of years. Secondly, the experts themselves in many cases cannot formally present the knowledge. The above difficulties in acquisition of knowledge can, however, be overcome by autonomously encoding knowledge through machine learning. The schematic view for elicitation of knowledge by the machine learning approach is presented in fig. 20.2.

The database in fig. 20.2 is extracted from experts or other reasoning systems. The machine learning unit grabs these data and attempts to acquire new knowledge out of it. The acquired knowledge is then transferred to the knowledge base for future usage. In some systems, the knowledge base need not be extended, but may be refined with respect to its internal parameters. For instance, certainty factor of the rules in a knowledge base may be refined based on the estimated certainty factors of proven case histories. A generic scheme for knowledge refinement is presented in fig. 20.3.

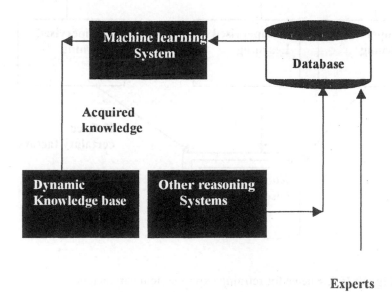

Fig. 20.2: Principles of automated knowledge acquisition.

Fig. 20.3 presents a scheme for automatic estimation of some parameters in an expert system. For instance, certainty factor of knowledge may be refined from their initial values and steady-state inferences of n number of proven case histories. The refinement should be carried out in a manner, so that steady-state inferences are consistent with the derived certainty factors. However, all n problems being similar, it is likely that some pieces of knowledge may be common to two or more number of knowledge bases. So, the resulting certainty factors of a common piece of knowledge may come up with different values from different knowledge bases. A scheme for normalization, therefore, has to be devised to take into account such certainty factors.

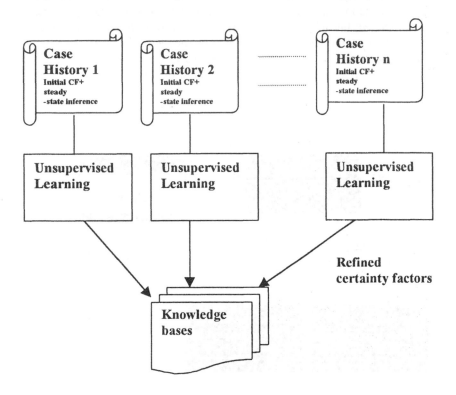

Fig. 20.3: A scheme for refining expert system parameters.

The unsupervised learning algorithm used in fig. 20.3 differs from one system to another depending on the type of knowledge representation and reasoning schemes. For example, if a Bayesian Belief network is employed to represent the knowledge base, the unsupervised learning network could be a Hopfield network. The nodes in the Hopfield net in the present context could represent the beliefs of evidences, while the weights could be the conditional probabilities, representing certainty factors of knowledge. Alternatively, if the knowledge-based systems are realized with Petri nets, then a Hebbian learning could be adopted to derive the steady-state certainty factors, represented by its weights.

20.5 Knowledge Refinement by Hebbian Learning

This section presents a new method for automated estimation of certainty factors of knowledge from the proven and historical databases of a typical reasoning system. Certainty factors, here, have been modeled by weights in a special type of recurrent fuzzy neural Petri net. The beliefs of the propositions, collected from the historical databases, are mapped at places of a fuzzy Petri net and the weights of directed arcs from transitions to places are updated synchronously following the Hebbian Learning principle until an equilibrium condition [15],[9] following which the weights no longer change further is reached. The model for weight adaptation has been chosen for maintaining consistency among the initial beliefs of the propositions and thus the derived steady-state weights represent a more accurate measure of certainty factors than those assigned by a human expert.

20.5.1 The Encoding Model

The process of encoding of weights consists of three basic steps, presented below:

- **Step-I :** A transition tr_i is enabled if all its input places possess tokens. An enabled transition is firable. On firing of a transition tr_i, its FTT t_i is updated using expression (20.8)[8], where places $p_k \in I$ (tr_i), n_k is the belief of proposition mapped at place p_k, and th_i is the threshold of transition tr_i.

$$t_i(t+1) = (\bigwedge_{1 \le k \le n} n_k(t)) \wedge u [(\bigwedge_{1 \le k \le n} n_k(t)) - th_i] \qquad (20.8)$$

Expression (20.8) reveals that if $\bigwedge_{1 \le k \le n} n_k > th_i$,

$$t_i(t+1) = \bigwedge_{1 \le k \le n} n_k(t)$$
$$= 0, \text{ otherwise.}$$

Step-II : After the FTTs at all the transitions are updated synchronously, we revise the fuzzy beliefs at all places concurrently. The fuzzy belief n_j at place p_j is updated using expression 20.9 (a), where $p_i \varepsilon O (tr_j)$ and by using 20.9(b) when p_i is an axiom, having no input arc.

$$n_i(t+1) = \overset{m}{\underset{j=1}{V}} \quad (t_j(t+1) \; w_{ij}(t)) \qquad\qquad 20.9(a)$$

$$= n_i(t) \text{ , when } p_i \text{ is an axiom} \qquad\qquad 20.9(b)$$

- **Step-III :** Once the updating of fuzzy beliefs are over, the weights w_{ij} of the arc connected between transition tr_j and its output place p_i are updated following Hebbian learning [10] by expression (20.10).

$$w_{ij} (t+1) = t_j (t+1) \; \wedge \; n_i(t+1) \qquad\qquad (20.10)$$

The above three step cycle for encoding is repeated until the weights become time-invariant. Such a time-invariant state is called equilibrium. The steady-state values of weights are saved for subsequent reasoning in analogous problems.

Theorem 20.1: *The encoding process of weights in an associative memory realized with FPN is unconditionally stable.*

Proof: Proof is simple and omitted for space limitation.   □

20.5.2 The Recall / Reasoning Model

The reasoning model of a recurrent FPN has been reported elsewhere [8-9]. During the reasoning phase, we can use any of these models including the new model proposed below.

The *reasoning / recall model in an FPN* can be carried out in the same way as in *the first two steps of the encoding model* with the following exceptions.

- While initiating the reasoning process, the known fuzzy beliefs for the propositions of a problem are to be assigned to the appropriate places. It is to be noted that in the encoding model the fuzzy beliefs of propositions were submitted using proven case histories.

- The reasoning model should terminate when the fuzzy beliefs associated with all propositions reach steady-state values, i.e., when for all places,

$n_i(t+1) = n_i(t)$ at $t = \min (t)$. The steady-state beliefs thus obtained are used to interpret the results of typical analogous reasoning problems. The execution of the reasoning model is referred to as belief revision cycle [8].

Theorem 20.2: *The recall process in an FPN unconditionally converges to stable points in belief space.*

Proof: Proof is simple and omitted for space limitation.   □

20.5.3 Case Study by Computer Simulation

In this study, we consider two proven case histories described by (Rule base I, Database I) and (Rule base II, Database II). The beliefs of each proposition in the FPNs (vide fig. 20.4 and 20.5) for these two case histories are known. The encoding model for associative memory presented above has been used to estimate the CFs of the rules in either cases. In case the estimated CF of a rule, obtained from two case-histories differs, we take the average of the estimated values as its CF.

    Case History I

    Rule base I :

            PR1 :   Loves(x,y), Loves(y,x) → Lover(x,y)
            PR2 :   Young(x), Young(y),
                    Opposite-Sex (x,y) → Lover(x,y)
            PR3 :   Lover(x,y), Male(x), Female(y) →
                    Marries(x,y)
            PR4 :   Marries(x,y) → Loves(x,y)
            PR5 :   Marries(x,y), Husband(x,y) → Wife(y,x)
            PR6 :   Father(x,z), Mother(y,z) → Wife(y,x)

    Database I :

            Loves (ram,sita), Loves(sita,ram),
            Young(ram), Young(sita), Opposite-
            Sex(ram,sita), Male(ram), Female(sita),
            Husband(ram,sita), Father(ram, kush), Mother
            (sita, kush)

While reasoning in analogous problems, the rules may be assigned with the estimated CFs, as obtained from the known histories. In this example, case-III is a new test problem, whose knowledge base is the subset of the union of the knowledge bases for case-I and case-II. Thus the CFs of the rules are known. In case-III, the initial beliefs of the axioms only are presumed to be known. The aim is to estimate the steady-state belief of all propositions in the network. Since stability of the reasoning model is guaranteed, the belief revision process is continued until steady- state is reached. In fact steady-state occurs in the reasoning model of case-III after 5 belief revision cycles. Once the steady-state condition is reached, the network may be used for generating new inferences.

Table 20.1: Parameters of case history I.

Initial Weights w_{ij}	$w_{71}=0.8$, $w_{72}=0.7$, $w_{93}=0.6$, $w_{14}=0.9$, $w_{13,5}=0.8$, $w_{13,6}=0.5$
Initial Fuzzy Beliefs n_i	$n_1=0.2$, $n_2=0.8$, $n_3=0.75$, $n_4=0.9$, $n_5=0.6$, $n_6=0.75$, $n_7=0.35$, $n_8=0.85$, $n_9=0.45$, $n_{10}=0.85$, $n_{11}=0.7$, $n_{12}=0.65$, $n_{13}=0.$
Steady-state weights after 4 iterations	$w_{71}=0.35$, $\qquad\qquad w_{72}=0.60$, $w_{93}=0.35$, $w_{14}=0.35$, $\qquad\qquad w_{13,5}=0.35$, $w_{13,6}=0.50$
$th_j =0$ for all transitions tr_j	

The FPN, given in fig. 20.4, has been formed using the above rule-base and database from a typical case history. The fuzzy beliefs of the places in fig. 20.4 are found from a proven historical database. The initial weights in the network are assigned arbitrarily and the model for encoding of weights is used for computing the steady-state value of weights (vide Table 20.1).

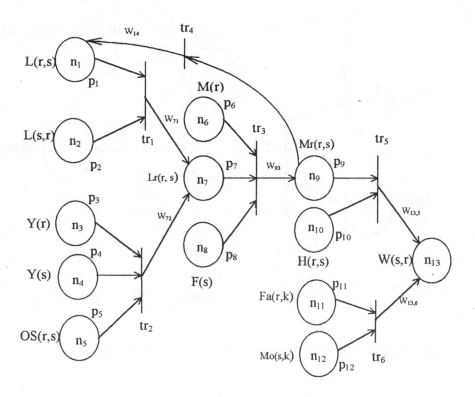

L= Loves, Y = Young, OS =Opposite-Sex, Lr = Lover, M = Male, F = Female,
Mr = Married, H = Husband, W = Wife, Fa = Father, Mo = Mother
r = Ram, s = Sita, k = Kush

Fig. 20.4: A FPN with initially assigned known beliefs and random weights.

Case History II

Rule base II :

PR1 :   Wife(y,x), Loves(x,z), Female(z) → Hates(y,x)
PR2 :   Husband(x,y), Loves(x,y) →̷ Wife(y,x)
PR3 :   Hates(z,x) → Loves(x,y)

Database II :

Husband(jadu,mala),   Loves(jadu,mala),
Loves(jadu,rita), Female(rita)

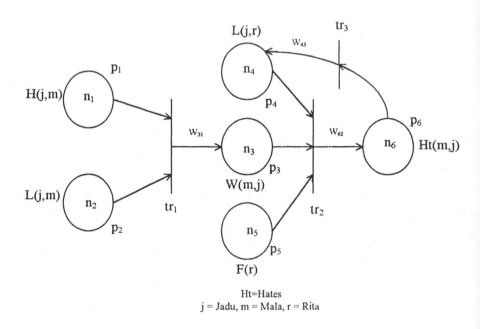

Ht=Hates
j = Jadu, m = Mala, r = Rita

Fig. 20.5: A second FPN with known initial beliefs and random weights.

The FPN of fig. 20.5 is formed with the rule base and database given above. The system parameters of the FPN of fig. 20.5 are presented in Table 20.2.

Table 20.2: Parameters of Case History 2.

Initial Weights w_{ij}	$w_{31}=0.75$, $w_{62}=0.95$, $w_{43}=0.8$
Initial Fuzzy Beliefs n_i	$n_1=0.8$, $n_2=0.7$, $n_3=0.1$, $n_4=0.2$, $n_5=0.9$, $n_6=0.3$
Steady-state weights after 3 iterations	$w_{31}=0.7$, $w_{62}=0.10$, $w_{43}=0.10$
$th_j =0$ for all transitions tr_j	

The Current Reasoning Problem

Now, to solve a typical reasoning problem, whose knowledge and databases are presented herewith, we need to assign the derived weights from the last two case histories. The reasoning model can be used in this example to compute the steady-state belief of the proposition: Hates (lata,askoke), with the given initial beliefs of all the propositions. The system parameters of the FPN in fig. 20.6 are presented in Table 20.3.

20.5.4 Implication of the results

The analysis of stability envisages that both the encoding and the recall model of associative memory are unconditionally stable. The time required for convergence of the proposed model is proportional to the number of transitions on the largest path (cascaded set of arcs) [9] in the network. The model could be used for determination of CF of rules in a KB by maintaining consistency among the beliefs of the propositions of known case histories.

Current Rule base:

PR1 :   Loves(x,y), Loves(y,x) → Lover(x,y)

PR2 :   Young(x),   Young(y),   OS(x,y)   →   Lover(x,y)

PR3 :   Lover(x,y),   Male(x),   Female(y)   →   Marries(x,y)

PR4 :   Marries(x,y) → Loves(x,y)

PR5 :   Marries(x,y), Husband(x,y) → Wife(y,x)

PR6 :   Father(x,z), Mother(y,z) → Wife(y,x)

PR7 :   Wife(y,x),   Loves(x,z),   Female(z)   →   Hates(y,x)

PR8 :   Hates(z,x) → Loves(x,y)

Current Database:

Loves(ashoke,lata), Loves(lata,ashoke), Young(asoke), Young(lata), Opposite-Sex(asoke, lata), Male(ashoke), Female(lata), Husband(ashoke,lata), Father(asoke, kamal), Mother(lata, kamal), Loves(ashoke,tina), Female(tina)

Table 20.3: Parameters of the current reasoning problem.

Initial weight w_{ij} taken from the steady-state CFs of corresponding rules from earlier case histories I & II shown in parenthesis	$w_{71}=w_{71}(I)=0.35$, $w_{72}=w_{72}(I)=0.60$, $w_{93}=w_{93}(I)=0.35$, $w_{14}=w_{14}(I)=0.35$, $w_{13,5}=w_{13,5}(I)=0.35$, $w_{13,6}=w_{13,6}(I)=0.50$, $w_{16,7}=w_{62}(II)=0.10$, $w_{15,8}=w_{43}(II)=0.10$
Initial Fuzzy Beliefs n_i	$n_1=0.4$, $n_2=0.8$, $n_3=0.75$, $n_4=0.85$, $n_5=0.65$, $n_6=0.9$, $n_7=0.3$, $n_8=0.7$, $n_9=0.3$, $n_{10}=0.95$, $n_{11}=0.65$, $n_{12}=0.6$, $n_{13}=0.25$, $n_{14}=0.55$, $n_{15}=0.35$, $n_{16}=0.40$
Steady-state Belief at place p_{16} for proposition Hates(l,a)	$n_{16}=0.10$
$th_j = 0$ for all transitions tr_j	

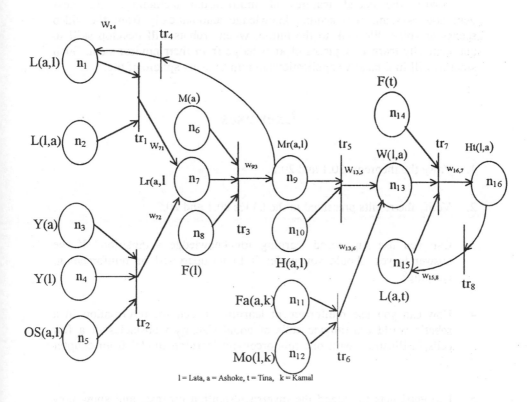

Fig.20.6: An FPN used for estimating the belief of Ht(l,a) with known initial belief and CFs.

20.6 Conclusions

Acquisition of knowledge itself is quite a vast field. The chapter introduced the principles of knowledge acquisition through examples. It demonstrated: how the opinion of a number of experts can be combined judiciously. It also presented an unsupervised model of knowledge acquisition that works on the principles of Hebbian learning.

The subject of knowledge acquisition is an active area of modern research in AI. Researchers are keen to use recent technologies such as inductive logic programming and the reinforcement learning for automated acquisition of knowledge. Fractals, which correspond to specialized mathematical functions of recursive nature, have also been employed recently in knowledge acquisition.

With the added features of multi-media technology, the next generation systems will acquire knowledge automatically from the video scenes or voice. We look to the future, when robots will develop skill to automatically learn new pieces of knowledge from their environment. Such systems will find massive applications in almost every sphere of life.

Exercises

1. Prove the theorems 20.1 and 20.2.

2. Verify the results presented in the tables 20.1 and 20.2.

3. Can you use supervised learning for knowledge acquisition? If the answer is yes, should you prefer it to unsupervised or reinforcement learning?

4. How can you use reinforcement learning to acquire information in a robot's world (partitioned cells of equal size with obstacles in a few cells)? Illustrate with one reinforcement learning model from chapter 13.

5. Use Petri nets to extend the inverse resolution process, and show how you can use it for knowledge acquisition. Use the elementary model of Petri nets, presented in chapter 8. [*open ended for researchers*]

References

[1]   Banks, D. and Carley, K., "Metric inference for social networks," *J. Classification*, vol. 11, pp. 121-149, 1994.

[2]   Banning, R. W., "Knowledge acquisition and system validation in expert systems for management," In *Human Systems Management*, vol. 4, Elsevier/North-Holland, New York, pp. 280-285, 1984.

[3]   Cooke, N. and Macdonald, J., "A formal methodology for acquiring and representing expert knowledge," *Proc. of the IEEE*, vol. 74, no. 10, pp. 1422-1430, 1986.

[4]   Degreef, P. and Breuker, J., "A case study in structured knowledge acquisition," *Proc. of IJCAI*, pp. 390-392, 1985.

[5]   Gaschnig, J., Klahr, P., Popple, H., Shortliffe, E. and Terry, A., "Evaluation of expert systems: Issues and case studies," *Building Expert Systems*, Hayes-Roth, F., Waterman, D. A. and Lenat, D. B., Eds., Addison–Wesley, Reading, MA, 1983.

[6]   Hart, A., *Knowledge Acquisition for Expert Systems*, McGraw-Hill, New York, 1986.

[7]   Hayes-Roth, F., Waterman, D. and Leant, D., *Building Expert Systems*, Addison-Wesley, Reading, MA,1983.

[8]   Konar, A. and Mandal, A. K., " Uncertainty management in expert systems using fuzzy Petri nets," *IEEE Trans. on Knowledge and Data Engg.*, vol. 8, no. 1, pp. 96-105, 1996.

[9]   Konar, A. and Pal, S., Modeling cognition with fuzzy neural nets, In *Fuzzy Systems Theory: Techniques and Applications*, Leondes, C. T., Ed., Academic Press, New York, 1999.

[10]   Kosko, B., "Fuzzy associative memory systems," In *Fuzzy Expert Systems*, Kandel, A., Ed., Addison-Wesley, Reading, MA, Dec. 1986.

[11]   Looney, C. G., "Fuzzy Petri nets for rule based decision making,"*IEEE Trans. on Systems, Man and Cybernetics*, vol. SMC-18, no. 1, Jan./ Feb., 1988.

[12]   McGraw, K. and Seale, M., "Knowledge elicitation with multiple experts: Considerations and techniques," *Artificial Intelligence Review*, vol. 2, no. 1, Jan. / Feb., 1988.

[13]   McGraw, K. L. and Harbison-Briggs, K., *Knowledge Acquisition: Principles and Guidelines*, Prentice-Hall, Englewood Cliffs, NJ, pp. 280-321,1989.

[14]   Mittal, S. and Dym, C., "Knowledge acquisition from multiple experts," *The AI Magazine*, vol. 7, no. 2, pp. 32-37, 1985.

[15]   Paul, B., Konar, A. and Mandal, A. K., "Estimation of certainty factor of rules by Associative Memory realised with Fuzzy Petri Nets," *Proc. of IEEE Int. Conf. on Automation, Robotics and Computer Vision*, Singapore, 1996.

[16]   Prerau, D., " Knowledge acquisition in the development of a large expert system," *AI Magazine*, vol. 8, no. 2, pp. 43-52, 1987.

[17]   Rush, R. and Wallace, W. A., " Elicitation of knowledge from multiple experts using network inference, *IEEE Trans. on Knowledge and Data Engg.*, vol. 9, no. 5, pp. 688-696, 1997.

21

Validation, Verification and Maintenance Issues

The chapter provides a detailed overview of the latest tools and techniques required for 'performance evaluation' of intelligent reasoning systems. Performance evaluation serves two main objectives. First, one can test: whether the designed system performs satisfactorily for the class of problems, for which it is built. This is called validation. Secondly, by evaluating the performance, we can determine whether the tools and techniques have been properly used to model "the expert". This is called verification. Besides validation and verification, the third important issue is "the maintenance", which is required to update the knowledge base and refine the parameters of an expert system. The chapter covers all these issues in sufficient detail.

21.1 Introduction

Validation, verification and maintenance are the three widely used terms commonly referred to to evaluate and improve the performance of a knowledge based system. 'Validation' and 'verification', though having a finer level of difference in their usage, are being used interchangeably in the existing literature of expert systems. Strictly speaking, 'validation' refers to 'building the right system' [3]-[4] that truly resembles the system intended to be built. In other words, validation corresponds to testing the performance of the system, and suggests reformulation of the problem characteristics and concepts based on the deviation of its performance from that of the ideal system. The 'ideal system', in general, could be the expert himself or an already validated system.

'Verification', on the other hand, deals with implementational issues. For instance, it includes redesigning the organization of the knowledge base and refinement of the pieces of knowledge so that the performance of the expert system is comparable to that of an ideal system.

Buchanan [8] first proposed the life cycle of expert systems with regard to the various stages of their development. For illustrating the scope of the validation and verification schemes in an expert system, we slightly extend this model (vide fig. 21.1). In this figure, the first four major steps are involved to represent the problem by an equivalent knowledge base, and the remaining two steps are employed to improve the performance of the system through feedback cycles. To illustrate the figure, let us consider the domain of 'circuit problems'. An electrical circuit consists of components like resistors, inductors, capacitors, transistors and electrical conductors (wire).

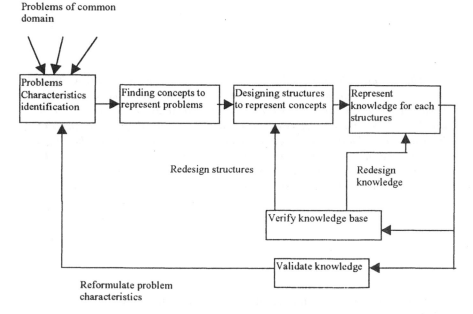

Fig. 21.1: Life cycle of an expert system.

Here the problem characteristic is to determine current through a component or the voltage drop across a component and the specifications for the problem include voltage drop across or current through a different component. Once the (input) specifications and the (output) requirements of the problems are known, we have to search concepts that can represent the problems. In case of electrical circuits, Kirchoff's laws, Ohm's laws, Thevenin's theorem, reciprocity theorem, superposition theorem, etc. correspond to concepts that may be used to represent the problem. The next task is to design structures to organize knowledge. In electrical circuit problems, 'finding structures' means determining the appropriate rules (laws/theorems) to represent the problem. For instance, in a simple loop, when the current through a passive component is known, and the value of the component is also known, we can employ Ohm's law to find the drop across it. When the currents through some of the components and /or voltage drops across one or more components are known and there exist a number of loops, we could employ Kirchoff's law or Maxwell's loop equation. Thus the structures in the present context could be of the following type.

IF <input specification> & <output requirement> & <configuration>
THEN use <rule/ theorem>.

Here, the input specification and the output requirement correspond to voltage and currents supplied and to be evaluated respectively. The 'configuration' in the present context may refer to number of loops (one or more). A look at fig. 21.2 will be useful to understand the structure used to formulate concepts.

1. *IF <voltage/current > & <voltage/current > & <single loop>*
   THEN apply <Kirchoff's voltage law (KVL) >.

2. *IF <voltage across a component> & <current through that component>*
   &
   <undefined number of loops>
   THEN apply <Ohm's law>.

3. *IF <voltage/current> & <voltage/current> & <many loops>*
   THEN apply <Maxwell's loop equation / Thevenin's theorem>.

Fig. 21.2: Instances of the structure representing a concept.

It may be added here that the structure describing the concepts need not be always IF-THEN statements. Structure and pointers, filler and slots or any other AI techniques commonly used to represent knowledge could equally represent it. After the construction of the structure is over, its component, here, for instance, the THEN part of each rule has to be encoded by some pieces of knowledge. Ohm's law, for example, can be formalized as follows:

IF V= (voltage across a component) &
  Z= (impedance of the component) &
  I= (current through the component)

THEN $V := I * Z$.

Kirchoff's voltage law (KVL) in a loop can be represented as follows:

IF (no. of circuit component in a loop = n)   &
  (voltage across and current through Z_1 are V_1 and I_1)   &
  (voltage across and current through Z_2 are V_2 and I_2)   &

  .   .
  .   .
  .   .

  (voltage across and current through Z_n are V_n and I_n)   &
  (total driving emf in a loop = V)

THEN $(V = V_1 + V_2 + \ldots\ldots + V_n)$ OR
  $(V = I_1 * Z_1 + I_2 * Z_2 + \ldots + I_n * Z_n)$.

After the construction of the knowledge base, the performance of the system in fig. 21.1 is evaluated through verification and validation of the system. A question that then naturally arises is how to compare the performance of the proposed system with an 'ideal' one. One way to solve the problem is to call the expert, based on whose knowledge the system is built. Suppose the expert selects superposition theorem to solve a problem but the expert system built with his expertise cannot identify 'superposition' theorem as the best choice for solving the problem. Obviously, there is a problem with the system design. Most likely there exist some loopholes in the identification of problem characteristics. So, reformulation of the problem characteristics through validation of the system could overcome this problem. In case reformulation of the new problem characteristics cannot be inferred by the validation procedure, then the structures used to represent concepts or the knowledge base could be re-designed. This is done by invoking a verification procedure.

Apart from validation and verification, another significant issue in improving the performance of an expert system lies with its maintenance. The knowledge base of an expert system is gradually enriched with new pieces of knowledge. Unless there is some formalization of entering new rules, inconsistencies in various forms may creep into the system. Further, there must be some sort of partitioning in the knowledge base; otherwise, the system will suffer from inefficient pattern matching cycle.

The chapter will cover various issues relating to validation, verification and maintenance of knowledge-based systems.

21.2 Validation of Expert Systems

Validation of an expert system involves comparing the performance of the system with an ideal system (or the expert) and then updating the knowledge base and the inferential procedures, depending on the deviation of the expert system performance from that of the ideal system. Validating a system, thus, not only means checking its performance with the expert, but it also includes the procedures to adjust the systems' parameters. A schematic view for validation of an expert system is presented in fig. 21.3.

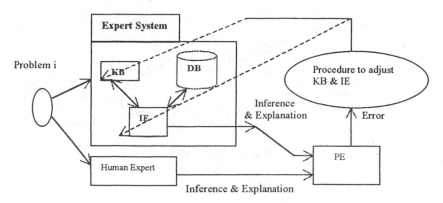

Fig. 21.3: Validating an expert system.

The expert system in fig. 21.3 comprises of three modules: the database (DB), the knowledge base (KB) and the inference engine (IE). The IE interprets the DB with the KB and thus derives inferences and presents explanation to the performance evaluator (PE). A human expert is also given the same problem, to infer its conclusion and present a suitable explanation for its conclusion. If the conclusions generated by the expert system deviates from the inferences by the expert, the explanation traced by the system and the expert may be compared to determine the disparity in rule firing. Let the state-space graph corresponding to the system and the expert be like those in fig. 21.4.

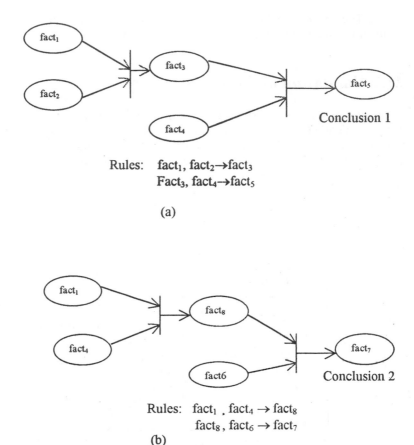

Rules: fact$_1$, fact$_2$→fact$_3$
         Fact$_3$, fact$_4$→fact$_5$

(a)

Rules: fact$_1$. fact$_4$ → fact$_8$
         fact$_8$, fact$_6$ → fact$_7$

(b)

Fig. 21.4: State-space graph used for explanation tracing:
(a) by the expert system, (b) by the expert.

A comparison of the two state-space graphs (Petri nets) in fig. 21.4 reveals that the conclusions derived by the expert are different from that derived by the system. The explanation traced by them also differs considerably. For instance, when fact$_3$ is derived by the system from fact$_1$ and fact$_2$, the expert derives fact$_8$ from fact$_1$ and fact$_4$. Since {fact$_1$, fact$_2$, fact$_4$ and fact$_6$} are available in the database, one possible reason of excluding the firing of the rule: fact$_1$, fact$_4$→fact$_8$ by the system is because of the lack of this knowledge from the KB of the expert system. The validation scheme should, therefore, ensure inclusion of this rule in the knowledge base of the expert system.

21.2.1 Qualitative Methods for Performance Evaluation

Validation of an expert system is carried out by qualitative means. Qualitative Validation employs subjective comparison of performances [4]. This section reviews a few methods for qualitative validation for expert systems. Among the qualitative validation schemes, the most popular is Turing test [7] and sensitivity analysis.

Turing Test: This test is named after Alan Turing, a pioneering AI researcher. Turing's view on intelligence can be stated as follows. A problem solving machine is intelligent, if a person trying to assess the intelligence of the machine cannot determine its identity, while working with a human problem solver and a computing machine. This method of testing intelligence of a machine is called Turing test. Turing's test has successfully been applied to evaluate performance of expert systems like MYCIN and ONCOCIN [4].

For assessing the performance of an expert system by Turing's test, the expert based on whose expertise the system is built and the expert system are assigned common problems. The assessment of the system performance is then compared with that of a human expert by another expert, without disclosing the performer's identity. If assessments can be measured objectively, then statistical techniques may be employed to measure the variations or consistency between the performance of the expert and the system. Examples of objective measurement of performance include number of correct inferences or the level of performance in grade points like excellent, good, fair, poor, etc. Turing's test, however, is not applicable to systems, where finding objective measurement of performance itself is a complex problem and thus is not amenable.

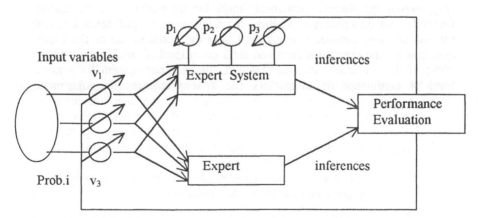

Fig. 21.5: Validating an expert system performance through sensitivity analysis.

Sensitivity Analysis: 'Sensitivity' in electrical engineering means change in response of a system, when there is even a small change in its input excitation. Thus an electrical system M_1 is more sensitive than the system M_2, if M_1 when compared to M_2 responds to smaller changes in the input. Sensitivity in expert systems, however, has a wider meaning. It corresponds to a change in inferences for a change in input variables or parameters of the expert system. Fig. 21.5 describes the scheme for validating an expert system performance through sensitivity analysis.

Let the value of the input variables supplied by the problem be $v_1 = \underline{v}_1$, $v_2 = \underline{v}_2$ and $v_3 = \underline{v}_3$ and the internal parameters (like certainty factors, conditional probabilities, etc.) of the expert system be $p_1 = \underline{p}_1$, $p_2 = \underline{p}_2$, $p_3 = \underline{p}_3$. Suppose, the inferences derived by the expert system match the inferences drawn by the expert for the above settings of input variables and parameters. The performance evaluator now adjusts each variable/parameter one by one by a small amount over their current value and observes the change in response of the system and the expert. If the responses are closer (w.r.t. some evaluator), then the design of the expert system is satisfactory; otherwise the system has to be updated with new knowledge or its inference engine has to be re-modeled. Sensitivity analysis, though a most powerful qualitative method for performance evaluation, has not unfortunately been exploited to any commercial expert systems. We hope that in the coming decade, it will play a significant role in performance evaluation of commercial systems.

21.2.2 Quantitative Methods for Performance Evaluation

When the responses of an expert system can be quantified (numerically), we may employ quantitative (statistical) tools for its performance evaluation. Generally, for quantitative evaluation of performance, a **confidence interval** for one or more measure is considered, and the performance of the expert system w.r.t. the expert is ascertained for a given level of confidence (95% or 99%). If the confidence interval obtained, for an expert system for a given level of confidence, is satisfactory the system requires no adjustment; otherwise we need to retune the knowledge base and the inference engine of the system. An alternative way of formulation is to define a hypothesis H_0 and test whether we would accept or reject the hypothesis w.r.t. a predefined performance range. The hypothesis H_0 may be of the following form.

> **H_0:** *The expert system is valid for the acceptable performance range under the prescribed input domain.*

Among the quantitative methods, the most common are paired t-test [4] and Hotelling's one sample T^2 test.

Paired t-test: Suppose $x_i \in X$ and $y_i \in Y$ are two members of the sets of random numbers X and Y. Let X and Y be the inferences derived by the expert system and the expert respectively. Also assume that for each x_i there exists a unique corresponding y_i. The paired t-test computes the deviation $d_i = x_i - y_i$, $\forall i$ and evaluates the standard deviation S_d and mean $\underline{d}$ for d_i, $1 \le i \le n$, where there exist n samples of d_i. A confidence interval for derived d is now constructed as follows:

$$\underline{d} - t_{n-1,\alpha} \le d \le \underline{d} + t_{n-1,\alpha}$$

where $t_{n-1,\alpha}$ is the value of the t-distribution with n-degrees of freedom and a level of confidence α. The hypothesis H_0 is accepted, if $d = 0$ lies in the above interval.

One main difficulty of realizing the t-test is that the output variables of the system have to be quantified properly. The method of quantification of the output variables for many systems, however, is difficult. Secondly, paired t-test should be employed to expert systems having a single output variable x_i (and y_i for the expert), which may be obtained for n cases. It may be added that for multivariate (i.e., systems having more than one variable) responses, t-test should not be used, as the variables x_i may be co-related and thus the judgement about performance evaluation may be erroneous. Hotelling's one sample T^2-test may be useful for performance evaluation for multivariate expert systems.

Hotelling's one sample T^2-test: Suppose that the expert system has m output variables. Thus for k set of input variables, there must be k output vectors, each having m scaler components. The expert, based on whose reference the performance of the system will be evaluated, should also generate k output vectors, each having m components. Let the output vectors of the expert system be $[X_i]_{m \times 1}, 1 \le i \le k$ and the same generated by the expert be $[Y_i]_{m \times 1}, 1 \le i \le m$. We now compute error vector $E_i = X_i - Y_i$, $1 \le \forall i \le k$ and the mean (%) of error vectors E_i, $1 <= i <= k$. The one sample T^2-test is then employed to check whether the % is significantly different from the null vector [4]. If the difference is not significant, then the hypothesis H_0 is acceptable.

21.2.3 Quantitative Method for Performance Evaluation with Multiple Experts

The last section discussed the methods for evaluating performance of an expert system based on the opinion of a single expert. This section explores the possibility of measuring the performance of an expert system based on multiple   expert responses. The main problem here is to measure the consistency among the experts, often called *inter-observer reliability*. The *'inter-class correlation co-efficient'* is a common measure for consistency among the experts. After a correlation co-efficient is estimated, we can use a related statistic to compare the joint expert agreement with the system [4]. If the experts' opinions are categorical variables like good, average, poor, etc., rather than continuous variables, *Kappa statistics* may be used to measure their composite reliability and then a related statistic to compare their joint agreement with the system agreement [4].

21.3   Verification of Knowledge Based System

Errors creep into knowledge based systems in various stages. First during knowledge acquisition phase, experts miss many points to highlight to the knowledge engineers. Secondly, the knowledge engineer misinterprets the expert's concept and encodes the semantics of expert's concept incorrectly. Thirdly, there is a scope of programming error. Lastly, as the knowledge base is developed in incremental fashion over several years, the entry of inconsistency, redundancy and self-referencing (circularity) in the knowledge base cannot be ignored. The knowledge base of an expert system, thus, needs to be verified before it is used in commercial expert systems.

Graph theoretic approaches [9-10] have been adopted for eliminating various forms of shortcomings from the knowledge-base. One such well-known graph is the Petri net, which has already been successfully used in knowledge engineering for dealing with imprecision of data and uncertainty of knowledge. In this section, we will use Petri nets for verification of knowledge base. It may be noted that various models of Petri nets are being used in knowledge engineering, depending on the type of application. The model of Petri nets that we will use here is an extension of the model presented in chapter 8. For the sake of convenience of the readers, we represent a set of knowledge by a Petri net (fig. 21.6).

Knowledge base:

1.   *Ancestor (X, Y) ← Parent (X,Y).*
2.   *Ancestor (X, Y) ← Parent (X,Z), Ancestor (Z,Y).*
3.   *Parent (d, j).*
4.   *Parent (j, m).*

The Petri net shown in fig. 21.6 consists of a set of places P, a set of transitions Tr, arc functions A and two tokens at place p_2 given by

$P = \{p_1, p_2\}$
$Tr = \{tr_1, tr_2, tr_3, tr_4\}$
$A = \{<X,Y>, <Z,Y>, <X,Z>\}$.
Tokens at $p_2 = \{(d, j), (j, m)\}$.

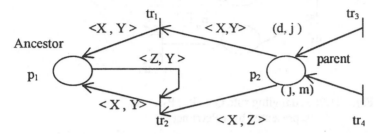

Fig. 21.6: A Petri net representation of a knowledge base.

The place p_2, transition tr_1 and place p_1 in sequence describes rule 1 in the given knowledge base. Place p_2, place p_1, transition tr_2 and consequent place p_1 together represent rule 2 of the system. The third and the fourth rules are presented by tokens of place p_2.

We now describe different forms of incompleteness of the knowledge base and their representation by Petri nets.

21.3.1 Incompleteness of Knowledge Bases

We will formalize three basic forms of inconsistency discussed below.

1. **Dangling condition:** Given a rule: $Q \leftarrow P_1, P_2, ..., P_n$, where P_i and Q are predicates; the arguments of the predicates are not presented here for brevity. Now, if any of the premises/antecedents of the rule are absent from the database and consequent part of all rules, then that rule can never be fired. Such a rule is called a **dangling rule**.

A dangling rule is represented by a Petri net in fig. 21.7. Here, the predicate P_1 of the antecedent part of rule: $Q \leftarrow P_1, P_2, ..., P_n$, is absent from both the available database and the consequent part of any rule. Consequently, P_1 can never be generated by firing of any rules. As a result, the last rule also cannot fire because of the non-existence of P_1 from the database. Many 'recognition-act' cycles are unnecessarily lost because of the presence of such dangling rules in the knowledge base.

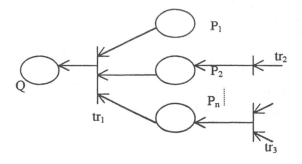

Fig. 21.7: A dangling rule $Q \leftarrow P_1, P_2 \dots P_n$
represented by a Petri net.

2. **Useless Conclusion:** When the predicates in the consequent part of a rule are absent from the antecedent part of all rules in the knowledge base, then that consequent predicate is called a **useless conclusion** [9-10]. In the Petri net, shown in fig. 21.8, R is a useless conclusion as no other rules use it as an antecedent predicate.

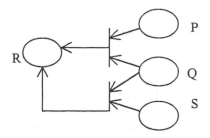

Fig. 21.8: R is a useless conclusion.

3. **Isolated Rules:** A rule is said to be isolated, if all the predicates in its antecedent part are dangling conditions and its predicate in the consequent part is a useless conclusion. Fig. 21.9 describes an isolated rule.

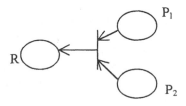

Fig. 21.9: An isolated rule represented by a Petri net.

21.3.2 Inconsistencies in Knowledge Bases

The following four types of inconsistencies are discussed in this section.

1. **Redundant Rule:** A rule is called **redundant** in a set of rules S, if all the predicates in the antecedent part of more than one rule in S are identical, the order of the predicates in the antecedent part of the rule being immaterial. For example, one of the following two rules is redundant.

$$Rule\ 1: R \leftarrow P_1, P_2, P_3.$$
$$Rule\ 2: R \leftarrow P_2, P_3, P_1.$$

It is to be noted that the Petri net representation of both the above rules is identical.

2. **Subsumed Rule:** If the consequent part of two rules includes some predicate, and the antecedent predicates of one rule are the subset of the antecedent predicates of the other rule, then the latter rule is called 'subsumed'. In the following set of rules, rule 2 is subsumed.

$$Rule\ 1: P \leftarrow Q, R.$$
$$Rule\ 2: P \leftarrow Q, R, S.$$

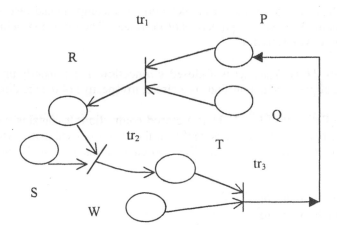

Fig. 21.10: Representation of a circular rule.

3. **Circular Rules:** A set of rules together is called circular, if their Petri net representation includes one or more cycles. The following are the examples of circular rules.

$$Rule\ 1: R \leftarrow P, Q$$
$$Rule\ 2: T \leftarrow R, S$$
$$Rule\ 3: P \leftarrow T, W$$

The Petri net corresponding to the above rules is given in fig. 21.10. Since it contains a cycle (through tr_1, tr_2 and tr_3), the corresponding rules are circular.

4. **Conflicting Rules:** Two rules are called conflicting if their antecedent part contains the same predicates, but the consequent parts are mutually exclusive. For example, the following two rules having some antecedent literal and different consequent literal are conflicting.

$$Rule\ 1: Bus(X) \leftarrow Four\text{-}wheeler(X), carries\text{-}passenger(X)$$
$$Rule\ 2: Car(X) \leftarrow Four\text{-}wheeler(X), carries\text{-}passenger(X)$$

It may be noted that conflict arises in two rules because of incorrect specification. For instance, to describe car (or bus) some other attributes of them are needed, rather than the common attributes only.

21.3.3   Detection of Inconsistency in Knowledge Bases

We shall formalize a scheme for the detection of incomplete and inconsistent rules by using the structural properties of Petri nets. The following definitions in this regard will be useful.

Definition 21.1: A subnet is a **closed connection**, if its underlying graph consisting of places and transitions is a circle, irrespective of the arc direction.

The Petri net of fig. 21.11 is a closed connection. It contains n closed patterns of places and transitions that form the cycles (closed path), when the direction of the arcs are ignored. The closed patterns are listed below for convenience.

i) p_1, tr_1, q, tr_2, p_1,
ii) p_2, tr_1, q, tr_2, p_2,
....
....
n) p_n, tr_1, q, tr_2, p_n.

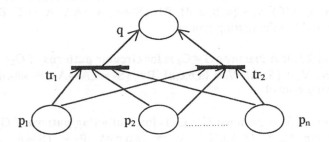

Fig. 21.11: A closed connection.

A generic form of closed pattern thus is given by place$_i$, transition$_j$, place$_k$, transition$_r$, place$_i$. Formally, if we denote

> a place by 'a',
>
> the arc from a place to a transition by 'b',
>
> the transition by 'c' and
>
> the arc from a transition to a place by 'd',

then we could represent the closed connection of fig. 21.11 by (abcdadcb) (abcdadcb)....(abcdadcb) where a sub-string in each pair of parenthesis describes a closed connection in the pattern. It may, however, be added that a certain order is implied here, when counting the closed connections.

Definition 21.2: Given two special strings α = abcd and β= adcb, the resulting pattern,

i) $(\alpha\beta)^n$, n>0 corresponds to redundant or subsumed rules,
ii) α^n, n>0 corresponds to circular rules, and
iii) $(\alpha^2\beta^2)^n$, n > 0 corresponds to conflicting rules,

where $\alpha\beta$ denotes concatenation of α and β and x^n denotes concatenation of x with itself n times.

We now present a grammar for syntactic recognition of the different patterns. Let G = (V$_n$, V$_t$, P, S) where the grammar G has four tuples; V$_n$ = a set of non-terminals, V$_t$ = a set of terminals, P = a set of production/re-write rules, S = a starting symbol. The following definitions of the grammar are useful for recognizing four distinct types of patterns.

Definition 21.3: A grammar $G = G_1$ is for **redundant or subsumed pattern** if $V_n = \{S, A, B, C\}$, $V_t = \{a, b, c, d\}$, $P = \{S \rightarrow A, A \rightarrow AA, A \rightarrow BC, B \rightarrow abcd, C \rightarrow adcb\}$ and S is the starting symbol.

Definition 21.4: A grammar $G = G_2$ is for **circular patterns**, if $G_2 = \{V_n, V_t, P, S\}$ where $V_n = \{S, A\}$, $V_t = \{a,b,c,d\}$, $P = \{S \rightarrow A, A \rightarrow AA, P \rightarrow abcd\}$ and S is the starting symbol.

Definition 21.5: A grammar $G = G_3$ is for **conflicting pattern** if $G_3 = \{V_n, V_t, P, S\}$, where $V_n = \{S,A,B,C\}$, $V_t = \{a,b,c,d\}$, $P = \{S \rightarrow A, A \rightarrow AA, A \rightarrow BBCC, B \rightarrow abcd, C \rightarrow adcb\}$ and S is the starting symbol.

Given a set of rules, they may be represented by Petri nets and the strings corresponding to the Petri net may be evaluated. Now, to check the type of inconsistency in the Petri net, each of the above three grammars has to be evolved one by one to generate languages for testing whether the string generated belongs to the language $L(G_i)$ of grammar G_i. If yes, then the type of inconsistency can be determined by the corresponding grammar. If no, then the process is repeated with the next grammar type, until the membership of the string is $L(G_i)$, $\exists i$ can be correctly identified. Zhang and Nguyen [9] designed a prototype software named PREPARE that includes the above procedure for determining inconsistency in knowledge base. For detecting incompleteness, the Petri net representation of rules provided them additional advantage, as the dangling condition, useless conclusions and isolated rules could be easily traced from the network itself.

21.3.4 Verifying Non-monotonic Systems

The last section was devoted to detect inconsistency and incompleteness of monotonic rule-based systems. Recently, Chang, Stachowitz and Combs [1] designed a scheme for verifying non-monotonic systems by detecting inconsistency and incompleteness of knowledge. We employ a new predicate $ab(X)$, which denotes X is abnormal, to overcome the '**qualification problem**'. It may be remembered that qualification problem deals with qualifying all possible exceptions of a fact. For instance, to state that Fly (birds), we need to qualify that Has-no-broken-wings(X), Has-power-of-flying(X), Not-penguin(X), etc. So, to represent birds fly, we add the qualifications as follows.

$$\text{Bird } (X) \wedge \text{Has-power-of-flying } (X) \wedge \text{Not-penguin } (X) \rightarrow \text{Fly } (X).$$

Since such qualifications have no finite limit in realistic cases, it would be convenient, if we write the above statement as presented below.

$$\text{Bird } (X) \wedge \neg \, ab(X) \rightarrow \text{Fly } (X)$$

A rule may also include a number of abnormal predicates. For instance, one simple rule to test the capability of car driving of X is given below.

Adult (X) $\wedge$ Car (Y) $\wedge$ $\neg$ab (X) $\wedge$ $\neg$ab (Y)$\rightarrow$Can-drive (X,Y).

We would now prefer to write ',' instead of '$\wedge$' in the L.H.S. of the '$\rightarrow$' operator. Thus the above rule can be expressed as

Adult (X), Car(Y), $\neg$ab(X), $\neg$ ab(Y) $\rightarrow$ Can-drive(X,Y).

In this section, *we would use a closed-world model for all predicates, except the ab predicates and postpone the commitment of the closed-world model for the ab predicates until we do not see other evidences* [1]. We now illustrate the above principle by an example. Consider the following knowledge base.

    1. Tall persons are normally fast-runners.
    2. Diseased persons are normally not fast-runners.
    3. John is a tall person.
    4. John is diseased.

The above pieces of facts and knowledge are represented below in non-monotonic logic using ab predicates.

    1. Tall(X), $\neg$ab$_1$(X)$\rightarrow$ Fast-runner(X).
    2. Diseased (X), $\neg$ ab$_2$(X) $\rightarrow$ $\neg$ Fast-runner(X).
    3. Tall (john).
    4. Diseased (john)

Now, postponing commitment to ab$_1$ and ab$_2$ predicates, we by resolution principle find

    $\neg$ ab$_1$ (john) $\rightarrow$ Fast-runner (john)
    $\neg$ ab$_2$ (john) $\rightarrow$ $\neg$ Fast-runner (john).

Now, unless we know anything about $\neg$ab$_1$(john) and $\neg$ab$_2$ (john), these two contingent facts are not inconsistent. In fact, we would assume ab$_1$ and ab$_2$ to be false, unless we discover anything about them. But, it is to be noted that setting ab$_1$ and ab$_2$ false yields a contradiction Fast-runner (john) and $\neg$ Fast-runner (john). To solve this problem, we have to rank the level of ab$_1$ and ab$_2$. Suppose, we assume ab$_2$ is more likely to be false than ab$_1$, then we can infer $\neg$Fast-runner (john) is more likely to be true than Fast-runner (john).

We now define a strategy for redundancy checking.

Redundancy checking: If rank of ab_2 < rank of ab_1. Consequently, the rule having $\neg ab_1$ in the L.H.S. of '$\rightarrow$' operator will be redundant. With reference to our previous example, for the same reason, we preferred

$$\neg ab_2(john) \rightarrow \neg Fast\text{-}runner\ (john)$$
over $\neg ab_1(john) \rightarrow Fast\text{-}runner\ (john)$.

Here, the last rule is redundant. We now present the principle for inconsistency checking in non-monotonic systems.

Inconsistency checking: Suppose we want to check inconsistency between the rules:

$$\neg\ ab_2(john) \rightarrow \neg Fast\text{-}runner\ (john)$$
and $\neg ab_1(john) \rightarrow Fast\text{-}runner\ (john)$.

If the rank of ab_2 and ab_1 are equal, then the above rules are inconsistent, else they are consistent.

Lastly, we conclude this section with the principle of incompleteness checking in non-monotonic systems.

Incompleteness checking: A non-monotonic knowledge base in many cases cannot produce correct inferences because of incompleteness, i.e., lack of rules or fact. For instance, consider the following rules and facts.

1. Match (X), Struck (X), $\neg ab_1$ (X)$\rightarrow$Lighted (X).
2. Match (a).
3. Wet (a).
4. Struck (a).

Here, from rule 1, 2 and 4 we define:

$$\neg ab_1\ (a) \rightarrow Lighted\ (a)$$

which, however, is false as Wet (a) is true. In fact, this occurs due to incompleteness of knowledge that

Match (a), Wet (a), $\neg ab_2$ (a)$\rightarrow \neg$Lighted (a).

To overcome the incompleteness of knowledge, we may add this rule to the knowledge base with a setting of

$$\text{rank of } ab_2 < \text{rank of } ab_1.$$

21.4 Maintenance of Knowledge Based Systems

Maintenance of knowledge based systems has not been given much priority during the last two decades of historical development in AI. The need for maintenance of expert systems is recently felt, as many of the commercial expert systems nowadays require updating of their knowledge bases on a regular basis. It may be noted that an augmentation of the knowledge base causes a significant decrease in inferential throughput and thus efficiency of the system. Designing the knowledge base in a structured manner and partitioning it into smaller modules thus help improving the efficiency of an expert system. An obvious question naturally arises: can we partition the knowledge base arbitrarily? The answer to this is obviously in the negative. The knowledge base is generally partitioned in a manner such that the related rules belong to a common partition.

The issues to be addressed in this section include i) effects of knowledge representation on maintainability, ii) difficulty of maintenance of systems built with multiple knowledge engineers, iii) difficulty in maintaining the data and control flow in reasoning.

21.4.1 Effects of Knowledge Representation on Maintainability

Production systems have some inherent advantages of knowledge representation. First of all, it is the simplest way to encode knowledge and thus the experts themselves can play the role of knowledge engineers by directly coding their expertise. This reduces the scope of errors in semantic translation of 'human expertise' into machine intelligence. This idea, in fact, motivated expert system developers to use production systems. It has been noted recently that typical production systems offer no resistance to the entry of inconsistency to the knowledge base. Thus for maintenance of knowledge in production systems, we require to 'verify' the knowledge base to reduce the scope of inconsistency. This is explained in fig. 21.12.

The knowledge acquisition system in fig. 21.12 generates new pieces of knowledge, which are subsequently added to the existing knowledge base. The knowledge base is now verified to check the existence of contradiction or incompleteness in it. If methods for eliminating inconsistency and incompleteness are known, that must be executed to overcome this problem.

Lee and O'Keefe [2] recently made experiments to study the effect of knowledge representation on the maintainability of expert systems. They observed that the time required to update knowledge by production systems approach is a bare minimum. They, however, concluded that a frame is more efficient than production systems in connection with the maintenance of knowledge. This is due to the reason that for augmentation of knowledge, a few pointers have to be attached only to the existing frame systems, whereas the complete rules have to be encoded in production systems. Thus from the point of view of memory requirement, frame is undoubtedly more efficient than production systems.

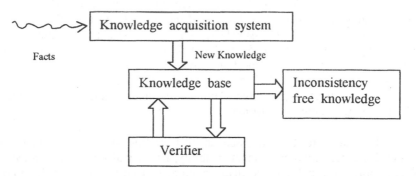

Fig. 21.12: Augmentation of knowledge, followed by verification, makes the knowledge base free from inconsistency.

21.4.2 Difficulty in Maintenance, When the System Is Built with Multiple Knowledge Engineers

An expert system, during the developmental cycle, is maintained by several knowledge engineers. The more the number of engineers, the more difficult is the maintenance problem. The problem is more severe, when languages like LISP are used for realization of the system, besides expert system shells or object oriented paradigms. This is obvious because each knowledge engineer has his own way of developing programs; so uniformity of coding may be lost, when the system is built by many knowledge engineers.

21.4.3   Difficulty in Maintaining the Data and Control Flow

The prototype model system of the expert system should be designed in a manner that the 'flow of control' is not affected by the augmentation of the knowledge bases. This is possible in relational model-based programs. On the other hand, in object-oriented systems, the flow of control does not have a clear path and is thus difficult to maintain. In relational models the flow of control of data and knowledge is unique, depending on their position in the program.

21.4.4 Maintaining Security in Knowledge Based Systems

'Security' [5], like other software, is equally useful in expert systems. It prevents unauthentic use / change of system resources. For instance, if there is no password check, anyone can change the knowledge base. A clever programmer may also affect the inference engine. Thus while updating knowledge, maintaining the security is a fundamental requirement in expert systems technology.

21.5 Conclusions

Performance evaluation of knowledge base systems is a fundamental issue of knowledge management. In this chapter we discuss different means for verifying the knowledge base. The key issues of knowledge verification are to keep the knowledge base free from incompleteness and inconsistency. For verifying non-monotonic systems the most important issue is to rank the level of the ab predicates in the same knowledge base. In my opinion, here the binary logic fails to yield a stable inference. In fact, multi-level logic may be used to overcome this difficulty of non-monotonic systems. Currently, special emphasis is given on the maintenance of the knowledge-based systems. This is useful as every year around 800-1000 new pieces of knowledge [7] are being added to expert systems. While augmenting the knowledge base, checking for inconsistency should be carried out to keep the system running smoothly over several years. Security of knowledge based systems should also be maintained to prohibit the unauthorized access to the system resources.

Exercises

1. Given the following knowledge base R, determine which of the rules in set R is dangling. R = {IF A, B THEN C, IF C THEN D}, Database = {A, C}.

2. Which is the useless conclusion in problem 1? Is there any isolated rule in the above system?

3. Determine the type of inconsistencies in the following knowledge bases (KB)

KB1= { IF A THEN B, IF B THEN A},
KB2= {IF A, B THEN C, If A THEN C},
KB3= { IF A, B THEN C, IF A, B THEN D}

4. Draw the Petri nets to illustrate the following type of rules and then detect the type of inconsistencies by using the Grammars, presented in the text:

 (a) redundant rules,
 (b) circular rules,
 (c) conflicting rules.

5. Examine inconsistency of the following system, if any:

 Man(X), Woman (Y), $\neg$ ab(X), $\neg$ ab (Y) $\rightarrow$ Loves (X, Y)
 Man (X), Woman (Y), Sister-of (Y, X) $\rightarrow\neg$ Loves (X, Y)

References

1. Chang, L. C., Stachowitz, A. R. and Combs, B. J., "Validation of nonmonotonic knowledge-based systems," in *Knowledge Engineering Shells: Systems and Technique*, Bourbakis, G.N., Ed., World Scientific, Singapore, 1993.

2. Lee, S. and O'Keefe, M. R., "The effect of knowledge representation on maintainability of knowledge-based systems," *IEEE Trans. on Knowledge and Data Engg.*, pp.173-178, Feb. 1996.

3. McGraw, L. K. and Briggs, H. K., *Knowledge Acquisition: Principles and Guidelines*, Ch.11, pp. 299-325, Prentice-Hall, Englewood Cliffs, NJ, 1989.

4. O'Keefe, M. R. and Balci, O., "Validating expert system performance," *IEEE Expert*, pp.81-89, Winter 1987.

5. O'Leary, D., "Expert System Security", *IEEE Expert*, pp.59-69, June 1990.

6. O'Leary, J. T., Goul, M., Moffilt, E. K. and Radwan, E. A., "Validating expert systems," *IEEE Expert*, pp. 51-58, June 1990.

7. Prearau, S. D., Gunderson, S. A., Reinke, E. R. and Adler, R. M., "Maintainability techniques in developing large expert systems", *IEEE Expert*, pp.71-80, June 1990.

8. Stefik, M., *Introduction to Knowledge Systems*, Morgan Kaufmann, San Mateo, CA, pp. 355-357, 1995.

9. Zhang, D. and Nguyen, D., "A tool for knowledge base verification," In *Knowledge Engineering Shells: Systems and Techniques*, Bourbakis, G. N., Ed., World Scientific, Singapore, 1993.

10. Zhang, D. and Nguyen, D., "PREPARE: A tool for knowledge-base verification," *IEEE Trans. Knowledge and Date Engineering*, vol.6, no.6, pp. 983-989, Dec.1994.

22

Parallel and Distributed Architecture for Intelligent Systems

An analysis of AI programs reveals that there exists a scope of massive parallelism in various phases of reasoning and search. For instance, the search problem in a given state-space can be subdivided and allocated to a number of processing elements. Further, many modules of the reasoning programs can also be realized on a parallel and distributed architecture. Generally, the parallelism in a reasoning program can appear at the following three levels: the knowledge representation level, the compilation and control level and the execution level. The chapter provides a brief introduction to the architecture of intelligent machines with special reference to representation and execution level parallelism in heuristic search, production systems and logic programming.

22.1 Introduction

As AI applications move from laboratory to the real world and AI software grows in complexity, the computational cost and throughput are becoming increasingly important concerns. The conventional Von Neuman machines are not suitable for AI applications because they are designed mainly for sequential, deterministic and numeric processing. The architectural features of AI machines depend largely on the tools and techniques used in AI and their applications in real world systems. The issues of designing efficient AI machines can be broadly classified into the following three levels:

1. Representational level
2. Control and Compilation level
3. Execution / Processor level.

The choice of an appropriate technique for knowledge representation and its efficient utilization are main concerns of representational level. Control and compilation level mainly deals with detection of dependencies and parallelism in the algorithms or programs for the problem and scheduling and synchronization of the parallel modules in the program. Execution level deals with maximization of the throughput by special architectures like CAM [8] for efficient search and matching of literals and fullest utilization of all the processors in the system.

22.2   Salient Features of AI Machines

An AI machine, in general, should possess the following characteristics [2]:

a) Symbolic processing: An AI machine should have the potential capability of handling symbols in the phase of acquisition of knowledge, pattern matching and execution of relational operation on symbols.

b) Nondeterministic computation: In a deterministic system, the problem states and their sequence of occurrences (or dependence) are known. In an AI problem, the occurrence of a state at a given time is unpredictable. For example, in a production system, which rule will be fired at a particular problem state cannot be predicted before arrival of the state. Such systems are usually called nondeterministic, and require special architecture for efficient and controlled search in an unknown space.

c) Dynamic execution: Because of nondeterministic nature of the AI computation, the size of data structures cannot be predicted before solving the problems. Static allocation of memory before execution of programs is, thus, not feasible for many AI programs. Dynamic allocation, which means creation of appropriate data structures as and when needed and returning the unnecessary data structures after execution of a module of a program, is preferred in AI programs. Besides dynamic allocation of memory, deadlocked tasks should also be dynamically allocated to different processors and communication topology should also be dynamically altered.

d) Massive scope of parallel and distributed computation: In parallel processing of deterministic algorithms, a set of necessary and independent tasks are identified and processed concurrently. This class of parallelism is called AND- parallelism. The large degree of nondeterminism in AI programs offers an additional source of parallel processing. Tasks at a non-deterministic decision point can be processed in parallel. The later class is called OR-parallelism. The above kinds of parallelism will be discussed in detail later.

Besides parallelism, many of the AI problems that include search and reasoning can be realized on a distributed architecture. System reliability can be improved to a high extent by such a distributed realization of the AI tools and models. The throughput and the reliability of the system thus can be enhanced jointly by fragmenting the system on a parallel and distributed architecture.

e) Knowledge management: Knowledge is an important component in reducing the complexity of a given problem. The richer is the knowledge base, the lesser is the complexity in problem solving. However, with the increase of the knowledge base, the memory requirement also increases and, thus, partitioning of the knowledge base is required. Besides storage of knowledge, its automated and efficient acquisition is also an important issue. While designing architectures of AI machines, management of knowledge for its representation and acquisition should be considered.

f) Open architecture: AI machines should be designed in a way, so that it can be readily expanded to support modification or extension of algorithms for the given problems.

The following sections elucidate various functional forms of parallelism in heuristic search and reasoning and illustrate the scope of their realization on physical computing resources.

22.3 Parallelism in Heuristic Search

In chapter 4 we covered the A* algorithm and the IDA* algorithm for heuristic search on OR graphs. The A* algorithm selects nodes for expansion based on the measure of $f = g + h$, where g and h denote the cost of generating a node (state) n and the predicted cost of reaching the goal from n respectively. The IDA* algorithm, on the other hand, selects a node n for expansion as long as the cost f at node n is within a pre-defined threshold. When no solution is found within the pre-defined threshold, it is enhanced to explore further search on the search space.

Because of non-determinism in the search process, there exists ample scope to divide the search task into possibly independent search spaces and each search sub-task may be allocated to one processor. Each processor could have its own local memory and a shared network for communication of messages with other processors. Usually there exist two common types of machines for intelligent search. These are *i) Single Instruction Multiple Data (SIMD) and ii) Multiple Instruction Multiple Data (MIMD) machines* [9]. In a SIMD machine, a host processor (or control unit) generates a single instruction at a definite interval of time and the processing elements work synchronously to execute that instruction. In MIMD machines, the processors, instructed by different controllers, work asynchronously. In this section, we present a new scheme for parallel realization of heuristic search on SIMD machines, following Mahanti and Daniels [6].

Mahanti and Daniels considered a SIMD architecture (fig. 22.1) with n processors: P_1, P_2, ..., P_n , each having a list for data storage. This list should have the composite features of PUSH and POP operations like those in a stack and DELETION operation like that in a queue. A host processor (controller) issues three basic types of commands to other processors (also called processing elements). These are

i)   *Balance the static load (on processors),*
ii)   *Expand nodes following guidelines and*
iii)   *Balance the dynamic loads (on processors).*

The static load balancing is required to provide each processor at least with one node. This is done by first expanding the search tree and then allocating the generated nodes to the processors, so that each get at least one node. Each processor can now expand the sub-tree rooted at one of the supplied nodes. The expansion of the sub-trees is thus continued in parallel. The expansion process by the processors can be done by either of two ways: **i) Partial Expansion (PE)** and **ii) Full Expansion (FE)**. The algorithms for these expansions along with the corresponding traces are presented in fig. 22.2- 22.5. During the phase of expansion, some processors will find many generated nodes, while some may have limited scope of expansion. Under this circumstance, the dynamic load balancing is required. Now, the host processor identifies the individual processors as **needy, wealthy** and **content** based on their possession of the number of nodes in their respective lists. A wealthy processor, that has many nodes, can donate nodes to a needy processor, which has no or fewer nodes. The transfer of nodes from the lists of wealthy processors is generally done from the rear end. The readers may note the importance of DELETION operation (of Queue) at this step. A content processor has a moderate number of nodes and thus generally does not participate in the process of transfer of nodes. The principle of node transfer has been realized in [6] by two alternative approaches: **i) stingy sharing** and **ii) Generous sharing**. The algorithms describing the sharing of nodes among processors and their traces will be presented shortly, vide fig. 22.6-22.7.

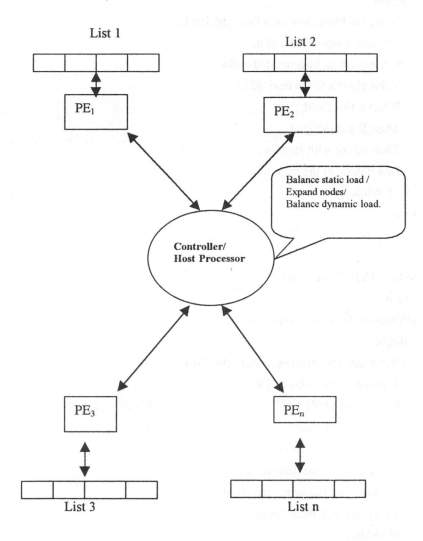

Fig.22.1: A SIMD architecture for IDA*.

Procedure Partial-Expansion

    Begin

      While the list is not empty **do**

      Begin

        Delete the front element n from the list L;

        Generate a new child c of n;

        If n has yet an ungenerated child

        Then place n at the front of L;

        If $f(c) \leq$ threshold

        Then If c is the goal

        Then return with solution;

        Else enter c at the front of L;

      End While;

    End.

Procedure Full-Expansion

    Begin

      While the list is not empty do

      Begin

        Delete the front element n from the list L;

        Generate a new child c of n;

        If $f(c) \leq$ threshold

        Then If c is the goal

        Then return with solution;

        Else enter c at the front of L;

        If n has yet an ungenerated child

        Then place n at the front of L;

      End While;

    End.

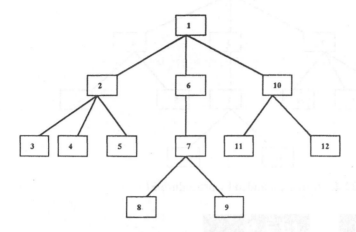

Fig.22.2: A tree expanded following the ascending order of nodes using the PE algorithm.

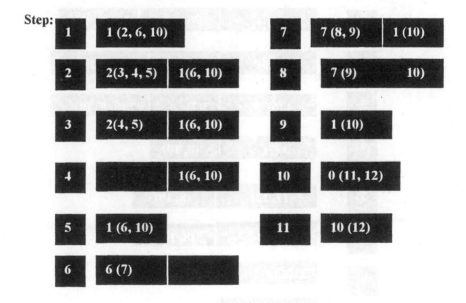

Fig. 22.3: Trace of the procedure PE on the tree of fig. 22.2 for a particular processor, where the elements of the list p (c_1, c_2, ...,c_n) represents a node p with its ungenerated children c_1, c_2, ...c_n.

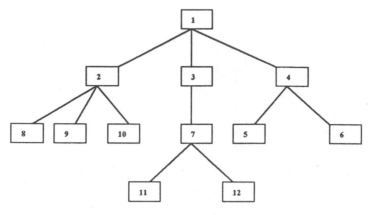

Fig.22.4: A tree expanded by procedure FE.

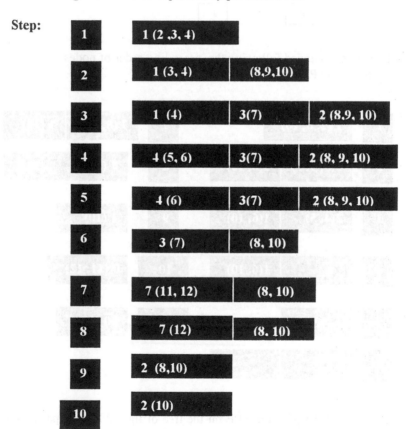

Fig. 22.5: Trace of the procedure FE on the tree of fig. 22.4.

Procedure Stingy-Sharing

   Begin

   Repeat

    Tag the empty lists as needy;

    Tag the lists containing multiple nodes as wealthy;

    For every wealthy list

     If its corresponding needy list exists

     Then donate a node from the rear end of the

      Wealthy list and put it at the front of the needy list;

    Until no lists are tagged needy or wealthy;

   End.

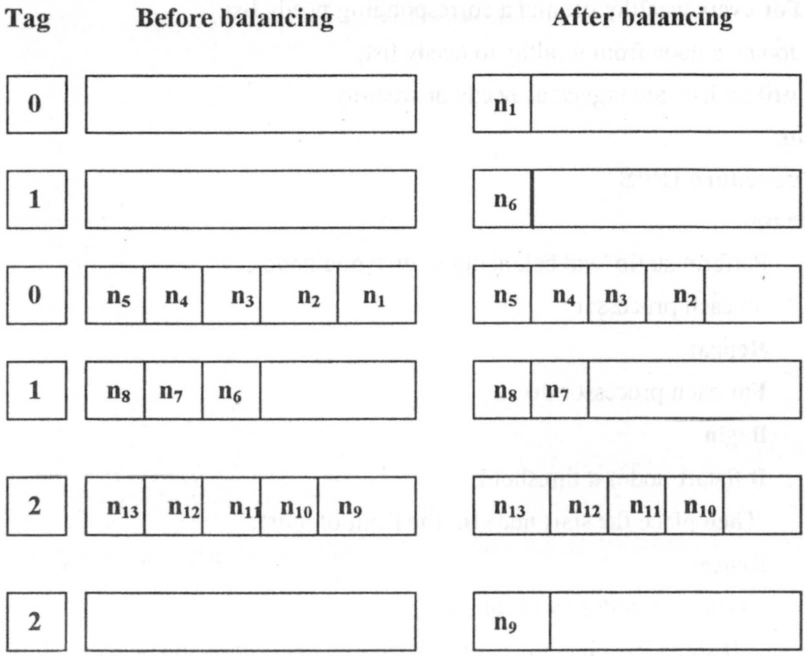

Tag	Before balancing	After balancing

Tag	Before balancing	After balancing
0		n_1
1		n_6
0	n_5 n_4 n_3 n_2 n_1	n_5 n_4 n_3 n_2
1	n_8 n_7 n_6	n_8 n_7
2	n_{13} n_{12} n_{11} n_{10} n_9	n_{13} n_{12} n_{11} n_{10}
2		n_9

Each wealthy list donates an element from its rear end to its corresponding
(having same tag number) needy list.

Fig. 22.6: One step of load balancing by procedure stingy sharing.

Procedure Generous-Sharing

Begin

Floor: = $\lfloor$ average list-size$\rfloor$;

 Ceil: = $\lceil$ average list size $\rceil$;

Repeat

   For every list with less than min (B-1, floor) nodes,

   tag the list as needy; // B = bound of needy lists//

   For every list with more than min (B, ceil) nodes, tag it wealthy;

   If no lists are tagged as needy **Then** tag all lists with

   min (B-1, floor) nodes as needy

  Else If no lists are tagged wealthy **Then** tag all lists

  with ceil nodes wealthy;

   For every wealthy list and a corresponding needy list,

   donate a node from wealthy to needy list;

 Until no lists are tagged as needy or wealthy;

End.

Procedure IPPS

Begin

1.   Perform static load balancing to give one node

       to each processor;

2.   **Repeat**

      For each processor **do**

      Begin

      If f(start node) $\leq$ threshold

       Then place the start node in the front of a list;

      Repeat

         Generate nodes by PE or FE;

          If some lists are empty apply dynamic load balancing

          by either of the two procedures mentioned above;

       Until the goal is found or all lists are empty;

2.   Update threshold value;

       Until the goal is identified;

   End.

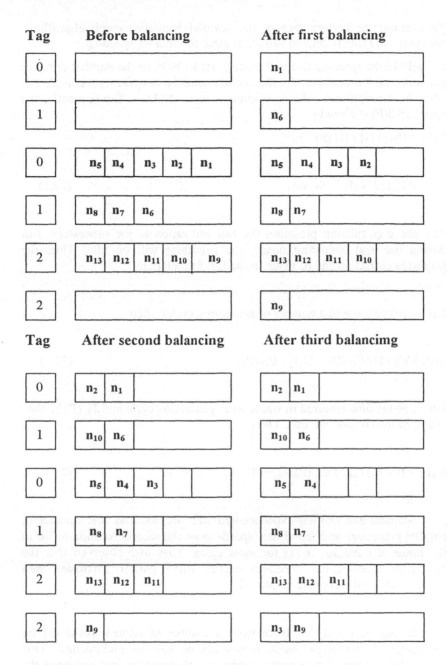

Fig. 22.7: Steps of GS load balancing with floor =2 and ceil =3.

To evaluate the performance of the parallel heuristic search algorithms, Mahanti and Daniels defined two alternative formula for speed-up.

Let $S(P)$ be the speed-up due to P processors; let $N(P)$ be the number of nodes generated on P processors. So, the average generation cycle time is $N(P) / P$. Thus for a uniprocesor, the generation time is $N(1) /1$. Consequently, the speed up $S(P)$ is given by

$$S(P) = [N(1) /1] / [N (P) / P]$$

$$= P N(1) / N (P) = S_1 \text{ ,say.} \qquad (22.1)$$

The above calculation presumes the full utilization of the processors. But during the load balancing phase, the processor will be idle. Thus the following correction can be made in the last formulation.

Let $C_G (P)$ be the actual number of generation cycles, then

$$S(P) = C_G (1) / C_G (P) = N(1) / C_G (P) \qquad (22.2)$$

Let t_G be the time required in single node generation cycle and $T_L (P)$ be the averaged load balancing time. Then

$$S (P) = [C_G (1) t_G] / [C_G (P) (t_G + T_L (P) / C_G (P)) \qquad (22.3)$$

Mahanti and Daniels experimented with 4K, 8K and 16K number of parallel processors and found the speed-up by the second formula to lie in the range of (2/3 to 3 /4) for most cases. They also observed that the maximum experimental speed-up occurs when partial expansion and generous load balancing are employed.

It may be noted that there exist a number of related works [9] on realization of intelligent search techniques on parallel architecture. This book, which is aimed at a wider audience, unfortunately does not have the scope to report them.

22.4 Parallelism at Knowledge Representational Level

Distributed representation of knowledge is preferred for enhancing parallelism in the system. A Petri net, for example, is one of the structural models, where each of the antecedent and the consequent clauses are represented by places and the if-then relationship between the antecedent and consequent clauses are represented by transitions. With such representation, a clause denoted by a place may be shared by a number of rules. Distribution of fragments of a knowledge on physical units (here places and transitions) enhances the degree of fault tolerance of the system. Besides Petri nets, other connectionist approaches for knowledge representation and reasoning include neural nets, frames, semantic nets and many others.

22.4.1 Parallelism in Production Systems

A production system consists of a set of rules, one or more working memory and an inference engine to manipulate and control the firing sequence of the rules. The efficiency of a production system can be improved by firing a number of rules concurrently. However, two rules where the antecedents of the second rule and the consequents of the first rule have common entries are in pipeline and therefore should not be fired in parallel. A common question, which may be raised: is how to select the concurrently firable rules. A simple and intuitive scheme is to allow those rules in parallel, which under sequential control of firing yield the same inferences. For a formal approach for identifying the concurrently firable rules, the following definitions [3] are used.

Let L_i and R_i denote the antecedents and consequents of production rule PR_i. Further, let

$$K_i = L_i \cap R_i. \tag{22.4}$$

The following definitions are useful to identify the concurrently firable rules in a Production system.

Definition 22.1: If the consequents of rule PR_1 have elements in common with the elements added to the database by firing of rule PR_2, i.e.,

$$R_1 \cap (R_2 - K_2) \neq \varnothing, \tag{22.5}$$

then rule PR_1 is said to be **output dependent** on rule PR_2 [3].

Definition 22.2: If the antecedents of rule PR_1 have elements in common with the part of data elements eliminated as a result of firing of rule PR_2, i.e.,

$$L_1 \cap (L_2 - K_2) \neq \varnothing, \tag{22.6}$$

then rule PR_1 is called **input dependent** on rule PR_2.

Definition 22.3: If persistent part of two rules PR_1 and PR_2 have common elements, i.e.,

$$K_1 \cap K_2 \neq \varnothing, \tag{22.7}$$

then one rule is said to be **interface dependent** to the other rule.

Definition 22.4: If the antecedents of rule PR_1 have elements in common with the elements added to the database due to firing of rule PR_2, i.e.,

$$L_1 \cap (R_2 - K_2) \neq \varnothing, \tag{22.8}$$

then rule PR_1 is called **input-output dependent** to rule PR_2.

Definition 22.5: If the consequents of rule PR_1 have elements in common with the data elements eliminated due to firing of rule PR_2, i.e.,

$$R_1 \cap (L_2 - K_2) \neq \varnothing, \tag{22.9}$$

then rule PR_1 is said to be **output-input dependent** to rule PR_2.

Definition 22.6: Two rules PR_1 and PR_2 are said to be **compatible**, if they are neither input dependent nor output-input dependent in both directions.

Definition22.7: Firing of rules PR_1 followed by PR_2 or vice versa and concurrent firing of PR_1 and PR_2 are **equivalent** from the point of view of changes in database, if the rules are compatible.

The list of concurrently firable rules of a Production system can be easily obtained following definition 22.7 with the help of a matrix called parallelism matrix, defined below.

Definition 22.8: In a Production system with n rules, a binary square matrix P of dimension (n x n) is called a **parallelism matrix**, if

$p_{ij} = 0$, when rule PR_i and rule PR_j are compatible

$\quad = 1$, otherwise.

The largest set of compatible rules can be easily determined by inspecting the following property of the P matrix.

Property: When $p_{ij} = p_{ji} = 0$, rules PR_i and PR_j are compatible.

Example 22.1: Let us consider a production system with 6 rules having the following P matrix:

$$P = \begin{pmatrix} 1 & 0 & 1 & 0 & 0 & 0 \\ 0 & 1 & 1 & 0 & 1 & 0 \\ 1 & 1 & 1 & 0 & 1 & 0 \\ 0 & 0 & 0 & 1 & 1 & 0 \\ 0 & 1 & 1 & 1 & 1 & 1 \\ 0 & 0 & 0 & 0 & 1 & 0 \end{pmatrix}$$

From the above matrix, the following set of compatible rules can be determined by using the property of the matrix, stated above.

$S1 = \{ PR_3, PR_4, PR_6 \}$

$S2 = \{ PR_1, PR_2, PR_4, PR_6 \}$

$S3 = \{ PR_1, PR_5 \}$.

For efficient execution of a rule-based system, the elements in a set of compatible rules should be mapped onto different processing elements. If the mapping of the compatible rules onto different processing elements is not implemented, the resulting realization may cause a potential loss in parallelism. While mapping rules onto processing elements, another point that must be taken into account is the communication overhead between the processing elements. When two rules are input dependent or input-output dependent, they must be mapped to processing elements, which are geographically close to each other, thereby requiring less communication time.

22.4.2 Parallelism in Logic Programs

In this section, the scope of parallelism in Logic Programs and especially in PROLOG is discussed briefly. A Logic program, because of its inherent representational and reasoning formalisms, includes four different kinds of parallelisms. These are AND parallelism, OR parallelism, Stream parallelism and Unification parallelism [2], [10].

a) AND -Parallelism

Consider a logic program, where the body of one clause consists of a number of Predicates, also called AND clauses, which may be unified with the head of other clauses during resolution. Generally, the resolution of the AND clauses is carried out sequentially. However, with sufficient computing resources, these resolutions can be executed concurrently. Such parallelism is usually referred to as AND parallelism. It is the parallel traversal of AND sub-trees in the execution tree [2]. For example, let us consider the following program.

1. Parent (Mo, Fa, X) ←Father (Fa, X), Mother (Mo, X).
2. Mother (jaya, tom) ←
3. Mother (ipsa, bil) ←
4. Father (asit, tom) ←
5. Father (amit, bil) ←

Query : ← Parent (Mo, Fa, bil)

The AND parallelism in the above Logic Program is demonstrated, vide fig.22. 8.

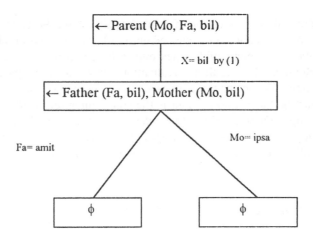

Fig. 22.8: Demonstration of AND parallelism, where concurrent resolution of two AND clauses of one clause (rule) with two other clauses takes place.

If the AND-subtrees of the execution tree are executed in parallel, the resulting unification substitutions may sometimes be conflicting. This is often referred to as **binding conflict**. For example, in the following Logic Program

     1. F(X)   ← A (X), B (X).
     2. A (1) ←
     3. B (2) ←

the literals A(X) and B(X) in the body of the first clause are resolvable with the heads of the second and the third clauses concurrently. However, the binding variable X takes the value {1 / X, 2 / X}, which are conflicting. In an AND-parallelism, where the goals/sub-goal, that include shared variables are allowed to be resolved independently in parallel, is generally called **Unrestricted AND-parallelism**. Some sort of synchronization of shared variables and some method of filtering from the set of variable bindings are required for correctly answering the query in such systems. Since considerable runtime overhead is introduced for implementing such a scheme, AND-parallelism is allowed, only when variable bindings are conflict-free. Such AND-parallelism is referred to as **restricted AND-parallelism** [2]. For detection of the conflict freedom of the variable bindings, a program annotation is used to denote which goals/sub-goals produce or consume variable values.

b) OR- Parallelism

In a sequential PROLOG program, each literal in the body of a clause is unified in order with the head of other clauses during the resolution steps. For example, consider the following program:

1. main ← A (X), P(X).
2. A(1) ←   b, c, d .
3. A(2) ←   e, f, g.
4. A (3) ←   h, i.
5. P (3) ←   p, q.

Here, to satisfy the goal, main, one attempts to unify the first sub-goal (principal functor) A (x) with A (1) and after setting X=1, one starts searching for P (1) in the head of other clauses but unfortunately fail to get so. The same process is then repeated for X=2, but unfortunately P(2) is not available in the head of some clauses. Finally, the goal main is satisfied by unifying A(X) and P (X) with heads A(3) and P(3) of the last two clauses.

However, given sufficient computing resources, it is possible to perform the unification of A(X) with A(1), A(2) and A(3) in parallel (vide fig. 22.9). Such concurrent unification of A(X) with OR- clauses A(1), A(2) and A(3) is called OR-parallelism. The prime difficulty with OR- parallelism, with respect to the last example, is the propagation of the correct bindings of variables to P(X). This, however, calls for some knowledge about the existence of P(3) as a head of some clauses. Perhaps, by maintaining concurrency of AND as well as OR-parallelism with the help of synchronization signals, this could be made possible in future PROLOG machines.

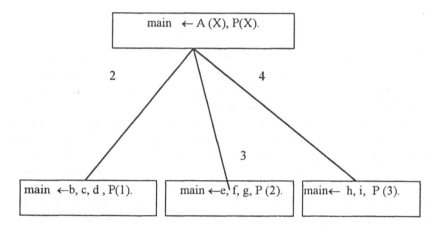

Fig. 22.9: Demonstrating the concurrent resolution of a clause with
           three OR clauses.

c) Stream Parallelism

Stream parallelism occurs in PROLOG, when the literals pass a stream of variable bindings to other literals, each of which is operated on concurrently. Literals producing the variable bindings are called **producers**, while the literals that use these bound values of variables are called **consumers**. As an example, consider the following logic program :

1. main   ← Int (N), Test (N), Print(N).
2. Int (0) ←
3. Int (N) ← Int (M), N is M + 1.

In the above program, Int (N) produces the value of N, which is passed on to Test (N) and Print (N) in succession. However, the procedure for Test (N) and Print(N) may be executed with old bindings of N, while new bindings of N may be generated concurrently for alternative solutions. Such parallelism is referred to as **Stream Parallelism**. Stream parallelism has similarity with pipelining. Here, Test(N) and Print(N) could be like one process, and Int(N) is another process. The former waits for the latter for data streams. There lies the similarity with pipelining.

d) Unification Parallelism

In a sequential PROLOG program, if a predicate in the body of a clause contains a number of arguments, then during unification of that predicate with the head of another clause, each argument is matched one by one. However, with adequate resources, it is possible to match the multiple arguments of the predicate concurrently with the corresponding positioned terms in a head clause. The parallelism of matching of variables of a predicate with appropriate arguments of other predicates is generally referred to as **Unification Parallelism**. As an example, consider the following logic program.

1. main $\leftarrow$ A (X,Y), ...

2. A (f (b,c).g(d)) $\leftarrow$

Here, the instantiation of X = f (b, c) and Y = g(d) should be done concurrently in a PROLOG machine that supports Parallel Unification.

22.5 Parallel Architecture for Logic Programming

It is clear from our discussion in the last section that 4 different types of parallelisms may co-exist in a logic program. Petri nets, which we have introduced in chapter 8, can be employed to represent these parallelisms by a structural approach. This section begins with an n-tuple definition of a Petri net. Next, the scope of parallelism in logic programs by Petri net models will be introduced. The principles of forward and backward firing of a Petri net will then be outlined. An algorithm for concurrent resolution of multiple clauses will be constructed subsequently with the Petri net models. Finally a schematic logic architecture of the system will be developed from the algorithm mentioned above. An analysis of the time estimate will then be covered to compare the relative performance of the architecture with respect to SLD-resolution, the fundamental tool for a PROLOG compiler.

22.5.1 The Extended Petri Net Model

An Extended Petri Net (EPN) [1], which we could use for reasoning under FOL, is an n-tuple, given by

EPN = { P, Tr, D, f, m, A, a, I, o } where

$P = \{p_1, p_2,......, p_m\}$ is a set of places,

$Tr = \{tr_1, tr_2,......, tr_m\}$ is a set of transitions,

$D = \{d_1, d_2,......, d_m\}$ is a set of predicates,

$P \cap Tr \cap D = \phi$,

Cardinality of P = Cardinality of D,

f: $D \rightarrow P^{\infty}$ represents a mapping from the set of predicates to the set of places,

m: $P \rightarrow (x_i, ..., y_i, X, ...,Y)$ is an association function, represented by the mapping from places to terms, which may include both constant(s), variable(s) and function of variables,

A: $(P \rightarrow Tr) \cup (Tr \rightarrow P)$ is the set of arcs, representing the mapping from the places to the transitions and vice versa,

a: $A \rightarrow (X, Y,...., Z)$ is an association function of the arcs, represented by the mapping from the arcs to terms,

I: $Tr \rightarrow P$ is a set of input places, represented by the mapping from the transitions to their input places,

0: $Tr \rightarrow P$ is a set of output places, represented by the mapping from the transitions to their output places.

Example 22.2: A Petri net consists of places, transitions, tokens at places and are functions. A detailed representation of the given rules with Petri nets is presented in fig. 22.10.

Given rules:

Son (Y, Z), Daughter (Y, Z) ← Father (X, Y), Wife (Z, X) .
Father (r, l) ←
Wife (s, r) ←
¬ Daughter (l, s) ←

It may be noted that the first rule includes two literals in the head, while the last rule includes negated literal in the head, both of which are violated in standard logic programs, but valid in predicate logic.

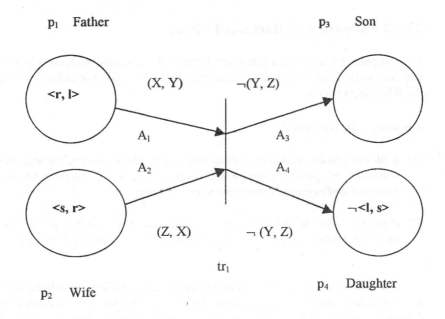

Fig. 22.10: Parameters of an EPN used to represent knowledge in predicate logic.

Here, $P = \{p_1, p_2, p_3, p_4\}$,
$Tr = \{tr_1\}$,
$D = \{$Father, Wife, Son, Daughter $\}$,
$f($Father $) = p_1$, $f($Wife$) = p_2$, $f($Son$) = p_3$, $f($Daughter$) = p_4$,

$m(p_1) = \ <r, l>$, $m(p_2) = \ <s, r>$, $m(p_3) = \ < \phi >$, $m(p_4) = \ \neg <l, s>$
initially and can be computed subsequently through unification of predicates in the process of resolution of clauses.

$A = \{A_1, A_2, A_3, A_4\}$,
$a(A_1) = (X, Y)$, $a(A_2) = (Z, X)$, $a(A_3) = \neg (Y, Z)$,
$a(A_4) = \neg (Y, Z)$ are the arc functions,
$I(tr_1) = \{p_1, p_2\}$, and $0(tr_2) = \{p_3, p_4\}$.

It is to be noted that if- then operator of the knowledge has been represented in the figure by tr_1 and the antecedent-consequent pairs of knowledge have been denoted by input (I) – output (0) places of tr_1. Moreover, the arguments of the predicates have been represented by arc functions.

22.5.2 Forward and Backward Firing

For computing a token at a place, the transition associated with the place is required to be fired. Firing of a transition, however, calls for satisfaction of the following criteria.

A transition tr_i is enabled,

(1) if all the places excluding at most one empty place, associated with the transition tr_i , possess appropriately signed tokens, i.e., positive literals for input and negative literals for output places.

(2) if the variables in the argument of the predicates, associated with the input and output places of the transition, assume consistent bindings.

It means that a variable in more than one arc function, associated with a transition, should assume unique value, which may be a constant or renamed variable [7] and should not be contradictory. For instance, if variable X assumes a value 'a' in one arc function and 'b' in another arc function of the same transition, then they are inconsistent. The bindings of a variable X in an arc function are evaluated by setting its value to the same positioned term in the token, located in the connected place.

When a transition is enabled, we can 'fire' the transition. After firing, tokens are generated as a result of resolution and are transferred to its input/output place. It is to be noted that the value of the token at these places is decided by the corresponding arc functions.

In the case of forward firing, the empty place is on the output side, whereas in the case of backward firing, it must be at the input side of the transitions.

Examples of forward and backward firing

The concept of forward and backward firing in Petri nets is illustrated below with examples.

Forward firing: Let us consider the Petri net shown in fig. 22.10. Here the variables in the arc function corresponding to the arcs A_1, A_2, A_3 and A_4 are given by (X, Y), (Z, X), ¬(Y, Z) and ¬(Y, Z) respectively. The places connected through these arcs contain tokens < r, 1 >, < s, r >, < φ > and <¬1,s> respectively. Thus the instantiated values of arc function (X, Y) are computed from its associated place p_1 and is given by a set: {r /X, 1 /Y}. Similarly, we get the other instantiations for the variables of the arc functions corresponding to arc A_2 and A_4. The resulting instantiations are

{ s / Z, r/ X} and

{1 / Y, s / Z}.

For the determination of the common variable bindings, we take an intersection of the value of X from the first and the second sets: {r /X, 1 /Y} and { s / Z, r/ X} respectively, which yields X = r. Similarly, the value of Y is evaluated from the first and the third set, which is found to be Y =1 and the value of Z is from the second and the third set is computed as Z = s.

It may be noted that in case the result of an intersection is a null set, we say that the corresponding variable bindings from the different arc functions are inconsistent.

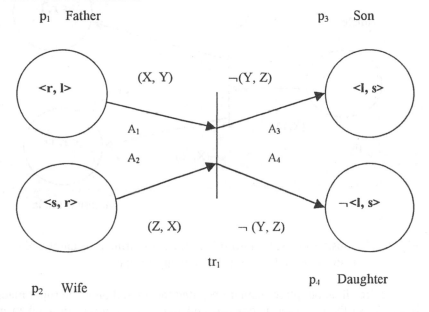

Fig. 22.11: The Petri net of fig. 22.10 after forward firing of transition tr_1.

After the bindings of the variables are evaluated by the above method, the value of the tokens are computed by assuming an oppositely signed arc function and the results are entered into all places associated with the fired transition. If the generated token already exists in a place, it need not be entered further. In the present context, the only new token is generated for place p_3, which is given by $\neg\{\neg (Y, Z)\} = <l, s >$. This has been entered in place p_3 of fig. 22.11.

Backward reasoning: For backward reasoning we consider the following knowledge base.

Paternal-uncle $(X, Y) \vee$ Maternal-uncle $(X, Y) \leftarrow$ uncle (X, Y)
$\neg$Paternal- uncle $< r, 1 > \leftarrow$
$\neg$ Maternal- uncle $< r, 1 > \leftarrow$

The Petri net representation of the above knowledge base is presented in fig. 22.12.

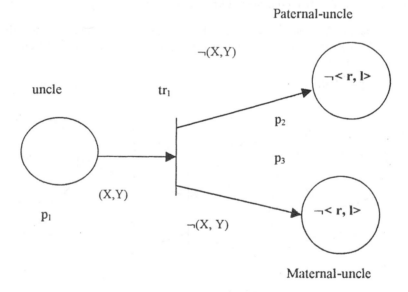

Fig. 22.12: Illustrating the backward firing: Before firing p_1 is empty; after firing a new token $\neg< r, 1 >$ will appear in p_1.

Here all output places contain negated tokens and only one input place is empty; so the transition is enabled and thus fires resulting in a token at place p_1. The value of a token in the present context is also evaluated through consistency analysis, as in the case of forward firing. The resulting consistent variable binding here is {r /X, 1 /Y}. Thus the new token at place p_1 is found to be $\{\neg (X, Y)\} = (\neg < r, 1 >)$. Note that there is no negation in the arc. But, you have to assume an opposite sign to the arc function for computing the token. Thus we assumed $\{\neg (X, Y)\}$ and hence $= (\neg < r, 1 >)$.

22.5.3 Possible Parallelisms in Petri Net Models

The Petri net model presented in this chapter supports AND, OR, stream and unification parallelisms, as outlined below.

AND - Parallelism

Consider the following three clauses, where clause 1 can be resolved with clause 2 and 3 concurrently. From the previous section, we call such parallel resolution of AND clauses A(X) and B(X) the AND–parallelism. Such type of parallelism can also be represented and realized by Petri nets as shown in fig. 22.13.

$$F(x) \leftarrow A(x) , B(X) \qquad (1)$$
$$A(1) \leftarrow \qquad (2)$$
$$B(1) \leftarrow \qquad (3)$$

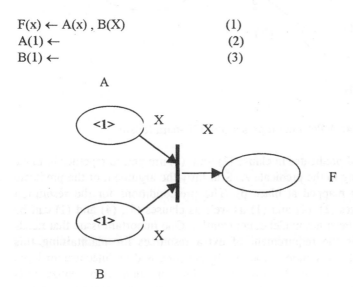

Fig. 22.13: A Petri net representing AND-Parallelism.

OR - Parallelism

Consider the following set of rules, where the predicate A(X) in clause 1 can be unified with the predicates A(1) and A(2) in clause 2 and 3.

$$F(x) \leftarrow A(X), B(Y) \qquad (1)$$
$$A(1) \leftarrow \qquad (2)$$
$$A(2) \leftarrow \qquad (3)$$
$$B(1) \leftarrow \qquad (4)$$

Since the resolution of the two OR-clauses (2) and (3) are done with clause 1 concurrently, we may refer to it as OR-parallelism. OR-parallelism can be realized easily with Petri net models. For instance, the above program when represented with a Petri net takes the form of fig. 22.14.

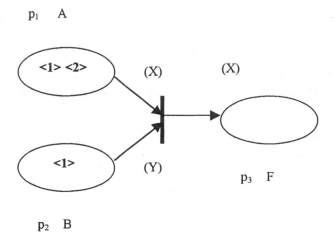

Fig. 22.14: A Petri net representing OR-parallelism.

The argument of predicates in clause (2) and (3) are placed together in place p_1, corresponding to the predicate A. Similarly the argument of the predicate in clause (4) is mapped at place p_2. The pre-conditions for the resolution process of clauses (2), (4) and (1) as well as clauses (3), (4) and (1) can be checked on the Petri net model concurrently. One important issue that needs mention here is the requirement of extra resources for maintaining this concurrency of the resolution process. In our proposed architecture for logic programs (to be presented shortly), we did not attempt to realize OR-Parallelism. However, with a little addition of extra logic circuits this could be implemented easily.

Stream Parallelism

Stream-parallelism is often referred to as a special form of OR-parallelism. This parallelism, as already stated, has similarity with the pipelining concept. In pipelining, processes that depend on the data or instructions produced by other process are active concurrently. Typical logic programs have inherent pipelining in the process of their execution. For instance, in the logic program presented below:

Grandson $(X, Z) \leftarrow$ Son(X,Y) , Father (Z,Y)	(1)
Father $(Y, X) \leftarrow$ Son(X, Y) , Male (Y)	(2)
Son $(r, d) \leftarrow$	(3)
Son $(l, r) \leftarrow$	(4)
Son $(c, l) \leftarrow$	(5)
Male $(d) \leftarrow$	(6)
Male $(r) \leftarrow$	(7)
Male $(l) \leftarrow$	(8)

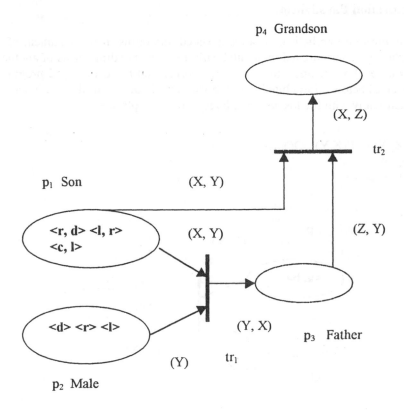

Fig. 22.15: Representing stream parallelism by Petri nets.

the result produced by clause (2) may be used by clause (1) to generate new solutions. Thus, while clause (1) is resolved with clause (2) and clause (4), the clause (2) will be resolved with clause (3) and clause (6). Thus the pipelining of clauses in the process of resolution exists in a logic program. However, it requires some formalism to continue this concurrent resolution in an automated manner.

Petri net models, presented in this chapter, can however be used to solve this problem. For example, the Petri net of fig. 22.15 demonstrates that until the token at place p_3 is computed, we cannot generate a token at place p_4. Thus the generation of a token at place p_4 is pipelined with that of place p_3. Further, when token generation at place p_4 and p_3 are made concurrently, so it is likely that transition tr_1 and tr_2 may fire concurrently when proper tokens are available at the input places of both transitions. Stream-parallelism has been realized in the architecture, to be presented shortly.

Unification Parallelism

In unification parallelism, as already stated, the terms in the argument of a predicate are instantiated in parallel with the corresponding terms of another predicate. For instance, the Petri net corresponding to the logic program presented below allows binding of the variables X and Y in the arc function concurrently with the tokens a and b respectively at place P.

$$R\ (Z, X) \leftarrow P(X,Y)\ ,\ Q(Y,Z) \qquad\qquad (1)$$
$$P\ (a, b) \leftarrow \qquad\qquad (2)$$
$$Q\ (b, c) \leftarrow \qquad\qquad (3)$$

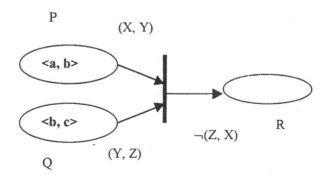

Fig. 22.16: A Petri net to illustrate the scope of parallel unification.

22.5.4 An Algorithm for Automated Reasoning

We shall use the following notations in the algorithm for automated reasoning.

Current-bindings (c-b) denote the set of instantiation of all the variables associated with the transitions.

Used-bindings (u-b) denote the set of union of the current-bindings up to the last transition firing.

Properly signed token means positive tokens for input places and negative tokens for output places.

The maximum value of the variable **no-of-firing** in the algorithm is set to number of transitions. This is because of the fact that number of transition denotes the largest possible reasoning path and the algorithm cannot generate new bindings after this many iterations.

Procedure Automated Reasoning
Begin
For each transition do
Par Begin
 used-bindings:= Null;
 Flag:= true; // transition not firable.//
Repeat
 If at least all minus one number of the input and
 output places possess properly signed tokens
 Then do
 Begin
   determine the set: current-bindings of all the
   variables associated with the transition;
 If (a non-null binding is available for all the variables) AND
   (current-bindings is not a subset of used-bindings)
  Then do Begin
    Fire the transition and send tokens to the input
    and the output places using the set current-bindings
    and following the arc functions with a presumed opposite sign;
    Update used-bindings by taking union with current-bindings;
    Flag:= false; //record of transition firing//
    Increment no-of-firing by 1;
  End

   Else Flag:= true;
  End;
  Until no-of firing = no-of-transition;
  Par End;
  End.

The above algorithm has been applied to the Petri net of fig. 22.17 and its trace is presented in table 22.1.

Table 22.1: Trace of the algorithm on example net of fig. 22.17

Time slot	Tran .	Set of c-b	Set of u-b	Flag=0, if c-b$\not\subset$ u-b =1, if c-b= u-i$\neq\{\phi\}$
First	tr_1	{r/x,d/y,a/z}	{{ϕ}}	0
cycle	tr_2	{n/x,a/y}	{{ϕ}}	0
Second	tr_1	{r/x,n/y,a/z}	{{r/x,d/y,a/z}}	0
cycle	tr_2	{d/x,a/y}	{{n/x, a/y}}	0
Third cycle	tr_1	{r/x,d/y,a/z}/ {r/x,n/y,a/z}	{{r/x,d/y,a/z}, {r/x,n/y,a/z}}	1
	tr_2	{n/x,a/y}/ {d/x,a/y}	{{n/x, a/y}, {d/x, a/y}}	1

22.5.5 The Modular Architecture of the Overall System

It is evident from the discussion in section 22.5.3 that most of the possible forms of parallelism of logic programs can be represented by Petri nets. Section 22.5.4 described a scheme for concurrent resolution of multiple clauses in a logic program. This section includes a mapping from Petri nets to logic circuits for efficient realization of the algorithm presented in the last section on a high speed inference engine.

Before executing the program, a compiler, specially constructed for this purpose, is employed to parse the given program for syntax analysis. On successful parsing, the variables used in the programs are mapped onto a specialized hardwired unit, called Transition History File (THF) register. The compiler also assigns the value of the variables, hereafter called tokens, at specialized hardwired units, called Place Token Variable Value Mapper (PTVVM). The sign of the arc function variables is also assigned to the PTVVM by the compiler.

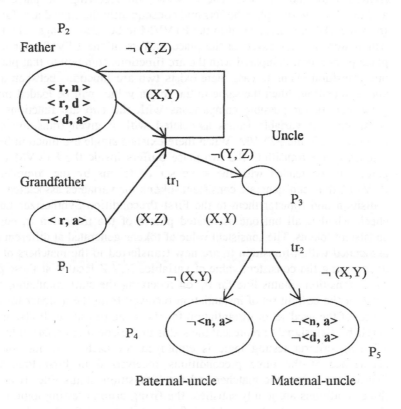

Fig. 22.17: An illustrative Petri net with initial token assignments used to verify the algorithm automated-reasoning.

The architecture of the proposed system consists of six major modules:

(i)   Transition History File (THF) for transition tr_i, $1 \leq \forall i \leq n$;

(ii)   Place Token Variable Value Mapper (PTVVM) for place p_j , $1 \leq \forall j \leq m$;

(iii)   Matcher for transition tr_i , $1 \leq \forall i \leq n$;

(iv)   First pre-condition synthesizer, realized with AND-OR Logic for transition tr_i , $1 \leq \forall i \leq n$;

(v)   Transition Status File (TSF) for transition tr_i , $1 \leq \forall i \leq n$;

(vi)   Firing criteria testing logic for transition tr_i , $1 \leq \forall i \leq n$

When a start command is issued from the power line, the Transition History Files for each transition generates the place names associated with that transition for the PTVVM. The PTVVM, on receiving the place names, addresses its internal place buffers and consequently the signed arc functions from the THF are propagated to the PTVVM to be used as tags. The tokens, which were already saved in the place buffers of the PTVVM for a given place p_j, are now compared with the arc functions common to that place and one transition. Thus in case there exists two arc functions between a place and a transition, then the same initial token value will be loaded into two place buffers for possible comparisons with the two arc functions. The consistency of variable values associated with a given transition can be checked inside the PTVVM. When there exists a single arc function between a place and a transition, the two place buffers inside the PTVVM for that place will be loaded with the same set of tokens by the compiler. The PTVVM thus can generate consistent tokens for variables associated with a transition and transfer them to the First Precondition Synthesizer Logic to check whether all but one associated places of the transition tr_i possesses consistent tokens. The consistent value of tokens generated at different places associated with a transition tr_i are now transferred to the matchers of tr_i for determining the consistent value of variables X,Y,Z located at these places. The Transition Status File for tr_i, on receiving the start command, checks whether the current set of instantiation received from the matcher for tr_i is a subset of the used set of instantiation for the same transition. It also issues a single bit flag signal to represent the status of the condition referred to above. A firing criteria testing logic is employed for each tr_i to test the joint occurrence of the three preconditions received from First Precondition Synthesizer Logic, the matcher and the Transition Status File. If all these three conditions are jointly satisfied, the firing criteria testing logic issues a 'fire tr_i' command to the matcher, informing it to transfer the value of new tokens for each place associated with tr_i to the PTVVM and the value of the current-bindings set to the Transition Status File. The process is continued until the number of transition firings becomes equal to the no-of-transitions.

22.5.6 The Time Estimate

An SLD resolution generally takes two clauses, one at a time. Thus for a logic program comprising of n clauses, we require (n-1) no. of resolutions. If T_2 denotes the time required for resolution of two clauses, then the time requirement for resolving all the n clauses, taking two at a time, becomes

$$T_{SLD} = (n-1)\, T_2 \qquad\qquad (22.10)$$

A typical logic program, however, includes a number of concurrently resolvable clauses. Let us assume that a logic program comprising of n clauses includes m_1, m_2 ,........., m_k no. of concurrently resolvable clauses. So the total number of resolutions required for such a program

$$= \{n - (m_1 + m_2 + \ldots + m_k) - 1\} + 1 \times k$$

$$= n - \Sigma \, m_i + (k-1)$$

$$\cong n - \Sigma \, m_i + k \tag{22.11}$$

Let us now assume that the time required for resolving m_i no. of clauses is T_i. Thus the total time required (T_u) for resolving ($n - \Sigma \, m_i + k$) no. of clauses becomes

$$T_u = (n - \Sigma \, m_i) \, T_2 + \sum_{i=1}^{k} T_i \tag{22.12}$$

If the time T_i could be set to T_2, the time required for binary resolution, then the resulting time for resolving ($n - \Sigma \, m_i + k$) number of clauses becomes T_E.

$$T_E = (n - \Sigma \, m_i) \, T_2 + \sum_{i=1}^{k} T_2$$

$$= (n - \Sigma \, m_i) \, T_2 + k \, T_2$$

$$= n - (\Sigma \, m_i - k) T_2 \tag{22.13}$$

It is evident that $T_E < T_{LSD}$ if $(\Sigma \, m_i - k) > 1$. The larger the value of $(\Sigma \, m_i - k)$, the better is the relationship $T_E \ll T_{SLD}$.

In our proposed architecture we kept provisions for resolving 5 clauses one at a time [1]. Thus if $m_i \le 5$, T_i remains constant. The schematic architecture describing the major modules and their pipelining is presented in fig. 22.18.

The logic architecture of the above system is available in [1]. An analysis of the logic architecture reveals that the PTVVM requires approximately 25 T_c, where T_c is the time period of the system clock. The THF requires approximately 1 memory access. The matcher requires 10 gate delay, while the FPS requires approximately 2 gate delay.

For the sake of convenience, the pipelining and parallelism of different modulus is presented in fig. 22.18. An estimation of the cycle time for firing a transition is computed in fig. 22.19 following the pipelining of the stages in fig. 22.18. It is evident from fig. 22.19 that the TSF works in parallel with the pipelined stages comprising of THF, PTVVM and Matcher or THF, PTVVM and FPS in sequence. The matcher and the FPS start working only when the PTVVM completes the current task. The time required for the matcher being larger than that of FPS we consider the time required by the matcher only following THF and PTVVM. It is also clear from the figure that the matcher finishes its tasks at time e, which is larger than the time d required by the TSF. An additional 2 gate delay followed by a few memory access cycles plus gate delay makes the total cycle time equal to g, which is the effective cycle time for firing a transition. The time requirement for each unit is shown in the fig. 22.19 itself and needs no further clarification.

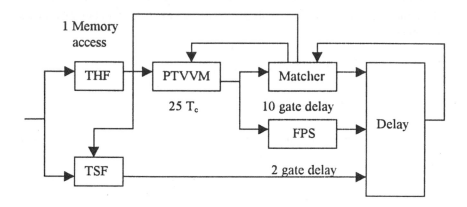

Fig. 22.18: Pipelining among the major modules of the architecture for the execution of FOL programs; each circuit corresponds to one transition.

Since a number of transitions are concurrently firable, it is expected that the execution of a complete logic program will require an integer multiple of this transition firing cycle time. Assuming a clock cycle time equal to T_c and ignoring smaller gate delays, it has been found that the total cycle time required for firing a transition is 25 T_c.

Thus assuming a 100M-Hz clock frequency, the cycle time for one transition is equal to 25×10^{-8} S = 250 nS (nanosecond). Now for the sample logic program represented by the Petri net of fig. 22.17, the total time for executing the program thus becomes 3 transition firing cycle = 3 x 250nS = 750 nS.

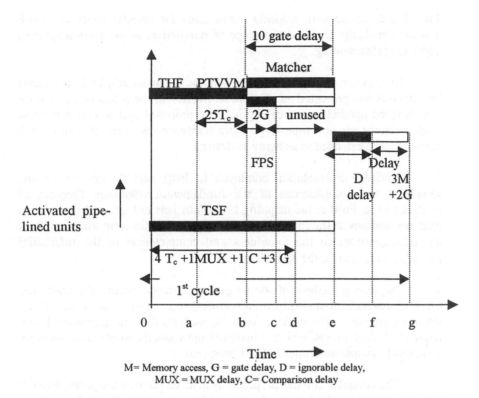

M= Memory access, G = gate delay, D = ignorable delay,
MUX = MUX delay, C= Comparison delay

Fig. 22.19: Cycle time estimation for the proposed parallel architecture.

To have an idea of the timings, let us consider the execution of a logic program on a PC. On the other hand, if the above logic is realized on a PC with a 100 megahertz clock, the system has to execute 8 resolutions for a total number of 9 clauses. For efficient execution of a PROLOG program, the predicate names and their arguments are to be stored in memory locations. Thus for unification of two predicates with n arguments, we require as memory $(2n+2)$ memory access time + $(2n+2)/2$ comparisons. Now, for resolving two clauses, each having m predicates, we require m^2 comparisons. Assuming $m = m_{max} = 4$ and n=9 as an example, we found a total time estimate to be = $8 \{2 \times 9+2 \} M + 8 \{(2 \times 9+2)/2 + (4)^2 \}C + 9$ stack handling + large compilation time= $160 M + 208 C + 9$ stacks + compilation, which is much larger than $(25 T_c) \times p$, where p is the number of concurrent transition firing cycles.

22.6 Conclusions

The chapter started with a SIMD architecture for parallel heuristic search and then gradually explored the scope of parallelism in production systems and logic programming.

The chapter emphasized a new scheme for analyzing logic programs by Petri nets and presented an efficient realization of the proposed scheme on a high speed parallel inference engine. The proposed system can resolve at most 5 clauses one at a time. With extra hardware resources, this limit of 5 clauses can be extended to as many as desired.

Principles of resolution employed in Petri nets do not violate the soundness and completeness of the fundamental resolution theorem of predicate logic. Further the mapping from Petri nets to logic architecture too does not add any extra constraints to the resolution theorem and thus does not pose questions to the soundness and completeness of the inferential procedures realized on the architecture.

The timing analysis of the proposed inference engine elucidates the basis of parallelism and pipelining among the various modulus of the inference engine. It is evident from the analysis that the transition firing requires 25 clock cycles, which is insignificantly small compared to memory access and comparison times in SLD programs.

The compiler for the proposed system serves two purposes. First it acts as a parser to the syntax of a given logic program. Secondly, it maps the variables, constants and predicates onto different modules of the architecture and initializes the flags of the system. The construction of the compiler and the VLSI testing of the architecture is under progress.

The complete realization of the parallel inference engine will serve as a new type database machine. Furthermore, it will be able to realize 'datalog programs' on efficient inference engines and demonstrate an alternate means to answer queries on high speed architecture.

Exercises

1. Determine which of the following rules are input dependent, output dependent, interface dependent, input-output dependent and output-input dependent on others. Hence, determine the compatible rules.

Rule 1: p, q, r → s, u, t
Rule 2: s, q, t →w, v, p
Rule 3: p, r, u →w, m, n
Rule 4: w, p, m→ s, v, n
Rule 5: r, t, w → v, p, s

Also construct the parallelism matrix P for the above rules.

2. Consider a processing element P_i, which is growing a tree, presented in fig. 22.20.

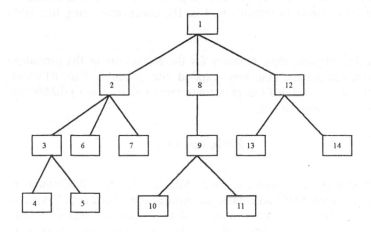

Fig. 22.20: A tree being expanded by the PE algorithm.

Show the snapshots of the list maintained by processor P_i by using procedure PE.

3. Number the nodes in the tree of fig. 22.20 in order of their generation by processor P_i following the procedure FE and also show the snapshots of the list maintained by the processor.

4. Illustrate the stingy sharing and generous sharing with your own example lists. Consider 8 lists: S_1 through S_8.

5. Construct a Petri net to represent the following pieces of knowledge.

Rule 1: P (X), Q(X) → R(X)
Rule 2: R(X), W(Y) →S(X, Y)

Rule 3: $S(X, Y) \rightarrow Z(Y, X)$
Rule 4: $\neg Z(b, a) \leftarrow$
Rule 5: $W(b) \leftarrow$
Rule 6: $Q(a) \leftarrow$

6.   Apply procedure automated-reasoning on the Petri net you constructed in problem 5 to compute the goal: $P(X) \rightarrow?$.

7.   Did you observe any parallelism in the firing of transitions in problem 6? If no, why?

8.   Extend the Petri net you constructed in problem 5, so as to demonstrate parallelism in firing of transitions. List the concurrent firing like table 22.1.

9.   Design the complete logic circuitry for the realization of the procedure automated-reasoning. You may expand the modules like PTVVM, matcher, etc. or design from grassroots level [*open ended problem for term paper or dissertation*].

References

[1]   Bhattacharya, A., Konar, A. and Mandal, A. K., "Realization of predicate logic based automated reasoning using Petri nets," *Proc. of 3rd Int. Conf. on Automation, Robotics and Computer Vision*, Singapore, pp. 1896-1900, 1994. Revised and extended version is to be submitted to *IEEE Trans. on Knowledge and Data Engg.*

[2]   Halder, S., *On the design of efficient PROLOG machines*, M.E. dissertation, Jadavpur University, 1992.

[3]   Ishida, T., "Parallel rule firing in production systems," *IEEE Trans. On Knowledge and Data Engg.*, vol. 3, no. 1, March 1991.

[4]   Kumar, V., Gopalkrishnan, S. P. and Kanal, N. L., Eds., *Parallel Algorithms for Machine Intelligence and Vision*, Springer-Verlag, Berlin, pp. 42-65, 1990.

[5]   Konar, A., *Uncertainty Management in Expert Systems Using Fuzzy Petri Nets*, Ph. D. dissertation, Jadavpur University, 1994.

[6]   Mahanti, A. and Daniels, J. C., "A SIMD approach to parallel heuristic search," *Artificial Intelligence*, vol. 60, pp. 243-282, 1993.

[7]   Murata, T. and Yamaguchi, H., " A Petri net with negative tokens and its application to automated reasoning," *Proc. of the 33rd Midwest Symposium on Circuits and Systems*, Calgary, Canada, pp. 762-765, 1990.

[8]   Naganuma, J., Ogura, T. and Yamada, I. S., " High speed CAM based architecture for a PROLOG machine (ASCA)," *IEEE Trans. on Computers*, vol. 37, no.11, Nov. 1988.

[9]   Powley, C. and Korf, E. R., "SIMD and MIMD parallel search" in *Proc.AAAI Spring Symposium on Planning and Search*, Stanford, CA, pp. 49-53, 1989.

[10]   Wah, B. W. and Li, J. G., "A survey on the design of multiprocessing systems for artificial intelligence applications," *IEEE Trans. on Systems, Man and Cybernetics*, vol. 19, no. 4, July / August 1989.

23

Case Study I: Building a System for Criminal Investigation

A new hierarchical scheme for criminal investigation, based on multi-sensory data including voice, fingerprint, facial image and incidental description, has been outlined in this chapter. The proposed system first attempts to classify a given fingerprint of a suspect into any of the six typical classes and then determines the fingerprint that matches best under the class. In case the fingerprint matching fails, an automated scheme for matching facial images is called for. On successful match, the suspect could be identified. In case the matching results in poor performance, the speaker identification scheme that works with phoneme analysis is utilized. When no conclusion about the suspect could be detected by voice, incidental description is used as the resource for criminal investigation. The incidental description based model is realized on a Petri net that can handle imprecision and inconsistency of data and uncertainty of knowledge about the criminology problem.

23.1 An Overview of the Proposed Scheme

The available techniques of automated criminal investigation generally rely on fingerprint and facial imagery of suspects. Unfortunately, classification of fingerprints by pixel-wise matching is tedious and the feature based schemes often lead to misclassification and hence improper matching. Moreover, facial imagery undergoes changes with aging and mood of the persons and thus matching of facial images in many occasions fails to identify the suspects. The chapter utilizes two more sets of information such as voice and incidental description to uniquely identify the suspects by a hierarchical approach. The steps of matching used in suspect identification follows a linear sequence, vide fig. 23.1.

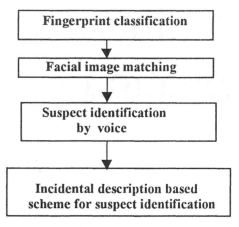

Fig. 23.1: A hierarchical scheme for suspect identification.

The matching of fingerprints of a suspect with a standard database, recorded by the investigation departments, requires two main steps. In the first step, the fingerprints of the suspect are classified into any one of the six typical classes, based on the number and location of two features named core and delta points [15]. After classification, the fingerprint of the suspect is compared with all members under that class by an algorithm of image matching, which is also used for matching facial images of the suspects. If the fingerprint matching is satisfactory, no further steps are required. Otherwise, the algorithm for matching facial images is invoked.

The 'image matching' algorithm attempts to partially match the facial image of the suspects with known images. The main trick in this matching lies in 'fuzzy membership-distance product', which keeps track of the important features in human faces and their relative distances. The matching scheme has other advantages of size and rotational invariance. This means that the matching scheme is insensitive to variation of image sizes or their angular rotation on the facial image plane. In case facial image matching also fails to identify the suspects, a voice classification scheme may be employed to check whether the suspect is a marked criminal of known voice.

The voice classification requires prior training instances. The input and the output training instances in the present context are speech features and recorded suspect number respectively. We trained a multi-layered feed-forward neural net with the known training instances. The training is given offline by the well-known back-propagation algorithm. During the recognition phase, only the speech features of the suspect are determined and supplied to the input of the neural net. A forward pass through the network generates the output signals. The node with the highest value in the output layer is considered to have correspondence with the suspect. In case these tests are inadequate for identification of the suspects, the incidental description is used to solve the problem.

The incidental description includes facts like Loved (jim, mita), Had-strained-relations-between (jim, mita) and may contain both imprecision and inconsistency of facts. We used a simplified model of fuzzy Petri net, presented in chapter 10, to continue reasoning in the presence of the above types of incompleteness of the database. The reasoning system finally identifies the culprit and gives an explanation for declaring the person as the culprit. The proposed system was tested with a number of simulated criminology problems. The field testing of the system is under progress.

The next section covers image matching as it has been used in both fingerprint and face identification from raw images.

23.2 Introduction to Image Matching

Fuzzy logic has been successfully used for matching of digital images [2], [3]. However, the methods of matching adopted in these works are computationally intensive and sensitive to rotation and size variation of images. Further, the existing matching techniques, which search a reference image among a set of images, often fail to identify the correct image in the presence of noise. The present work attempts to overcome these limitations by a new approach using the concept of 'fuzzy moments' [2].

In this work, a gray image has been partitioned into n^2 non- overlapped blocks of equal dimensions. Blocks containing regions of three possible characteristics, namely, 'edge', 'shade' and 'mixed-range' [14], are then identified and the sub-classes of edges based on their slopes in a given block are also estimated. The degree of membership of a given block to contain edges of typical sub-classes, shades and mixed-range is measured subsequently with the help of a few pre-estimated image parameters like average gradient, variance and difference of maximum and minimum of gradients. Fuzzy moment, which informally means the membership-distance product of a block $b[i, w]$ with respect to a block $b[j, k]$, is computed for all $1 \le i, w, j, k \le n$. A feature called 'sum of fuzzy moments' that keeps track of the features and their relative distances is used as *image descriptors*. The descriptors of an image are compared subsequently with the same ones of other images. We used an Euclidean distance measure to determine the distance between the image descriptors of two images. To find the best matched image among a set of images, we compute the Euclidean distance of the image descriptors of the reference image with all the available images. The image with the smallest Euclidean distance is considered to be the 'best matched image'.

This section has been classified into six sections. Section 23.2.1 is devoted to the estimation of fuzzy membership distributions of a given block to contain edge, shade and mixed-range. A scheme for computing the fuzzy moments and a method for constructing the image descriptors are presented in section 23.2.2. Estimation of Euclidean distances between descriptors of two different images is also presented in this section. An algorithm for image matching is presented along with its time complexity analysis in section 23.2.3. The insensitivity of the matching process to rotation and size variation of image is discussed in section 23.2.4. The simulation results for the proposed matching algorithm are presented in section 23.2.5. Implications of the results are included in section 23.2.6.

23.2.1 Image Features and Their Membership Distributions

A set of image features such as edge, shade and mixed-range and their membership distribution are formally defined in this section.

Definition 23.1: An **edge** is a contour of pixels of large gradient with respect to its neighbors in an image.

Definition 23.2: A **shade** is a region over an image with a small or no variation of gray levels.

Definition 23.3: A **mixed-range** is a region excluding edges and shades on a given image.

Definition 23.4: A linear edge segment that makes an angle α with respect to a well defined line (generally the horizontal axis) on the image is said to be an **edge with edge-angle** α. In this chapter we consider edges with edge-angle $-45°$, $0°$, $45°$ and $90°$.

Definition 23.5: Fuzzy membership distribution $\mu_Y(x)$ denotes the degree of membership of a variable x to belong to Y, where Y is a subset of a universal set U.

Definition 23.6: The **gradient [5]** at a pixel (x, y) in an image here is estimated by taking the square root of the sum of difference of gray levels of the neighboring pixels with respect to pixel (x, y).

Definition 23.7: The **gradient difference** (G_{diff}) within a partitioned block is defined as the difference of maximum and minimum gradient values in that block.

Definition 23.8: The gradient average (G_{avg}) within a block is defined as the average of the gradient of all pixels within that block.

Definition 23.9: The **variance** (σ^2) **of gradient** is defined as the arithmetic mean of square of deviation from mean. It is expressed formally as

$$\sigma^2 = \Sigma(G-G_{avg})^2 P(G)$$

where G denotes the gradient values at pixels, and P(G) [7] represents the probability of the particular gradient G in that block.

Fuzzy Membership Distributions: Once the features of the partitioned blocks in an image are estimated following the above definitions, the same features may be used for the estimation of membership value of a block containing edge, shade and mixed-range.

For these estimations we, however, require the membership distribution curves, describing the degree of a block containing edge, shade and mixed-range. These distributions have been assumed intuitively by common sense reasoning. To illustrate the concept of intuitive guess of the membership functions, we present example 23.1.

Example 23.1: In this example we demonstrate the membership curve for a block to contain edge, shade or mixed-range w.r.t. σ^2. In order to keep the method of estimation of membership values in a simplified way, we describe the intuitive curves by standard mathematical functions. For example, the membership curves for edge, shade and mixed-range w.r.t. σ^2 may be described as

$\mu_{shade}(b_{j\,k}) = e^{-a\,x^2}$, where $x = \sigma^2$, $a \gg 0$

$\mu_{edge}(b_{j\,k}) = 1 - e^{-b\,x^2}$, where $x = \sigma^2$, $b > 0$

$\mu_{mixed\text{-}range}(b_{j\,k}) = cx^2 / (d + ex^2 + fx^3)$, where $x = \sigma^2$, and c, d, e, f > 0.

The membership distributions with respect to G_{diff} and G_{avg} have also been defined analogously. Table 23.1 below presents the list of membership functions used in this chapter.

Table 23.1: Membership functions for features.

PARAMETERS	Mixed Range Membership	Edge Membership	Shade Membership
G_{avg}	$\dfrac{(\eta x^2)}{(\rho + \theta x^2 + \varphi x^3)}$ $\eta, \rho, \theta, \varphi > 0$	$1 - e^{-8\,x}$	$e^{-a\,x^2}$
G_{diff}	$\dfrac{\alpha x^2}{(\beta + \lambda x^2 + \delta x^3)}$ $\alpha, \beta, \lambda, \delta > 0$	$1 - e^{-bx^2}$	e^{-ax^4}
σ^2	$\dfrac{cx^2}{(d + ex^2 + fx^3)}$ c,d,e,f > 0	$1 - e^{-bx^2}$ b>0	$e^{-a\,x}$ a≫0

The membership values of a block b[j, k] containing edge, shade and mixed-range can be easily estimated if the parameters and the membership curves are known. The fuzzy production rules, described below, are subsequently used to estimate the degree of membership of a block b [j, k] to contain edge (shade or mixed-range) by taking into account the effect of all the three parameters together.

Fuzzy Production Rules: A fuzzy production rule is an If-Then relationship representing a piece of knowledge in a given problem domain. For the estimation of fuzzy memberships of a block b [j, k] to contain, say, edge, we need to obtain the composite membership value from their individual parametric values. The If-Then rules represent logical mapping functions from the individual parametric memberships to the composite membership of a block containing edge. The production rule PR1 is a typical example of the above concept.

PR1: *IF*   *(G_{avg} > 0.142)*   *AND*
                 (G_{diff}>0.707)   *AND*
                 ($\sigma^2 \cong 1.0$)
      THEN (the block contain edges).

Let us assume that the G_{avg}, G_{diff} and σ^2 for a given partitioned block have found to be 0.149, 0.8 and 0.9 respectively. The $\mu_{edge}(b_{j\,k})$ now can be estimated first by obtaining the membership values $\mu_{edge}(b_{j\,k})$ w.r.t. G_{avg}, G_{diff} and σ^2 respectively by consulting the membership curves and then by applying the fuzzy AND (minimum) operator over these membership values. The single valued membership, thus obtained, describes the degree of membership of the block b [j, k] to contain edge. For edges with edge-angle α, we use the membership curves and obtain the composite membership of a block containing edge with edge-angle α by ANDing the membership of a block containing an edge with the membership of its having an edge angle α.

The composite degree of membership of a block containing shade and mixed-range has been computed similarly with the help of more production rules, the format of which are similar to that of PR1.

23.2.2 Fuzzy Moment Descriptors

In this section, we define fuzzy moments and evaluate image descriptors based on those moments. A few definitions, which will be required to understand the concept, are in order.

Definition 23.10: **Fuzzy shade moment** $[M_{i\,w}]_{shade}^{jk}$ is estimated by taking the product of the membership value $\mu_{shade}(b_{j\,k})$ (of containing shade in the block b[j, k]) and normalized Euclidean distance $d_{i\,w,\,j\,k}$ of the block b[j, k] w.r.t. b[i, w]. Formally,

$$[M_{i\,w}]_{shade}^{jk} = d_{i\,w,\,j\,k} \times \mu_{shade}(b_{j\,k}) \tag{23.1}$$

Fuzzy mixed-range and edge moments with edge-angle α are also estimated using definition 23.10 with only replacement of the term "shade" by appropriate features.

Definition 23.11: The **fuzzy sum of moments (FSM), for shade** S_{iw}, w.r.t. block b[i, w] is defined as the sum of shade moments of the blocks where shade membership is the highest among all other membership values. Formally,

$$S_{i\,w} = \sum_{\exists jk} d_{i\,w,j\,k} \times \mu_{shade}(b_{j\,k}) \qquad\qquad (23.2)$$

where $\mu_{shade}(b_{j\,k}) \geq \mu_x(b_{j\,k})$, $x\,\varepsilon$ set of features.

The FSM of the other features can be defined analogously following expression (23.2).

After the estimation of fuzzy membership values for edges with edge-angle α, shades and mixed-range, the predominant membership value for each block and the predominant feature are saved. The FSMs with respect to the predominant features are evaluated for each block in the image. For each of six predominant features (shade, mixed-range and edges with edge-angle $-45°$, $0°$, $45°$ and $90°$) we thus have six sets of FSMs. Each set of FSM (for example the FSM for shade) is stored in an one dimensional array and is sorted in a descending order. These sorted vectors are used as descriptors for the image.

For matching a reference image with a set of known images, one has to estimate the image descriptors for the known images. Normally, the image descriptors for a known set of images are evaluated and saved prior to the matching process. The descriptors for the reference image, however, are evaluated in real time when the matching process is invoked. The time required for estimation of the descriptors, therefore, is to be reduced to an extent, whatever possible, to keep the matching process executable in real time.

The matching of images requires estimation of Euclidean distance between the reference image with respect to all other known images. The measure of the distance between descriptors of two images is evident from definition 23.12.

Definition 23.12: The **Euclidean distance, $[E_{i,j}]_k$** between the corresponding two k-th sorted FSM descriptor vectors V_i and V_j of two images i and j of respective dimensions $(n \times 1)$ and $(m \times 1)$ is estimated first by ignoring the last $(n - m)$ elements of the first array, where $n > m$ and then taking the sum of square of the elemental differences of the second array with respect to the modified first array having m elements.

It may be added that the elements of the second and the modified first array are necessarily non-zero.

Definition 23.13: The measure of distance between two images, hereafter called **image distance,** is estimated by taking exhaustively the Euclidean distance between each of the two similar descriptor vectors of the two images and then by taking the weighted sum of these Euclidean distances.

Formally, the distance $D_{r\,y}$ between a reference image r and a image y is defined as

$$D_{r\,y} = \sum \beta_k \times [E_{i\,j}]_k$$

where the suffix i and j in $[E_{i\,j}]_k$ corresponds to the set of vectors V_i for image r and V_j for image y, for $1 \leq I, j \leq 6$.

For identifying the best matched image (among the set of known images) with respect to the reference image, one has to estimate the image distance D_{ry} where y $\in$ the set of known images and r denotes the reference image. The image Q for which the image distance $D_{r\,q}$ is the least among all such image distances is considered the **best matched image.**

23.2.3 Image Matching Algorithm

The major steps of the image matching are presented in fig. 23.2 and the details are included in procedure Image-matching, given below:

Procedure Image-matching (IM_1, IM_2,........,IM_{m+1})
    //IM_1 = reference image //
   Begin
     For p:=1 to m+1 **do begin**
        Partition IM_p into non-overlapping blocks of $(n \times n)$ pixels;
        // estimation of parameters and membership values //
        For block:=1 to n^2 **do begin**
            Find-parameters $(f(x,y), G_{avg}, G_{diff}, \sigma^2)$;
            Find-membership $(G_{avg},$
            $G_{diff}, \sigma^2, \mu_{edge}(block), \mu_{shade}(block), \mu_{MR}(block));$
        End For;
        // Sum of moment computation //
        For i:=1 to n **do begin**
        For w :=1 to n **do begin**
        k:=n $\times$ (i -1)+ w; // Mapping from 2-d with indices (i,w) to 1-d with
                index k //
        Find-moment-sum($S_{i\,w}$, $MR_{i\,w}$, $E0_{i\,w}$, $EP45_{i\,w}$, $E45_{i\,w}$, $E90_{i\,w}$);
           $S_P[k]:=S_{i\,w}$;
           $E45_P[k]:=E45_{i\,w}$;
           $EP45[k]:=EP45_{i\,w}$;
           $E0_P[k]:=E0_{i\,w}$;

$E90_P[K]:=E90_{i\,w};$

$MR_P[K]:=MR_{i\,w};$

End For;

End For;

Sort(S_P , MR_P , $E45_P$, $EP45_P$, $E0_P$, $E90_P$);

// This procedure sorts arrays S_P , MR_P, etc. into descending order and
places the resulting   vectors into corresponding arrays //

// Image identification from Euclidean distance //

  For p:=2 to (m+1) **do begin**

    // p= an index to represent image //

    $Euclid_P :=0;$ $Euclid_1 :=0;$

      For $X_P \in$ { S_P, MR_P , $E0_P$, $E45_P$, $EP45_P$, $E90_P$} and

          $X_1 \in \{S_1$, MR_1 , $E0_1$, $E45_1$, $EP45_1$, $E90_1\}$ in order **do begin**

          Find-distance (X_P ,X_1 ,d_{X_P});

          $Euclid_P :=[Euclid_P +d^2_{X_P}]^{1/2}$;

      End For;

        IF $Euclid_P > Euclid_{P-1}$ then

            image:= p -1 ;

          ELSE   image :=p;

  End For;

  End For;

End.

23.2.4 Rotation and Size Invariant Matching

In order to keep the matching process free from size and rotational variance of
the reference image, the following strategies have been used.

1. The Euclidean distances used for estimation of fuzzy moments are
   normalized with respect to the diagonal of the image, which is assumed to
   have a unit distance. Thus the Euclidean distance between each two blocks
   of an image are normalized with respect to the image itself. This
   normalization of distances keeps the matching process insensitive to the
   size variation of the images.

2. The descriptor vectors are sorted so as to keep the blocks with most
   predominant features at the beginning of the array, which only participate
   subsequently in the matching process. Thus the matching process is free
   from rotational variance of the reference image. The insensitivity of
   matching process to size and rotational variance of the reference image has
   also been proved through computer simulation.

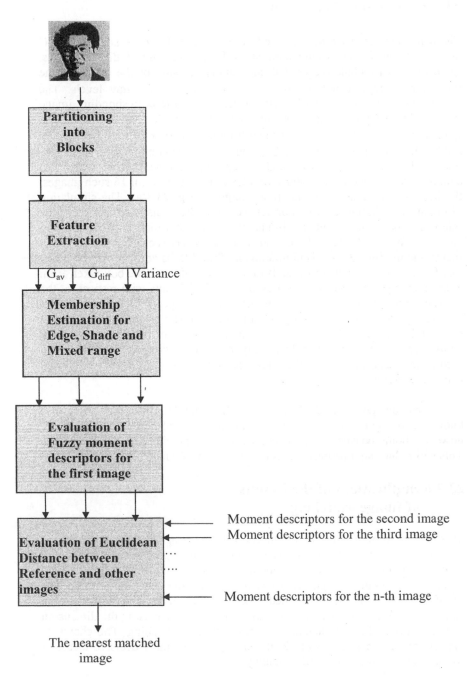

Fig. 23. 2: An outline of the image matching scheme.

23.2.5 Computer Simulation

The technique for matching presented in this chapter has been simulated in C language under C^{++} environment on an IBM PC \ AT with 16 digital images, some of which include the rotated and concise version of the original. The images were represented by 64×64 matrices with 32 gray levels. The simulation program takes each image in turn from the corresponding image files and computes the gradients for each pixel in the entire image. The gradient matrix for the image is then partitioned into 16×16 non-overlapping sub blocks. Parameters like G_{avg}, G_{diff} and σ^2 are then computed for each sub-block. The fuzzy edge, shade and mixed-range moment vectors are also computed subsequently. The above process is repeated for all 16 such images, the first one being the reference (boy image) in fig. 23.3(a). The simulation program then estimates the normalized Euclidean distance between the reference image and all other subsequent images, of which the first four are shown in fig. 23.3(b)-(e). The Euclidean distance between the reference boy image and the fig. 23.3 (d) is minimum and found to be zero. It may be noted that fig. 23.3 (b), which corresponds to the image of the same boy taken from a different angle, has an Euclidean distance of 1.527 units with respect to the reference boy image. The Euclidean distance of images 23.3 (c) and 23.3(e) with respect to 23.3(a) being large enough of the order of 4.92 and 5.15 units respectively proves the disparity of matching. The rotational and size invariance of the proposed matching algorithm is evident from the resulting zero image distance between the reference boy image and the size-magnified and rotated version of the same image.

It may be added that the feature extraction and descriptor formation for a known set of images, which were performed in the real time by our program, however should be carried out offline before the matching process is invoked. This will reduce significantly the time required for the matching process.

23.2.6 Implications of the Results of Image Matching

This section introduced a new concept for matching of digital images by estimating and comparing the fuzzy moments with respect to each partitioned block of images. The proposed method is free from size and rotational variance and requires insignificantly small time for the matching process. The smaller the size of the partitioned block in the image, the higher is the computational time for matching. On the other hand, increasing the dimension of the partitioned blocks hampers the resolution of matching. The choice of the size of each partitioned block, therefore, is a pertinent decisive factor in connection with the process of matching.

The fuzzy membership functions used in this context have been chosen intuitively. There exists ample scope of selecting appropriate membership functions in a judicious manner. Selection of membership functions that cause the least error in the process of matching is yet to be identified. This is an open problem to the best of the author's knowledge to date.

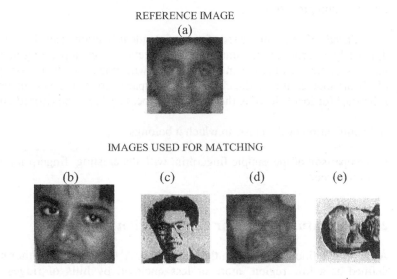

REFERENCE IMAGE
(a)

IMAGES USED FOR MATCHING

(b) (c) (d) (e)

Fig. 23.3: Matching of a reference image (a) with images (b) through (e).

23.3 Fingerprint Classification and Matching

Fingerprints are graphical flow-like ridges [11] present on human fingers. Their formation depends on the initial conditions of the embryonic mesoderm from which they develop [15]. Each fingerprint is a map of ridges and valleys (explained later) in the epidermis layer of the skin. The ridge and valley structure form unique geometric patterns that act as a basis for classification.

Fingerprints have long been used for personnel identification and criminal investigation, and also in applications such as access control for high security installations, credit card verification and employee identification because of the following features:

i) Uniqueness, i.e., fingerprint images of two individuals are not alike.

ii) Permanence, i.e., fingerprint images of an individual do not change throughout the life span.

The main reason for the popularity of automatic fingerprint identification is to speed up the matching (searching) process. Manual matching of fingerprints is a highly tedious task because the matching complexity is a function of the size of the image database which can vary from a few hundred records to several million records, which takes several days in some cases. The manual classification method makes the distribution of records uneven resulting in more work for commonly occurring fingerprint classes. These problems can be overcome by automating the fingerprint based identification process.

Speed of an automated fingerprint identification can be increased drastically by grouping the images into different classes depending upon their features, so that searching can be done only with images of that class, instead of all the images thus reducing the search space. So, whenever an image is submitted for identification, the following processes are to be carried out:

i) identification of the class, to which it belongs

ii) comparison of the sample fingerprint with the existing fingerprint images of that class.

23.3.1 Features Used for Classification

A **ridge** is defined as a line on the fingerprint. A **valley**, on the other hand, is defined as a low region, more or less enclosed by hills of ridges. Each fingerprint is a map of ridges and valleys in the epidermis layer of the skin. Ridge and valley structure form unique geometric patterns. In a fingerprint, the ridges and valley alternate flowing in a local constant direction. A closer analysis of the fingerprint reveals that the ridges (or valleys) exhibit anomalies of various kinds such as ridge bifurcation, ridge endings, short ridges and ridge cross over [15]. These features are collectively called **minutiae** and these minutiae have a pattern that is unique for each fingerprint. The directions of the ridges, the relative positions of the minutiae and the number of ridges between any pair of minutiae are some of the features that uniquely characterize a fingerprint. Automated fingerprint identification and verification systems that use these features are considered minutiae based. Vast majorities of contemporary automated fingerprint identification and verification systems are minutiae based systems.

In this work, we however considered singular points to classify the fingerprints. Two distinct types of singular points have been used to identify fingerprints. These are **core** and **delta** points. Fig. 23.4, presented below, describes these singular points.

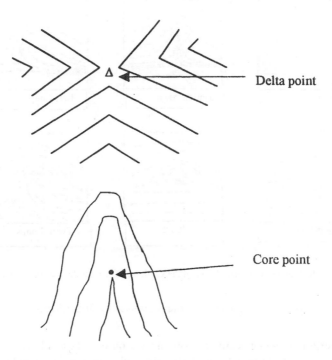

Fig. 23.4: The delta and the core points in fingerprint images.

23.3.2 Classification Based on Singular Points

A point is detected as an ordinary point, core point or delta point by computing the poincare index. **Poincare index** is computed by summing up the changes in the direction angle around the curve, when making a full counterclockwise turn around the curve in a directional image [15]; the directional angle turns 0 degree, +180 degree, -180 degree during this trip. A point is called ordinary if the angle has turned 0 degree, core if it has turned +180 degree and delta if it has turned –180 degree.

Depending upon the number of core-delta pairs they can be categorized into Arch (core-delta pair =0), Tented arch / Loop (core-delta pair = 1) or Whorl /Twin loop (core-delta pair =2). If the number of core-delta pair is greater than two then further processing is to be done to identify the exact class.

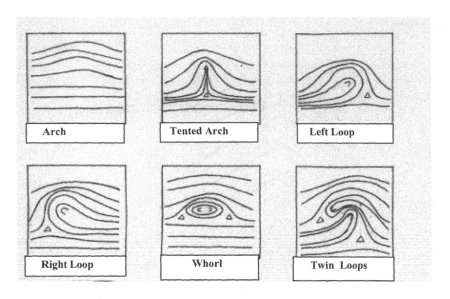

The triangles represent delta and the semicircles/ circles denote core points.

Fig. 23.5 (a): The six typical classes of fingerprints.

Distinction between Tented Arch and Loops: These classes contain one core-delta pair. Suppose the core and delta points are connected by an imaginary line. In tented arch class, this line orientation is along the local direction vectors, while in a loop image type the line intersects local directions transversely.

Let β be the slope of the line connecting the core and delta points. Let $\alpha_1, \alpha_2 \ldots \alpha_n$ be the local direction angles on the line segment. If the averaged difference $(1/n) \Sigma [\alpha_i - \beta]$ is less than a threshold, say (0.9), then the image is classified as a tented arch; otherwise it is a Loop.

Distinction between the Whorl and the Twin Loops: As mentioned these classes contain two core-delta pairs. Now, we try to connect the core points along the direction vector. If the effort fails, it is said to belong to Twin class; else it belongs to Whorl class.

Distinction between the Left and the Right Loops: Further classification of Loop into Right loop or Left loop is possible based on whether the delta point is located to the left or right of the core point. This discrimination is done as follows. Starting from a core point and moving along the direction vectors, the delta point remains to the left in a Left loop image and to the right in a Right loop image [16].

Thus we have altogether 6 different fingerprints (fig. 24.5(a)), which can be classified by using the classification tree presented in fig. 24.5 (b).

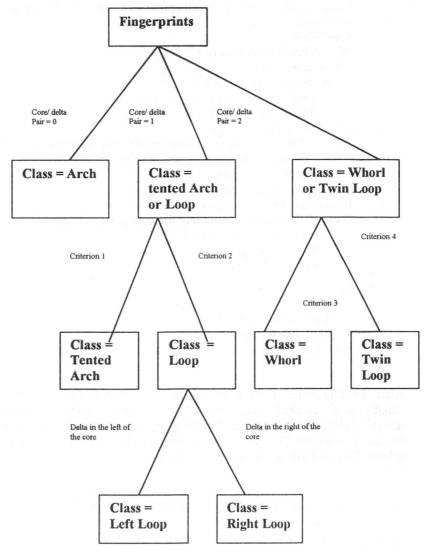

Criterion 1: Line joining the core and the delta is in the local direction, Criterion 2: Line joining the core and the delta crosses the local direction, Criterion 3:Line joining the core points is along the local direction, Criterion 4: Line joining the core points crosses the local direction vector.

Leaves, denoted by dark shaded nodes, are the final classes.

Fig. 23.5(b): The fingerprint classification tree used in the simulation.

Procedure find-fingerprint-class;
   Begin
   1   **While** number of core plus delta points exceed 2, use Gaussian
       smoothing
       End While;
   2   **If** number of core-delta pair = 0
       Then class = arch;
        Else If number of core-delta pair = 1
            Then class = tented arch or loop;
            Else class = whorl or twin loop;
   3   **If** number of core-delta pair = 1
       Then   **If** the line joining the core and delta point intersects
             the line of local direction transversely
            Then class = loop;
            Else class = tented arch;
   4   **If** number of core-delta pair =2
       Then If the straight line joining the core points do not
             touch the local direction
            Then class = whorl
            Else class = twin loop;

   5   **For** class $\in$ loop;
        If core is left to delta
        Then class = left loop;
        Else class = right loop.
       End For;
  End.

After the fingerprint is classified into one out of the above six possible classes, it is searched among the set of fingerprint databases under the identified class. If a suitable match is found by the image matching algorithm, discussed in section 23.2.3, then the suspect could be detected; otherwise the scheme for facial image matching is called for.

23.4 Identification of the Suspects from Voice

During phonation of the voiced speech, the vocal tract is excited by the periodic glottal waveform generated by the vocal cords. The periodicity of this waveform is called **pitch period**, which depends upon the vocal tract shape that varies with the speakers. The shape of the vocal tract uniquely determines the sounds that are produced and characterized by natural frequencies (or formants) which correspond to the resonant frequencies of the acoustic cavity [17]. Since formants depend upon the longitudinal cross-section of the vocal tract, which is different for various speakers, formant frequency is the unique parameter for the speaker identification [13].

For the production of the nasal sounds (such as / m/ and /n/), the vocal tract is blocked at some point determined by the identity of the nasal consonant and the valum is moved to connect the nasal tract to the vocal tract. The nasal sounds are radiated through the nostrils. The power spectrum of the acoustic radiation produced during phonation of the nasal consonants provides a good clue to the speaker identity [20]. So we find the average power spectral density in different range of frequencies for the word "nam" in the Bengali sentence "*dinubabur desher nam phalakata*".

23.4.1 Extraction of Speech Features

The feature extraction scheme that was used in past includes a set of measurements at 10 to 20 millisecond intervals throughout the utterances. This approach has the following limitations. First limitation is regular and the rapid sampling of the voice signal with the characterizing measurement produces large sets of the data that have a high degree of redundancy. Secondly, a given set of parameters is not optimally suited to every segment of an utterance. Some of them are of no use during many spoken sounds, as in the case of the fundamental frequency during voiceless intervals.

An alternative scheme of feature extraction uses characteristics averaged over time such as long term spectra; here parameters are measured over specific context or over long enough intervals to take the benefits of context dependence. It leads to much smaller data sets, and virtually eliminates the effect of the timing differences, but it excludes a large class probability of the speech data having useful speaker dependent effects.

The most efficient approach, which is used in this work, is to perform some degree of segmentation and recognition of the linguistic components of the speech signal. In most of the speaker identification applications, it is reasonable to assume that a known phrase or sentence is used for the speech samples. In that case the speech segmentation and recognition would not be difficult. Ideally the system designer would be free to record an advantageous set of phonemes, which could be easily segmented. The ability to find its way about the utterances allows the system to locate certain interesting points of the speech events and then they are extracted to get the appropriate parameters at each of these points.

The speaker dependent information could be easily extracted from the pre-specified signal which in this context is a Bengali sentence: "*dinubabur desher nam phalakata*'. The given sentence was digitized by a sonogram (model no. 5500), maintained by the Indian Statistical Institute, Calcutta and the digital spectra is broken up in to four parts corresponding to *"dinubabur"*, *"desher"*, *"nam"*, and *"phalakata"*. The analysis of the complete sentence is done to acquire some useful speech parameters, which help in the identification of the speaker with the Speech Processing toolbox in MATLAB on a PENTIUM based system. The various features acquired by the MATLAB are:

1. Formant frequencies[18] (fundamental and the harmonics),

2. Peak power spectrum density (at different frequency ranges),

3. Average value of the spectral density of the word "nam" which is in the CVC context (Consonant followed by a Vowel followed by the Consonant) at different range of frequency, and

4. The pitch, which is estimated by counting the number of pulses per second by taking the digitized data of sampling frequency 10240 Hz.

A sample of estimation of the formant frequencies, from the digitized speech signal of a person, uttering the above sentence, is presented in fig. 23.6.

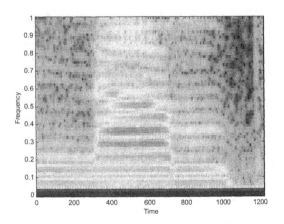

The dark horizontal shades correspond to the formant frequencies of specified duration. First 8- formant frequencies are visible in the figure.

Fig. 23.6: The speech spectrogram, showing the formant frequencies.

23.4.2 Training a Multi-layered Neural Net for Speaker Recognition

The second part is the identification of the speaker, which is done with the help of Artificial Neural Network (ANN). The ANN model is trained by the well known Back-propagation algorithm. Although any conventional approach of statistical classification can be used, we preferred the ANN model for its accuracy and simplicity. To reduce the number of iterations adaptive learning rate and momentum are added to the back-propagation, so that the error goal is achieved earlier and the network is not stuck to a shallow minimum in the error surface.

The input parameters for the training instances in the present context are the first 6 formant frequencies, maximum power spectral density, average power spectral density for two words and the pitch, while the output training instances correspond to the decoded binary signals, where a high output at a given output node corresponds to a given speaker. The training was given for 10 sets of training instances, each corresponding to one speaker by using the MATLAB neural net toolbox.

The neural network model with four layers having ten neurons in the first and the last layer and twelve neurons in the intermediate layers has been used for the computer simulation. The network uses LOGSIGMOID function, which generates the output between 0 and 1. The training is done by TRAINBPX, which uses both the momentum and the adaptive learning rate to speed up the learning. The learning rate and the number of iterations (epoch) are chosen, so as to get the desired error goal accurately. An adaptive learning rate requires some changes in the training procedure used by TRAINBP. First the initial network output and error are calculated. At each epoch new weights and biases are calculated using the current learning rate. New output and error are then calculated. If the new error exceeds the old error by more than a pre-defined ratio (typically 1.04), the new weights, biases, output and error are discarded, and the learning rate is decreased (typically by multiplying by 0.7); otherwise the new parameters are saved. If the new error is less than the old error, the learning rate is increased (typically by multiplying by 1.05).

Two layer sigmoid TRAINBPX adds momentum to back-propagation to decrease the probability that the network will get stuck in to a shallow minimum in the error surface and to decrease training times. The function also includes adaptive learning rate to further decrease the training time by keeping the learning rate reasonably high while ensuring the stability.

The training is given offline. Thus during recognition, only the speech features of the unknown person are extracted and are submitted to the input of the trained neural net. A forward pass through the network determines the signals at the nodes of the output layer. The node with the highest value corresponds to the suspect.

23.5 Identification of the Suspects from Incidental Descriptions

When fingerprint, facial image and voice matching are inadequate to identify the suspects, we have to rely on the incidental descriptions. The incidental descriptions are generally incomplete and sometimes contradictory, especially when received from multiple sources. In this section, we shall employ Fuzzy Petri nets (FPN) for reasoning about the suspects from the incomplete and the contradictory databases and a knowledge base, whose certainty also is not guaranteed.

Imprecision and inconsistency are the most common forms of incompleteness of information, from which the database of an expert system (ES) suffers. Non-monotonic logic has established its success for tackling inconsistency in database, whereas fuzzy logic had demonstrated its capability of reasoning with imprecise databases. The reasoning process of the two techniques, as reported in the current literature [6], is orthogonal to each other. In this chapter, a combined approach for tackling all the above forms of incompleteness of databases and uncertainty of knowledge bases has been proposed based on the model for belief-revision and the same formalisms have been applied for reasoning in a generic class of ES. The proposed technique for handling incompleteness in databases has been illustrated with reference to an ES for criminal investigation, designed by the author and realized in Pascal by his students Abhijit Das and Abhik Mukherjee (see Appendix A), while working for his (the author's) Ph.D. thesis at the Systems and Information laboratory of ETCE department, Jadavpur University, India. The system presented in this chapter is an extension of that expert system that includes facial images, speech and fingerprints also as inputs.

The proposed ES comprises of three main modules, namely, *i) the database (DB), ii) the knowledge base (KB) and iii) the inference engine (IE)*. A brief description of each of the modules with special emphasis on the management of imprecision and inconsistency of database will be presented in this chapter.

23.5.1 The Database

Fuzzy beliefs of the information, collected from sources with various degrees of authenticity levels, are first normalized using a set of intuitively constructed **grade of membership functions**, shown in fig. 23.7. The information with their normalized values is then recorded in a database in the form of a data-tree, to be described shortly.

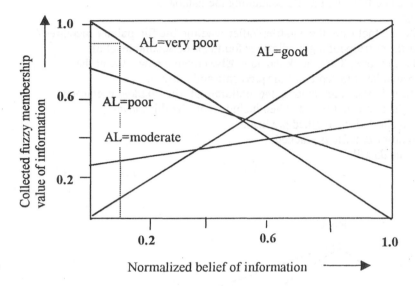

Fig.23.7: Normalization of fuzzy beliefs using membership functions.

As an illustration of the normalization process, it is observed from fig. 23.7 that an information with a fuzzy belief of 0.85, collected from a source with very poor authenticity level (AL), gives rise to a normalized belief of 0.12 (shown by dotted lines).

23.5.2 The Data-tree

The database in the proposed ES has been organized in the form of a data-tree (fig. 23.8) having a depth of 3 levels. The root simply holds the starting pointer, while the second level consists of all the predicates, and the third level contains the relevant facts corresponding to each predicate of the second level. The normalized beliefs corresponding to each fact are also recorded along with them at the third level. Such organization of the data-tree helps in efficient searching in the database. To illustrate the efficiency of searching, let us consider that there exist P number of distinct predicates and at most L number of facts under one predicate. Then to search a particular clause, say has-alibi (jadu) in the data-tree, we require P+L number of comparisons in the worst case, instead of P* L number of comparisons in a linear sequential search. Now, we present the algorithm for the creation of the data-tree.

Procedure create-tree (facts, fuzzy-belief);
Begin
create root of the data-tree;
open the datafile, i.e., the file containing the database;
Repeat
 i) Read a fact from the datafile; //after reading, the file pointer increases//
 ii) **If** the corresponding predicate is found at the second level of
   the data-tree, **Then** mark the node **Else** create a new node at the
   second level to represent the predicate and mark it;
 iii) Search the fact among the children of the marked predicate;
 iv) **If** the fact is not found, **Then** include it at the third level
   as a child of the marked predicate
Until end-of-datafile is reached;
 Close the datafile;
End.

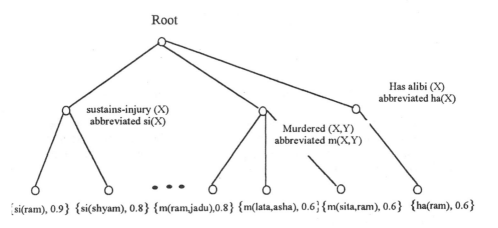

Fig. 23.8: The data-tree representation of the database.

23.5.3 The Knowledge Base

The KB of the proposed ES comprises of a set of default rules (DR) [1], a set of production rules (PR) and two working memory (WM) modules. DR are used to guess about the suspects. The typical structure [1], [4] of the DR is as follows:

$$\text{DR1:} \quad \frac{\text{SR (Y, X) OR (M (X)} \land \text{M (Y)} \land \text{L (X, w (Y)))}}{\text{S (Y, murderer-of(X))}}$$

which means that if (Y had strained relations with X) or ((X and Y are males) and (X loves wife of Y)) then unless proved otherwise, (suspect Y as the murderer of X).

The variables X, Y, etc. of a predicate are instantiated by users through a query language. A working memory WM2 in the proposed system [8] is used to hold all the instantiated antecedents of a default rule. If all the instantiated antecedents of a rule are consistent, then the rule is fired and the values bound to Y for the predicate suspect (Y, murderer-of (X)) are recorded. These values of Y together with a particular value of X (=x say) are used subsequently as the space of instantiation for each variable under each PR.

The list of suspects, guessed by the DR, is also used at a subsequent stage for identifying the suspect-nodes of the FPN. The process of collecting data from users through query language and representation of those data in the form of a tree, called default-data-tree, can be best described by the following procedure.

Procedure default-data-tree formation (no-of-predicates);
```
        //no-of-predicates denotes number of predicates
                in the entire set of default rules//
Begin
 Create root-node;
 i:= 1;
 While i ≤ no-of-predicates do
 Begin
  writeln "Does predicate pᵢ (X,Y) exist ?" ;
  readln (reply);
  If reply = no Then do
  Begin
  Create a predicate pᵢ (X, Y) at the second level
  under the root;
   Repeat
    writeln "predicate pᵢ (X,Y), X = ? ,Y = ?" ;
    readln (X, Y);
    If (X ≠ null ) and (Y ≠ null) Then
    Create a data node with the read values of X and Y
    under the predicate pᵢ ;
   Until (X = null ) and ( Y = null)
```

End;
  i : = i+ 1;
 End While;
End.

Once a default-data-tree is formed, the following procedure for suspect-identification may be used for identifying the suspects with the help of the file of default rules (DRs) and the default-data-tree.

Procedure suspect-identification (DRs, default-data-tree);
Begin
i : = 1;
Open the file of DRs;
Repeat
 If (all the predicates connected by AND operators OR at least
     one of the predicates connected by or operator in
     DR$_i$ are available as predicates in the default-data-tree)
 Then do
   Begin
       Search the facts (corresponding to the predicates) of
       the DR$_i$ under the selected predicates in the tree so
       that the possible variable bindings satisfy the rule;
       Fire the rule and record the name of suspects in a list;
   End;
     i : = i + 1;
 Until end-of-file is reached;
   Close the file of DRs;
End.

The structure of a production rule (PR 1) used in our ES is illustrated through the following example.

PR1: murdered (X, Y) :-
     suspect (Y, murderer-of(X)),
     has-no-alibi (Y),
     found-with-knife ((Y), on-the-spot-of-murder-of (X)).

Variables of each predicate under a PR, in general, have to be instantiated as many times as the number of suspects plus one. The last one is due to the person murdered. However, some of the variable bindings yield absurd clauses like has-alibi (ram) where 'ram' denotes the name of the person murdered. Precautions have been taken to protect the generation of such absurd clauses. The procedure for instantiating variables of PRs with the name of suspects and the person murdered is presented below. In this procedure, however, no attempts have been taken to prevent the generation of absurd clauses to keep the algorithm simple.

Procedure variable-instantiation-of-PRs (PRs, suspect, person-murdered);
Begin
 For i := 1 to no-of-PRs **do**
 Begin
 Pick up the PR $(X_1, X_2, ..., X_m)$;
 Form new PRs by instantiating $X_1, X_2, ...,X_m$ by elements from the set of
 suspects and the person murdered such that $X_1 \neq X_2 \neq ... \neq X_m$;
 Record these resulting rules in a list;
 End For;
End.

Once the instantiation of the variables in the production rules is over based on the procedure presented above, the facts of the resulting rules are searched in the static database, represented in the form of a data-tree. A working memory WM1 keeps track of the instantiated clauses of production rules, after these are detected in the data-tree. Moreover, if all the AND clauses of a PR are found in the data-tree, the rule is fired and the consequence of the PR is also recorded in the WM1. The information stored in WM1 is subsequently used for the encoding of the place-transition pairs (of FPN), corresponding to the PR. The WM1 is then cleared for future storage.

23.5.4 The Inference Engine

The inference engine (IE) of the said ES comprises of five modules, namely, i) the module for searching the antecedents of PRs on the data-tree, ii) the FPN formation module, iii) belief-revision and limit-cycle detection & elimination module, iv) non-monotonic reasoning module, and v) decision making [19] and explanation tracing module.

Searching antecedents of PR in the datatree: The following procedure describes the technique for searching the instantiated clauses of a PR in a data-tree.

Procedure search-on-data-tree (data-tree, key-data);
//searches the antecedents of PRs on data-tree//
Begin
Search the predicate corresponding to the key-clause in the
second level of the data-tree; **If** search is successful **Then** do
 Begin
 Search the key-clause among the children of the predicate under
 consideration;

If search is successful **Then** print "key found"
  Else print "key-clause not found";
End;
 Else print "key-predicate not found";
End.

Formation of FPN: The algorithm for FPN-formation that we present here has a difference with that we presented in chapter 10. The basic difference lies in the fact that here the production rules contain no variables, as they are already instantiated. This is required for Pascal implementation, as one cannot execute unification operation in Pascal, and thus has to depend on string matching. In chapter 10, we, however, considered non-instantiated rules and thus presumed the scope of unification of predicates within the programming language.

Procedure FPN-formation (list of PRs, data-tree);
//r = number of rules for which the searching has to be continued
  at a given iteration//
Begin
r : = no-of-instantiated-PRs;
Repeat
 temp : = 0;
 For i : = 1 to r **do**
  Begin
  pick up PR_i ;
  If all the antecedent clauses of PR_i are present in the data-tree
  Then do
  Begin
   i) store the antecedent-consequent pairs into the WM1
    for appending the FPN;
   ii) search the antecedent clauses, stored in WM1, on
    the FPN; **if** found, mark those places, **else** create new
    places and mark these places; create one transition
    and the consequent clauses and establish connectivity
    from the antecedents to the transition and from the transition
    to the consequent places;
   iii) append the data-tree by the consequent clauses;
   iv) eliminate PR_i from the list of instantiated PRs;
   v) r := r -1;
   vi) erase the contents of WM1;
  End;
 End For;
Until (r = 0) or (no further augmentation of the FPN takes place)
End.

23.5.5 Belief Revision and Limitcycles Elimination

The belief revision model we employed in criminal investigation is a simplified model we covered in chapter 10. In fact, we here used Looney's model [12] with the additional considerations that the FPN may include cycles. Unlike multi-valued distributions as used in chapter 10, we here considered single valued tokens, representing the belief of the proposition / predicates associated with the places in the FPN. The **belief revision cycle** in the present context includes updating of the fuzzy truth token values associated with the transitions, followed by updating of fuzzy beliefs at all places. The following expressions were used for updating truth tokens at transitions and beliefs at places.

For a given transition tr_q having input places $\{p_1, p_2, ..., p_v\}$ with the corresponding single valued tokens $\{n_1(t), n_2(t), ..., n_v(t)\}$ at time t, let

$$. \quad x = \bigwedge_{1 \leq j \leq v} \{ n_j(t) \}$$

Now, if x exceeds the threshold marked with transition tr_q, the **fuzzy truth token** (FTT) at transition tr_q, denoted by t_q, is computed as follows.

$$t_q(t+1) = x \wedge cf_q$$

$$= 0, \text{ otherwise}$$

where cf_q denotes the certainty factor of the rule, described by the transition.

Further, the **fuzzy belief** $n_i(t+1)$ at time $(t+1)$ is computed at place p_i, the common output place of transitions $tr_1, tr_2, ..., tr_u$ by the following expression.

$$n_i(t+1) = \bigvee_{1 \leq k \leq u} \{ t_k(t+1) \}$$

where $t_k(t+1)$ denotes the FTT of transition tr_k at time $(t+1)$.

If place p_i is an axiom (i.e., it is not an output place of any transition), then

$$n_i(t+1) = 0.$$

Let m be the number of transitions in the network. Now, it has been shown in [9] that there may be two situations after m number of belief revision cycles. First, the fuzzy beliefs at all places may reach an equilibrium (steady) state, following which there will be no more changes in beliefs. Alternatively, the beliefs at one or more places in the network may exhibit sustained oscillation.

We call this oscillation 'limitcycles'. No inferences can be derived from an FPN exhibiting limitcycles. For elimination of limitcycles, either more information has to be added to the system, or one has to delink the cycles carefully, so that resulting inferences are minimally disturbed. We employed the second scheme here, assuming that we always will not get more information. The following definitions are useful to explain the scheme for limitcycles elimination.

Definition 23.14: An **arc** represents a connectivity from a place to a transition and vice versa.

Definition 23.15: An arc is called **dominant**, if it carries the highest FTT among all the arcs incoming to a place, or the least fuzzy belief (LFB) among all the arcs incoming to an enabled transition (i.e., a transition having tokens at all its input places) provided the LFB does not exceed the certainty factor of the transition.

Definition 23.16: An arc is called **permanently dominant**, if it remains dominant for an infinite number of belief revision cycles.

Definition 23.17: Fuzzy gain of an acyclic single reasoning path between two places is computed by taking the minimum of the fuzzy beliefs of the places and the certainty factors of the transitions on that path. In case there exists parallel reasoning paths between two places, then the overall fuzzy gain of the path is defined as the maximum of the individual fuzzy gains. Fuzzy gain of a reasoning path, starting from axioms up to a given transition, is defined as the maximum of the fuzzy gains from each axiom up to each of the input places of the transition.

Limitcycles Elimination: It has already been discussed that existence of limitcycles in an FPN can be understood only after m number of belief revision cycles. An analysis [8] of FPN reveals that limitcycles occur because of permanent dominance of all the arcs on one or more cycles. The simplest way to eliminate limitcycles is to make at least one arc on those cycles non-dominant. Further, an input or output arc of a transition on a cycle can be non-dominant, if it has more than one input or output places. But how should one make an arc non-dominant? This calls for adjustment of threshold of some transitions on cycles, so that its output arc becomes permanently non-dominant. But what should be the value of the threshold that would cause permanent non-dominance of its output arc? One choice of threshold is the largest fuzzy belief on the reasoning path from the axioms to the selected transition.

A question that naturally arises: shall we not lose anything by adjusting thresholds of transitions? The answer to this is affirmative, if we can ensure that the inferences will be minimally disturbed in the worst case. This is possible in the present context, if we can adjust threshold to a transition that lies on a reasoning path with the least fuzzy gain. Further, the selected transition should not induce limitcycles in the neighborhood cycles. An algorithm [8] for elimination of limitcycles can be easily constructed based on the last two criteria.

23.5.6 Non-monotonic Reasoning in a FPN

In this section, a new technique for non-monotonic reasoning, which eliminates one evidence out of each pair of contradictory evidences from an FPN, will be presented. However, before presenting the technique, the possible ways by which non-monotonicity is introduced into a FPN will be illustrated.

Creeping of non-monotonicity into a FPN: Contradictory evidences may appear in a FPN because of i) inconsistency in a database, ii) inconsistency in production rules, and iii) uncertainty of information in a database. To illustrate how contradictory evidences and hence non-monotonicity are introduced into the FPN, we consider the following examples.

i) Inconsistency in database: To illustrate how inconsistency in database finds its way in the FPN, let us consider a database that includes the information, namely, has-alibi (ram), has-no-alibi (ram), has-precedence-of-murder (ram). Further assume that 'ram' is included in the list of suspects. The set of PRs before instantiation with the name of suspects and the person murdered are listed below:

PR1:   not-suspect (X) :- has-precedence-of-murder (X),
*   Has-alibi (X).*

PR2:   culprit (X) :- has-precedence-of-murder (X),
*   has-no-alibi (X).*

Now, first by instantiating the variable X in the above rules by 'ram' and then checking the antecedent clauses of the resulting rules in the database, we finally form the FPN, shown in fig. 23.9. In this fig. d_1, d_2, d_3, d_4 and d_5 represent has-alibi(ram), has-precedence-of-murder (ram), has-no-alibi(ram), not-suspect (ram) and culprit (ram) respectively. Thus the contradictory evidences has-alibi (ram) and has-no-alibi (ram) could enter into the FPN because of inconsistency in the database.

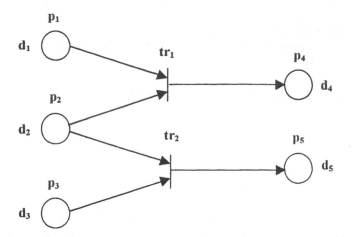

Fig. 23.9: Entry of inconsistency in an FPN because of inconsistent database.

ii) Inconsistency in PRs: To illustrate this, let us consider the database that contains the clauses named as loved (ram, sita), proposed-to-marry (lata, ram), loved (lata, ram). Moreover, assume that the name of suspects and the person murdered belong to the set S = {lata, sita, ram} and the rule-base includes

PR1: proposed-to-marry (X,Y) :- loved (X,Y).

PR2: hates (X,Y) :- proposed-to-marry (X,Y),
                  loved (Y,Z).

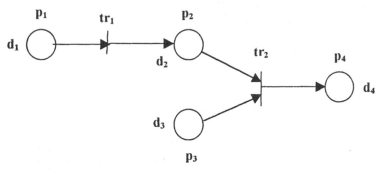

Fig. 23.10: Entry of inconsistency in an FPN due to incompleteness of rules.

The variables in the above rules are first instantiated with elements of the set S and then the clauses of the resulting rules are searched in the database for the formation of the FPN. The FPN of fig. 23.10, which has been formed in this way, includes contradictory evidences loved (ram, sita) and hates (ram, sita) because of the presence of inconsistency in PRs. d_1, d_2, d_3 and d_4 in fig. 23.10 represent loved (lata, ram), proposed-to-marry (lata, ram), loved (ram, sita) and hates (lata, ram) respectively.

iii) Lack of precision of facts in the database: To illustrate how lack of precision of facts causes non-monotonicity to appear in an FPN, let us consider a database comprising of the clauses named as loved (ram, sita), loved (madhu, sita), loses-social-recognition (sita), husband (ram), another-man (madhu) and the set of PRs that includes

RR1:   loses-social-recognition (W) :- loved (H,W),
*   loved (AM,W).*
PR2:   suicides (W) :- loses-social-recognition (W).
PR3:   suicides (W) :- tortured (H,W).
PR4:   tortured (H,W) :- not-loved (H,W).

where   H,W   and   AM   represent   husband-wife   and   another-man respectively.

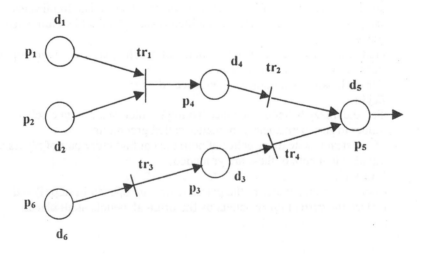

Fig. 23.11: Entry of inconsistency in an FPN because of lack of precision of facts.

Now, assume that the suspects and the person murdered belong to set $S1$ = {ram, sita, madhu}. The variables of the above rules are now instantiated with the elements of set $S1$ and the resulting rules are stored in a file. This file of instantiated rules together with the data-tree is now used for the formation of the FPN. One point that needs special mention at this stage is clarified as follows. Let us assume that the initial fuzzy belief for the information loved (ram, sita) is 0.001, i.e., very small. Under this circumstance, it is not unjustified to assume that not-loved (ram, sita) is true. However, we cannot ignore loved (ram, sita), since it is present in the database. So, because of uncertainty of the information loved (ram, sita), the pair of inconsistent information d_1 and d_6, defined below, both appear in the FPN of fig. 23.11. In fig. 23.11, d_1, d_2, d_3, d_4, d_5 and d_6 represent loved (ram, sita), loved (madhu, sita), tortured (ram, sita), loses-social-recognition(sita), suicides (sita) and not-loved (ram, sita) respectively.

23.5.7 Algorithm for Non-monotonic Reasoning in a FPN

Irrespective of the type of inconsistency, the following algorithm for non-monotonic reasoning may be used for continuing reasoning in the presence of contradictory evidences in an FPN [10].

Procedure non-monotonic reasoning (FPN, contradictory pairs)
Begin
i)   Open the output arcs of all the contradictory pair of information;
ii) Check the existence of limit cycles by invoking procedures for limitcycle detection; If limitcycles are detected, eliminate limitcycles, if possible; If not possible report that inconsistency cannot be eliminated and exit;
iii) Find the steady-state fuzzy beliefs of each place by invoking the steps of belief-revision procedure;
iv) **For** each pair of contradictory evidences do
     Begin
     Set the fuzzy belief of the place having smaller steady- state belief value to zero permanently, hereafter called grounding ;
     If the steady-state fuzzy belief of both the contradictory pair of places is equal, **Then** both of these are grounded;
     End For;
v)   Connect the output arcs of the places which are opened in step (i) and replace the current fuzzy beliefs by the original beliefs at all places, excluding the grounded places;
vi) Open the output arcs of transitions leading to grounded places;
     //This ensures that grounded places are not disturbed.//
     End.

23.5.8 Decision Making and Explanation Tracing

After execution of the non-monotonic reasoning procedure, the limit-cycle-detection and elimination modules are invoked for reaching an equilibrium condition in the FPN. The decision-making and explanation-tracing module is then initiated for detecting the suspect-node with the highest steady-state fuzzy belief among a number of suspect nodes in the network. This very suspect is declared as the culprit. The module then starts backtracking in the network, starting from the culprit node and terminating at the axioms and thus the reasoning paths are identified. Fuzzy gains are then computed for all the reasoning paths , terminating at the culprit-node. The paths with the highest fuzzy gain is then identified and information on this path is then used for providing an explanation to the user.

23.5.9 A Case Study

The ES presented above is illustrated in this section with reference to a simulated criminal investigation problem. The problem is to detect a criminal from a set of suspects and to give a suitable explanation for declaring the suspect as the criminal.

A set of default rules are first used to find the set of suspects. Let the set of suspects be S = {sita, ram,lata} and the name of the person murdered is 'sita'. The database for the system along with the abbreviated forms is given below.

Database	Abbreviation
Husband (ram, sita)	H(r, s)
Wife (sita, ram)	W(s, r)
Girl-friend (lata, ram)	GF (l, r)
Boy-friend (madhu, sita)	BF (m, s)
Loved (sita, ram)	L (s, r)
Loved (sita, madhu)	L (s, m)
Loved (ram, lata)	L (r, l)
Loved (lata, ram)	L (l, r)
Hated (sita, ram)	Ht (s, r)
Lost-social -recognition (sita)	LSR (s)
Commits-suicide (sita)	S (s)
Tortured (ram,sita)	T (r, s)
Had-strained-relations-with (ram, sita)	SR (r, s)
Murdered (ram, sita)	M (r, s)
Proposed-to-marry (ram, lata)	PTM (r, l)
Accepted-marriage-offer-from (lata, ram)	AMO (l, r)
Has-precedence-of-murder (lata)	HPM (l)
Murdered (lata, sita)	M (l, s)

The production rules for the system with the abbreviated predicates and variables, as defined in the database, are presented below.

Production rules:

1. LSR (W):-
   L(W, H),
   L (W, BF).

2. S (W):-
   LSR (W).

3. S (W):-
   T (H, W).

4. PTM (H, GF):-
   L (H, GF).

5. Ht (W, H):-
   PTM (H,GF).

6. AMO (GF, H):-
   L (GF, H),
   PTM (H, GF).

7. M (GF, W):-
   AMO (GF,H),
   HPM (GF).

8. T (H,W):-
   Ht (W,H).

9. T (H, W):-
   SR (H, W).

10. SR (H, W):-
    T (H, W).

11. M (H, W):-
    SR (H,W),
    PTM (H,GF).

The variables used in the above PRs are first instantiated with the elements of set S. The resulting set of rules, thus obtained, is then stored into a file. A data-tree corresponding to the proposed database is then formed. The data-tree together with the file of instantiated PRs is then used for the formation of the FPN as shown in fig. 23.12. The initial normalized fuzzy beliefs corresponding to each predicate are shown in fig. 23.12. It is assumed that the threshold associated with all transitions except tr_6 is zero. The threshold of tr_6 is set to 0.95 arbitrarily.

Through consultation with a dictionary, the system identifies a pair of contradictory evidences namely L(s, r) and Ht(s, r). The non-monotonic reasoning procedure, now, opens the output arcs of these two predicates and then detects the existence of limit cycles in the FPN. The limitcycles are then eliminated by setting the threshold of tr_4 to 0.9, by using the guidelines of limit cycle elimination. The steady-state fuzzy beliefs of the propositions L (s, r) and Ht(s, r) are then found to be 0.4 and 0.6 respectively. Therefore, according to the non-monotonic reasoning procedure, L (s, r) is grounded, the opened arcs are reconnected and the initial fuzzy beliefs at all places except at L (s, r) are returned. The initial thresholds are also returned to appropriate transitions.

The limit cycle detection procedure, which is then invoked, detects limitcycles. The limitcycle elimination module again eliminates limit cycles by setting the value of threshold for transition tr_4 to 0.9 [8].

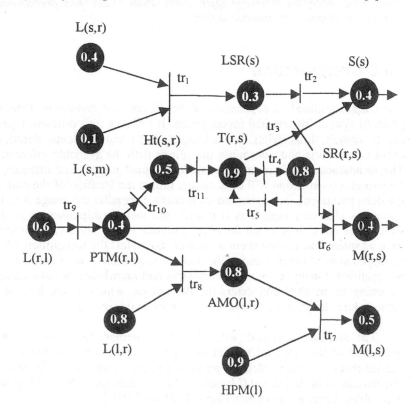

Fig. 23.12: The FPN used in case study.

The steady-state fuzzy beliefs in the network are then computed. Now, out of the three suspect nodes S (s), M (r, s) and M (l, s) the steady-state value of M (l, s) is found to be the highest (0.6). So, 'lata' is considered to be the culprit. The decision-making and explanation-tracing module then continues backtracking in the network, starting from M (l, s) node until the axioms are reached. The reasoning paths L(r, l)-PTM(r, l)-AMO(l, r)-M(l, s), HPM (l)-M(l, s) and L(l, r)-AMO(l, r)-M(l, s) are then identified. Fuzzy gain of all these three paths is then found to be 0.6. Thus the following conclusion and explanation is reported to the users.

Conclusion: Lata is the criminal.

Explanation: *Ram loved Lata. Ram proposed to marry Lata. Lata loved Ram. Lata accepted marriage offer from Ram. Lata has precedence of murder. Therefore, Lata murdered Sita.*

23.6 Conclusions

The chapter outlined an application of the AI and soft computing tools in a practical system for criminal investigation. It receives four different types of inputs, namely, fingerprints, facial images, voice and incidental description about the suspects and determines the culprit with the available information. The techniques used here are insensitive to small noises. For instance, the fingerprint classification will be accurate, unless the location of the core and the delta points only are corrupted with noise. Secondly, the image matching scheme is also very rugged as it records the membership-distance product, which does not change much for small gaussian noise. Thirdly, the formants have adequate information about a speaker. So, unless the recognition scheme is trapped at local minima during the training cycles, the correct speaker could be identified. Finally, even with noisy facts and knowledge, we can continue reasoning in an FPN and determine the culprit, which to the best of the author's knowledge is the first successful work in this domain.

The software included with this book demonstrates the scope of realization of the proposed system in practice. The interested students may execute them with the guidelines prescribed in the Appendix A. The software was developed in Pascal and C based on the considerations that the beginners in the subject may not have expertise in LISP or PROLOG.

Exercises

1. Given an image, whose gray values at pixel (x, y) is f(x, y). Develop a program to compute the x-gradient G_x (x, y) and y-gradient G_y (x, y) at pixel (x, y) by using

   G_x (x, y) = f (x+1, y) – f (x, y) and

   G_y (x,y) = f (x, y+1) – f (x, y)

at all feasible x, y. Then estimate local direction at pixel (x, y) by computing the angle θ with respect to the image x-axis by the expression

$$\theta = \tan^{-1}(G_y / G_x).$$

Now synthesize a binary (having two gray levels) fingerprint image on a graph paper and try to identify the core and delta points, using the above definition of local direction angles.

2. Extend the program used for image matching (see Appendix A) by computing the membership distance product of those blocks only, whose membership exceeds a given threshold, say 0.6. What benefit will you derive by doing so? How much computational time (complexity) do you save by doing so?

3. Within the narrow scope of a textbook, we just defined (in chapter 14) but could not explain in detail the weight adjustment policy using momentum. Read any textbook on artificial neural nets that covers the back-propagation algorithm. Implement the algorithm without momentum and with momentum.

4. Take 5 fingerprints of your friends and try to classify them by observation. Run the fingerprint classification program given in the CD (see Appendix A) and test whether you get the correct classification by the program.

5. Setting all threshold of transitions to be 0.1, and assuming that transition tr_5 is open from fig. 23.12, continue belief revisions, until steady-state is reached, both manually and by the supplied program in the CD (see Appendix A).
Determine the culprit in this context.

6. Does a purely cyclic net consisting of n places and n transitions, each transition having a single input and single output place, exhibit limitcycles? If yes, can you eliminate it by the suggested guidelines? If no, why?

7. Verify the results obtained in the case study in connection with non-monotonic reasoning on fig. 23.12.

8. Design some properties of an FPN by which you can say that the net will exhibit limitcycles for some ranges of initial fuzzy beliefs and thresholds. [*open ended problem for researchers*]

References

[1]   Bernerd, P., *An Introduction to Default Logic*, Springer-Verlag, Berlin, ch. 3, pp. 13-30, 1989.

[2]   Biswas, B., Konar, A. and Mukherjee, A. K., "Image matching with fuzzy moment descriptors," communicated to *Engineering Applications of Artificial Intelligence,* Pergamon, Elsevier, North Holland.

[3]   Dellepiane, S. and Vernazza, G., "Model generation and model matching of real images by a fuzzy approach," *Pattern Recognition*, vol. 25, no. 2, pp. 115-137, 1992.

[4]   Farreny, H. and Prade, H., "Default and inexact reasoning with possibility degrees," *IEEE Trans. on Systems, Man and Cybernetics*, vol. SMC-16, pp. 270-276, Mar.- Apr. 1986.

[5]   Gonzaledz, R. C. and Wintz, P., *Digital Image Processing*, Addison Wesley, Reading, MA, 1993.

[6]   Graham, I. and Jones, P. L., *Expert Systems: Knowledge, Uncertainty and Decision*, Chapman and Hall, London, 1988.

[7]   Pratt, W. K., *Digital Image Processing*, Wiley-Interscience Publications, New York, 1978.

[8]   Konar, A., *Uncertainty Management in Expert SystemsUusing Fuzzy Petri Nets*, Ph. D. thesis, Jadavpur University, 1994.

[9]   Konar, A. and Mandal, A. K., "Uncertainty management in expert systems using fuzzy Petri nets," *IEEE Trans. on Knowledge and Data Engineering*, vol. 8, no. 1, pp. 96-105, 1996.

[10]   Konar, A. and Mandal, A. K., " Non-monotonic reasoning in expert systems using fuzzy Petri nets," *Advances in Modeling and Analysis, AMSE Press*, vol. 23, no. 1, pp. 51-63, 1992.

[11]   Lee, H. C. and Gaensslen, R. E., *Advances in Fingerprint Technology*, CRC Press, Boca Raton, FL, 1994.

[12]   Looney, C. G., " Fuzzy Petri nets for rule based decision making," *IEEE Trans. on Systems, Man and Cybernetics*, vol. 18, no. 1, Jan-Feb. 1989.

[13]   Malmberg, B., *Manual of Phonetics*, North-Holland, Amsterdam.

[14]   Ramamurthi, B. and Gershe, A., "Classified vector quantization of images," *IEEE Trans. on Communication*, vol. Com 34, no. 11, pp. 1105-1115, Nov. 1986.

[15]   Ratha, N. K., Karu, K,. Chen, S. and Jain, A. K., "A real time matching system for large fingerprint databases," *IEEE Trans. on Pattern Analysis and Machine Intelligence*, vol. 18, no. 8, pp. 799-813, 1996.

[16]   Sasnur, K., *Minutiae Based Automated Fingerprint Identification System*, M.E. thesis, Jadavpur University, 1998.

[17]   Shukla, A., "Hindi vowel recognition using ANN," *J. of AMSE*, vol. 32, B, no. 1, 1995.

[18]   Sharma, V. V. S. and Yegnarayan, B., "A critical survey of ASR system," *J. of the Computer Society of India*, vol. 6, no. 1, pp. 9-19, Dec. 1975.

[19]   Waterman, A. D. and Hayes-Roth F., *Pattern Directed Inference Systems*, Academic Press, New York, 1976.

[20]   Zadgaoker, A. S. and Shukla, A., "Classification of Hindi consonants according to the place of articulation using ANN," *Int. J. Elect. Engg. Edu.*, Manchester, vol. 32, pp. 273-275, 1995.

24

Case Study II: Realization of Cognition for Mobile Robots

This is an applied chapter that demonstrates the scope of realization of the philosophical models of cognition on a mobile robot. A mobile robot has to generate a navigational plan in a given environment between pre-defined starting and goal points. The robot's environment includes many obstacles and thus finding the shortest path without touching the obstacles in many cases is an extremely complex problem. The complexity of the problem increases further, when the obstacles are dynamic and there exists many spatial constraints for the movement of the robots. The chapter considers the problems of navigational planning for both static and dynamic environments. It outlines the existing tools and techniques, currently available for handling these problems, and illustrates the scope of the soft computing tools through computer simulations for application in this domain.

24.1 Mobile Robots

Mobile robots are equipped with various types of sensors to recognize the world around them. In most of the commercial robots [1], [13], where navigation is required in a closed environment (like factory), ultrasonic sensors are adequate. For long range navigation of the order of 100 meters or more, laser range finders are occasionally employed in robots. Further, for determining the shape and structure of the objects in its territory, video cameras are mounted with the robot. The cameras grab color or gray images, which are processed by a computer placed within the robot. Fig. 24.1 describes the external features of the Nomad Super Scout II mobile robot, manufactured by the Nomadic Technologies, USA. It has 14 ultrasonic sensors, mounted around its entire periphery, covering a total of 360 degrees. A video camera is

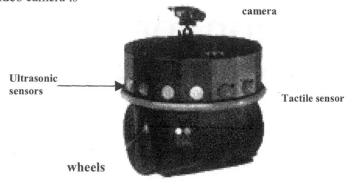

Fig. 24.1: The Nomad Super Scout II robot (collected from the public utilities of the internet site).

fixed with the robot to acquire the gray images around its world. The ring below the ultrasonic sensors, surrounding the robot, serves the purpose of tactile sensors. The robot has two large and one small wheels. The large wheels are driven by separate stepper motors to control the linear and circular motion of the robot. For instance, when both the motors are moved at equal speed, the robot moves forward; when the left motor moves, but the right is static, the robot moves in clockwise direction. Similarly, when the left wheel is static and the right is commanded to move, the robot moves anti-clockwise. The differential velocity of the motors thus can be used to move the robot along a prescribed direction. The third small back wheel is required for mechanical balancing, and it has no active control. We used the above robot for our experiments in Robotics and Vision Laboratory, Jadavpur University.

Let us now have a look at the electrical side of the robot. The robot contains 3 printed circuit boards altogether. At the bottom of the robot, there is a power board that supplies power to motors and the rest of the boards. The next board from the bottom is a controller card that controls the speed of the motors by sensing the target direction at an instant of time. It is realized in Nomad by a pulse width modulator that controls the duty cycles of the pulse trains applied at the digital input of the motor. The topmost card fixed inside the top cover of the robot is a Pentium mother board that provides the software interface of the robot with the users.

The robot can operate in three modes. First, users can control the movement of the robot by a joystick, connected to one serial port of the robot. Secondly, the robot can be linked with a color VGA monitor, a keyboard and a mouse for testing its operations like range measurement by sonar sensors or slow movement of the motors. This is done after the robot boots the Linux operating system and issues a robot prompt. Thirdly, the motion of the robot can be also controlled through a radio communication link. This is useful for practical utilization of the robot.

The robot can be used for various types of experiments, including control, motion planning and co-ordination. We however used it only for navigational planning in an environment occupied with obstacles. The algorithms that we tested on this robot, however, are presented here in generic form to satisfy a large audience, many of whom may not be familiar with this robot.

24.2 Scope of Realization of Cognition on Mobile Robots

We remember from our discussion in chapter 2 that the elementary model of cognition [14] includes three main cycles. Among these, the 'sensing-action' cycle is most common for mobile robots. This cycle inputs the location of the obstacles and subsequently generates the control commands for the motors to set them in motion. The second cycle passes through perception and planning states of cognition, while the third includes all possible states including sensing, acquisition, perception, planning and action [12]. Sensing here is done by ultrasonic sensors / camera or by both. The sensed range signals or images are saved for multi-sensory data fusion. The fusion of multi-sensory data is generally carried out by Dempster-Shafer theory, Bayesian belief networks [10],[25] and Kalman filtering [2], [18]. A robot can also construct high level knowledge from relatively lower level data and knowledge. This is referred to as perception. Construction of knowledge about its environment from range data and gray images helps it in generating its plan for action. For instance, a robot can plan its trajectory, if it knows its local environment. The local range information may be combined to construct global information about its world. There exists ample scope of building perception for robots, but very little of it could have been realized so far in practical systems.

The task planning by a robot becomes easier, when it has the requisite knowledge about its world. For example, determining the trajectory in a known world map is simpler, in contrast to an unknown environment. In an unknown environment, the robot has to execute only the sensing-action cycle. Automated methods for constructing knowledge by a robot about its environment are therefore a significant research problem in modern robotics. After the navigational planning is completed, the remaining task is to generate control commands to the actuators and final control elements for the execution of the plans.

Among the 5 states of cognition cycle, AI is required mainly in perception and planning. It is already mentioned that research on perception did not take any shape yet. In this chapter, we thus mainly concentrate on planning and co-ordination of a robot. It may be added here that the term 'co-ordination' here refers to co-ordination among sensors and arms of a robot, or it may equally include the co-ordination of multiple robots in complex task planning problems.

24.3 Knowing the Robot's World

For operation in a closed environment, a robot is required to know its environment for its safety and security and also for the reliability of its performance. A robot develops an elementary form of perception by visiting and sensing the 2-dimensional obstacles around it. There are many algorithms for construction of the robot's world map [20]. We, here, present a simple algorithm, devised and tested by my colleague Mr. S. Patnaik. The beauty of Patnaik's algorithm [23] lies in its ordered search and traversal of the robot around the obstacles. The principle of search and traversal are presented below with examples.

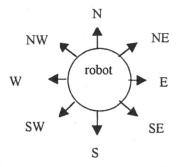

Fig. 24.2: The schematic view of a robot with 8 ultrasonic sensors.

Consider a robot having 8 ultrasonic sensors directed towards north (N), north east (NE), east (E), south east (SE), south (S), south west (SW), west (W) and north west (NW). Fig. 24.2 describes the schematic view of the proposed system. The ultrasonic sensors provide the robot with the location and direction of the obstacles within its field of view.

NORTH

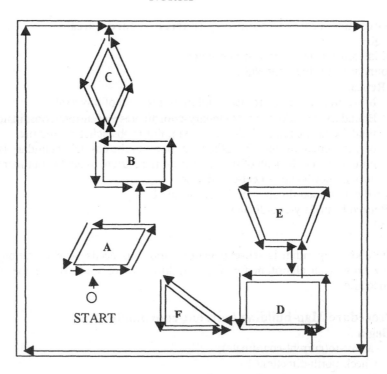

SOUTH

Dark shaded regions denote obstacles A through F.

Fig. 24.3: Depth first traversal by the robot for building its 2-D world map.

The algorithm to be proposed for map building starts searching an obstacle in the north direction of the current obstacle. If an obstacle is found in the north, the robot moves to the obstacle and turns around it (through a shortest path) to record its 2-dimensional shape. The robot then starts searching an obstacle in the north of the current obstacle. If an obstacle is found it recursively repeats the process, else it attempts to find an obstacle in the next possible direction in order of NE, E, SE, S, SW, W and NW. If no obstacle is found in any of these directions, then the robot has to backtrack to the previous obstacle or the starting position (wherefrom it moved) and start searching an obstacle in order of NE, E, SE, S, SW, W and NW. The process is repeated until the robot traverses all the obstacles and the inside boundary of its world.

Traversing around an obstacle and storing the boundary coordinates of the obstacle is one of the important steps in map building. We first discuss this algorithm and later use it as a procedure in the map-building algorithm.

Procedure Traverse-boundary (current-coordinates)
Begin
Initial-coordinate = current-coordinate;
Boundary-coordinates:= Null;
 Repeat
   Move-to (current-coordinate) and mark the path of traversal;
   Boundary-coordinates: = Boundary-coordinates ∪ {current-coordinate};
   Find the next point such that the next point is obstacle free and the
   perpendicular distance from the next point to the obstacle boundary is
   minimum and the path of traversal from the current to next is not marked;
   current-coordinate := next-coordinate;
 Until current-coordinate = initial-co-ordinate
 Return Boundary-coordinates;
End.

The above algorithm is self-explanatory and thus needs no elaboration. We now present an algorithm for map building, where we will use the above procedure.

Procedure Map-building (current-coordinate)
 Begin
    Move-to(current-coordinate);
    Check-north-direction( );
    If (new obstacle found) **Then do**
    Begin
      Current-obstacle = Traverse-boundary(new-obstacle-coordinate);
      Add-obstacle-list (current-obstacle); //adds current obstacle to list//
      Current-position = find-best-point (current-obstacle) // finding the best
       take off point from the current obstacle//
      Call Map-building (current-position);
    End
    Else do Begin
    Check-north-east-direction ( );
     If (new obstacle found) **Then do**

```
      Begin
        Current-obstacle =  Traverse-boundary( new-obstacle-coordinate );
        Add-obstacle-list ( current-obstacle );
        Current-position = find-best-point (current-obstacle);
        call Map-building (current-position);
      End;
  End;
  Else do Begin
   Check east direction( );
  //Likewise in all remaining directions//
  End
  Else backtrack to the last takeoff point on the obstacle
        (or the starting point);
End.
```

Procedure Map-building is a recursive algorithm that moves from an obstacle to the next following the depth first traversal criteria. The order of preference of visiting the next obstacle comes from the prioritization of the directional movements in a given order. The algorithm terminates by back-tracking from the last obstacle to the previous one and finally to the starting point.

24.4 Types of Navigational Planning Problems

The term "Planning" [9] refers to the generation of sequences of action in order to reach a given goal state from a predefined starting state. In the navigational planning problem, a robot has to identify the trajectories between a starting and a goal position within its environment. Depending on the nature of the planning problems, we classify it into two broad categories: (i) Offline and (ii) Online planning. Offline planning is concerned with determination of trajectories when the obstacles are stationary. It includes three classical types of problems such as (i) Obstacle Avoidance; (ii) Path Traversal Optimization; (iii) Time Traversal Optimization. The "Obstacle Avoidance" type of problems deals with identification of obstacle free trajectories between the starting and the goal point. The "Path Traversal Optimization" problem is concerned with identification of the paths having shortest distance between the starting and the goal point. The "Time Traversal Optimization" problem deals with searching a path between the starting and the goal point that requires the minimum time for traversal.

The second category of problem that handles the path planning and navigation concurrently in an environment full of dynamic obstacles is referred to as "Online Navigation and Planning" problem.

Depending on the type of planning problems the robot's environment is represented by a specified formalism such as tree, graph, partitioned blocks and the like. Thus before presenting a technique for handling the planning problems, we in most cases will represent the environment by some structure.

1	1	1	1	1	1	1	1	1	1	1	1	1	1	1	1
1	2	2	2	2	2	2	2	2	2	2	2	2	2	2	1
1	2	1	1	1	1	1	2	1	1	1	1	1	1	2	1
1	2	1	■	■	■	1	2	1	■	■	■	1	2	1	1
1	2	1	■	■	■	1	2	1	■	■	■	1	2	1	1
1	2	1	■	■	■	1	2	1	1	■	■	1	2	2	1
1	2	1	1	1	1	1	2	2	1	■	■	1	2	2	1
1	2	2	2	2	2	2	2	2	1	■	■	1	2	2	1
1	2	1	1	1	1	2	3	2	1	1	1	1	2	2	1
1	2	1	■	■	1	2	3	2	2	2	2	2	2	2	1
1	2	1	■	■	1	2	3	2	1	1	1	1	2	1	1
1	2	1	■	■	1	2	3	2	1	■	■	1	2	1	1
1	2	1	■	■	1	2	3	2	1	■	■	1	2	1	1
1	2	1	1	1	1	2	3	2	1	1	1	1	2	1	1
1	2	2	2	2	2	2	2	2	2	2	2	2	2	2	1
1	1	1	1	1	1	1	1	1	1	1	1	1	1	1	1

(a): Step 1

1															1
	2	2	2	2	2	2	2	2	2	2	2	2	2		
	2		■	■	■		2		■	■	■		2		
	2		■	■	■		2		■	■	■		2		
	2		■	■	■		2			■	■		2		
	2						2			■	■	2			
	2	2	2	2	2	2	2			■	■	2			
	2						3					2			
	2		■	■			3	2	2	2	2	2	2		
	2		■	■			3			■	■		2		
	2		■	■			3			■	■		2		
	2		■	■			3			■	■		2		
	2						3						2		
	2	2	2	2	2			2	2	2	2	2	2		
1															1

(b): Step 2

Fig. 24.4: Steps of constructing a Voronoi diagram.

24.5 Offline Planning by Generalized Voronoi Diagram (GVD)

The construction process of GVD [28] by Cellular Automata consists of two main steps. In step 1, the boundaries of each obstacle and the inner space boundary of the environment are filled in with a numeral say 1(One). As the distance from the obstacles and the inner boundary increases, the coordinates of the space will be filled in with gradually increasing numbers. The process of filling in the workspace, by numerals, is thus continued until the entire space is filled in. In step 2, the highest numerals are labeled and its neighboring coordinates containing the highest numerals or next to the highest are labeled. The process of labeling is continued until each obstacle is surrounded by a closed chain of labeled numerals. Such closed curves are retained and the rest of the numerals in the space are deleted. The graph, thus constructed, is called the GVD. An example shown above (vide fig. 24.4 (a) & (b)) demonstrates the construction process of GVD.

Once the construction of a Voronoi diagram is complete, the shortest path from the starting position and the goal position of the robot to the graph, depicting the Voronoi diagram, are evaluated. A heuristic search may now be employed to find the shortest path on this graph.

24.6 Path Traversal Optimization Problem

There exist many approaches to handle this type of problem. We will however limit our discussion to i) Quadtree based heuristic search and ii) Genetic algorithm based scheme.

24.6.1 The Quadtree Approach

The quadtree is a tree, in which each node in the tree will have four children nodes. We can represent the given 2- dimensional image map in the form of quadtree by a recursive decomposition. Each node in the tree represents a square block of the given image map. The size of the square block may be different from node to node. The nodes in the quadtree can be classified into three groups: i) free nodes, ii) obstacle nodes and iii) mixed nodes.

A **free node** is a node that has no obstacle in the square region represented by it. An **obstacle node** is a node whose square region is totally filled with obstacles. A **mixed node's square region** is partially filled with obstacles. For example, consider the image map given in fig. 24.5.

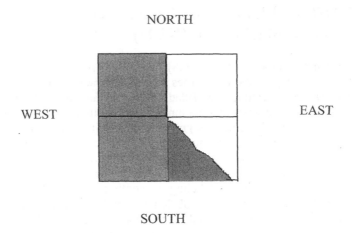

Fig. 24.5: Representation of a simple 2D world map, in which
the gray region represents obstacles.

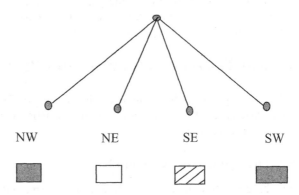

Fig. 24.6: The decomposition of 2D world map of fig. 24.5 into quadtree
nodes. The type of each node is represented by a small box, with
different fill pattern: gray color for obstacle nodes, white color
for free nodes, and hatched boxes for mixed nodes.

In the decomposition process of the 2-dimensional image map above, it
is first divided into four sub-square regions (four children nodes), namely
NW, NE, SW, SE according to directions. The square regions NW, SW are
fully occupied with obstacles (gray region). So they will come under
"Obstacle node" category. Square region NE is not having any obstacle region
in it. So it will come under "Free node " category. The square region SE is
partially filled with obstacles. So it will come under "Mixed node" category.
The quad tree representation of the above decomposition is shown in fig. 24.6.

For both free nodes and obstacle nodes there is no need to decompose them further. They will remain as leaves of the quadtree. Each mixed node must be recursively subdivided into four sub-quadrants, which form children of that node. The subdivision procedure is repeated continuously until either of the conditions mentioned below is satisfied.

i)   The node is either a free node or an obstacle node.

ii)   The size of the square region represented by the node's children is less than the size of the mobile robot.

The complete example for generating the quadtree for a given 2-dimensional image map is illustrated below. Consider the image map of fig. 24.7.

Fig. 24.7: A representative 2-D world map of a robot, containing obstacles, denoted by the shaded regions.

The gray areas in the figure are regions occupied by obstacles. For simplicity we have considered the square obstacles only. In the first stage of its decomposition, the image map is divided into four square regions of equal size as shown below (fig. 24.8). The image map itself is the root of the quadtree. Let us call it A.

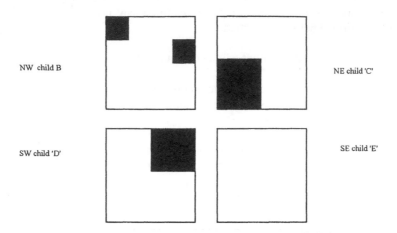

NW child B

NE child 'C'

SW child 'D'

SE child 'E'

Fig. 24.8: The decomposition of the 2-D world map presented in fig. 24.7.

In the above decomposition the child 'E' has no obstacle in its square region. So it is a free node and will not be decomposed further. It will remain as a leaf node of the quadtree. The remaining children 'B', 'C', and 'D' have some free space and some obstacle region; they will come under mixed nodes. The quadtree developed up to this stage (fig. 24.9) is as follows. The small square box under each node represents the status of the node. The white box represents a free node. The gray box represents the obstacle node. The box with hashed lines represents the mixed node. They have to be decomposed further, and the decomposition is shown in the fig. 24.10

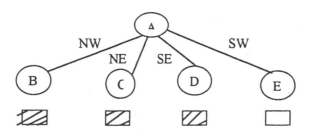

Fig. 24.9: The quadtree representation of the decomposition.

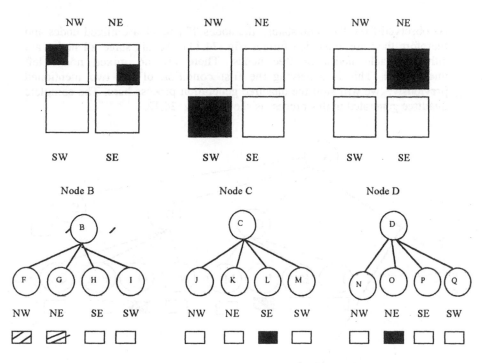

Fig. 24.10: The second stage decomposition of the 2-D world map given in fig 24.9.

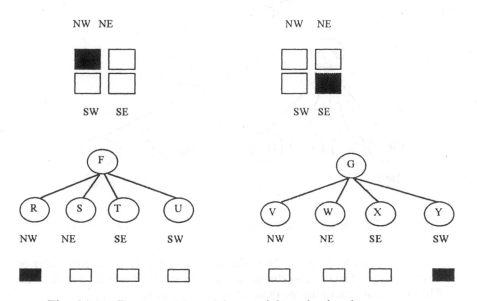

Fig. 24.11: Decomposition of the remaining mixed nodes.

As observed from the node status, the nodes 'F' and 'G' are mixed nodes and therefore they are expanded further (fig. 24.11). At this stage all nodes are either obstacle nodes or free nodes. There are no mixed nodes left unexpanded. This is satisfying the first condition of the two mentioned previously; we will stop the quadtree generation process here. The complete quadtree generated in this process is shown in fig. 24.12.

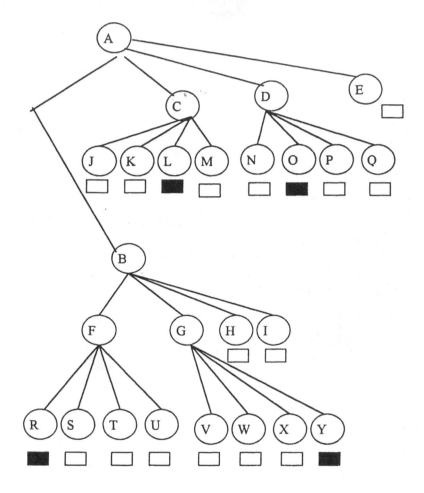

Fig. 24.12: The quadtree representation of the world map given in fig. 24.7.

We observe from fig. 24.12 that all leaf nodes are either free nodes or obstacle nodes. The nodes-status are denoted by small boxes under each node. The process of generating the quadtree is now complete.

It may be noted that the neighborhood relationship of the partitioned modules is not directly reflected in the tree we constructed. There is an interesting algorithm [27] to find the neighbors of a node in the tree, but we do not present it here, for it requires a lot of formalization. We just presume that the map is partitioned into modules and the robot can safely move from one module (leaves of the tree) to another. But how do we find the shortest path between two given modules? The heuristic search will give an answer to this question. In fact the A* algorithm that reduces the sum of the cost of traversal from a starting point to an intermediate point (module), and the predicted cost of reaching the goal from the intermediate point to the goal, is used in our simulation (see Appendix A) to select the intermediate points. Here, we used the Euclidean distance measure to estimate the distance between two partitioned modules.

24.6.2 The GA-based Approach

Michalewicz [16] first successfully applied GA [6] in navigational planning of mobile robots. In one of their recent papers [29], they considered the data structure of fig. 24.13(a), for the chromosome in path planning problem, where the first two field: x_i, y_i, $1 \leq \forall i \leq n$ denote the co-ordinate of the robot, the third field b_i denotes whether the point is on an obstacle and the fourth field denotes a pointer pointing to the next node to be visited.

Fig. 24.13(a): A chromosome, representing a path from the starting point (x_1, y_1) to the goal (x_n, y_n) through a number of points.

They considered the standard crossover operation that could be applied between two chromosomes and the cross-site could be selected as the region between two structures within a chromosome. Thus one can change the current sub-goal by executing a crossover operation. They also devised a number of mutation operations to improve smoother paths or to delete a point on the path to avoid its overlap with an obstacle. For details, the readers may refer to [30]. The fitness function in their GA-based planner comprises of three factors: i) finding the shortest path between the starting and the goal point, ii) the path should be smooth and iii) there must be sufficient clearance between the obstacles for the easy movement of the robot through the path.

In their evolutionary planner algorithm, Michalewicz considered a set of operators including crossover and mutations. An operator is selected based on its probability of occurrence and the operation with the highest probability is executed. The fitness evaluation function is then measured and proportional selection is employed to get the population in the next generation. A sample execution of their algorithm is presented in fig. 24.13(b) to illustrate the concept. Fig. 24.13(b) describes a static planner, i.e. possible paths when obstacles are stationary. We simulated this and found analogous results. Michalewicz et al. claimed that for dynamic obstacles they can use the same algorithm, but the algorithm may be required to be executed from intermediate sub-goals, especially in cases when an obstacle is close enough to the planned trajectory.

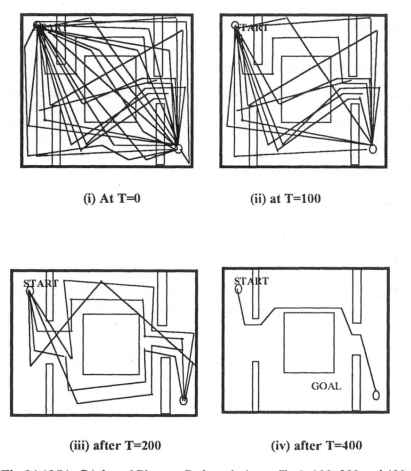

(i) At T=0   (ii) at T=100

(iii) after T=200   (iv) after T=400

Fig.24.13(b): GA-based Planner: Path evolution at T= 0, 100, 200 and 400 iterations.

24.6.3 Proposed GA-based Algorithm for Path Planning

In the previous planning algorithm [30], much time is wasted for planning a complete path, which later is likely to be disposed of. An alternative method for path planning in a dynamic environment is to select the next point and the path up to that point only in one genetic evolution. This is extremely fast and thus can take care of movable obstacles of speeds comparable to the robot.

Our first step in this path planning is to set up the initial population. For this purpose, instead of taking random coordinates, we have taken the sensor information into account and the coordinates obtained from those sensors are used to set up the initial population. With this modification it is assured that all the initial population are feasible, in the sense they are obstacle-free points and the straight line paths between the starting point and the selected next points are obstacle free.

Since in one genetic iteration, we plan the path up to the selected next point, the data structure to represent the chromosome becomes extremely simple, as presented below (fig. 24.14).

Fig. 24.14: Representation of the chromosome in our simulation.

Here (X_i, Y_i) is the starting point and (X_j, Y_j) is the one of the 2D points, obtained from the sensor information. All these chromosomes form the initial population. The next step is to allow crossover among the initial population. But what about the crossover point? If we choose the cross-site randomly, it is observed that most of the offsprings generated in the process of crossover are not feasible, as those paths may fall outside the 2D work space. So instead of binary crossover, we employed integer crossover. The crossover process is shown below. Consider the two chromosomes as shown in fig. 24.15 and the crossover point is set between the third and the fourth integer for every chromosome.

After making crossover between all pairs of initial population, we will get the new population. For this new population we will find the feasibility, i.e., they are reachable from the starting point by the straight line path or not. The next step of the algorithm is making the mutation. This will make fine tuning of the path, such as avoiding the sharp turns. In this process we select a binary bit randomly on the bit stream of the sensor coordinates and alter that binary bit value, such that the feasibility should not be lost for that chromosome.

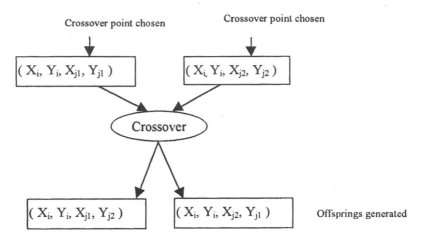

Fig. 24.15: The crossover operation used in the proposed algorithm.

Our next task is estimating the fitness of each and every chromosome of the total present population (both for the initial and new populations). Calculation of the fitness involves finding the sum of the straight line distance from the starting point (X_i, Y_i) to the coordinate (X_{j1}, Y_{j1}) obtained from the sensor information and the distance from (X_{j1}, Y_{j1}) to the goal point (X_g, Y_g).

Fitness of a chromosome $(X_i, Y_i, X_{j1}, Y_{j1}) =$

$1 / \{$(distance between $(X_i, Y_i,)$ and$(X_{j1}, Y_{j1})) +$

(distance between (X_{j1}, Y_{j1}) and $(X_g, Y_g))\}$

After finding the fitness value of the chromosomes, we will evaluate the best fit chromosome, i.e., for which the fitness is the best. In this case, the best fit chromosome represents the predicted shortest path from the starting point to the goal. We restrict the number of generations to one, since in the first generation itself we are getting a near optimal intermediate point to move. That third and fourth integer field of the best fit chromosome will become the next intermediate point to move. Then we update the starting point with this better point and the whole process of the GA, from setting up the initial population, is repeated, until the best fit chromosome will have its third and the fourth field equal to the x- and y-coordinates of the goal location. The algorithm is now formally presented below.

Procedure GA-path

// (x_i, y_i) = starting point;
(x_g, y_g) =goal point;   //
add-pathlist (x_i, y_i) ;
Repeat
  i) Initialization:
    Get sensor information in all possible directions
      $(x_{j1}, y_{j1}), (x_{j2}, y_{j2}), \ldots (x_{jn}, y_{jn})$.
    Form chromosomes like (x_i, y_i, x_j, y_j);
  ii) Crossover:
    Select crossover point randomly on the third and the fourth
    fields of the chromosome.
    Allow crossover between all chromosomes and get new
    population as
      $(x_i, y_i, x_{j1}, y_{j1}), (x_i, y_i, x_{j2}, y_{j2}), (x_i, y_i, x_{j1}{}^i, y_{j1}{}^i), (x_i, y_i, x_{j2}{}^{ii}, y_{j2}{}^{ii})$;
  iii) Mutation:
    Select a mutation point in bitstream randomly and
    complement that bit position for every chromosome.

  iv) Selection:
    Discard all chromosomes (x_i, y_i, x_j, y_j) from population
    whose line segment is on obstacle region
    For all chromosomes in population find fitness using
    Fitness$(x_i, y_i, x_j, y_j) = 1/ ((x_j - x_i)^2 + (y_j - y_i)^2 + (x_g - x_j)^2 + (y_g - y_j)^2)$;
    Identify the best fit chromosome $(x_i, y_i, x_{bf}, y_{bf})$;
    Add-pathlist(x_{bf}, y_{bf});
    $X_i = x_{bf}$; $y_i = y_{bf}$.
    End For,
  Until ($x_i = x_g$) and ($y_i = y_g$);
End.

A sample execution of the above algorithm is presented in fig. 24.16 and the details of how to execute the algorithm is available in Appendix A. A look at the Appendix reveals that the GA based evolution scheme can identify a path easily, when the obstacles are convex-shaped. For concave shaped obstacles, the evolution has to be continued quite a large number of iterations to find the goal point. This, of course, is not the limitation of our algorithm. It happens so for all GA-based path planning. To overcome this problem, one simple way is to follow the boundary of the concave obstacles, and when there is a 90 degree turn, again start the evolution. Thus GA may be used in an interleaved manner. We tested and found this to be satisfactory in most cases.

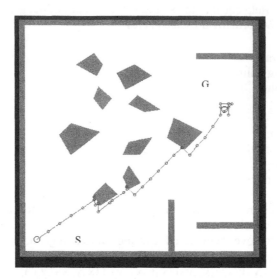

S= starting point, G= Goal point, Polygons are obstacles.

Fig. 24.16: Computer simulation of the proposed GA-based path planner.

24.7 Self-Organizing Map (SOM)

In a self-organizing feature map, a set of two dimensional points (n × n), called neurons, are mapped to a set of (l)   one dimensional structure (m × 1) such that $l \times m \times 1 \ll n^2$. The principle employed to obtain these feature vectors is outlined below for the convenience of the readers (for details, see chapter 14).

Algorithm for SOM
Assumption: Assume that each neuron on the 2D plane has a weight vector of m components;

    Randomize weight vectors of all neurons;
    Repeat
      a: For each neuron on space S = the 2D plane, find the Euclidean distance (error) between its weight vector and the feature vector;
      b: Find the neuron with minimum distance in space S, and select it as the center of mass of a small square 2-D space; Call this space S;
      c: Adjust components of the weight vectors of the neurons within S by the respective errors (weight := weight + α error; α = learning rate);
    Until S converges to a point

The above three steps are to be repeated for each input pattern.

The SOM algorithm has been successfully applied by Heikkonen et al. [8] to online navigational planning. An outline of the algorithm for online navigation is presented below.

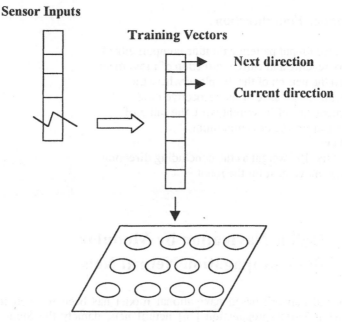

Fig. 24.17: Architecture of SOM showing the association between the sensory inputs and the control outputs.

Algorithm

Input: A set of l vector of 10 dimension, 8 of which denote the sensory reading for range of obstacles, 1 for current direction of movement and 1 for the next direction of movement

A set of such patterns is used as inputs;

A set of n^2 neurons with random weights on a 2D plane

Output: The weight corresponding to the next direction of movement in the previous vector.

The algorithm consists of two modules: Procedure training and Find-direction.

Procedure Training

Begin

  For each input pattern

  Call algorithm for SOM

End For;

End.

Once the training is over, the system will be able to determine the next direction of movement when all the nine components of the feature vector are supplied. Here the input sensory information is available for the determination of next direction of movement.

Procedure Find-direction
Begin
For a given input pattern with nine components of
sensory reading and current direction of movement
    Find the neuron of the 2D plane where the
    Euclidean distance of the respective nine
    components of its weight w.r.t. the same of
    the feature vector is minimum;
End For;
  Select the 10th weight as the concluding direction
  of next movement for the robot.
End.

24.8   Online Navigation by Modular Back-propagation Neural Nets

The neural network based navigational model has been realized in [15] by employing 3 back-propagation [26] neural nets, namely the Static Obstacle Avoidance (SOA) net, the Dynamic Obstacle Avoidance (DOA) net and the Decision Making (DM) net, shown in fig. 24.18. The SOA net receives sensory information and generates control commands pertaining to motions and direction. The DOA net, which works in parallel with the SOA net, generates directions of motions from the predicted motion of obstacles. The primary plans for motion generated by these two nets are combined by the AND logic to determine the common space of resulting motion. A final decision about the schedule of actions is generated by employing the DM net, which acts upon the resulting plan of motion generated by the AND logic.

Each of the SOA, DOA and DM models are realized on three layered neural networks, and are trained with well-known back-propagation algorithms (fig. 24.18 (a) and (b)). The sample training patterns for the SOA, DOA and DM nets are presented in tabular form (vide table 24.1 – 24.3), where the entries under inputs denote the cell numbers surrounding the Robot (vide fig. 24.18 (b)). The tables are self-explanatory and thus need no further elaboration. Readers, however, should try to explain the table on their own from the point of view of current position of static obstacles and predicted direction of movement of the dynamic obstacles.

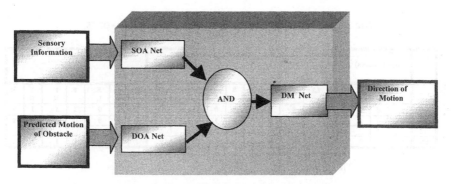

SOA= Static Obstacle Avoidance; DOA= Dynamic Obstacle Avoidance; DM= Decision Making

Fig.24.18 (a): Structure of the navigator.

13	14	6	7	8
12	5	1	4	9
11	3	MR	2	10

Fig.24.18 (b): The position of mobile robot (MR) and its surrounding cells numbered arbitrarily.

Table 24.1: Training pattern samples of SOA network.

N	INPUTS (Square Values)									OUTPUTS						
	11	12	13	14	6	7	8	9	10	L	TL	FL	F	FR	TR	R
1	0	0	0	1	1	1	1	0	0	1	1	0	0	0	1	0
2	0	0	0	0	1	1	1	0	0	1	1	0	0	0	1	0
.	.	.	.	.												
28	0	0	0	0	0	0	0	0	0	1	1	1	1	1	1	1

Inputs are position of obstacle at respective cell of the grid (1= presence and 0= absence of obstacle); Outputs are five action commands (TL= Turn Left; FL= Forward Left; F= Forward; FR= Forward Right; TR= Turn Right) and two obstacle-position indicator (L= Left; R= Right)

Table 24.2: Training pattern samples of DOA networks.

No	INPUT (Square values)													OUTPUT							
	11	3	12	5	13	14	6	7	8	4	9	2	10	L	T L	F L	F	F R	F R	T R	R
1	1	0	0	0	0	0	0	0	0	0	0	0	0	0	0	0	1	1		1	0
2	0	0	0	1	0	1	1	0	0	1	0	1	0	0	1	0	0	0		0	0
.	.		.	.	.	.	.	.	.	.	.	.	.	.	.	.	.			.	
55	0	0	0	0	0	0	0	0	0	0	0	0	0	1	1	1	1	1		1	1

The notation in this table has the same meaning as earlier.

Table 24.3: Training pattern samples of the decision-making network.

No	INPUT							OUTPUT						
	L	T L	F L	F	F R	T R	R	G	T L	F L	F	F R	T R	S T
1	0	1	0	0	0	0	0	0	1	0	0	0	0	0
2	0	0	0	1	0	0	0	0	0	0	1	0	0	0
.	.	.	.	.	.	.	.							
37	1	1	1	1	1	1	1	1	0	0	0	0	0	0

Outputs of Table 24.1 and 24.2 are ANDed and given as INPUT to table 24.3.
Outputs are decision of movement (G= Go; ST= Stop and the others are same as above)

The BP algorithm in our discussion was meant for a closed indoor environment [21]-[24]. Recently Pomerleau at *Carnegei Mellon University* designed an autonomous Land Vehicle called ALVINN that traverses the road using the Back-propagation algorithm [11], shown in fig.24.19. In their system, they employed a three layered neural net, with 1024 (32 × 32) neurons in the input layer, that grabs the (32 × 32) pixel-wise video image and generates one of the 30 commands at the neurons located at the output layers. The commands are binary signals, representing the angle changes in the movement. The figure below describes some of the control signal. A human trainer generates the control commands for a specific instance of an image. After training with a few thousand samples of instances, the network becomes able to generate the control commands autonomously from known input excitations.

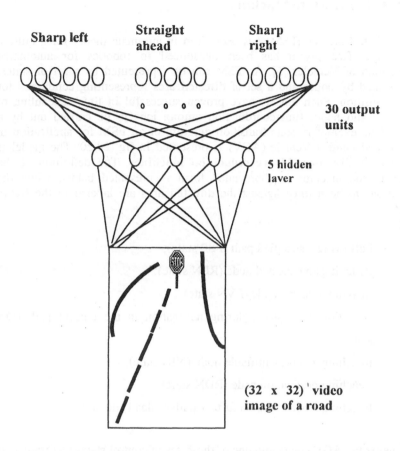

Fig. 24.19: The structure of the ANN showing the desired direction of
movement from (32 x 32) gray image.

24.9 Co-ordination among Sub-modules
in a Mobile Robot

The co-ordination in a mobile robot is required for autonomously handling a
set of concurrent as well as sequential tasks, pertaining to Sensing, Motion
and Control of various subsystems within the robot. A number of models of
coordination are available in the realm of artificial intelligence. The finite
state machines, A Language For Action (ALFA) and Timed Petri nets are a
few popular models. For lack of space, we only present the Finite State
machine and timed Petri net models for co-ordination in this chapter.

24.9.1 Finite State Machine

Finite State Machine (FSM) has been used in automata theory for a quite a long time. Recently it has been introduced to robotics for automated Navigation of Vehicles [5]. An FSM can be represented by a set of states, represented by nodes and a set of directed arcs representing conditions for state transition. Such models have proven successful in time scheduling of activities and hence time-coordination among jobs to be carried out by a robot. One exemplary state transition diagram of an FSM for application in Automated Guided Vehicle (AGV) is presented in fig. 24.20. The model is based on well-known **Moore sequential machines**. The definition of the states of the machine in abbreviated form is presented below, while the conditions to be marked against the directed arcs are labeled in the figure itself.

(i)   Following a specified path (FSP state);

(ii)   Reaching an expected node (REN state);

(iii)   Leaving a visited node (LVN state);

(iv)   Avoiding an obstacle during navigation in the current path (OA state);

(v)   Reaching a non-identifiable node (NIN state);

(vi)   Reaching destination node (RDN state);

(vii)   Reaching a node with no further path or dead end node.

Whenever the AGV enters into one of these seven control states its response is fixed according to the present state. It is represented below with respect to the respective inputs.

(i)   Move further along current path (MFCP)

(ii)   Turn to the next path (TNP)

(iii)   Move further away from visited node (MFVN)

(iv)   Generate alternate local path (GALP)

(v)   Generate new path from current path (GNP)

(vi)   Stop and send done signal to central controller (DONE)

(vii)   Stop and send help signal to central controller for alternate path (HELP)

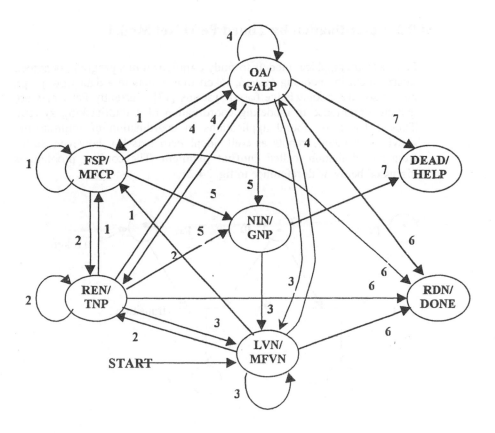

1= following a specified path already generated; 2= reaching at a known node; 3= moving further in the present path; 4= reached at a known node; 5= reaching at a unknown node; 6= reaching at the destination point; 7= reaching at a dead node, from where no path can be further generated

Fig. 24.20: State transition diagram of finite state machine for Automated Guided Vehicle Navigation.

One major drawback of FSM is its incapability to handle the notion of "real time". Consequently, from the current status of a state, one cannot have an idea about the time of occurrence of the state. To overcome this limitation, an alternative scheme for coordination was devised by incorporating time on Petri nets. Such a model of Timed Petri nets not only identify the time of occurrence of a state, but also in addition coordinates more than one activity also, concurrently.

24.9.2 Co-ordination by Timed Petri Net Model

Petri nets are graphical models to study parallelism in a program, deadlock management in operating system, token movements in a data flow graph and reasoning in knowledge based systems [17]. Currently Petri nets are gaining importance in modeling coordination of a multitasking system. Thus it has also found applications in coordination of multitasking activities in mobile robots as well. Ideal Petri nets [4], however, suffer from one limitation, called **conflict problem**. The conflict problem is discussed below with reference to fig. 24.21.

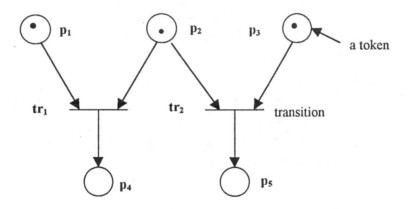

Fig. 24.21: A Petri net with transition tr_1 and tr_2, places p_1, p_2, p_3, p_4, p_5, where p_1, p_2 are input places of tr_1; p_2, p_3, are input places of tr_2. p_4, p_5 are output places of tr_1 and tr_2 respectively. Dots denote tokens in a place.

In this figure, the transition tr_1 and tr_2 could fire if each of them has a token at all its input places. However, if tr_1 fires earlier than tr_2, then according to the notion of Petri nets, p_4 will receive one new token and p_1, p_2 will lose their tokens. Consequently, transition tr_2 cannot fire as place P_2 does not possess a token. Thus either of these two transitions can fire only in this system. However none of the transitions tr_1 and tr_2 will fire, unless no additional condition is imposed regarding the priority of the firing of the transition. This problem of waiting in a loop is often referred to as *time conflict* in Petri nets theory. One way to restore this conflict is to attach time with the tokens and firing delay with the transitions. Such Petri nets that include the notion of time with tokens are called timed Petri nets. Fig. 24.22 describes a Timed Petri net, where the semantics of the places are presented in the figure itself. The firing conflict in a Petri net with transitions like tr_1 and tr_2 of fig. 24.21 is avoided by using the following strategy.

Strategy: *If the arrival time of the latest token in the input place of one transition + its firing delay < the arrival time of the latest token in the input place of the second transition (with common place) + its firing delay, then the former will fire.*

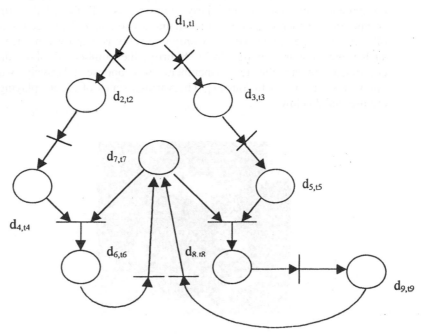

$d_{1,t1}$ = Sense surrounding by sensor and camera at time t1;
$d_{2,t2}$ = Build local map of surrounding from sonar sensor and compare with world map at time t2;
$d_{3,t3}$= Image processing and matching with image database at time t3;
$d_{4,t4}$ = Plan for obstacle free path at time t4;
$d_{5,t5}$ = Object is recognized (suppose that is a useful tool) at time t5;
$d_{6,t6}$ = Start from node n_i at time t6;
$d_{7,t7}$ = Stop at node n_{i+1} at time t7;
$d_{8,t8}$ = Start movement to keep the tool in pre specified place at time t8;
$d_{9,t9}$ = Place the tool at proper place at time t9;

Fig. 24.22: Timed Petri net showing the coordination of multitasking mobile robot.

24.10 An Application in a Soccer Playing Robot

The model of cognition cited in chapter 2 finds application in a wide domain of problems under robotics. Though only a few states of cognition are commonly used in most systems, there still exist a few limited applications, where all these are equally utilized.

We start this section with such an application of cognition in football tournaments by robots, popularly known as "ROBOCUP". In such competitions, generally a football tournament is organized between two teams of soccer playing robots. Till now only one robot is employed per team. One such robot is shown in fig. 24.23. Shortly, the number of robots that will coordinate among the team members and oppose reactively with the opponents will be set to four [7]. The various states of soccer playing robot are presented below.

Fig.24.23: A soccer playing robot, collected from the public utility of the internet site.

The player robot (hereafter player only) of a team, say A, has to recognize: i) the ball, ii) the team mates, iii) opposite players and iv) the goal post of the opponent team B. After taking each snap shot of the field, the sensory information acquired is saved in working memory of the robot. This helps the robot construct higher level knowledge from the acquired information. An example that illustrates the construction of knowledge is the automated learning behavior of the losing team from the possible goals of the winning team. For instance consider the teams consisting of fewer members only. Now if the opponent team scores successive goals from a fixed zone on the ground and the ball is unprotected, then the learning behavior of the losing team will help them to change their orientation. This could be formally represented by a learning rule [3] as follows:

Rule 1: If the opponent scores successive goals from the same spot,
      Then (there must be lack of coordination among team members) or
      (the ball is far away from the current position of the team members)

Rule 2: If the moving ball is far away from the defending team, then
      accelerate speed or change positions of the members.

So, the rule 3 that describes a pipelined action of rule 1 and 2 will be learned autonomously by the losing team.

Rule 3: If the opponent scores successive goals from the same spot then
      accelerate speed or change position of the members.

The state of perception of *Cognition Cycle* thus generated new knowledge from experience of the soccer playing robot.

The task of planning in the elementary model of cognition, as stated earlier, requires breaking a decision into a few pieces of sequential / parallel action. For instance to pass a ball to a teammate closer to the opponent's goalpost, a member of the former team has to plan to pass the ball through many of its teammates under a defensive environment. The decision of scheduling of the ball transfer from one teammate to another is a complex planning problem, which has been successfully realized in ROBOCUP tournaments [19], [27].

The most interesting and fascinating task in a football tournament, perhaps, is the coordination among the teammates and reactive opposition against the opponents. The passing of balls from one team member to another to lead the ball towards the goal post is an example of such a coordination problem. The reactive opposition is required to pass the ball through a defensive environment. The state of action in a soccer-playing robot is clear and thus requires no elaboration.

This is only an interesting example of mobile robots. There exist, however, quite a large number of industrial applications of mobile robots. A few of them includes factory automation, injurious system handling, factory supervision, material management and product planning. Mobile robots have applications also in fire fighting, mine clearing and nuclear waste disposal. The next decade will find mobile robots taking care of our day-to-day household activities.

Exercises

1. Devise an algorithm for automatically finding the neighbors of a node in a quadtree.

   [**Hints:** If two nodes in the tree are of the same size and have common parents, then they are neighbors.

   For a given node in the tree, find the path leading to the root. Propagate down-stream and select the mirror image nodes at the same level, i.e., if at level 3, the up-node is NE, then its mirror image down node is SW. The process is continued, until one visits the same labeled nodes, etc.]

2. What mutation operators can we use to make the paths smooth?

3. In fig. 24.22, assign time tagged tokens at the axioms (places having no input arcs) and generate tokens at all places for a complete traversal of the loops of the Petri net. List the concurrent and the sequential activities in a tabular form. Where is the scope of co-ordination in the net? Does it justify the scheme of co-ordination?

4. Can you modify the algorithm of map building to include the notion of minimum traversal? If so, how?

5. Run the program for GA-based path planning, supplied in the CD (see Appendix A to learn its use). What conclusions do you derive from the generated paths for convex and concave obstacles? Can you extend the program to improve the planning, when the obstacle has a concave geometry?

References

[1] Borenstain, J., Everett, H. R., and Feng, L., *Navigating Mobile Robots: Systems and Techniques*, A. K. Peters, Wellesley, 1996.

[2] Brown, R. G. and Hwang, Patrick Y. C., *Introduction to Random Signals and Applied Kalman Filtering*, John Wiley & Sons, New York, 1997.

[3]   Budenske, J. and Gini, M., "Sensor explication: Knowledge-based robotic plan execution through logical objects," *IEEE Trans. on SMC*, Part B, Cybernetics, vol.27, no.4, August 1997.

[4]   Freedman, P., "Time, Petri nets, and robotics," *IEEE Trans. on Robotics and Automation*, vol.7, no. 4, pp.417-433, 1991.

[5]   Fok, Koon-Yu and Kabuka, M. R., "An automatic navigation system for vision guided vehicles using a double heuristic and a finite state machine," *IEEE Trans. on Robotics and Automation,* vol.7, no.1, Feb. 1991.

[6]   Goldberg, D. E., *Genetic Algorithm in Search Optimization and Machine Learning*, Addition Wesley, Reading, MA, 1989.

[7]   Hedberg, S., "Robots playing soccer? RoboCup poses a new set of AI research challenges," *IEEE Expert*, pp. 5-9, Sept/Oct, 1997.

[8]   Heikkonen, J. and Oja, E., "Self-organizing maps for visually guided collision-free navigation," *Proc. of the Int. Joint Conf. on Neural Networks*, Vol. 1, 669-692, 1993.

[9]   Hutchinson, S. A. and Kak, A. C., "Planning sensing strategies in a robot work cell with Multi-sensor capabilities," *IEEE Trans. on Robotics and Automation,,* vol.5, no.6, 1989.

[10]   Krebs, B., Korn, B., and Burkhardt, M., "A task driven 3D object recognition system using Bayesian networks,", *Proc. of Int. Conf. on Computer Vision,* pp.527-532, 1998.

[11]   Kluge, K. and Thorpe, C., "Explicit models for road following," *IEEE Conf. on Robotics and Automation*, Carnegie Mellon University, Pittsburgh, Pennsylvania, 1989.

[12]   Konar, A. and Pal, S., "Modeling cognition with fuzzy neural nets" In *Fuzzy Systems Theory: Techniques and Applications*, Leondes, C. T., Ed., Academic Press, New York, 1999.

[13]   *Artificial Intelligence and Mobile Robots: Case Studies of Successful robot system*s, Kortenkamp, D., Bonasso, R. P. and Murphy, R., Eds., AAAI Press/ The MIT Press, Cambridge, MA, 1998.

[14]   Matlin, W. Margaret, *Cognition*, Hault Sounders, printed and circulated by Prism books, India, 1996.

[15]   Meng, H. and Picton, P. D., "Neural network for local guidance of mobile robots," *The Third Int. Conf. on Automation, Robotics and Computer Vision (ICARCV' 94)*, Nov.1994, Singapore.

[16]   Michalewicz, Z., *Genetic Algorithms + Data Structure = Evolution Programs*, Springer-Verlag, New York, 3rd edition, 1986.

[17]   Murata, T., "Petri nets: Properties, analysis and applications," *IEEE Proceedings*, vol.77, no.4, April, 1989.

[18]   Nicholas, A., *Artificial Vision for Mobile Robots*, The MIT Press, Cambridge, MA, 1991.

[19]   Nourbaksh, I., Morse, S., Becker, C., Balabanovic, M., Gat, E., Simons, R., Goodridge, S., Potlapalli, H., Hinkle, D., Jung, K., and Vanvector, D., "The winning robot from the 1993 robot competition," *AI Magazine*, vol. 14 (4), pp. 51-62, 1993.

[20]   Pagac, D., Nebot, E. M. and Durrant. W., H., "An evidential approach to map building for autonomous robots," *IEEE Trans. on Robotics and Automation*, vol.14, no.2, pp. 623-629, Aug. 1998.

[21]   Patnaik, S., Konar, A., and Mandal, A. K., "Visual perception for navigational planning and coordination of mobile robots," *Indian Journal of Engineers*, vol. 26, Annual Number '97, pp.21-37, 1998.

[22]   Patnaik, S., Konar, A., and Mandal, A. K., "Navigational planning with dynamic scenes," *Proc. of the Int. Conf. on Computer Devices for Communication (CODEC-98)*, pp. 40-43 , held at Calcutta, Jan., 1998.

[23]   Patnaik, S., Konar, A., and Mandal, A. K., "Map building and navigation by a robotic manipulator," *Proc. of Int. Conf. on Information Technology*, pp.227-232, TATA-McGraw-Hill, Bhubanes-war, Dec. 1998.

[24]   Patnaik, S., Konar, A., and Mandal, A. K., "Constrained hierarchical path planning of a robot by employing neural nets," *Proc. of the Fourth Int. Symposium on Artificial Life and Robotics* (AROB 4th '99), held at Japan, Jan. 1999.

[25]   Pearl, J., "Fusion, propagation and structuring in belief networks," *Artificial Intelligence*, Elsevier, vol. 29, pp. 241-288, 1986.

[26]   Rumelhart, D. E., Hilton, G. E., and Williams R. J., "Learning internal representations by error propagation," In *Parallel Distributed*

Processing, Rumelhart, D. E. and McClelland, J. L., Eds., vol. 1, The MIT Press, Cambridge, MA, pp. 318-362, 1986.

[27]   Samet, H., "Neighbor finding techniques for images represented by quadtrees," *Computer, Graphics and Image Processing*, vol. 18, pp. 37-57, 1982.

[28]   Takahashi, O. and Schilling, R. J., "Motion planning in a plane using generalized voronoi diagrams," *IEEE Trans. on Robotics and Automation*, vol.5, no.2, 1989.

[29]   Xiao, J., Michalewicz, Z., Zhang, L. and Trojanowski, K., "Adaptive evolutionary planner/ navigator for mobile robots," *IEEE Trans. on Evolutionary Computation* , vol.1, no.1, April 1997.

[30]   Xiao, J., "The evolutionary planner / navigator in a mobile robot environment," in *Handbook of Evolutionary Computing*, Back, J., Fogel, D. B. and Michalewicz, Z., Eds., IOP and Oxford University Press, New York, 1997.

24⁺

The Expectations from the Readers

This is the closing chapter of the book that emphasizes the practical utilization of the theories underlying in AI and Soft Computing. It expects more participation from the software developers and the hardware engineers for a faster and better realization of the theories in practice.

24.1⁺ The Expectations

Chapter 24 could be the concluding chapter of the book. But the author feels that he has a moral responsibility to say a few more words, without which the book remains incomplete.

The book presented various logical and mathematical tools for realizing cognition on machines. It aimed at a large readership and thus readers with different backgrounds will find it in different tastes, but hopefully most of them, after reading this book, will be able to shape their illusionary dreams about AI and Soft Computing. The book, however, is not meant just for that

747

purpose. It has a bigger role in reducing the gap between the subject and its readers.

There exists an unwritten agreement between any new subject and its readers. The readers have a great expectation from the subject and the subject demands its enrichment by its readers. Thus there is a closed loop, somewhat like 'the chicken before the egg' and 'the egg before the chicken'. But which one to start with? Obviously, the enrichment of the subject should be done first. In fact, the theoretical foundation of AI is rich enough, but the practical utilization of the theories remains limited. The author through his simple writing invites a larger readership including software developers, hardware engineers, researchers and students for a faster and better realization of the theories in practice.

Appendix A

How to Run the Sample Programs?

The appendix includes the sample runs of a few programs developed to illustrate the case studies in chapter 23 and 24. The programs were built in C, C++ and Pascal languages. The source code and the .exe files of these programs are included in the CD, placed inside the back cover of the book. Before executing the .exe files, save them in A drive, as the pathnames for the data files and the BGI file are given accordingly. Please be sure to include the associated data files or the BGI file in the floppy, when you execute a program. The associated file names are given in the program description below. The source codes can also be compiled by a Microsoft CPP / Turbo Pascal compiler and executed on a Pentium- based machine.

A.1 The Fpnreas1 Program

The Fpnreas1 program is a natural language interface program that inputs the database and knowledge base for the criminal investigation problem. The knowledge base has two modules: the default rules and the production rules.

The user has to submit both types of rules in a specific menu driven format. This is a Pascal program, which also orders the rules, if they are pipelined (i.e., consequent of one rule is an antecedent of another rule).

NOTE: SAVE THREE FILES fpnreas1, fpnreas2 and fpnreas3 TOGETHER IN A FLOPPY (A –Drive) before execution of the programs. The order of execution should be fpnreas1, fpnreas2 and fpnreas3 respectively.

The sample run of the program **Fpnreas1** is presented below.

B:\>fpnreas1

KEEP CAPS LOCK ON
ARE YOU READY (Y/N)? Y

ENTER generic filename without extension KONAR
MAIN MENU:

1-   Create / Append static database
2-   Create / Append production rules
3-   Create / Append default rules
4-   Arrange production rules properly
5-   Delete clauses from database
6-   Delete clauses from pr rule base
7-   Delete clauses from def. rule base
0-   Terminate

SELECT:
1
C to create, A to append: C
enter predicate and arguments
one by one for each clause

if some argument is inapplicable or redundant in the context
enter "END" as predicate to terminate

press any key to continue

Predicate: L
Argument1: S
Argument2: R

Predicate: L
Argument1: S
Argument2: M

Predicate: LSR
Argument1: S
Argument2: _

Predicate: S
Argument1: S
Argument2: _

Predicate: END

MAIN MENU:

1- Create / Append static database
2- Create / Append production rules
3- Create / Append default rules
4- Arrange production rules properly
5- Delete clauses from database
6- Delete clauses from pr rule base
7- Delete clauses from def. rule base
0- Terminate

SELECT:
2

general structure of a production rule:

antecedents without complementation
antecedents1: predicate, argument1, argument2
antecedents2: predicate, argument1, argument2
antecedents3: predicate, argument1, argument2
antecedents4: predicate, argument1, argument2
antecedents5: predicate, argument1, argument2

antecedents with complementation
antecedents1: predicate, argument1, argument2
antecedents2: predicate, argument1, argument2
antecedents3: predicate, argument1, argument2
antecedents4: predicate, argument1, argument2
antecedents5: predicate, argument1, argument2

conclusion
predicate, argument1, argument2

Threshold value

press any key to continue

further instructions

each rule must contain
atmost 5 antecedents without complementation, and
atmost 5 antecedents with complementation
if the actual no. of antecedents is less than 5.
Enter "END" as predicate
When all the antecedents of the rule are entered
use upper case letters to represent arguments
if some argument is inapplicable or redundant in the context
enter it as an underscore

press any key to continue
antecedents without complementation:
ANTECEDENT 1: Predicate: L
, ARGUMENT1 : X
, ARGUMENT2 : Y

ANTECEDENT 2: Predicate: L
, ARGUMENT1 : X
, ARGUMENT2 : Z

ANTECEDENT 3: Predicate: END

antecedents with complementation:
ANTECEDENT 1: Predicate: END

conclusion
Predicate: LSR
Argument1 : X
,Argument2: _

Threshold value: 0.7

More rules to create (y / n)? Y

antecedents without complementation:
ANTECEDENT 1: Predicate: LSR
, ARGUMENT1 : X
, ARGUMENT2 : _

ANTECEDENT 2: Predicate: END

antecedents with complementation:

ANTECEDENT 1: Predicate: END

conclusion
Predicate: S
Argument1 : X
,Argument2: _

Threshold value: 0.7

More rules to create (y / n)? N

MAIN MENU:

1- Create / Append static database
2- Create / Append production rules
3- Create / Append default rules
4- Arrange production rules properly
5- Delete clauses from database
6- Delete clauses from pr rule base
7- Delete clauses from def. rule base
0- Terminate

SELECT:
3
C to create, A to append: C
General structure of a production rule:

antecedents without complementation
antecedents1: predicate, argument1, argument2
antecedents2: predicate, argument1, argument2
antecedents3: predicate, argument1, argument2
antecedents4: predicate, argument1, argument2
antecedents5: predicate, argument1, argument2

antecedents with complementation
antecedents1: predicate, argument1, argument2
antecedents2: predicate, argument1, argument2
antecedents3: predicate, argument1, argument2
antecedents4: predicate, argument1, argument2
antecedents5: predicate, argument1, argument2

conclusion
predicate: SUSPECT, argument1, argument2

press any key to continue

further instructions

each rule must contain
atmost 5 antecedents without complementation, and
atmost 5 antecedents with complementation
if the actual no. of antecedents is less than 5.
Enter "END" as predicate
When all the antecedents of the rule are entered
use upper case letters to represent arguments
if some argument is inapplicable or redundant in the context
enter it as an underscore

press any key to continue
antecedents without complementation:
ANTECEDENT 1: Predicate: L
, ARGUMENT1 : X
, ARGUMENT2 : Z

ANTECEDENT 2: Predicate: L
, ARGUMENT1 : X
, ARGUMENT2 : Y

ANTECEDENT 3: Predicate: END

antecedents with complementation:
ANTECEDENT 1: Predicate: END

conclusion
Predicate: SUSPECT Argument1: S
Argument2: _

More rules to create (y / n)?N

MAIN MENU:

8-   Create / Append static database
9-   Create / Append production rules
10- Create / Append default rules
11- Arrange production rules properly
12- Delete clauses from database
13- Delete clauses from pr rule base
14- Delete clauses from def. rule base
1-   Terminate

SELECT:
4

Rules properly arranged

MAIN MENU:

15- Create / Append static database
16- Create / Append production rules
17- Create / Append default rules
18- Arrange production rules properly
19- Delete clauses from database
20- Delete clauses from pr rule base
21- Delete clauses from def. rule base
2- Terminate

SELECT:
0

A.2 The Fpnreas2 Program

It is also a natural language interface program, developed in Pascal, which asks for the name of persons involved in the criminology problem. The variables in the predicates of the production rules are then instantiated with the supplied values and the truth–falsehood (T / F) of the resulting predicates is asked. Thus from a number of instantiated predicates, the program can filter a few, which subsequently results in the rules with properly bound variables in the predicates. This is done so to realize unification in an indirect manner. Note that Pascal does not possess the instructions for unification, so one has to use its power of string matching.

The sample run of the program is presented below.

B:\> fpnreas2

KEEP CAPS LOCK ON. ARE YOU READY
Y

ENTER generic filename without extension KONAR

QUERY SESSION:

enter names of persons involved
enter "END" to terminate
Person1: S
Person2: R

Person3: M
Person4: END

enter T if true, F if false
L(S,S): F
L(S,R): T
L(S,M): T
L(R,S): F
L(R,R): F
L(R,M): F
L(M,S): F
L(M,R): F
L(M,M): F

suspects selected by DRs are:
S
R
M

Enter the name of the conclusion Nodes
Enter "END" to terminate
S
END

A.3 The Fpnreas3 Program

This also is a Pascal program that computes the belief for all predicates located at the places of the FPN. The belief revision cycles are continued, until the steady-state is reached, following which there will be no more changes in beliefs at a place. The program can also construct an FPN from the rules and data supplied in Fpnreas1 program. The program finally determines the culprit place from many possible suspects and gives an explanation for the selection of the culprit.

The sample run of the program is presented below.

B:\> fpnreas3

net to display (y / n)? Y
Nodes displayed in the following fashion

Level	POSITION_FROM_LEFT	NODE
1	1	H: This is
nethead (_, _)		
2	1	I: L (S,R)
3	1	T: LSR (S,_)
4	1	N: LSR(S,_)
5	1	T: S (S,_)
6	1	N: S (S,_)
7	1	F: S (S,_)
3	2	T: LSR (S,_)
2	2	I: L(S, M)
3	3	T: LSR (S,_)
3	4	T: LSR(S,_)

palette:2

Here it draws the FPN, the top part of that network we considered in case study in chapter 23. The figure is not good. Readers may try to improve it.

Initialization Session:

which model to use (2 OR 3): 2

I: L (S, R): Initial FTT: 0.4
I: L (S, M): Initial FTT: 0.1
N: LSR (S, _): Initial FTT: 0.4
N: S (S, _): Initial FTT: 0.3
F: S (S, _): Initial FTT: 0.3

No Of updations: 10

Stability obtained after 2 updations

CONCLUSION:
conclusion no.1: S (S,_): FFT= 4.0000000000E -01

Conclusion1:
paths
path no: 1
path gain: 0.400
I: L (S, R): FTT: 0.400
T: LSR (S, _): FTT: 0.400
N: LSR (S, _): FTT: 0.400
T: S (S, _): FTT: 0.400
N: S (S, _): FTT: 0.400

F: S (S, _): Initial FTT: 0.400

node type: F
Predicate :S
Argument1: S
Argument2: _
Node found

Choose palette no: 2

Here it draws the belief distribution of the place for S (s). This part is not at all good, and I request the readers to spend some time in improving the graphics part of the program.

A.4 The Imagemat Program

This program, written in C, is used to match a reference image with a given set of images and the best matched image has to be identified. It is designed based on the concept of fuzzy moment descriptors in chapter 23. You can submit image files boy1.dat, boymus.dat (boy with moustache) and boyglass.dat (boy with spectacles) included in the floppy diskette. Note that one file must be the reference image. The other images with which the reference is to be matched should include the copy of the reference image.

NOTE: Keep the data files boy1.dat, boymus.dat and boyglass.dat along with the program Imagemat in a floppy before execution.

The sample run of the program is presented below.

no. of pictures to compare = 3

no. of subblocks in each picture
(make sure you are giving perfect square) = 16
enter name of file in which d.image of 1 picture available = boy1.dat
enter number of rows or columns in image = 64

taking the gray values (in rowwise) form from file

calculating gradient matrix

dividing the picture into subblocks
number of subblocks = 16

calculating the statistical parameters

calculating membership values
enter name of file in which d.image of 2 picture available = boy1.dat

enter number of rows or columns in image = 64

taking the gray values (in rowwise) form from file

calculating gradient matrix

dividing the picture into subblocks
number of subblocks = 16

calculating the statistical parameters
calculating membership values
enter name of file in which d.image of 3 picture available = boymus.dat

enter number of rows or columns in image = 64

taking the gray values (in rowwise) form from file

calculating gradient matrix

dividing the picture into subblocks
number of subblocks = 16

calculating the statistical parameters
calculating membership values

The best matched picture is in the position = 2
The best matched picture is in file boy1.dat
The distance between best matched picture boy1.dat and reference picture
boy1.dat is 0.000000

A.5 The Finger Program

This program, written in C, determines the class of a given fingerprint. When
the number of core-delta pairs is greater than 2, the fingerprint image has to
pass through an average, median or a Gaussian filter, included in the program.

NOTE: The following files: fin1, cim1.mat, fin2, cim2.mat, fin3, cim3.mat,
fin5 and cim5.mat should be saved with the exe file or source code of finger
together on a floppy before execution.

The sample run of the program finger is presented below.

Input the filename without extension
fin1
give the output file to store integer data
cim1.mat
1-Avg, 2-Median, 3-Gaussian
1
it belongs to arch type

input the filename without extension
fin2
give the output filename to store integer data
cim2.mat
1-Avg, 2-Median, 3-Gaussian
1
it belongs to tented arch

input the filename without extension
fin3
give the output filename to store integer data
cim3.mat
1-Avg, 2-Median, 3-Gaussian
1
it belongs to left loop

input the filename without extension
fin5
give the output filename to store integer data
cim5.mat
1-Avg, 2-Median, 3-Gaussian
1
it belongs to twin category

A.6 The Cimgmat Program

This program is a minor extension of the 'imagemat program' we discussed earlier. Once a fingerprint image is classified by the 'finger' program, we match the classified fingerprint with all possible members within its class by the present program.

NOTE: Save the cim*..mat files with the .exe and source files of the cimgmat program in A-drive before execution.

A sample run of the program is presented below.

C > File currently in progress = cim7.mat pos =1
Taking the gray values (in row wise) from file
Calculating gradient matrix
 Dividing the picture into subblocks
Number of subblocks =16
Number of rows in subblock =16

Calculating statistical parameters
Calculating membership values

File currently in progress = cim7.mat pos =2
Taking the gray values (in row wise) from file
Calculating gradient matrix
 Dividing the picture into subblocks
Number of subblocks =16
Number of rows in subblock =16

Calculating statistical parameters
Calculating membership values

File currently in progress = cim8..mat pos =3
Taking the gray values (in row wise) from file
Calculating gradient matrix
 Dividing the picture into subblocks
Number of subblocks =16
Number of rows in subblock =16

Calculating statistical parameters
Calculating membership values

File currently in progress = cim2..mat pos =4
Taking the gray values (in row wise) from file
Calculating gradient matrix
 Dividing the picture into subblocks
Number of subblocks =16
Number of rows in subblock =16

Calculating statistical parameters
Calculating membership values

Best matched picture is in pos = 2
Best matched picture is in file cim7.mat
The distance between the matched picture cim7.mat
And reference picture cim7.mat is =0.0000

A.7 The Gapath Program

This program employs Genetic Algorithm to plan a trajectory between a given starting and a goal point.

NOTE: Save the program and the BGI file on a floppy drive (A:) before execution of the program.

The sample run of the program for two special cases is given below.

When the goal is within convex obstacle:

The workspace coordinate are (50,50) and (450,450)
Enter starting x (between 50 & 450): 70 (press enter)
Enter starting y (between 50 & 450): 70 (press enter)
Enter sensing range (between 10 & 50): 30 (press enter)
Enter robot step size (5 to 30) : 20 (press enter)

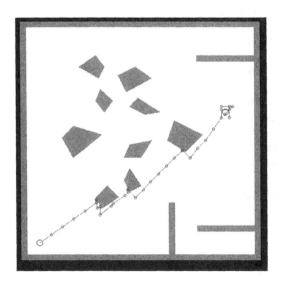

Time taken to search path: 0.549451
Distance covered = 604.238 units

When the goal is within concave obstacle:

The workspace coordinate are (50,50) and (450,450)
Enter starting x(between 50 & 450): 70 (press enter)
Enter starting y(between 50 & 450): 70 (press enter)
Enter sensing range (between 10 & 50): 30 (press enter)
Enter robot step size (5 to 30) : 20 (press enter)

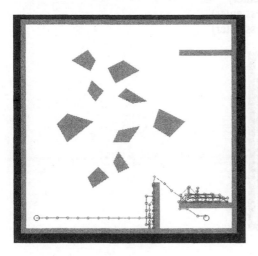

Time taken to search path: 1.868
Distance covered = 1710.915 units

A.8 The Quadtr1 Program

This program partitions a given workspace by Quadtree approach and determines the trajectory of the robot by employing heuristic search on the tree.

NOTE: Save the BGI file with the program on an A-drive before running it.

A sample run of the program is given below.

Workspace coordinate range in between (80,80) and (400,400)
//diagonal vertices//
The obstacle are lying in the following coordinates
 i.e, coordinates of diagonal of the square region (x1, y1) to (x2,y2)

Obstacle1: (80,80) to (120,120)
Obstacle2: (200,120) to (240,160)
Obstacle3: (240,160) to (320,240)

Obstacle4: (160,240) to (240,320)

You have to enter the (x,y) coordinates of the starting position
of the robot within the workspace and should not lie on the
obstacle zone

Set goal x_location: 380 (Press enter)
Set goal y_location: 380 (Press enter)
Set starting x_location: 130 (Press enter)
Set starting x_location: 130 (Press enter)

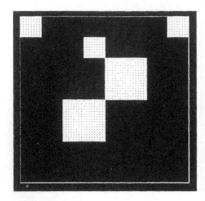

(Press enter)

Current Node : 120,120 and 160,160

Total Neighbours
120,80 and 160,120 ------- status: empty node
160,120 and 200,160 ------- status: empty node
80,160 and 160,240 ------- status: empty node
80,120 and 120,160 ------- status: empty node

number of neighbours =4

path selected
160,120 and 200,160

Total Neighbours
160,80 and 200,120 ------- status: empty node
200,120 and 240,160 ------- status: occupied node
160,160 and 240,240 ------- status: empty node
120,120 and 160,160 ------- status: empty node

number of neighbours =4

path selected
160,160 and 240,240

Total Neighbours
240,160 and 320,240 ------- status: occupied node
160,240 and 240,320 ------- status: occupied node
80,160 and 160,240 ------- status: empty node
160,120 and 200,160 ------- status: empty node
200, 120 and 240,160 ------- status: occupied node

number of neighbours =5

path selected
80,160 and 160,240
Total Neighbours
160,160 and 240,240 ------- status: empty node
80,240 and 160,320 ------- status: empty node
80,120 and 120,160 ------- status: empty node
120,120 and 160,160 ------- status: empty node

number of neighbours =4

path selected
80,240 and 160,320

Total Neighbours
80,160 and 160,240 ------- status: empty node
160,240 and 240,320 ------- status: occupied node
80,320 and 160,400 ------- status: empty node

number of neighbours =3

path selected
80,320 and 160,400

Total Neighbours
80,240 and 160,320 ------- status: empty node
160,320 and 240,400 ------- status: empty node

number of neighbours =2

path selected
160,320 and 240,400

Total Neighbours
160,240 and 240,320 ------- status: occupied node
240,240 and 400,400 ------- status: empty node
80,320 and 160,400 ------- status: empty node

number of neighbours =3

path node selected
240,240 and 400,400

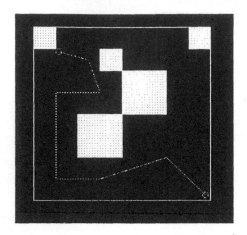

A.9 The World Program

This C++ program builds a map of the obstacles and the room boundary by first determining the room boundary and then employing the depth first traversal following an ordered priority based directed traversal.

NOTE: Save the BGI file with the program in A-drive.

The sample run of the program is presented below.

The workspace is rectangular and co-ordinates are (85,85), (85,395), (395,85) & (395,395)

Starting position of robot x(enter value between 85 - 395): 100

Starting position of robot y(enter value between 85 - 395): 380

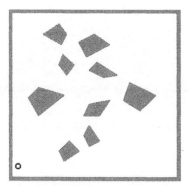

(a)   Robot entering the room

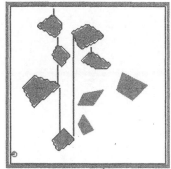

(b) After Visiting 6th Obstacle

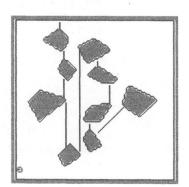

(c) After Visiting 10th Obstacle

(d) Obstacle boundaries stored

Appendix B

Derivation of the Back-propagation Algorithm

This appendix presents the derivation of the back-propagation algorithm, covered in chapter 14. We derive it for two cases, with and without non-linearity in the output layered nodes.

B.1 Derivation of the Back-propagation Algorithm

The back-propagation algorithm is based on the principle of gradient descent learning. In gradient descent learning, the weight $w_{p,q,k}$ which is connected between neurons p in layer (k-1) with neuron q at the k-th (output) layer is hereafter denoted by W_{pq} for simplicity. Let E be the Euclidean norm of the error vector, for a given training pattern, produced at the output layer. Formally,

$$E = (1 / 2) \sum_{\forall r} (t_r - \text{Out}_r)^2$$

where t_r and Out_r denote the target (scaler) output and the computed output at node r in the output layer.

The gradient descent learning requires that for any weight W_{pq},

$$W_{pq} \leftarrow W_{pq} - \eta(\partial E / \partial W_{pq}) \qquad (B.1)$$

where η is the learning rate.

The computation of $\partial E / \partial W_{pq}$, however, is different in the last layer from the rest of the layers. We now consider two types of output layers: output layer with and without non-linearity of neurons.

When the neurons in the output layer contains no non-linearity,

$$Out_q = Net_q = \sum W_{pq}.Out_p \qquad (B.2)$$
$$\forall p \text{ in the}$$
$$\text{penultimate layer}$$

Now, $\partial E / \partial W_{pq}$

$$= (\partial E / \partial Out_q) \ (\partial Out_q / \partial W_{pq})$$

$$= -(t_q - Out_q) \ Out_p \qquad (B.3)$$

Consequently, $W_{pq} \leftarrow W_{pq} - \eta(t_q - Out_q) \ Out_p$ [by (B.1)] $\qquad (B.4)$

Denoting, $(t_q - Out_q)$ by δq we have

$$W_{pq} \leftarrow W_{pq} + \eta.\delta q. \ Out_p \qquad (B.5)$$

Now, we consider the case, when the neurons in all the layers including the output layer contain sigmoid type non-linearity. The network structure with two cascaded weights is given in fig. B.1.

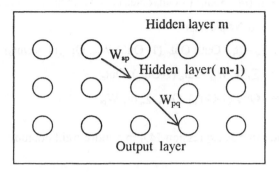

Fig. B.1: Defining the weights in a feed-forward topology of neurons.

Here, $Out_p = 1/(1+e^{-Net_p})$

$Net_p = \Sigma_r W_{rp}.Out_r$

where the index r corresponds to neurons in the hidden layer m.

$Out_r = 1/(1+e^{-Net\ r})$

$Net_r = \Sigma_i W_{ir}.Out_i$

where w_{ir} are the weights connected to neuron r from its preceding layer.

Now, for the output layer with sigmoid type non-linearity we have

$\partial E/\partial W_{pq}$

$= (\partial E/\partial Out_q)\ (\partial Out_q/\partial Net_q)\ (\partial Net_q/\partial W_{pq})$

$= -(t_q - Out_q)\ Out_q\ (1 - Out_q)\ Out_p$

$= - \{ (t_q - Out_q)\ Out_q\ (1 - Out_q) \}\ Out_p$

$= -\delta q.\ Out_p$ (say).

The readers may now compare this result with that given in fig. 14.6, where this is written as $\delta_{q,k} Out_{p,j}$.

Now, we compute the updating rule for W_{sp}:

$\partial E / \partial W_{sp}$

$= \sum_r (\partial E / \partial Out_r) \; (\partial Out_r / \partial Net_r) \; (\partial Net_r / \partial Out_p). \; (\partial Out_p / \partial Net_p)$

$\quad (\partial Net_p / \partial W_{sp})$

$= \sum_r - (t_r - Out_r) \; Out_r \; (1 - Out_r) \; W_{pr}. \; Out_p \; (1 - Out_p). \; Out_s$

$= - \sum_r \delta r \; W_{pr} \; Out_p \; (1 - Out_p). \; Out_s$

$= - Out_p.(1 - Out_p). \; Out_s \; \sum_r \delta r. \; W_{pr}$

which is the same as expression 14.4 in a more rigid notation.

Appendix C

Proof of the Theorems of Chapter 10

This appendix presents the proof of the theorems stated in chapter 10. The new equations derived in the chapter are numbered C.1 through C.10. Here we will use some of the equations of chapter 10 as well.

Proof of theorem 10.1: From the definition of the associated places of m_{ij} it is clear that if m_{ii} is 1 in matrix M_k, then place p_i is reachable from itself. Further, $M_k = (P \circ Q)^k$ indicates that the diagonal $p_i \in IRS^k (p_i)$. Therefore, place p_i lies on a cycle through k-transitions. Since it is true for any diagonal element of M_k, all the places corresponding to the diagonal elements having value 1 of matrix M_k will lie on cycles, each having k-transitions on the cycle. This is all about the proof. $\square$

Proof of theorem10.2: For an oscillatory response, the fuzzy belief at a place should increase as well as decrease with respect to time. However, since $N(t+1) \geq N(t)$ for all integer $t \geq 0$ (vide expression10.7), hence, the components of $N(t)$ cannot exhibit oscillatory response. Further, as components of $N(0), T(0),$ Th and R are bounded between 0 and 1, and $N(t+1)$ is derived by using fuzzy AND/ OR operators over R, $N(t)$ and $T(t)$, hence, components of $N(t)$ remains bounded. Thus fuzzy beliefs at places being bounded and free from oscillation exhibit stable behavior. This completes the first part of the theorem.

Now to prove the second part, we first rewrite expression (10.5) in the following form :

$$T(t+1) = T(t) \wedge [R \circ (Q' \circ N^c(t))^c] \wedge U [R \circ (Q' \circ N^c(t))^c - Th] \qquad (C.1)$$

$$N(t+1) = N(t) \vee P' \circ T(t+1) \qquad [(10.5), \text{ rewritten }]$$

Now, we compute $T(1)$ and $T(0)$ and $N(1)$ and $N(0)$ by expression (C.1) and (10.5) respectively and find

$$N(1) = N(0) \vee P' \circ T(1) \qquad\qquad (C.2)$$

Now computing $T(2)$ and $N(2)$ by expression (C.1) and (10.5) it is found

$$T(2) \leq T(1) \qquad\qquad (C.3)$$

$$\text{and} \quad N(2) = N(1) \vee P' \circ T(2)$$
$$= N(0) \vee P' \circ T(1) \vee P' \circ T(2) \qquad [\text{ by } C.2]$$
$$= N(0) \vee P' \circ T(1) \qquad\qquad [\text{ by } C.3]$$
$$= N(1)$$

Thus it can be easily shown that $N(t+1) = N(t)$ for all integer $t \geq 1$. Therefore, steady-state is reached with only one belief revision step. Hence, the theorem follows. □

Proof of theorem10.3: Unconditional stability of the model can be proved following the steps analogous to the proof of theorem 10.2.

To prove the second part of the theorem, let us assume that the equilibrium condition is reached at time $t = t^*$. Thus, by definition of theorem 10.2

$$N(t^* + 1) = N(t^*) \quad (= N^*, \text{ by statement of the theorem }) \qquad (C.4)$$

Now, expression (10.8) satisfies expression (C.4) when

$N^* \geq P' o [\{ R o (Q o N^{*c})^c \} \wedge U \{ R o (Q o N^{*c})^c - Th \}]$ (C.5)

Further, if $R o (Q o N^{*c})^c \geq Th$, all the components of U vector being unity, it can be dropped from expression (C.5). Thus we get expression (10.10).□

Proof of theorem 10.5: Q'_{fm}, by definition, is a TPC matrix used in forward reasoning under IFF implication relationship, with elements $q_{ij} = 1$ when $p_i \in O(tr_j)$ and $q_{ij} = 0$, otherwise. P'_{bm}, on the other hand, is a PTC matrix used in back-directed reasoning under IFF implication relationship with elements $p_{ji} = 1$ when $p_i \in I(tr_j)$, otherwise $p_{ji} = 0$. Thus for all i, j

$P'_{bm} = (Q'_{fm})^T$. Analogously, $Q'_{bm} = (P'_{fm})^T$ can also be proved. □

Proof of theorem 10.6: We will use lemma 1 and 2, listed below, to prove this theorem.

Lemma 1: *The distributive law of product holds good with respect to fuzzy composition operator*, i.e.,

$$A o [B \ V C] = (A o B) \ V (A o C), \qquad (C.6)$$

where A is a (n x m) fuzzy matrix and B and C are either fuzzy matrix or vectors of compatible dimensions.

Proof: Proof, being simple, is omitted

Lemma 2: De Morgan's law holds good for fuzzy matrices A and B, i.e.,

$$(A \ominus B) = (A^c \ o \ B^c)^c \qquad \text{(C.7(a))}$$

and $\quad A o B = (A^c \ominus B^c)^c$ (C.7(b))

where $\ominus$ denotes fuzzy OR-AND composition operator, which plays the role of AND-OR composition operator, with replacement of AND by OR and OR by AND operators.

Proof: Proof is simple and hence omitted.

Now, let us consider the reciprocity relations given by expressions (10.19) and (10.22). Since expression (10.19) is nothing but an identity of N_f, it is valid for any arbitrary fuzzy vector N_f. Assume N_f to be a vector with only one element equal to 1 and the rest are zero. Further, to keep the proof brief, let us consider that N_f is of (3 x 1) dimension.

Thus, we get

$$\begin{bmatrix} 1 \\ 0 \\ 0 \end{bmatrix} = Q'_{fm}{}^T \circ R_{fm} \circ \left(Q'_{fm} \circ \begin{bmatrix} 0 \\ 1 \\ 1 \end{bmatrix} \right)$$

$$= Q'_{fm}{}^T \circ R_{fm} \circ \begin{bmatrix} q_{12}{}^c \wedge q_{13}{}^c \\ q_{22}{}^c \wedge q_{23}{}^c \\ q_{32}{}^c \wedge q_{33}{}^c \end{bmatrix}$$

$$(\text{C.8})$$

analogously ,

$$\begin{bmatrix} 0 \\ 1 \\ 0 \end{bmatrix} = Q'_{fm}{}^T \circ R_{fm} \circ \begin{bmatrix} q_{11}{}^c \wedge q_{13}{}^c \\ q_{21}{}^c \wedge q_{23}{}^c \\ q_{31}{}^c \wedge q_{33}{}^c \end{bmatrix}$$

$$(\text{C.9})$$

and

$$\begin{bmatrix} 0 \\ 0 \\ 1 \end{bmatrix} = Q'_{fm}{}^T \circ R_{fm} \circ \begin{bmatrix} q_{11}{}^c \wedge q_{12}{}^c \\ q_{21}{}^c \wedge q_{22}{}^c \\ q_{31}{}^c \wedge q_{32}{}^c \end{bmatrix}$$

$$(\text{C.10})$$

where q_{ij} are the elements of Q'_{fm} matrix. Now, combining (C.8), (C.9) and (C.10) we find

$$\begin{pmatrix} 1 & 0 & 0 \\ 0 & 0 & 0 \\ 0 & 0 & 0 \end{pmatrix} \lor \begin{pmatrix} 0 & 0 & 0 \\ 0 & 1 & 0 \\ 0 & 0 & 0 \end{pmatrix} \lor \begin{pmatrix} 0 & 0 & 0 \\ 0 & 0 & 0 \\ 0 & 0 & 1 \end{pmatrix}$$

$$= Q'_{fm}{}^T \, o \, R_{fm} \, o \, \left\{ \begin{pmatrix} q_{12}{}^c \land q_{13}{}^c & 0 & 0 \\ q_{22}{}^c \land q_{22}{}^c & 0 & 0 \\ q_{32}{}^c \land q_{32}{}^c & 0 & 0 \end{pmatrix} \lor \right.$$

$$\begin{pmatrix} 0 & q_{11}{}^c \land q_{13}{}^c & 0 \\ 0 & q_{21}{}^c \land q_{23}{}^c & 0 \\ 0 & q_{31}{}^c \land q_{33}{}^c & 0 \end{pmatrix} \lor$$

$$\begin{pmatrix} 0 & 0 & q_{11}{}^c \land q_{12}{}^c \\ 0 & 0 & q_{21}{}^c \land q_{22}{}^c \\ 0 & 0 & q_{31}{}^c \land q_{32}{}^c \end{pmatrix}$$

$$\Rightarrow I = Q'_{fm}{}^T \, o \, R_{fm} \, o \, \begin{pmatrix} q_{12}{}^c \land q_{13}{}^c & q_{11}{}^c \land q_{13}{}^c & q_{11}{}^c \land q_{12}{}^c \\ q_{22}{}^c \land q_{23}{}^c & q_{21}{}^c \land q_{23}{}^c & q_{21}{}^c \land q_{22}{}^c \\ q_{32}{}^c \land q_{33}{}^c & q_{31}{}^c \land q_{33}{}^c & q_{31}{}^c \land q_{32}{}^c \end{pmatrix}$$

(by lemma 1)

$$\Rightarrow I = Q'_{fm}{}^T \, o \, R_{fm} \, o \, [\, Q'_{fm}{}^c \, \Theta \, I^c \,]$$

$$= Q'_{fm}{}^T \, o \, R_{fm} \, o \, (\, Q'_{fm} \, o \, I)^c \qquad \text{(by lemma 2)}$$

$$\Rightarrow I = Q'_{fm}{}^T \, o \, R_{fm} \, o \, [\, Q'_{fm}{}^c \, \Theta \, I^c \,]$$

$$= Q'^T_{fm} \; o \; R_{fm} \; o \; (Q'_{fm} \, o \, I \,)^c \qquad\qquad \text{(by lemma 2)}$$

Now, extending the above operations for an $((n \, . \, z) \times 1)$ N_f vector (vide section III B), the same results can be easily obtained. Considering expression (10.22), an identity of T_f, expression (10.23(b)) can be proved analogously.

INDEX

I

J

K

L

U

V

W

T - #0210 - 101024 - C0 - 234/156/44 [46] - CB - 9780849313851 - Gloss Lamination